# Modern Applied U-Statistics

BICENTENNIAL 1807 WILEY 2007 BICENTENNIAL

## THE WILEY BICENTENNIAL–KNOWLEDGE FOR GENERATIONS

Each generation has its unique needs and aspirations. When Charles Wiley first opened his small printing shop in lower Manhattan in 1807, it was a generation of boundless potential searching for an identity. And we were there, helping to define a new American literary tradition. Over half a century later, in the midst of the Second Industrial Revolution, it was a generation focused on building the future. Once again, we were there, supplying the critical scientific, technical, and engineering knowledge that helped frame the world. Throughout the 20th Century, and into the new millennium, nations began to reach out beyond their own borders and a new international community was born. Wiley was there, expanding its operations around the world to enable a global exchange of ideas, opinions, and know-how.

For 200 years, Wiley has been an integral part of each generation's journey, enabling the flow of information and understanding necessary to meet their needs and fulfill their aspirations. Today, bold new technologies are changing the way we live and learn. Wiley will be there, providing you the must-have knowledge you need to imagine new worlds, new possibilities, and new opportunities.

Generations come and go, but you can always count on Wiley to provide you the knowledge you need, when and where you need it!

William J. Pesce
PRESIDENT AND CHIEF EXECUTIVE OFFICER

Peter Booth Wiley
CHAIRMAN OF THE BOARD

# Modern Applied U-Statistics

**Jeanne Kowalski**
*Division of Oncology Biostatistics*
*Johns Hopkins University*
*Baltimore, MD*

**Xin M. Tu**
*Department of Biostatistics and Computational Biology*
*University of Rochester*
*Rochester, NY*

BICENTENNIAL 1807 WILEY 2007 BICENTENNIAL

**WILEY-INTERSCIENCE**
**A John Wiley & Sons, Inc., Publication**

Published by John Wiley & Sons, Inc., Hoboken, New Jersey.
Published simultaneously in Canada.



Wiley Bicentennial Logo: Richard J. Pacifico

***Library of Congress Cataloging-in-Publication Data is available.***

ISBN 978-0-471-68227-1

Printed in the United States of America.

10 9 8 7 6 5 4 3 2 1

# Contents

# Preface

This book is an introduction to the theory of U-statistics and its modern applications by in-depth examples that cover a wide spectrum of models in biomedical and psychosocial research. A prominent feature of the book is its presentation of U-statistics as an integrated body of regression-like models, with particular focus on longitudinal data analysis. As longitudinal study designs are increasingly popular in today's research and textbooks on U-statistics theory that address such study designs are as yet non-existent, this book fills a critical void in this vital research sector. By integrating U-statistics models with regression analyses, the book unifies the two classic dueling paradigms — the U-statistics based non-parametric analysis and model-based regression analysis — to present the theory and application of U-statistics in an unprecedented broad and comprehensive scope.

The book is self-contained, although the content of the text does require knowledge of the classic statistical inference theory at a level comparable to the book by Casella and Berger (1990) and familiarity with statistics asymptotics or large sample theory. As U-statistics are not presented as an isolated entity, but rather as natural extensions of single-response based regression models to multisubject-defined functional response models, the book is structured to reflect the constant alternation between U-statistics and regression analysis. In Chapter 1, we review the theory of statistics asymptotics, which forms the foundation for the development of later chapters. In Chapter 2, we discuss regression analysis for cross-sectional data, contrast the classic likelihood-based approach with the modern estimating equation based distribution-free inference framework and explore the pros and cons of these two dueling paradigms. In Chapter 3, we introduce univariate U-statistics and discuss their properties. In Chapter 4, we revert back to regression and discuss inference for regression models introduced in Chapter 2 within the context of longitudinal data. We focus on distribution-free inference and discuss the popular generalized estimating equation (GEE) and weighted GEE (WGEE). In Chapter 5, we return to the discussion of U-statistics by extending the theory in Chapter 3 to multivariate U-statistics for applications to longitudinal data analysis. In Chapter 6, we introduce a new class of function response models as a unifying paradigm for regression by bringing together U-statistics based models discussed in Chapters 3 and 5 and regression models covered in Chapters 2 and 4 under a single regression-like modeling framework and discuss inference by introducing a new class of U-statistics based GEE (UGEE) and WGEE (UWGEE) under complete and missing data scenarios.

We opted to use a dedicated website,

http://www.cancerbiostats.onc.jhmi.edu/Kowalski/ustatsbook, as a venue to post real data applications and software, and to continuously update both to reflect current interest and timely research. We consider this preferable to the traditional approach of including real data examples in the book, since application interests change rapidly in today's research environment and real data examples will likely become obsolete much sooner than examples that focus on modeling principles. Likewise, we have only included some key references in the book. With the worldwide web and powerful search engines, such as Google, enabling the retrieval of references and related information at unprecedented speed and with only a few key strokes, static media, such as books, are no longer the best choice for documenting and finding references for research on a given topic.

This book may be used as a text for a one-semester topic course on U-statistics by focusing on Chapters 3, 5, and 6. This approach assumes that students have a course on generalized linear models (GLM), longitudinal data modeling, and distribution-free inference with GEE and WGEE. Alternatively, one can precede such a topic course by using Chapters 1, 2, and 4 either as a primary or secondary textbook in a one-semester course on GLM and longitudinal data analysis. The book is intended for second and third year graduate students in a Biostatistics or Statistics department, but may also serve as a self-learning text or reference for researchers interested in the theory and application of U-statistics based models.

Many people have contributed to this book project. We are very grateful to Mr. Yan Ma, Ms. Cheryl Bliss-Clark, Dr. Changyong Feng, Ms. Haiyan Su, Dr. Wan Tang and Ms. Qin Yu in the Department of Biostatistics and Computational Biology at the University of Rochester, and to Ms. Hua Ling Tsai and Mrs. Alla Guseynova in the Division of Biostatistics at the Sidney Kimmel Comprehensive Cancer Center at Johns Hopkins for their help with reading parts of the manuscript and helping make numerous corrections during the various stages of the manuscript development, as well as website design, construction, and development. We are especially indebted to Ms. Bliss-Clark, who painstakingly proofread the entire book numerous times and helped rewrite many parts of the book to improve the presentation, to Mr. Ma, who diligently read drafts of all chapters of the book and made countless corrections and changes to help eradicate typos and technical errors, and to Dr. Wan Tang, who selflessly shared with us his knowledge and expertise in the theory of semiparametric models for longitudinal data and helped edit many sections and exercises of the book. We would not have finished the book without the tenacious help and dedication from these three marvelous individuals.

Finally, we would like to thank Mr. Steve Quigley from Wiley for his unrelenting support of this book project, and Ms. Jacqueline Palmieri and Ms. Melissa Yanuzzi for their careful review of the manuscript and assistance with the publication of the book.

Jeanne Kowalski
Baltimore, Maryland
Johns Hopkins University

Xin Tu
Rochester, New York
University of Rochester

October 2007

# Chapter 1

# Preliminaries

This chapter serves to provide a systematic review of the concepts and tools that are fundamental to the development of the asymptotic and U-statistics theories for inference of statistical models introduced in the book. Since the majority of concepts and techniques discussed build upon the theory of probability, it is necessary to introduce and discuss them within the progression from the theory of probability to statistics asymptotics. This chapter comprehensively covers a spectrum of topics in probability theory in order to form a foundation for the development of statistics asymptotics utilized throughout the book. We distinguish *statistics asymptotics* or the large sample behavior of statistics from other asymptotics in that statistics asymptotics studies the convergence of random variables and vectors, as opposed to a sequence of fixed quantities, as in the study of asymptotics. In the interest of space, most technical results will be presented without proof (some left as exercises). Readers familiar with statistics asymptotics may opt to skip this chapter except for Section 1.1 without loss of continuity, and instead use the materials contained within as a reference when reading subsequent chapters.

The concepts and techniques presented focus on investigating the large sample behavior of statistics, particularly within the context of parameter estimates that arise from statistical models. Because of the frequent difficulties encountered in applying and extending such techniques, we begin this chapter by motivating discussion through examples that require asymptotic theory to address a wide range of problems arising in biomedical and psychosocial research. We then highlight the fundamental concepts and results that underlie the foundation of statistics asymptotics and its applications to inference of statistical models. Among the topics reviewed, those pertaining to boundedness and convergence of sequences of independently and identically distributed, or i.i.d., random variables play a particularly important role in the theoretical development of subsequent chapters.

## 1.1 Introduction

Asymptotic theory has an indispensable role in the inference of modern statistical models. What is statistics asymptotic theory? In brief, it is the study of the behavior of statistics, particularly model estimates, as the sample size approaches infinity. What is the advantage to applying such large sample theories? First, statistics asymptotic theory provides a powerful and unified theoretical framework to systematically simplify otherwise algebraically messy and often intractable computations in order to enable inference based on parameters of interest. Second, statistics asymptotics provides a theoretical basis for studying the properties of estimates of statistical models, such as bias and efficiency (or precision) to aid in the development of robust and optimal estimates, as well as to select such estimates among competing alternatives. Third, statistics asymptotics enables the study of distributions of statistics arising from statistical models and, as such, provides a framework in which to develop inference for such models to facilitate both data and power analysis.

In this section, we illustrate the importance of statistical asymptotic theory with some prototype models that are highlighted throughout the book. These models will be frequently visited in later chapters, along with their extensions to address more complex data types and study designs.

### 1.1.1 The Linear Regression Model

Regression models are widely used to model an outcome of interest, or dependent variable, as a function of a set of other variables, or independent variables, in almost all areas of research, including biomedicine, epidemiology, psychology, sociology, economics and public health. The dependent variable, or response, may be continuous or discrete. The independent variables are also referred to as explanatory variables, predictors, or covariates, depending on their roles in the statistical model. A clarification on the taxonomy used to delineate different types of independent variables is presented in Chapter 2. When the dependent variable is continuous, linear regression is the most popular choice for modeling a relationship with other variables.

Consider a study with $n$ subjects. Let $y_i$ denote a continuous response and $\mathbf{x}_i = (x_{i1}, x_{i2} \ldots, x_{ip})^\top$ a $p \times 1$ vector of independent variables from the $i$th subject $(1 \leq i \leq n)$. The linear regression for modeling $y_i$ as a function of $\mathbf{x}_i$ is defined by the following model statement:

$$y_i = \mathbf{x}_i^\top \boldsymbol{\beta} + \epsilon_i, \quad \epsilon_i \sim \text{i.i.d. } N\left(0, \sigma^2\right), \quad 1 \leq i \leq n \tag{1.1}$$

where $\boldsymbol{\beta} = \left(\beta_1, \beta_2, \ldots, \beta_p\right)^\top$ is the vector of parameters, $\epsilon_i$ denotes the error term in modeling $y_i$ through $\mathbf{x}_i^\top \boldsymbol{\beta}$, $N\left(\mu, \sigma^2\right)$ denotes a normal distribution with mean $\mu$ and variance $\sigma^2$. In (1.1), $\epsilon_i$ ~i.i.d. $N\left(0, \sigma^2\right)$ means that $\epsilon_i$ are independently and identically distributed with the same normal distribution $N\left(0, \sigma^2\right)$. The linear function $\eta_i = \mathbf{x}_i^\top \boldsymbol{\beta}$ is often called the linear predictor. In most applications, $x_{i1} \equiv 1$ so that the linear predictor of the model includes an intercept, that is,

$$\eta_i = \beta_1 + x_{i2}\beta_2 + \cdots + x_{ip}\beta_p$$

Given the distribution assumption for the error term $\epsilon_i$, the maximum likelihood method is widely used to estimate and make inference about $\boldsymbol{\beta}$. Let

$$\mathbf{y} = (y_1, \ldots, y_n)^\top, \quad X = (\mathbf{x}_1, \ldots, \mathbf{x}_n)^\top$$

Then, the maximum likelihood estimate (MLE) of $\boldsymbol{\beta}$ is $\widehat{\boldsymbol{\beta}} = \left(X^\top X\right)^{-1} X^\top \mathbf{y}$ (see Chapter 2 for details). Further, under the assumption of a constant matrix $X$, the sampling distribution of $\widehat{\boldsymbol{\beta}}$ is a multivariate normal (see Chapter 2 for details):

$$\widehat{\boldsymbol{\beta}} = \left(X^\top X\right)^{-1} X^\top \mathbf{y} \sim N\left(\boldsymbol{\beta}, \sigma^2 \left(X^\top X\right)^{-1}\right) \tag{1.2}$$

Note that we do not distinguish *estimate* and *estimator* and will use *estimate* consistently throughout the book.

The assumption of a constant $X$ is reasonable for some studies, especially controlled experiments. For example, in a study to compare two treatments with respect to some response of interest $y$, we first determine the sample size for each treatment and then assign patients to one of two treatment conditions. We can use two binary indicators to code the treatment received by each patient as:

$$x_{ik} = \begin{cases} 1 \text{ if treatment } = k \\ 0 \text{ if otherwise} \end{cases}, \quad k = 1, 2, \quad 1 \leq i \leq n \tag{1.3}$$

With the sample size $n_k$ $(k = 1, 2)$ for each treatment fixed, the design matrix $X$ with $\mathbf{x}_i = (x_{i1}, x_{i2})^\top$ is a constant and $\beta_1$ and $\beta_2$ in the linear predictor $\eta_i = \beta_1 x_{i1} + x_{i2}\beta_2$ represent the mean response of $y$ within each treatment condition. The MLE $\widehat{\boldsymbol{\beta}}$ in this particular case reduces to $\widehat{\boldsymbol{\beta}} = \left(\widehat{\beta}_1, \widehat{\beta}_2\right)^\top = (\overline{y}_{1\cdot}, \overline{y}_{2\cdot})^\top$, where $\overline{y}_{k\cdot} = \frac{1}{n_k}\sum_{i=1}^n x_{ik} y_i$ and $n_k = \sum_{i=1}^n x_{ik}$ $(k = 1, 2)$. The sampling distribution of $\widehat{\boldsymbol{\beta}}$ in (1.2) also simplifies to (see exercise):

$$\frac{\sqrt{n}}{\sigma^2}\left(\widehat{\boldsymbol{\beta}} - \boldsymbol{\beta}\right) \sim N\left(\boldsymbol{\beta}, \Sigma_\beta = \begin{pmatrix} \frac{n}{n_1} & 0 \\ 0 & \frac{n}{n_2} \end{pmatrix}\right) \tag{1.4}$$

When $\sigma^2$ is known, the above distribution describes the variability of the estimates $\overline{y}_{k\cdot}$ of the mean responses of the two treatment groups. In most applications, however, $\sigma^2$ is unknown. By estimating $\sigma^2$ using the following pooled sample variance,

$$\widehat{\sigma}^2 = \frac{1}{n-2}\sum_{i=1}^n (y_i - \widehat{\eta}_i)^2 = \frac{1}{n-2}\sum_{i=1}^n \left[y_i - \left(\widehat{\beta}_1 x_{i1} + \widehat{\beta}_2 x_{i2}\right)\right]^2$$

we can use the bivariate $t$ distribution (see Chapter 2 for details),

$$\frac{\sqrt{n}}{\widehat{\sigma}^2}\left(\widehat{\boldsymbol{\beta}} - \boldsymbol{\beta}\right) \sim t\left(\mathbf{0}, \begin{pmatrix} \frac{n}{n_1} & 0 \\ 0 & \frac{n}{n_2} \end{pmatrix}, n - p\right) \tag{1.5}$$

for inference about $\boldsymbol{\beta}$. For example, we can use (1.4) or (1.5) to test whether there is any between-group difference, that is, $\beta_1 - \beta_2 = 0$, depending on whether $\sigma^2$ is known or estimated.

However, in many applications, especially in observational studies arising in epidemiological and related research, the response $y_i$ and independent variables $\mathbf{x}_i$ are typically observed concurrently. For such study designs, it is not sensible to hold $\mathbf{x}_i$ fixed when considering the sampling variability of $y_i$. Even with controlled randomized trials, such as the example for comparing two treatment conditions discussed above, it is not always possible to have a constant design matrix. For example, if the between-treatment difference is significant in that example, we may want to know whether such a difference varies across subgroups defined by different demographic and comorbid conditions such as age, gender, and race. Such *moderation* analyses have important implications in biomedical and psychosocial research for determining treatment specificity and deriving optimal treatment regimes

(Baron and Kenny, 1986). Within the context of the study for comparing two treatments, if we want to test whether age moderates the treatment effect, we can fit the model in (1.1) with the linear predictor, 0

$$\eta_i = \beta_1 x_{i1} + x_{i2}\beta_2 + x_{i3}\beta_3 + x_{i2}x_{i3}\beta_4$$

where $x_{i3}$ is the covariate of subject's age and $x_{ik}$ $(k = 1, 2)$ are defined in (1.3). If $\beta_4 \neq 0$, then age is a moderator. Since moderation analyses are often post-hoc (after treatment difference is established) and moderators can be discrete as well as continuous variables, it is practically impossible to fix their values prior to subjects' recruitment.

When $X$ varies from sample to sample, the matrix $\left(X^\top X\right)$ in the sampling distribution (1.2) changes as a function of the sampling process. The sampling-dependent variance matrix of the estimate not only violates the definition of a sampling distribution, but more importantly, it leaves open the question of whether (1.2) is valid for inference about $\boldsymbol{\beta}$. This issue is addressed in greater detail in Chapter 2 by utilizing the theory of statistics asymptotics.

The normal distribution in the linear regression (1.1) places a great restriction on applications of the model to real study data for inference. In recent years, many distribution-free modeling alternatives have been developed to address this critical limitation. These alternative approaches do not assume any parametric distribution for the response and thus apply to a much wider class of data types and distributions arising in real study applications. In the absence of a parametric assumption imposed on the distribution of the response, likelihood-based approaches, such as the maximum likelihood, cannot be applied for inference of model parameters; in this case, statistics asymptotics plays a critical role in the development of alternative approaches to inference. We will discuss such distribution-free regression models and inference procedures for cross-sectional study designs in Chapter 2 and longitudinal study designs in Chapter 4.

### 1.1.2 The Product–Moment Correlation

Correlation analysis is widely used in a range of applications in biomedical and psychosocial research for assessing rater reliability, gene expression relations, precision of diagnosis, and accuracy of proxy outcomes (Barnhart et al. 2002; King and Chinchilli, 2001; Schroder et al. 2003; Kowalski et al. 2004; Kowalski et al. 2007; Tu et al. 2007). Although various types of correlation have been used, the product–moment correlation is the most popular and is a major focus within the context of correlation analysis throughout this book. Thus, the word *correlation* is synonymous to the product–moment correlation within this book unless otherwise noted.

Like regression, correlation analysis is widely used to assess the strength of a linear relationship between a pair of variables. In contrast to linear regression, correlation analysis does not distinguish between a dependent and an independent variable and is particularly useful for modeling dynamic relationships among concurrent events and correlates of medical and psychiatric conditions, such as heart disease and depression, and for assessing rater agreement in diagnoses, test–retest reliability in instrument development and fidelity of psychotherapy intervention.

Let $(x_i, y_i)$ denote an i.i.d. sample of bivariate continuous outcomes $(1 \leq i \leq n)$. The product–moment correlation is defined as:

$$\rho = \frac{Cov(x, y)}{\sqrt{Var(x) Var(y)}} \tag{1.6}$$

where $Cov(x, y)$ denotes the covariance between $x$ and $y$ and $Var(u)$ the variance of $u$ ($u = x$ or $y$). If we assume a bivariate normal distribution for the i.i.d. pairs, $(x_i, y_i)$, the maximum likelihood estimate of $\rho$ is readily derived as:

$$\widehat{\rho} = \frac{\sum_{i=1}^{n} (x_i - \overline{x}_\cdot)(y_i - \overline{y}_\cdot)}{\sqrt{\sum_{i=1}^{n} (x_i - \overline{x}_\cdot)^2 \sum_{i=1}^{n} (y_i - \overline{y}_\cdot)^2}}$$

where $\overline{u}_\cdot = \frac{1}{n}\sum_{i=1}^{n} u_i$ ($u = x$ or $y$). The above estimate, known as Pearson's correlation, is widely used as a measure of association between two variables.

However, unlike linear regression, the exact distribution of $\widehat{\rho}$ no longer has a known parametric form. Statistics asymptotic theory is required to study the sampling variability of $\widehat{\rho}$ for large samples. More generally, as in the discussion of the linear regression model, the normal assumption may not apply to outcomes arising from real study data, making it necessary to develop inference procedures for $\rho$ without the bivariate normal assumption imposed on the data distribution.

A key difference between linear regression discussed in Section 1.1.1 and the product–moment correlation defined in (1.6) is that linear regression models the conditional (on covariates) mean response, or first order moment of $y$, given $x$, while the product–moment correlation centers on cross-variable variability between $x$ and $y$, as measured by the second order moments of $x$ and $y$. This difference is not confined to linear regression and in fact, most regression models including the more general *generalized linear models* are based upon modeling the mean or first order moment of a response of interest (see Chapter 2 for details). Thus, the correlation model represents a significant departure from this mainstream modeling paradigm, in terms of modeling high order moments. Of note, the difference in modeling of moments carries over to drastically different inference paradigms. Indeed, as we demonstrate in further detail in Chapters 3, the typical approach to asymptotics-based inference theory developed for mean-based models, such as linear regression, does not in general apply to higher order moment based models. Although standard asymptotic approaches as typified by direct applications of the law of large numbers and central limit theorem may still be applied to study the sampling variability of the Pearson estimate $\widehat{\rho}$ (see Section 1.7), such applications require much more work both computationally and analytically. A more elegant solution is to utilize the theory of U-statistics. As we discuss in Chapters 5 and 6, such an alternative approach also allows us to generalize this popular measure of association to longitudinal study designs, as well as to effectively address missing data.

### 1.1.3 The Rank-Based Mann–Whitney–Wilcoxon Test

When comparing two treatment groups, one can use the linear regression model as discussed in Section 1.1.1 provided that the data are normally distributed. In the presence of severe departure from normality, especially when the data distribution is highly skewed, the Mann–Whitney–Wilcoxon rank sum test is a popular alternative to regression. This test compares the distributions of the response of interest between two groups based on the rankings of the values of the response so that no analytical distribution is imposed upon the response, at least in regard to its shape or form. Thus, the test applies to a much wider range of data distributions.

Consider a study with two treatment groups with $n_k$ subjects constituting the $k$th sample $(k = 1, 2)$. Let $y_{ki}$ be two i.i.d. samples from some continuous response of interest $(1 \leq i \leq n_k,\ k = 1, 2)$. We are interested in testing whether there is any difference in mean response between the two samples. However, unlike the regression setting, we do not assume analytic form for the distributions of the two samples, except that they differ by a location shift, that is, if $F_1$ is the cumulative distribution function (CDF) of $y_{1i}$, the CDF of $y_{2j}$ is given by: $F_2(y) = F_1(y - \theta)$, where $\theta$ is some unknown parameter (see Section 1.5 for definition of CDF). The Mann–Whitney–Wilcoxon test for the equality of distributions, that is, $\theta = 0$, is based on rank scores that are created as follows.

First, the responses from the two groups $y_{ki}$ are combined and ordered from the smallest to the largest. If there are ties among the observations, they are arbitrarily broken. The ordered observations are then assigned rank scores based on their rankings. For tied observations, average rankings are assigned to them. For example, if $n_1 = 5$ and $n_2 = 6$, and the observations of $y_{1i}$ and $y_{2j}$ are given by

$$y_{1i} : 4,\ 7,\ 3,\ 15,\ 1\ ; \quad y_{2j} :\ 10,\ 14,\ 9,\ 3,\ 17,\ 24$$

then the ordered observations of combined groups when arranged from the smallest (left) to the largest (right) are given as follows:

$$1,\ 3,\ 3,\ 4,\ 7,\ 9,\ 10,\ 14,\ 15,\ 17,\ 24$$

The rank scores of the ordered observations are the rankings of the ordered sequence starting from 1 for the smallest observation with tied observations assigned the average rankings, that is,

$$1,\ \frac{2+3}{2},\ \frac{2+3}{2},\ 4,\ 5,\ 6,\ 7,\ 8,\ 9,\ 10,\ 11$$

The rank scores $R_{ki}$ for the two groups are given by

$$R_{11} = 4,\ R_{12} = 5,\ R_{13} = \frac{2+3}{2},\ R_{14} = 9,\ R_{15} = 1$$
$$R_{21} = 7,\ R_{22} = 8,\ R_{23} = 6,\ R_{24} = \frac{2+3}{2},\ R_{25} = 10,\ R_{26} = 11$$

Note that for two continuous variables $y_{1i}$ and $y_{2j}$, the probability of having a tie either within the variable or between the variables is 0 (see Section 1.5). Although an improbable event in theory, it is possible to have tied observations in practice because of limited precision in measurement.

Wilcoxon (1945) and Mann–Whitney (1947) independently developed two statistics for testing the equality of the distributions of $y_{ki}$ based on the rank scores $R_{ki}$ for the observations from the two groups. More specifically, Wilcoxon proposed the rank sum statistic:

$$\text{Wilcoxon rank sum test :} \quad W_n = \sum_{i=1}^{n_1} R_{1i}$$

Note that the sum of the rank scores $\sum_{j=1}^{n_2} R_{2j}$ from the second group may also be used as a statistic. However, it is readily shown (see exercise) that $W_n + \sum_{j=1}^{n_2} R_{2j} = \frac{n(n+1)}{2}$, where $n = n_1 + n_2$. Thus, only one of the sums of the rank scores can be used as a statistic. The test initiated by Mann–Whitney is

$$\text{Mann–Whitney test :} \quad U_n = \sum_{i=1}^{n_1} \sum_{j=1}^{n_2} I_{\{y_{2j} - y_{1i} \leq 0\}} \tag{1.7}$$

where $I_{\{u \leq 0\}}$ is a set indicator with $I_{\{u \leq 0\}} = 1$ if $u \leq 0$ and 0 if otherwise. Since $W_n = U_n + \frac{n_1(n_1+1)}{2}$ in the absence of ties (see exercise), the two tests are equivalent (see Chapter 3). Throughout this book, we will use the Mann–Whitney form of the test and refer to it as the Mann–Whitney–Wilcoxon rank sum test.

To use the test in (1.7), we must find the sampling distribution of $U_n$. Although it is possible to find the exact distribution (see Chapter 3), we are also interested in the behavior of the statistic in large samples, since such behavior will enable the study of efficiency and extend the test to a more general setting (see Chapter 5 and 6). However, unlike the linear regression model and product–moment correlation, it is more difficult to study the asymptotic behavior for such a statistic, as it cannot be expressed in the usual form of an independent sum of random variables. In Chapter 3, we will carefully characterize the asymptotic behavior of this statistic and generalize it in Chapters 5 and 6 to address design issues arising in modern study trials by using the theory of U-statistics.

## 1.2 Measurability and Measure Space

Measurable sets and measures of such sets are the most basic elements of the theory of probability. In this section, we review these fundamental concepts.

### 1.2.1 Measurable Space

Let $\Omega$ be a sample space and $\mathcal{F}$ some collection of subsets of $\Omega$. A $\sigma$-field or $\sigma$-algebra $\mathcal{F}$ is a collection of subsets of $\Omega$, which has the following defining properties:

1. $\phi \in \mathcal{F}$ (empty set is contained in $\mathcal{F}$).
2. If $A \in \mathcal{F}$, then $A^c \in \mathcal{F}$ ($\mathcal{F}$ is closed under complementation).
3. If $A_i \in \mathcal{F}$, then $\bigcup_{i=1}^{\infty} A_i \in \mathcal{F}$ ($\mathcal{F}$ is closed under *countable* unions).

Note that under (2) $\Omega \in \mathcal{F}$ iff (if and only if) $\phi \in \mathcal{F}$. Thus, (1) may be equivalently stated as:

1'. $\Omega \in \mathcal{F}$ (empty set is contained in $\mathcal{F}$)

The pair $\mathcal{S} = (\Omega, \mathcal{F})$ is called a measurable space. So, a measurable space is a sample space equipped with some $\sigma$-field. For any sample space $\Omega$, there always exist two special $\sigma$-fields. One is $\mathcal{F} = \{\Omega, \phi\}$ and the other is $\mathcal{F} = \{\text{all subsets of } \Omega \text{ including } \phi \text{ and } \Omega\}$, representing the smallest and largest $\sigma$-field, respectively. The latter is often called the power set of $\Omega$.

**Example 1**. Consider rolling a die. There are six possible outcomes, 1, 2, ...,6, the die will register when thrown. Let

$$\Omega = \{1, 2, 3, 4, 5, 6\}, \quad \mathcal{F} = \{\text{all subsets of } \Omega \text{ including } \Omega \text{ and } \phi\}$$

Then, $\Omega$ is the sample space containing all possible outcomes from a toss of the die and the power set $\mathcal{F}$ is a $\sigma$-field. Thus, $\mathcal{S} = (\Omega, \mathcal{F})$ is a measurable space.

**Example 2**. In Example 1, consider a collection of subsets of $\Omega$ defined by:

$$\mathcal{F}_2 = \{\phi, \Omega, \{1, 3, 5\}, \{2, 4, 6\}\}$$

It is readily checked that $\mathcal{F}_2$ is also a $\sigma$-field. Thus, $\mathcal{S} = (\Omega, \mathcal{F}_2)$ is again a measurable space. When compared to $\mathcal{F}$ in Example 1, this $\sigma$-field has fewer subsets. In other words, $\mathcal{F}_2$ is contained in $\mathcal{F}$, that is, $\mathcal{F}_2 \subset \mathcal{F}$.

**Example 3**. Let $\Omega = R$, the real line, and

$$\mathcal{F}_2 = \{\phi, \Omega, (-\infty, 0], (0, \infty)\}, \quad \mathcal{F}_3 = \{\phi, \Omega\}$$

Then, $\mathcal{F}_2$ and $\mathcal{F}_3$ are both $\sigma$-fields. Clearly, $\mathcal{F}_3 \subset \mathcal{F}_2$.

The $\sigma$-fields in all three examples above are quite simple, since each has only a finite number of subsets. In general, it is impossible to enumerate the members in a $\sigma$-field if it contains infinitely many subjects. One way of constructing a $\sigma$-field in this case is to use the following result.

**Proposition.** The intersection of $\sigma$-fields is a $\sigma$-field.

Let $\mathcal{A}$ denote a collection of subsets of $\Omega$. To construct a $\sigma$-field including all subsets in $\mathcal{A}$, we can take the intersection of all the $\sigma$-fields containing $\mathcal{A}$. Since the *power set* of $\Omega$ (the collection of all subjects of $\Omega$) is always a $\sigma$-field (see exercise), the intersection always exists. This smallest $\sigma$-field is also called the $\sigma$-field generated by $\mathcal{A}$, denoted by $\sigma(\mathcal{A})$.

**Example 4**. Let $\Omega = R$ and the Borel $\sigma$-field $\mathcal{B}$ be the $\sigma$-field generated by any one of the following class of intervals:

$$\begin{aligned} &\{(-\infty, a]; a \in R\}, \quad \{(a, b]; a, b \in R\}, \quad \{(a, b); a, b \in R\} \\ &\{(a, \infty); a \in R\}, \quad \{[a, \infty); a \in R\}, \quad \{(-\infty, a); a \in R\} \end{aligned} \tag{1.8}$$

More generally, let $\Omega = R^k = \{(x_1, ..., x_k)\,; x_i \in R, 1 \le i \le k\}$ be the $k$-dimensional Euclidean space. The Borel $\sigma$-field $\mathcal{B}^k$ is the $\sigma$-field generated by any one of the following class of subsets:

$$\left\{\otimes_{i=1}^{k}(-\infty, a_i]\,; a_i \in R\right\}, \quad \left\{\otimes_{i=1}^{k}(a_i, b_i]\,; a_i, b_i \in R\right\} \tag{1.9}$$
$$\left\{\otimes_{i=1}^{k}(a_i, b_i)\,; a_i, b_i \in R\right\}, \quad \left\{\otimes_{i=1}^{k}(a, \infty)\,; a_i \in R\right\}$$
$$\left\{\otimes_{i=1}^{k}[a_i, \infty)\,; a_i \in R\right\}, \quad \left\{\otimes_{i=1}^{k}(-\infty, a_i)\,; a_i \in R\right\}$$

where $\otimes_{i=1}^{k} A_i = A_1 \otimes A_2 \cdots \otimes A_k$ denotes the Cartesian product of sets such that $\{(x_1, ..., x_k) : x_i \in A_i, 1 \le i \le k\}$.

The Borel $\sigma$-field contains all open and closed sets. For example, let $O \subset R^k$ be some open set. For any point $a \in O$, the defining properties of an open set ensure that we can find an open interval $O_a \subset O$ that contains $a$. Since the set of rational numbers is dense, we can find a rational number in $O_a$ to index such an open interval. The collection of all such open intervals indexed by a subset of the rational numbers $Q$ covers all points in $O$. Since $Q$ is countable, $O$ is a countable union of open intervals. It thus follows from (3) of the definition of $\sigma$-field that $O \in \mathcal{B}^k$. To show that any closed set is in $\mathcal{B}^k$, simply note that the complement of a closed set is an open set.

The $\sigma$-field is an abstract notion of information and is a building block of probability theory and its vast applications in all areas of research. In many experiments and studies, we may not be interested in every single outcome in the sample space and we can use $\sigma$-field to represent the information pertaining to our interest. For example, if we only care whether the outcome registered by a toss of a die is an odd or even number, the simpler $\sigma$-field $\mathcal{F}_2$ in Example 2 suffices to communicate this interest. The defining properties of a $\sigma$-field ensure that we can perform meaningful operations regarding the pieces of information in a $\sigma$-field as well.

**Example 5**. In Example 2, let $\mathcal{A} = \{\{1, 3, 5\}\}$. This collection has a single subset communicating the interest of observing an odd outcome when tossing the die. However, $\mathcal{A}$ does not contain sufficient information to describe all possible outcomes, since the event of even outcomes, $\{2, 4, 6\}$, is not in $\mathcal{A}$. The $\sigma$-field generated by $\mathcal{A}$, $\mathcal{F}_2 = \{\phi, \Omega, \{1, 3, 5\}, \{2, 4, 6\}\}$, has all the information needed to describe all potential outcomes in terms of the event of interest.

For relatively simple sample spaces like the one in the above example, we can readily construct a $\sigma$-field containing the events of interest. For more complex sample spaces, especially those with an infinite number of elements such as $R$, it is difficult to assemble such a $\sigma$-field through enumerating all its subsets. However, for any events of interest represented by a collection of subsets $\mathcal{A}$, we can always find a $\sigma$-field containing $\mathcal{A}$, such as the one generated by $\mathcal{A}$, $\sigma(\mathcal{A})$, as guaranteed by the Proposition.

### 1.2.2 Measure Space

Given a measurable space $\mathcal{S} = (\Omega, \mathcal{F})$, let $\mu$ define a mapping or function from $\mathcal{F}$ to $R \cup \{\infty\}$, with the following properties:

1. $\mu(A) \geq 0$ for any $A \in \mathcal{F}$.
2. (*Countable additivity*) If $A_i \in \mathcal{F}$ is a disjoint sequence for $i = 1, 2, \ldots$, then

$$\mu\left(\bigcup_{i=1}^{\infty} A_i\right) = \sum_{i=1}^{\infty} \mu(A_i)$$

If $\Omega$ has a finite number of elements, the countable additivity property becomes $\mu\left(\bigcup_{i=1}^{n} A_i\right) = \sum_{i=1}^{n} \mu(A_i)$. Such a mapping $\mu$ is called a measure. A measure space is defined by adding $\mu$ to $\mathcal{S}$ and denoted by the triplet $\mathcal{M} = (\Omega, \mathcal{F}, \mu)$.

Thus, a measure is defined for each element of the $\sigma$-field $\mathcal{F}$, that is, subsets rather than each individual element of the sample space $\Omega$. This distinction is especially important when $\Omega$ has an *uncountable* number of elements such as the real line $R$.

If a $\sigma$-field represents the collection of information to describe some outcome of interest, a measure then allows us to quantify the pieces of information in the $\sigma$-field in a certain fashion according to our interest. For example, in the theory of probability, such a *probability* measure is used to quantify the occurrence of event as represented in a $\sigma$-field in the study of random phenomena (see Section 1.4 for the definition of probability measure).

**Example 6.** Within the context of Example 1, define a set function $\mu$ as follows:

$$\mu(A) = |A|, \quad A \in \mathcal{F}$$

where $|A|$ denotes the number of elements (*singletons*) in $A$. Then, $\mu$ is a measure. This measure $\mu$ assigns 1 to each singleton.

In general, a measure may be constructed by assigning a value to each singleton, and the measure of a subset is then the sum of the measures of all its members. The countable additivity of such a measure follows directly from the additivity of arithmetic addition. For example, to check the countable additivity property (or finite additivity for this example, since $\Omega$ is finite), consider $A = \{1, 2\} = \{1\} \cup \{2\}$. Then, we have

$$\mu(A) = \mu(\{1\}) + \mu(\{2\}) = 2$$

Since $\mu(A)$ counts the number of elements in $A$, it is called the *count* measure. Note that by restricting $\mu$ to $\mathcal{F}_2$ in Example 2, $\mu$ becomes a measure for the measurable space $\mathcal{S} = (\Omega, \mathcal{F}_2)$.

**Example 7**. Consider the number of surgeries performed in a hospital over a period of time. It is difficult to put an upper bound on such an outcome. As a result, this outcome of interest has a theoretical range from 0 to $\infty$. Thus, an appropriate sample space for the number of surgeries is the *countable* space containing the non-negative integers, $\Omega = \{0, 1, 2, \ldots\}$. Let $\mathcal{F}$ be the $\sigma$-field containing all subsets of $\Omega$ including $\Omega$ itself and $\phi$, that is, the power set of $\Omega$.

The measure defined in Example 6 may be extended to the countable space. It is readily checked that $\mu(A \cup B) = \mu(A) + \mu(B)$ for any $A, B \in \mathcal{F}$ with $A \cap B = \phi$. In addition, the countable additivity also holds true. Thus, $\mu$ is a well-defined measure. However, unlike the count measure in Example 6, $\mu(\Omega) = \infty$. Since $\Omega$ is a countable union of measurable sets with finite measure (e.g. $\Omega = \bigcup_{k=0}^{n} \{k\}$), $\mu$ is called a *$\sigma$-finite measure.*

For a sample space with a finite or countable number of elements, we can define a measure by first limiting it for each element, as in Examples 5–7. For a sample space with an uncountable number of elements, such as $R$, this approach generally does not work. For example, for the Borel measurable space $\mathcal{S} = (R, \mathcal{B})$, $\mathcal{B}$ contains all different types of intervals as well as points. It is impossible to develop a measure by first defining its values for all individual points of the sample space $R$. This is because a proper interval such as $(0, 4)$ contains an uncountable number of points and as such it is not possible to extend such a point-based function to a measure that is defined for proper intervals using the defining countable additivity property. It is also generally difficult to define a measure by assigning its values to all subsets in a $\sigma$-field directly because such assignments must satisfy the relationship given by the following theorem.

**Theorem 1**. Given a measure space $\mathcal{M} = (\Omega, \mathcal{F}, \mu)$, we have

1. (*Continuity from below*) If $A_n \in \mathcal{F}$, $A \in \mathcal{F}$ and $A_n \uparrow A$, then $\mu(A_n) \uparrow \mu(A)$.

2. (*Continuity from above*) If $A_n \in \mathcal{F}$, $A \in \mathcal{F}$ and $A_n \downarrow A$, then $\mu(A_n) \downarrow \mu(A)$, provided that at least one $\mu(A_n)$ is finite.

3. (*Continuity subadditivity*) If $A_i \in \mathcal{F}$, then $\mu\left(\bigcup_{i=1}^{\infty} A_i\right) \leq \sum_{i=1}^{\infty} \mu(A_i)$.

4. (*Rule of total measure*) If $\{C_i \in \mathcal{F}; i = 1, 2, \ldots\}$ is a partition of $\Omega$, then $\mu(A) = \sum_{i=1}^{\infty} \mu(A \cap C_i)$.

Note that $A_n \uparrow A$ ($A_n \downarrow A$) means that $A_1 \subseteq A_2 \subseteq \ldots \subseteq A_n \subseteq \ldots$ and $A = \bigcup_{i=1}^{\infty} A_i$, while $\mu(A_n) \uparrow \mu(A)$ ($\mu(A_n) \downarrow \mu(A)$) implies that $\mu(A_n)$ increases (decreases) to $\mu(A)$.

Because of these difficulties, a measure for a $\sigma$-field is generally constructed by first defining it for a class of subsets that generates the $\sigma$-field. For example, if a class of subsets $\mathcal{A}$ is a *ring*, that is, $A \cup B \in \mathcal{A}$ (closed under pairwise union) and $A \cap B^c \in \mathcal{A}$ (closed under relative complement) for any $A, B \in \mathcal{A}$, then any measure defined on $\mathcal{A}$ may be uniquely extended to the $\sigma$-field generated by it. This is known as the *Caratheodory's extension theorem* (e.g., Ash, 1999).

**Example 8**. Consider again the measurable space $S = \left(R^k, \mathcal{B}^k\right)$. Since $\mathcal{B}^k$ is generated by subsets of the form $\mathcal{A} = \left\{\otimes_{i=1}^{k} (a_i, b_i]\, ; a_i, b_i \in R\right\}$ and $\mathcal{A}$ is a ring, it follows that we can create a measure $\mu$ on $\mathcal{B}^k$ if we first define it for such sets. For example, if we measure each $\otimes_{i=1}^{k} (a_i, b_i]$ by its *volume* $\prod_{i=1}^{k} (b_i - a_i)$, we obtain the *Lebesgue* measure for $\mathcal{B}^k$.

**Example 9**. In Example 8, let $k = 1$ and consider $S = (R, \mathcal{B})$. Let $F$ be some non-decreasing, right continuous function $F$ on $R$. For intervals of the form $(a, b]$, let $\mu((a, b]) = F(b) - F(a)$. Then, by Example 8, $\mu$ can be uniquely extended to the Borel $\sigma$-field $\mathcal{B}$. In particular, if $F(x) = \exp(-\alpha x)$ if $x \geq 0$ and 0 if otherwise, where $\alpha$ is a known positive constant, this $F$-induced measure gives rise to the exponential distribution (see Section 1.4 for details about distribution functions).

## 1.3 Measurable Function and Integration

Functions and their integration form the foundation of mathematical science. The *Riemann integral* is the first rigorous definition of integration of a function over an interval in $R$ or subset in $R^k$ and has been widely used in applications, including statistics. However, Riemann integration has several limitations. First, as the definition of Riemann integral relies on $R^k$, it is difficult to extend the notion of integration to general measurable spaces. Second, within the Euclidean space, a wide class of functions of theoretical interest is not Riemann integrable, that is, the integral does not exist. In this section, we review measurable functions and their integrals, which address the fundamental limitations of Riemann integration.

### 1.3.1 Measurable Functions

Consider two measurable spaces, $\mathcal{S}_1 = (\Omega_1, F_1)$ and $\mathcal{S}_2 = (\Omega_2, F_2)$. A mapping from $\Omega_1$ to $\Omega_2$, $T : \Omega_1 \to \Omega_2$ is measurable $F_1/F_2$ if

$$T^{-1}A \in F_1, \quad \text{for any } A \in F_2 \tag{1.10}$$

where $T^{-1}A = \{\omega \in F_1, T(\omega) \in A\}$ is the inverse image of $A \in F_2$ in $F_1$. If $\mathcal{S}_2 = (R, \mathcal{B})$, $T$ is called a *real measurable function* and often denoted by $f$.

Note that $T^{-1}$ should not be interpreted as the inverse mapping of $T$, since $T$ is not necessarily a one-to-one function. Note also that unlike measures that are defined as set functions based on $\sigma$-fields, $T$ is a point-based function defined for each element of the sample space $\Omega_1$.

Let $\sigma(T)$ denote the $\sigma$-field generated by $T$, that is,

$$\sigma(T) = \sigma\left\{T^{-1}A, \quad \text{for any } A \in F_2\right\}$$

Then, $T$ is measurable $\sigma(T)/F_2$. Thus, for any mapping $T$, we can always make it measurable with respect to some $\sigma$-field.

**Example 1.** Let $\mathcal{S}_1 = \mathcal{S}_2 = (R, \mathcal{B})$. Let $f$ be a continuous real function from $\mathcal{S}_1$ to $\mathcal{S}_2$. Consider any open set $O \in F_2 = \mathcal{B}$. By the definition of a continuous function, $f^{-1}O$ is also an open set in $F_1 = \mathcal{B}$. Thus, $f$ is measurable.

**Example 2.** Consider the measurable spaces $\mathcal{S}_1 = (R, \mathcal{B})$ and $\mathcal{S}_2 = (R, F_2)$ in Example 3 of Section 1.2. Define the identity mapping $T$ from $R$ to itself, that is, $T(x) = x$ for any $x \in R$. Then, $T$ is measurable $\mathcal{B}/F_2$. This follows since for any $A \in F_2$, $A$ is either $\phi$, $(-\infty, 0]$, $(0, \infty)$ or $R$, all of which are elements of $\mathcal{B}$. Thus, $T^{-1}A = A \in \mathcal{B}$ for any $A \in F_2$. Note that it is readily checked that the reverse is not true, that is, $T$ is not measurable $F_2/\mathcal{B}$.

It is inconvenient if not impossible to check measurability of a mapping by the defining condition in (1.10). For example, even for a very simple $\sigma$-field $F_2$ and a mapping in Example 2, we must inspect every element $A$ of $F_2$ to ensure that $T^{-1}A \in \mathcal{B}$. When $F_2$ contains an infinite number of subsets, such as $\mathcal{B}$, it is not possible to perform such a task. Fortunately, we can use the following results to circumvent the difficulties.

**Theorem 1.** Let $\mathcal{S}_k = (\Omega_k, F_k)$ denote three measurable spaces $(1 \leq k \leq 3)$. Let $\mathcal{A} \subset F_2$, a collection of subsets contained in $F_2$, and $F_2 = \sigma(\mathcal{A})$. Let $G$ be a $\sigma$-field for $\mathcal{S}_1$. Let $T_1$ and $T_2$ be measurable mappings from $\mathcal{S}_1$ to $\mathcal{S}_2$ and from $\mathcal{S}_2$ to $\mathcal{S}_3$, respectively. Then,

1. If $T_1^{-1}A \in \mathcal{F}_1$ for each $A \in \mathcal{A}$, then $T_1$ is measurable $\mathcal{F}_1/\mathcal{F}_2$.
2. If $T_1$ is measurable $\mathcal{F}_1/\mathcal{F}_2$ and $\mathcal{F}_1 \subset G$, then $T_1$ is measurable $G/\mathcal{F}_2$.
3. If $T_1$ is measurable $\mathcal{F}_1/\mathcal{F}_2$, and $T_2$ is measurable $\mathcal{F}_2/\mathcal{F}_3$, then the composite mapping from $\Omega_1$ to $\Omega_3$, $T_2 \circ T_1$ is measurable $\mathcal{F}_1/\mathcal{F}_3$.

Thus, by Theorem 1/1, we can check measurability of a mapping $T_1$ from $\Omega_1$ to $\Omega_2$ by limiting the defining condition in (1.10) to a collection of subsets generating the $\sigma$-field $\mathcal{F}_2$. In addition, we can use Theorem 1/2 and 1/3 to construct new mappings for a given purpose.

**Example 3**. In Example 2, let $\mathcal{A} = \{(-\infty, 0]\}$. Then, $\mathcal{A} \subset \mathcal{F}_2$. Consider again the identity mapping $T(x) = x$ for any $x \in R$. It is readily checked that $T^{-1}A = (-\infty, 0] \in \mathcal{B}$. Since $\mathcal{F}_2 = \sigma(\mathcal{A})$, it follows from Theorem 1/1 that $T$ is measurable $\mathcal{B}/\mathcal{F}_2$. In comparison to Example 2, we only checked the measurability of $T$ with respect to a single subset $A = (-\infty, 0]$ in $\mathcal{F}_2$.

**Example 4**. Let $\mathcal{S}_1 = \mathcal{S}_2 = \mathcal{S}_3 = (R, \mathcal{B})$. Since $f(x) = \sin(x)$ and $g(x) = |x|$ are both continuous functions, it follows from Example 2 that $f$ is measurable $\mathcal{F}_1/F_2 = \mathcal{B}/\mathcal{B}$ and $g$ is measurable $\mathcal{F}_2/F_3 = \mathcal{B}/\mathcal{B}$. It follows from Theorem 1/3 that $|\sin(x)| = g(f(x)) = g \circ f(x)$ is measurable $\mathcal{F}_1/F_3 = \mathcal{B}/\mathcal{B}$. In other words, $|\sin(x)|$ is a real measurable function on $R$.

**Theorem 2**. Let $\mathcal{S} = (\Omega, \mathcal{F})$ denote a measurable space. Let $f$ and $g$ be real measurable functions. Then,

$$\max(f, g), \quad \min(f, g), \quad f \pm g, \quad f \cdot g, \quad \frac{f}{g} \text{ (if } g \neq 0)$$

are all measurable.

### 1.3.2 Convergence of Sequence of Measurable Functions

Limit and convergence are key to studying the behavior of measurable functions and their integrations. In this section, we discuss the notion of convergence of sequence of measurable functions and in particular review two popular modes of convergence.

Consider a measurable space $S = (\Omega, \mathcal{F})$. For each $n$ $(\geq 1)$, let $f_n$ be a real measurable function defined on $S$, that is, $f_n$ is a mapping from $S$ to $(R, \mathcal{B})$ and $\mathcal{F}/B$ measurable. For such a sequence of measurable functions, we can always construct two sequences of functions, $g_n = \max_{1 \leq k \leq n} f_k$ and $h_n = \min_{1 \leq k \leq n} f_k$ $(n = 1, 2, \ldots)$. By Theorem 3 of Section 1.3, both $g_n$ and $h_n$ are $\mathcal{F}/\mathcal{B}$ measurable. In addition, both sequences are monotone increasing in the sense that $g_n \leq g_{n+1}$ and $h_n \geq h_{n+1}$. Thus, for each element $\omega \in \Omega$, the limits $\lim_{n\to\infty} g_n$ and $\lim_{n\to\infty} h_n$ exist if $\infty$ is considered to be a "real" value. We denote these limits by $\sup_{n\geq 1} f_n = \lim_{n\to\infty} g_n$ and $\inf_{n\geq 1} f_n = \lim_{n\to\infty} h_n$.

Based on these two limit functions we may further define two sequences of functions, $\sup_{n\geq k} f_n$ and $\inf_{n\geq k} f_n$ ($k \geq 1$). It is straightforward to check that these two sequences are monotone (see exercise). Thus, their limits exist, denoted as $\limsup f_n$ and $\liminf f_n$, respectively. It is easy to show that $f_n$ converges iff $\limsup f_n = \liminf f_n$ and the limit is denoted by $\lim_{n\to\infty} f_n$. The following asserts that all these functions constructed are also measurable (see exercise).

**Theorem 3.** Let $\mathcal{S} = (\Omega, \mathcal{F})$ be a measurable space. Let $f_n$ be a sequence of real measurable functions. Then,

1. The following functions,

$$\max_{1\leq n\leq k} f_n, \quad \min_{1\leq n\leq k} f_n, \quad \sup_{n\geq 1} f_n, \quad \inf_{n\geq 1} f_n$$

$$\limsup f_n = \lim_{n\to\infty} \sup_{k\geq n} f_k, \quad \liminf f_n = \lim_{n\to\infty} \inf_{k\geq n} f_k$$

are all measurable.

2. The set $\{\omega : \limsup f_n = \liminf f_n\}$ is measurable. In particular, if $\lim_{n\to\infty} f_n$ exists, then the limiting function is measurable.

As will be seen in Section 1.3.3, we can alter the values of a real measurable function in a set of measure 0 without changing the integral of the function. Thus, two functions $f$ and $g$ that differ on a set of measure 0, that is, $\mu(\{x : f(x) \neq g(x)\}) = 0$, are not distinguishable by integration and are said to be equivalent *almost everywhere* (a.e.). By applying this notion of equivalence to a sequence of measurable functions, we say that $f_n$ converges to $f$ almost everywhere, $f_n \to f$ a.e., if the set on which $f_n$ does not have a limit has measure 0, that is,

$$\mu(\{\omega : \limsup f_n \neq \liminf f_n\}) = 0$$

Thus, under a.e. convergence, $f_n$ can have points on which $\lim_{n\to\infty} f_n$ does not exist, but the set of such points must have a measure of 0.

The a.e. convergence is defined for each element $\Omega$. Another measure of convergence for $f_n$ is to look at the difference between $f_n$ and its limit $f$ as quantified by $\mu$ directly. More specifically, $f_n$ is said to converge to $f$ *in measure*, denoted by $f_n \to_\mu f$, if for any $\delta > 0$, the deterministic sequence, $d_{n,\delta} = \mu[|f_n - f| > \delta]$, converges to 0 as $n \to \infty$. Stated formally, for any $\delta$, $\epsilon > 0$, we can find some integer $N_{\epsilon,\delta}$ such that

$$d_{n,\delta} = \mu(|f_n - f| > \delta) < \epsilon, \quad \text{for all } n \geq N_{\epsilon,\delta}$$

As will be seen in Section 1.6, this measure of convergence is widely used in statistics asymptotics to study large sample behaviors of statistics and/or model estimates.

The two measures of convergence for sequences of functions are not equivalent as the following example demonstrates.

**Example 5.** Let $f_{nk}$ be functions from $\mathcal{S} = (R, \mathcal{B})$ to itself as defined by

$$f_{nk}(x) = \begin{cases} 1 \text{ if } \frac{k-1}{n} \leq x \leq \frac{k}{n} \\ 0 \text{ if otherwise} \end{cases}, \quad 1 \leq k \leq n$$

Let $\mu$ be the Lebesgue measure on $S$. Then, the sequence, $f_{11}$, $f_{21}$, $f_{22}$, $f_{31}$, $f_{32}$, $f_{33}$,..., converges to function 0 in measure since $\mu(|f_{nk} - 0| > \delta) \leq \frac{1}{n} \to 0$ for any $\delta > 0$. However, the sequence does not converge a.e. and in fact, it converges nowhere since it may take values 0 and 1 for sufficiently large $n$.

On the other hand, consider $f_n(x) = I_{\{x>n\}}$ $(n \geq 1)$. Then, $f_n(x) \to 0$ for all $x \in R$. However, since $\mu(|f_n - 0| > \delta) = \mu(x > n) = \infty$ for any $\delta > 0$ $(\delta < 1)$, $f_n$ does not converge in measure.

In the above example, $\mu(R) = \infty$. This is not incidental, as in the construction of the second part of the example, as the following theorem indicates.

**Theorem 4.** Let $\mathcal{S} = (\Omega, \mathcal{F})$ be a measurable space. If $\mu(\Omega) < \infty$, then convergence a.e. implies convergence in measure.

Proof. Let $f_n(x) \to f(x)$ a.e. Then, since

$$\{\omega : f_n(\omega) \nrightarrow f(\omega)\} = \bigcup\nolimits_{\delta>0} \left\{\omega : \limsup_{n\to\infty} |f_n(\omega) - f(\omega)| > \delta\right\}$$

it follows that $\mu(\{\omega : \limsup_{n\to\infty} |f_n(\omega) - f(\omega)| > \delta\}) = 0$ for any $\delta > 0$. From the continuity properties of measure (with finite total measure), $\mu(\{\omega : \sup_{k=n}^{\infty} |f_n(\omega) - f(\omega)| > \delta\}) \to 0$. Thus

$$\mu(\{\omega : |f_n(\omega) - f(\omega)| > \delta\}) \to 0$$

and $f_n(x) \to f(x)$ in measure.

Thus, a.e. convergence is generally stronger than convergence in measure.

### 1.3.3 Integration of Measurable Functions

In Riemann calculus, the definition of integral relies critically on the metric of the Euclidean space. For example, consider integrating a function $f$

over an interval $[a, b]$. By partitioning the interval into $n$ subintervals, $a = a_0 < a_1 < ... < a_n = b$, we obtain a *Riemann sum* $\sum_{i=1}^{n} f(x_i)(a_i - a_{i-1})$ with $x_i \in \Delta_i = [a_{i-1}, a_i]$ $(1 \le i \le n)$. If such a sum converges as $\delta_n = \max_{1 \le i \le n}(a_i - a_{i-1}) \to 0$ and the limit does not depend on the selection of $x_i$ and $a_i$, $f$ is *Riemann integrable*, with $\int_a^b f(x)dx$ denoting the limit as the *Riemann integral* over $[a, b]$. The metric structure of $R$ is seen to be fundamental to the definition of the integral, since for a general measure space $\mathcal{M} = (\Omega, \mathcal{F}, \mu)$, it is unclear how the partition would be performed on $\Omega$. Further, even within $\mathcal{M} = (R, \mathcal{B}, \mu)$, many functions are not *Riemann integrable*. For example, let $A$ be the set of all irrational numbers within $[0, 1]$ and $f(x) = I_A$. Then, $f(x)$ is measurable since $A \in \mathcal{B}$. However, the *Riemann sum* $\sum_{i=1}^{n} f(x_i)(a_i - a_{i-1})$ does not have a limit; it is 0 or 1 if $x_i$ are all rational or all irrational numbers.

Integration for a general measurable function $f$ from $M_1 = (\Omega, \mathcal{F}, \mu)$ to $M_2 = (R, \mathcal{B}, \mu)$ is defined instead by partitioning $R$ in the image space $M_2$. For example, if $f$ is bounded, that is, $c \le f(\omega) \le d$ for all $\omega \in \Omega$, partition $[c, d]$ into $n$ subintervals, $c = c_0 < c_1 < ... < c_n = d$. For each $\Lambda_i = [c_{i-1}, c_i]$, its preimage $A_i = f^{-1}\Delta_i$ is a measurable set in $\mathcal{F}$ with the measure $\mu(A_i)$. Thus, the sum $\sum_{i=1}^{n} c_i\mu(A_i)$ is well defined. If this sum converges as $\lambda_n = \max_{1 \le i \le n}(c_i - c_{i-1}) \to 0$ and the limit is independent of the partition, $f$ is *integrable with respect to* $\mu$ and the integral is denoted by $\int_\Omega f d\mu$. Thus, the definition of integration with respect to a general measure space differs only in how the partition is performed. However, this difference is fundamental as it addresses the two key limitations of Riemann integration.

More rigorously, we define integration, denoted by $\int$, in several steps according to the complexity of the function.

1. If $f$ is an indicator function of a measurable set $A \in \mathcal{F}$, that is, $f = I_A$, we define the integral of $f$ by $\int_\Omega f d\mu = \mu(A)$.

2. For a *simple* measurable function (a linear combination of indicator functions), $f = \sum_{i=1}^{n} c_i I_{A_i}$, where $c_i \in R$ and $A_i \in \mathcal{F}$, define $\int_\Omega f d\mu = \sum_{i=1}^{n} c_i\mu(A_i)$. The integral so defined is unique. In other words, it can be shown that if $f$ can be expressed by a different linear combination, say $f = \sum_{i=1}^{n} d_i I_{B_i}$ with $d_i \in R$ and $B_i \in \mathcal{F}$, $\sum_{i=1}^{n} d_i\mu(B_i) = \sum_{i=1}^{n} c_i\mu(A_i)$ (see exercise).

3. For $f \ge 0$, we first find a monotone increasing sequence of nonnegative simple measurable $f_n$ that converges to $f$ and then define its integral as the limit of the sequence of the integrals of $f_n$.

More specifically, partition $\Omega$ into $n^2 + 1$ parts according to the values of $f$ as follows:

$$f_n(\omega) = \begin{cases} \frac{i-1}{n} \text{ if } \frac{i-1}{n} \leq f \leq \frac{i}{n}, \text{ for } i = 1, 2...n^2 \\ n \quad \text{if } f \geq n \end{cases} \tag{1.11}$$

Then, $f_n$ is a monotone increasing sequence of non-negative simple measurable functions and $f = \lim_{n\to\infty} f_n$ (see exercise). Define $\int_\Omega f d\mu = \lim_{n\to\infty} \int_\Omega f_n d\mu$. As in 2, it can be shown that the integral does not depend on the choice of $f_n$ and thus the integral is uniquely defined (see exercise).

4. For general $f$, write $f = f^+ - f^-$, where $f^+ = \max(f, 0)$ and $f^- = \max(-f, 0)$ are the positive and negative parts of $f$ and both are nonnegative. The integral of $f$ is defined as $\int_\Omega f d\mu = \int_\Omega f^+ d\mu - \int_\Omega f^- d\mu$, provided that at least one of $\int_\Omega f^+ d\mu$ and $\int_\Omega f^- d\mu$ is finite.

The integration of a measurable function $f$ on a measurable set $C \in F$ is defined as $\int_A f d\mu = \int_\Omega f I_A d\mu$. If $\mathcal{M} = (R, \mathcal{B}, \mu)$ and $\mu$ is the Lebesgue measure, $\int_A f d\mu$ is called the *Lebesgue* integral and is often denoted by $\int_A f(x)\,dx$.

From the definition, it is clear that $\int_\Omega f d\mu = 0$ if $f = 0$ a.e.. Thus, functions that differ on set of measure 0 have the same integrals.

**Example 6**. For the measure space in example 7 of Section 1.2, any function on it is measurable. The integral of a real function $f$ on a subset $A$ is actually the sum of its values of points in $A$, that is,

$$\int_A f d\mu = \sum_{i \in A} f(i) = \sum_{(f(i)>0)\cap A} f(i) - \sum_{(f(i)<0)\cap A} f(i)$$

For a finite subset $A$, the above is well defined. If $A$ contains infinitely many points, $\sum_{(f(i)>0)\cap A}$ and $\sum_{f(i)<0} f(i)$ may both be $\infty$, in which case $\int_A f d\mu$ does not exist.

**Example 7**. Consider again $f(x) = I_A$, where $A$ consists of all irrational numbers in $[0, 1]$. As noted earlier, this function is not Riemann integrable as it "oscillates" too much between 1 and 0. However, since the complement of $A$, the set of rational numbers in $[0, 1]$, is a countable union of points with measures 0, $\mu(A) = 1$. Thus,

$$\int_{[0,1]} f d\mu = \int_{[0,1]} I_A d\mu = \mu(A) = 1$$

and $f$ is Lebesgue integrable.

In general, it is difficult to compute the integral of a measurable function defined on a general measure space. As in Riemann calculus, the change of variable formula is very useful in converting such an integration problem into the more concrete Euclidean space.

Let $M_1 = (\Omega_1, \mathcal{F}_1, \mu_1)$ and $M_2 = (\Omega_2, \mathcal{F}_2, \mu_2)$ be two measure spaces and $h$ be a measurable mapping from $M_1$ to $M_2$. If $f$ is a real measurable function on $M_2$, then the composite $g = f \circ h$ is a real measurable function from $M_1$ to $S = (R, B)$. If $\mu_2$ is induced by $\mu_1$ by $h$, then the integral of $f$ with respect to $M_2$ can be computed from the integral of $g$ with respect to $M_2$ and vice versa, as the following theorem asserts.

**Theorem 5** (*Change of variable formula*). Under the above assumptions, we have

$$\int_{\Omega_1} g d\mu_1 = \int_{\Omega_1} f \circ h d\mu_1 = \int_{\Omega_2} f d\mu_2$$

The equality is understood as if one side of the equation exists, then both exist and agree with each other.

The theorem is most useful when $M_2 = (R, B, \mu)$. In this case, integration over an abstract measure space is changed to integration on the familiar space $R$. We will illustrate applications of Theorem 5 with respect to calculations of moments of random variables in Section 1.5.

Most functions used in statistics are Riemann integrable. A real-valued function $f$ on $[a, b]$ is Riemann integrable iff it is bounded and continuous almost everywhere (that is the measure of the set of non-continuity points is 0). For such functions, the Lebesgue and Riemann integrals are equal to each other. Thus, integration of such functions with respect to the Lebesgue measure can be carried out by the usual Riemann integration calculus using powerful tools such as Newton-Leibnitz and integrate by parts formulas.

The Fubini theorem may be used when integrating functions over $R^k$. The theorem states that if the absolute value of the function is integrable, then the Lebesgue integral over $R^k$ may be computed through iterated integrals, and the order of iteration does not matter. For example, if $|f(x, y)|$ is integrable over $R^2$, then,

$$\int_{R^2} f(x, y) d\mu = \int_R \left( \int_R f(x, y) dx \right) dy = \int_R \left( \int_R f(x, y) dy \right) dx$$

### 1.3.4 Integration of Sequences of Measurable Functions

As seen in Section 1.3.3, functions that are equal to each other almost everywhere have the same integral. Thus, if a sequence of measurable functions, $f_n \to f$ a.e., that is, $\mu(\{x : f_n(x) \to f(x)\}^c) = 0$, we would expect that the sequence of the integral of the sequence, $\int f_n d\mu \to \int f d\mu$, as $n \to \infty$. This is indeed true under some sensible assumptions. Listed below are some important properties with respect to integration of functions and sequences of functions.

1. (*Linearity*) If $f$ and $g$ are integrable, then $\int (af + bg)\, d\mu = \int af d\mu + \int bg d\mu$ for any $a, b \in R$.

2. If $f_i$ are integrable, then $\int (\sum_{i=1}^{\infty} f_i)\, d\mu = \sum_{i=1}^{\infty} \int f_i d\mu$.

3. If $f$ is an integrable function and $A_i$ are disjoint measurable subsets, then $\int_{\bigcup_{i=1}^{\infty} A_i} f d\mu = \sum_{i=1}^{\infty} \int_{A_i} f d\mu$.

4. $\int f d\mu \geq 0$ if $f$ is integrable and $f \geq 0$ a.e. The equality is achieved if and only if $f = 0$ a.e.

5. (*Fatou's Lemma*) If $f_n \geq 0$ and are integrable, $\liminf_{n\to\infty} \int f_n d\mu \geq \int (\liminf_{n\to\infty} f_n) d\mu$.

6. (*Monotone Convergence Theorem*) If $f_n$ are integrable and $0 \leq f_n \uparrow f$ a.e., then $\int f_n d\mu \uparrow \int f d\mu$.

7. (*Dominated Convergence Theorem*) If $f_n$ are integrable, $f_n \to f$ a.e., $|f_n| \leq g$ a.e. for all $n$ and $\int g d\mu < \infty$, then $\int f_n d\mu \to \int f d\mu$.

The special case of (7) in which $g$ is constant and the integration is over a subset with finite measure, is called the *bounded convergence theorem*. Part (5)–(7) are particularly useful in measure and probability theories as they permit the change of the order of integration and limit. In statistic asymptotic theory, we often face the problem of integrating a sequence of *random variables* (a special class of measurable functions), which is difficult to carry out because of the known measure $\mu$. However, such sequences often converge to a simple limit, the *normal random* variable, and integration of such a limit is easy to compute. We discuss random variables and convergence of sequences of such variables in Section 1.6.

Let $\mathcal{M} = (\Omega, \mathcal{F}, \mu)$ be a measure space and $f$ be some real measurable function defined on $\mathcal{M}$. Then, by applying the integral properties above, it is readily shown that $\upsilon(E) = \int_E f d\mu$ is a measure for any $E \in \mathcal{F}$ (see exercise). In addition, $\upsilon$ is *absolutely continuous* with respect to $\mu$ in the sense that $\mu(A) = 0$ implies $\upsilon(A) = 0$ for any $A \in \mathcal{F}$. The theorem below claims that the reverse is also true.

**Theorem 6** (*Radon–Nikodym*). If $\upsilon$ and $\mu$ are $\sigma$-finite measures $\mathcal{F}$ and $\upsilon$ is absolutely continuous with respect to $\mu$, then there is a $g \geq 0$ such that $\upsilon(E) = \int_E g d\mu$. The function $g$ is called the Radon–Nikodym derivative of $\upsilon$ relative to $\mu$.

From the Radon–Nikodym theorem, all absolute continuous measures relative to the usual Lebesgue measure on the Borel $\sigma$-field $\mathcal{B}$ have the form $\upsilon(E) = \int_E g d\mu$ for $E \in \mathcal{B}$. Let $F(x)$ be an antiderivative of $g$, that is,

$F'(x) = \frac{d}{dx}F = g(x)$. Then, the measure $\upsilon$ on $\mathcal{B}$ is determined by $F(x)$ and vice versa since $\upsilon((a,b]) = F(b) - F(a)$.

In Section 1.2, we discussed how to construct a measure on $M = (R, \mathcal{B}, \mu)$ from a nondecreasing, right continuous function $F(x)$. This $F$-induced measure is known as the Lebesgue–Stieltjes measure. There are three basic types of $F(x)$.

1. $F(x)$ is a step function. The jumps can only occur at a countable number of points (see exercise). The $F$-induced measure at such points is positive and 0 elsewhere. Thus, such a measure concentrates on these jump points and is called the discrete measure. A discrete measure is not absolute continuous relative to the Lebesgue measure.

2. $F(x)$ is absolutely continuous. For such a function, its derivative $f(x)$ exists a.e., and further, $F(x) = F(a) + \int_a^x f(t)dt$. The corresponding measure $\upsilon$ is given by $\upsilon(E) = \int_E f(x)\,dx$.

3. $F(x)$ is singular. For such a function, $F_s'(x)$ exists and $F_s'(x) = 0$ a.e., but $F(x)$ is in general not an antiderivative of $F_s'(x)$, unless $F(x)$ is a constant. The corresponding measure is called a singular measure on the Borel $\sigma$-field $B$. In general, it has a positive measure for a subset with Lebesgue measure 0, but no point has a positive measure. The singular measure is not absolutely continuous relative to the Lebesgue measure.

The following theorem states that a general $F(x)$ can be decomposed as a sum of the three basic types of functions.

**Theorem 7.** A right, continuous function $F(x)$ on $M = (R, \mathcal{B}, \mu)$ has the following decomposition:

$$F(x) = F_a(x) + F_s(x) + F_d(x)$$

Proof. The measure of a point $a$ is $F(a) - F(a+)$. Thus, a point in $R$ has a positive measure if and only if $F$ is discontinuous at the point.

Let $F_d(x) = F(x) - F(x^+) > 0$. Then, the number of such points $x$ is countable and $F(x) - F_d(x)$ is continuous (see exercise). Its derivative, $f(x) = (F(x) - F_d(x))'$, exists a.e.. Let $F_a(x)$ be an anti-derivative of $f(x)$, that is a function such that $F_a(x) = F_a(a) + \int_a^x f(t)\,dt$. Then, $F_a(x)$ is absolutely continuous, and $F(x) - F_d(x) - F_a(x)$ is a singular function.

Note that all three terms are not necessary to appear in a decomposition. For statistical applications, $F(x)$ is either discrete or absolutely continuous, or sometimes a combination of the two. Singular measures are usually of theoretical interest.

## 1.4 Probability Space and Random Variables

### 1.4.1 Probability Space

A probability space, $\mathcal{P} = (\Omega, \mathcal{F}, P)$, is a special case of measure space when the measure $P$ is a probability measure satisfying $P(\Omega) = 1$. Thus, as in the case of general measure space, a probability measure is defined for each element of $\mathcal{F}$, that is, subsets rather than each individual element of $\Omega$. Likewise, $P$ satisfies the defining as well as the additional properties of a measure given by Theorem 1 of Section 1.2.

**Example 1**. In Example 6 of Section 1.2, we can normalize the count measure to obtain a probability measure $P$ as follows:

$$P(k) = \frac{1}{6}, \quad 1 \leq k \leq 6$$
$$P(A \cup B) = P(A) + P(B), \quad \text{for } A \cap B = \phi \quad A, B \in \mathcal{F}$$

For example, to check the *countable additivity* property, or rather *finite additivity* since $\Omega$ is finite, consider $A = \{1, 2\} = \{1\} \cup \{2\}$. Then, we have

$$P(A) = P(1) + P(2) = \frac{2}{6}$$

Note that by restricting $P$ to $\mathcal{F}_2$ in Example 2, $P$ becomes a probability measure for the measurable space $\mathcal{S} = (\Omega, \mathcal{F}_2)$.

In general, if the total measure of the sample space is finite, we can normalize it into a probability space, that is, rescale the original measure $\mu$ by a factor $\frac{1}{\mu(\Omega)}$, where $\mu(\Omega)$ is the total measure of the sample space.

**Example 2**. In example 7 of Section 1.2, we discussed a measure space with countable elements, $\Omega = \{0, 1, 2, \ldots\}$, together with a count measure. Since the total measure of the sample space is $\infty$, the count measure cannot be normalized to form a probability measure. To define a probability measure $P$, we may assign to each outcome $k$ of $\Omega$ a value $p_k = P(k)$ satisfying:

$$p_k \geq 0, \quad \sum_{k=1}^{\infty} p_k = 1 \tag{1.12}$$

Since $\sum_{k=1}^{\infty} p_k$ is a convergent series, $p_k$ must converge to 0 as $k \to \infty$ (see exercise). Although various choices of $p_k$ are available, the most popular is to define $P$ as follows:

$$P(k) = \frac{\mu^k}{k!} \exp(-\mu), \quad \mu > 0, \quad k = 0, 1, \ldots \tag{1.13}$$

It is readily checked that $P$ satisfies the defining properties of a measure (see exercise). Thus, $P$ is a well-defined probability measure and is known as the Poisson probability law.

If we want to construct an absolute continuous probability measure $P$ with respect to the Lebesgue measure, then by the Radon–Nikodym theorem, $P(-\infty, \infty) = \int_{-\infty}^{\infty} f(x)\, dx$ for some non-negative function $f(x)$, and the requirement of total probability measure of the sample space =1 means that $\int_{-\infty}^{\infty} f(x)\, dx = 1$.

**Example 3.** Consider the positive function, $f(x) = \frac{1}{\sqrt{2\pi}} \exp\left(-\frac{x^2}{2}\right)$. It is readily checked that $\int_{-\infty}^{\infty} f(x)\, dx = 1$. Define a function $P$ for each interval of the type $(-\infty, a]$ by

$$P((-\infty, a]) = \int_{-\infty}^{a} \frac{1}{\sqrt{2\pi}} \exp\left(-\frac{x^2}{2}\right) dx$$

Since $A_n = (-\infty, n] \uparrow A$, it follows from Theorem 1/1 of Section 1.2 that

$$\begin{aligned} P((-\infty, \infty)) &= \lim_{n\to\infty} P((-\infty, n]) = \lim_{n\to\infty} \int_{-\infty}^{n} \frac{1}{\sqrt{2\pi}} \exp\left(-\frac{x^2}{2}\right) dx \\ &= \int_{-\infty}^{\infty} \frac{1}{\sqrt{2\pi}} \exp\left(-\frac{x^2}{2}\right) dx \end{aligned}$$

As $\frac{1}{\sqrt{2\pi}} \exp\left(-\frac{x^2}{2}\right)$ is continuous and bounded, the integral can be completed by Riemann calculus. It is readily shown that $\int_{-\infty}^{\infty} \frac{1}{\sqrt{2\pi}} \exp\left(-\frac{x^2}{2}\right) dx = 1$ (see exercise). Thus, $P$ is a probability measure for $\mathcal{S} = (R, \mathcal{B})$.

**Example 4.** Consider the measurable space $\mathcal{S} = (R, \mathcal{F}_2)$ defined in Example 3 of Section 1.1. Let

$$P_2(A) = \begin{cases} 0 \text{ if } A = \phi \\ \frac{1}{2} \text{ if } A = (-\infty, 0]\,, \text{ or } A = (0, \infty) \\ 1 \text{ if } A = \Omega \end{cases}$$

Then, $P_2$ is a probability measure for the measurable space $\mathcal{S} = (R, \mathcal{F}_2)$. Here, the sample space $R$ is uncountable, but $\mathcal{F}_2$ contains a finite number of subsets. The probability measure $P_2$ is defined for each of the four subsets of $\mathcal{F}_2$.

### 1.4.2 Random Variables

A random variable, $X$, is a real measurable function from a probability space, $\mathcal{S} = (\Omega, \mathcal{F})$, to the measurable space, $\mathcal{R} = (R, \mathcal{B})$. Since intervals of the form $(-\infty, a]$ $(a \in R)$ generate $\mathcal{B}$, it follows from Theorem 1 of Section 1.3 that $X$ is a random variable iff

$$X^{-1}((-\infty, a]) = \{X \leq a\} \in \mathcal{F}, \quad \text{for any } a \in R \tag{1.14}$$

The above often serves as the definition of a random variable.

Note that we can replace $(-\infty, a]$ in (1.14) by any one class of the intervals discussed in Example 4 of Section 1.2. For example, $X$ may also be defined by the intervals of the form $[a, \infty)$:

$$X^{-1}([a, \infty)) = \{X \geq a\} \in \mathcal{F}, \quad \text{for any } a \in R$$

For a random variable $X$, we denote by $\sigma(X)$ the $\sigma$-field generated by $X$.

**Example 5**. Consider Example 1 of Section 1.2 for the possible outcomes resulting from tossing a die and let $X$ be the number of dots registered on the die when tossed. Then, $X$ is a random variable defined on the measurable space $\mathcal{S} = (\Omega, \mathcal{F})$ introduced in that example. To see this, consider intervals of the form $(-\infty, a]$. It is readily checked that

$$\{X \leq a\} = \begin{cases} \Phi & \text{if } a < 1 \\ \{1, 2, \ldots, \lceil a \rceil\} & \text{if } 1 \leq a \leq 6 \\ \{1, 2, \ldots, 6\} & \text{if } a > 6 \end{cases} \tag{1.15}$$

where $\lceil a \rceil$ denotes the largest integer function, that is, $\lceil a \rceil$ is an integer with $a - 1 < \lceil a \rceil \leq a$. Since subsets of the form $\{1, 2, \ldots, \lceil a \rceil\}$ are all in $\mathcal{B}$, (1.15) shows that $X$ is a random variable. In addition, it is readily checked that $\sigma(X) = \mathcal{F}$.

**Example 6**. Within the context of Example 2 of 1.2, let $X = 1$ if the number of dots registered in a toss of the die is an even number and 0 if otherwise. Since

$$\{X \leq a\} = \begin{cases} \Phi & \text{if } a < 0 \\ \{1, 3, 5\} & \text{if } 0 \leq a < 1 \\ \Omega & \text{if } a \geq 1 \end{cases}$$

and $\{1,3,5\} \in \mathcal{F}_2$, it follows that $X$ is a random variable. Also, it is readily checked that $\sigma(X) = \{\{1,3,5\}, \{2,4,6\}, \Phi, \Omega\} = \mathcal{F}_2$, the $\sigma$-field $\mathcal{F}_2$ defined in that example.

Note that $X$ is also a random variable with respect to the measurable space $\mathcal{S} = (\Omega, \mathcal{F})$ in Example 1 of Section 1.1. This follows from Theorem 1/2 of Section 1.3 since $\sigma(X_2) \subset \mathcal{F}$.

**Example 7**. Consider a measurable space, $\mathcal{S} = (\Omega, \mathcal{F})$. Let $X(\omega) = c$, a real function that maps every element $\omega \in \Omega$ to a constant $c \in R$. Then,

$$\{X \leq a\} = \begin{cases} \Phi & \text{if } a < c \\ \Omega & \text{if } a \geq c \end{cases}$$

Thus, $\sigma(X) = \{\Phi, \Omega\}$, the smallest $\sigma$-field. Thus, $X$ is a random variable. This example shows that any constant is a random variable.

**Example 8**. Let $f$ be a continuous function from $\mathcal{S}_1 = (R, \mathcal{B})$ to $\mathcal{S}_2 = (R, \mathcal{B})$. Then, for any open set $O \in F_2 = \mathcal{B}$, $f^{-1}O \in F_1 = \mathcal{B}$. Thus, $f$ is a random variable defined on the measurable space $\mathcal{S}_1 = (R, \mathcal{B})$.

### 1.4.3 Random Vectors

Random vectors are defined in the same way. A $k$-dimensional random vector, $\mathbf{X} = (X_1, \ldots, X_k)^\top$, is a measurable mapping from a measurable space, $\mathcal{S} = (\Omega, \mathcal{F})$ to $\mathcal{R}^k = (R^k, \mathcal{B}^k)$, where $R^k$ denotes the $k$-dimensional Euclidean space and $\mathcal{B}^k$ the associated Borel $\sigma$-field. Since rectangles of the form $(-\infty, a_1] \times \ldots \times (-\infty, a_k]$ generate $\mathcal{B}^k$, a random vector is also defined as:

$$\{X_1 \leq a_1, \ldots, X_k \leq a_k\} = \bigcap_{i=1}^{k} \{X_i \leq a_i\} \in \mathcal{F}, \quad a_1, \ldots, a_k \in R \qquad (1.16)$$

As in the case of a random variable, we may replace the rectangles in (1.16) by other classes of rectangles such as $[a_1, \infty) \times \ldots \times [a_k, \infty)$ and even mixtures of different classes of rectangles such as

$$(-\infty, a_1] \times [a_2, \infty) \times \ldots \times [a_{k-1}, \infty) \times [a_k, \infty)$$

**Example 9**. If $\mathbf{X} = (X_1, \ldots, X_k)^\top$ is a random vector, then any of its components $X_i$ is a random variable. For example, let $a_{nj} = n$ for all $j$ except $j = i$. Then, as

$$A_n = \{X_1 \leq n, \ldots, X_i \leq a_i, \ldots, X_k \leq n\} \in \mathcal{F}$$

it follows that $\{X_i \le a_i\} = \bigcup_{n=1}^{\infty} A_n \in \mathcal{F}$.

**Theorem 1**. Let $\mathcal{S} = (\Omega, \mathcal{F})$ denote a measurable space.

1. Let $f_i : \Omega \to R$ $(1 \le i \le k)$ and $\mathbf{g} = (f_1, \ldots, f_k)^\top$. Then, $\mathbf{g}$ is measurable $\mathcal{F}/\mathcal{B}^k$ if and only if $f_i$ is measurable $F/\mathcal{B}$.

2. If $f$ is a continuous mapping from $R^k$ to $R^m$, then $f$ is measurable $\mathcal{B}^k/\mathcal{B}^m$.

3. If $\mathbf{X}$ is a $k$-variate random vector and $f$ is a continuous (or measurable) mapping from $R^k$ to $R^m$, then $f(\mathbf{X})$ is measurable $F/R^m$.

Thus, by Theorem 1 the reverse in Example 10 is also true: if each of the components of $\mathbf{X}$ is a random variable, then $\mathbf{X}$ is a random vector. In addition, Theorem 1 generalizes the results in Example 9 to a general case of random vector.

**Example 10**. Consider randomly drawing a point from the unit disk in $R^2$. The sample space is $\Omega = \{(x, y) : x^2 + y^2 \le 1\}$. Since $\Omega \in \mathcal{B}^2$, $(X, Y)$ is a random vector with respect to $\mathcal{F} = \{A; A = \Omega \bigcap B, B \in \mathcal{B}^2\}$, the $\sigma$-field $\mathcal{B}^2$ restricted to $\Omega$. It follows from Theorem 1 that the $X$-coordinate, $X$, of $(X, Y)$ is a random variable. It is easy to check that if $a < -1$, $\{X \le a\} = \Phi$, and if $a > 1$, then $\{X \le a\} = \Omega$.

**Example 11**. Let $\mathcal{S} = (\Omega, \mathcal{F})$ be a measurable space. Let $f(X, Y) = X + Y$ be a mapping from $R^2$ to $R$. If $X$ and $Y$ are random variables, then it follows from Theorem 1 that $Z = f(X, Y) = X + Y$ is also a random variable. Similarly, $Z = XY$, $Z = X/Y$ $(Y \ne 0)$ and $Z = X - Y$ are also random variables.

**Example 12**. Let $X_i$ be a finite sequence of $n$ random variables. By Theorem 3 of Section 1.3, $\max_{1 \le i \le n} X_i$ and $\min_{1 \le i \le n} X_i$ are both measurable. It follows from Example 11 that the range of the sequence, $\max_{1 \le i \le n} X_i - \min_{1 \le i \le n} X_i$, is also a random variable.

**Example 13**. Let $X_1$ and $X_2$ be two random variables. Then, $\{X_1 < X_2\}$ is a measurable set and thus the indicator $I_{\{X_1 < X_2\}}$ is a random variable.

We can utilize the results in Example 11 to show this. Since $Y = X_1 - X_2$ is a random variable by Example 12, the set $\{X_1 < X_2\} = \{Y < 0\}$ is measurable.

Alternatively, we can show directly that $\{X_1 < X_2\}$ is a measurable set. Since the set of rational numbers $Q$ is dense, it follows that (see exercise):

$$\{X_1 < X_2\} = \bigcup_{r \in Q} \{X_1 < r\} \cap \{r < X_2\} \tag{1.17}$$

For each $r \in Q$, $\{X_1 < r\}$ and $\{r < X_2\}$ are both measurable. Since $\mathcal{B}$ is closed under intersection and countable unions, the conclusion follows.

**Example 14**. Let $\mathbf{X} = (X_1, X_2, \ldots, X_k)^\top$ be a random vector. Then, the ordered vector:

$$\left(X_{(1)}, X_{(2)}, \ldots, X_{(k)}\right), \quad \text{with } X_{(1)} < X_{(2)} < \cdots < X_{(k)} \tag{1.18}$$

is also a random vector, where $\{(1), (2), \ldots, (k)\}$ is some permutation of the integer set $\{1, 2, \ldots, k\}$.

Let $\mathbf{i} = \{i_1, i_2, \ldots, i_k\}$ index a permutation of $\{1, 2, \ldots, k\}$ and $P^k$ denote the collection of all such permutations. Then, $\mathbf{X_i} = (X_{i_1}, X_{i_1}, \cdots, X_{i_k})^\top$ is a random vector and thus the subset $\mathbf{B_i} = \{X_{i_1} \leq X_{i_1} \leq \cdots \leq X_{i_k}\}$ is measurable. Then, for any $a_1, \ldots, a_k \in R$ (see exercise),

$$\begin{aligned} A = \left\{X_{(1)} \leq a_1, \ldots, X_{(k)} \leq a_k\right\} &= \bigcap_{l=1}^{k} \left\{X_{(l)} \leq a_l\right\} \\ &= \bigcup_{\mathbf{i} \in P^k} \left[\mathbf{B_i} \bigcap \left(\bigcap_{j=1}^{k} \{X_{i_j} \leq a_j\}\right)\right] \in \mathcal{B}^k \end{aligned} \tag{1.19}$$

The measurability of $A$ follows.

Note that since $a_i$ are not ordered, it is necessary to intersect $\mathbf{B_i}$ with $\bigcap_{j=1}^{k} \{X_{i_j} \leq a_j\}$ in (1.19) to ensure the identity since otherwise, we only have $A \subseteq \bigcup_{\mathbf{i} \in P^k} \left(\bigcap_{j=1}^{k} \{X_{i_j} \leq a_j\}\right)$. An alternative is to define $b_j = \min_{j \leq l \leq k} a_l$. Then, it is readily checked that (see exercise)

$$\begin{aligned} \bigcap_{j=1}^{k} \{X_{i_j} \leq a_j\} &= \bigcap_{j=1}^{k} \{X_{i_j} \leq b_j\}, \\ \bigcup_{\mathbf{i} \in P^k} \bigcap_{j=1}^{k} \{X_{i_j} \leq b_j\} &\subseteq \bigcap_{l=1}^{k} \{X_{(l)} \leq a_l\} \end{aligned} \tag{1.20}$$

from which it follows that $A = \bigcup_{\mathbf{i} \in P^k} \left[\left(\bigcap_{j=1}^{k} \{X_{i_j} \leq b_j\}\right)\right] \in \mathcal{B}^k$

The ordered vector in (1.18) is a rearrangement of the original $\mathbf{X}$ in the order from the smallest to the largest. If $X_i$ are i.i.d. random variables $(1 \leq i \leq n)$, then $\left(X_{(1)}, X_{(2)}, \ldots, X_{(k)}\right)$ is known as the order statistic.

# 1.5 Distribution Function and Expectation

## 1.5.1 Distribution Function

Let $\mathcal{P} = (\Omega, \mathcal{F}, P)$ be a probability space and $X$ a random variable defined on the measurable space $(\Omega, \mathcal{F})$. The *cumulative distribution function* of $X$ or CDF, denoted by $F(\cdot)$, is defined by

$$F(a) = P(X \le a), \quad \text{for } a \in R \tag{1.21}$$

For any value $a \in R$, $F(a)$ yields the probability of the event that $X \le a$.

The CDF has the following properties

1. $P(a < X \le b) = F(b) - F(a)$.
2. (*Continuity from above*) $F(-\infty) = \lim_{a \to -\infty} F(a) = 0$.
3. (*Continuity from below*) $F(\infty) = \lim_{a \to \infty} F(a) = 1$.

Proof.

1. This follows from:

$$\begin{aligned} F(b) &= P(-\infty < X \le b) = P(\{-\infty < X \le a\} \cup \{a < X \le b\}) \\ &= P(-\infty < X \le a) + P(a < X \le b) \\ &= F(a) + P(a < X \le b) \end{aligned}$$

2. Without the loss of generality, assume $a \downarrow -\infty$. Let $n = \lceil |a| \rceil$, where $\lceil \cdot \rceil$ denotes the largest integer function. It follows that

$$P(A_n) \le F(a) \le P(B_n) \tag{1.22}$$

Let

$$A_n = \{-\infty < X \le -n\}, \quad B_n = \{-\infty < X \le -n + 1\}$$

Then, as $n \to \infty$, $A_n \downarrow \phi$ and $B_n \downarrow \phi$ and thus by the continuity properties of measure (Theorem 1 of Section 1.2),

$$\lim_{n \to -\infty} P(A_n) = \lim_{n \to -\infty} P(B_n) = P(\phi) \tag{1.23}$$

It follows from (1.22) and (1.23) that $\lim_{a \to -\infty} F(a) = 0$.

Note that the continuity properties of measure only apply to a countable sequence. As $A_a = \{-\infty < X \le a\}$ is not such a sequence, it is necessary to construct the countable sequences $A_n$ and $B_n$ before applying the continuity properties of measure.

3. This follows from an argument similar to 2.

Since $\mathcal{P} = (\Omega, \mathcal{F}, P)$ is a measure space and $X$ a real measurable function from $\mathcal{P} = (\Omega, \mathcal{F}, P)$ to the measurable space $\mathcal{S} = (R, \mathcal{B})$, it follows from the discussion in Section 1.3.3 that the CDF defined in (1.21) is the value of the $X$-induced measure for the interval $(-\infty, b]$,

$$F(b) = \mu_X((-\infty, b]) = P\left(X^{-1}(-\infty, b]\right)$$

Following the discussion for general measures on $\mathcal{S} = (R, \mathcal{B})$, $F$ also generates the same measure as $\mu_X$. Further, $F$ can be decomposed into three parts consisting of a step function, an absolutely continuous function (with respect to the Lebesgue measure) and a singular function. In statistical applications, $F$ is either a step or an absolutely continuous function, depending on whether $X$ represents a discrete or continuous outcome. In the discrete case, $F$ can have at most a countable number of discontinuity points $x_k$, that is,

$$f_k = F(x_k) - F\left(x_k^-\right) > 0, \quad k = 1, 2, \ldots$$

Thus, $F$ in this case is completely determined by $\{f_k; k = 1, 2, \ldots\}$, which is called the *probability distribution* or *mass* function. If $F$ is absolutely continuous, $F(a) = \int_{-\infty}^{a} f(x)dx$. In this case, the *probability density function* (PDF), $f(x)$, is often used to identify the distribution. Following the discussion in Section 1.3 for general real measurable functions, the probability of a single point for a continuous $X$ is zero, $\mu_X(\{x\}) = 0$ and the PDF is unique up to a set of Lebesgue measure 0. Note that both the probability distribution and probability density functions are real measurable functions with respect to the Borel $\sigma$-field.

Through $X$ and $F$, we can calculate the probability of any event of interest in the probability space $\mathcal{P} = (\Omega, \mathcal{F}, P)$. This high level of abstraction makes it possible to study any random phenomenon through the single *induced* probability space $\mathcal{P}' = (R, \mathcal{B}, \mu_X)$ without referring to the nature and context of the original probability space $\mathcal{P}$. For this reason, we often denote the probability $P\left(X^{-1}A\right)$ of any event $A \in \mathcal{B}$ simply by $\Pr(X \in A)$, that is,

$$\Pr(X \in A) = \mu_X(A) = P\left(X^{-1}A\right)$$

**Example 1.** Consider a constant random variable $X = c$, where $c$ is a known constant. Since

$$F(a) = \Pr(X \leq a) = \begin{cases} 0 \text{ if } a < c \\ 1 \text{ if } a \geq c \end{cases}$$

$F(a)$ is a step function with a jump of size 1 at $a = c$, which is known as the *degenerate* distribution. The induced probability measure in this case is given by

$$\mu_X(A) = \begin{cases} 1 \text{ if } c \in A \\ 0 \text{ if } c \notin A \end{cases}$$

Thus, a constant random variable induces a probability measure $\mu_X$ on $\mathcal{S} = (R, \mathcal{B})$ with a point mass at $c$.

**Example 2**. Consider the probability space in Example 2 of Section 1.4, $\mathcal{P} = (\Omega, \mathcal{F}, P)$, where $\Omega = \{0, 1, 2, \ldots\}$, $\mathcal{F}$ is the power set of $\Omega$ and $P$ satisfies (1.12). Let $X = k$ be a real function from $\Omega$ to $R$. Then, $X$ is a random variable (see exercise). The CDF $F$ is a step function with jumps, $F(k) - F(k^-) = p_k$, at $k$. If $\Omega$ is viewed as a subset of $R$, the induced probability measure $\mu_X$ is identical to $P$. Nonetheless, they are conceptually different since $\mu_X$ is defined for a completely different measurable space $\mathcal{S} = (R, \mathcal{B})$. If $P$ satisfies (1.13), $X$ is said to have the Poisson distribution.

**Example 3**. Non-negative integrable functions can be used to construct distribution functions for continuous random variables. For example, consider a non-negative function $f(x) = I_{\{0<x<1\}}$. As $\int_{-\infty}^{\infty} f(x)\,dx = 1$, $f(x)$ is a PDF. The corresponding CDF is $F(x) = xI_{\{0<x<\leq 1\}}$. Since $f(x) = 1$, a constant over $(0, 1)$, it is called the *uniform* distribution $U[0, 1]$.

As another example, let $f(x) = \exp\left(-\frac{(x-\mu)^2}{2\sigma^2}\right)$. Since $\int_{-\infty}^{\infty} f(x)\,dx = \sqrt{2\pi\sigma^2}$, the normalized $g = \frac{1}{\sqrt{2\pi\sigma^2}} f(x)$ is a PDF of a *normal* distribution $N(\mu, \sigma^2)$ with mean $\mu$ and variance $\sigma^2$.

Finally, consider $f(x) = \alpha \exp(-\alpha x) I_{\{x \geq 0\}}$ with $a > 0$. Then, it follows that $\int_{-\infty}^{\infty} f(x)\,dx = 1$, yielding the PDF of an *exponential* distribution. The corresponding CDF is $F(x) = (1 - \exp(-\alpha x)) I_{\{x \geq 0\}}$.

**Example 4**. Let $F(a)$ be the CDF of some random variable $X$ defined on a probability space $\mathcal{P} = (\Omega, \mathcal{F}, P)$. Define the inverse function of $F$ as follows:

$$F^{-1}(q) = \inf\{x; F(x) \geq q\}, \quad \text{for any } 0 < q < 1 \tag{1.24}$$

If $F(x)$ is strictly increasing, the above reduces to

$$F^{-1}(q) = \{x; F(x) = q\}, \quad \text{for any } 0 < q < 1$$

For any $0 < q < 1$,

$$\{F(x) \geq q\} = \left[F^{-1}(q), \infty\right) \in \mathcal{B}$$

Since intervals of the form $[a, \infty)$ generate $\mathcal{B}$, $F(x)$ is $\mathcal{B}$ measurable when viewed as a real function from $\mathcal{S} = (R, \mathcal{B})$ to itself. Thus, $Y = F(X)$

is a measurable function $\mathcal{F}/\mathcal{B}$ and thus a random variable in its own right. Similarly, we can show that $F^{-1}(Y)$ is also a random variable (see exercise).

**Example 5.** (*Probability integral transformation*). Let $X$ be a continuous random variable. Then, it is readily checked that

$$F\left(F^{-1}(q)\right) = q, \quad \text{for any } 0 < q < 1 \tag{1.25}$$

Note that the above is generally not true for a discrete random variable. In fact, in general, we have

$$F\left(F^{-1}(q)\right) \geq q, \quad \text{for any } 0 < q < 1$$

It follows from (1.25) that

$$\begin{aligned} F_Y(y) &= \Pr(Y \leq y) = \Pr(F(X) \leq y) = \Pr\left(X \leq F^{-1}(y)\right) \\ &= F\left(F^{-1}(y)\right) = y \end{aligned}$$

Thus, $Y = F(X)$ is a uniform random variable between $[0, 1]$.

**Example 6.** In Example 3, let $U$ be a uniform random variable on $[0, 1]$. Then, it follows from Example 4 that $X = F^{-1}(U)$ is a random variable with the CDF given by $F$. This result is useful for generating random variables with particular distributions, such as in simulation studies.

For example, to generate a random variable with an exponential distribution, $F(x) = 1 - \exp\left(-\frac{x}{3}\right)$, we first generate values $u$ from the uniform $U[0, 1]$ and then generate $x$ through the inverse relationship: $x = F^{-1}(u) = -3\log(1-u)$.

### 1.5.2 Joint Distribution of Random Vectors

Let $\mathbf{X} = (X_1, \ldots, X_k)^\top$ be a random vector defined on the measurable space $(\Omega, \mathcal{F})$. The joint CDF of $\mathbf{X}$ is defined by

$$\begin{aligned} &F(\mathbf{a}) = \Pr(\mathbf{X} \leq \mathbf{a}) = \Pr(X_1 \leq a_1, ..., X_k \leq a_k) \\ &\text{for } \mathbf{a} = (a_1, a_2, ..., a_k)^\top \in R^k \end{aligned}$$

For any value $\mathbf{a}$, the function $F(\mathbf{a})$ yields the probability of the event $\{\mathbf{X} \leq \mathbf{a}\}$. As in the random variable case, for continuous $\mathbf{X}$, we can also define the probability density function (PDF), $\mathbf{f}(\mathbf{a}) = \frac{\partial^k F(\mathbf{a})}{\partial a_1 \ldots \partial a_k}$, which is again unique up to a set of Lebesgue measure 0. For discrete $\mathbf{X}$, we can similarly define the probability distribution function,

$$\begin{aligned} f_{i_1, \ldots i_k} &= \Pr(X_1 = x_{i_1}, ..., X_k = x_{i_k}) \\ &= F(x_{i_1}, ..., x_{i_k}) - F\left(x_{i_1}^-, ..., x_{i_k}^-\right) > 0 \end{aligned}$$

where $\{x_{i_1}, ..., x_{i_k}\}$ indexes all points with a positive mass $\{i_j \geq 1, 1 \leq j \leq k\}$.

**Example 7**. Let $X_i$ denote the number of surgeries performed in a hospital in year $i$ and $\mathbf{X} = (X_1, X_2)^\top$ $(i = 1, 2)$. Then, $F(k, m) = \Pr(X_1 \leq k, X_2 \leq m)$ is the probability that the hospital performed at most $k$ and $m$ surgeries in year 1 and 2, while $f_{k,m} = \Pr(X_1 = k, X_2 = m)$ is the probability that it performed exactly $k$ and $m$ surgeries $(k, m = 1, 2, \ldots)$.

Similar to the random variable case, the CDF of a random vector has the following properties:

1. $P(\mathbf{a} < \mathbf{X} \leq \mathbf{b}) = F(\mathbf{b}) - F(\mathbf{a})$.
2. $\lim F(\mathbf{a}) \to 0$ if $a_i \to -\infty$ for some $i$ $(1 \leq i \leq k)$.
3. $F(\infty) = \lim_{\mathbf{a}\to\infty} F(\mathbf{a}) = 1$ with $\mathbf{a} \to \infty$ denoting $a_i \to \infty$ for all $i$ $(1 \leq i \leq k)$.

As Example 9 of Section 1.4 shows, if $\mathbf{X} = (X_1, \ldots, X_k)^\top$ is a random vector, then any of its components $X_i$ is a random variable. In addition, let $F_i$ denote the CDF of $X_i$, which is known as a *marginal distribution* of $\mathbf{X}$. Then, we have

4. $\lim F(\mathbf{a}) \to F_i(a_i)$ if the $i$th component of $\mathbf{a}$ is fixed at $a_i$ while all the rest of $\mathbf{a}$ approach $\infty$.

It follows from (4) that marginal distributions are completely determined by the joint CDF of $\mathbf{X}$. The reverse is generally not true, except for a special class of *independent* random variables. To discuss such random variables, we first review the notion of *independence* within a collection of measurable sets as well across such collections.

Two measurable sets $A$ and $B$ are said to be independent if $\Pr(A \cap B) = \Pr(A)\Pr(B)$. A finite or countable collection of such sets, $\mathcal{A} = \{A_i; i \geq 1\}$, is defined to be independent if any finite subcollection of $\mathcal{A}$ is independent, that is, $\Pr\left(\bigcap_{i\in J} A_i\right) = \prod_{i\in J} \Pr(A_i)$, where $\prod$ denotes product and $J$ some finite subset of the integers $1, 2, \ldots$. Likewise, two collections of measurable sets $\mathcal{A}$ and $\mathcal{B}$ are independent if for any $A \in \mathcal{A}$ and $B \in \mathcal{B}$, $A$ and $B$ are independent. A finite or countable sequence of such collections $\{\mathcal{A}_n, n \geq 1\}$ is independent if any finite subsequence is independent, that is, $\{\mathcal{A}_k, k \in J\}$ is independent for any $J$ indexing a finite subsequence. Let $\{\mathcal{F}_n, n \geq 1\}$ with $\mathcal{F}_n = \sigma(\mathcal{A}_n)$ be a sequence of $\sigma$-fields generated by $\{\mathcal{A}_n, n \geq 1\}$. The following shows that independence on $\{\mathcal{A}_n, n \geq 1\}$ carries over to the sequence of $\sigma$-fields it generates.

**Proposition**. If the finite or countable sequence $\{\mathcal{A}_n, n \geq 1\}$ is independent, then $\{\mathcal{F}_n, n \geq 1\}$ is also an independent sequence.

**Example 8**. Consider $\Omega = \{1, 2, 3, 4, 5, 6\}$, where each point is assigned a probability $\frac{1}{6}$. Let $A = \{1, 2, 3\}$, $B = \{1, 2, 3\}$ and $C = \{3, 4, 5, 6\}$. Then, it is readily checked that

$$\Pr(A) = \frac{1}{2}, \quad \Pr(B) = \frac{1}{2}, \quad \Pr(C) = \frac{4}{6}$$

$$\Pr(A \cap B \cap C) = \Pr(A)\Pr(B)\Pr(C) = \frac{1}{6}$$

But, $A$, $B$, and $C$ are not independent since $\Pr(A \cap B) \neq \Pr(A)\Pr(B)$. Thus, it is important to check all subcollections for independence.

**Example 9.** A subset $A$ is independent with itself iff $\Pr(A) = 0$ or 1. This follows from $\Pr(A) = \Pr(A \cap A) = \Pr(A)\Pr(A)$.

A finite or infinite sequence of random variables $\{X_i; i \geq 1\}$ is independent if the sequence of the $\sigma$-fields $\{\sigma(X_i); i \geq 1\}$ is independent. By the Proposition, we can equivalently define such an independence through any sequence of collections of sets that generates $\{\sigma(X_i); i \geq 1\}$. For example, since $\{X_i \leq a_i\}$ generates $\sigma(X_i)$, the sequence $\{X_i; i \geq 1\}$ is independent iff $\{X_i \leq a_i; i \geq 1\}$ is a collection of independent sets for any $a_i \in R$ $(i \geq 1)$. If $\{X_i \leq a_i; i \geq 1\}$ is an independent sequence and $X_i$ all have the same CDF $F$, it is called an *independently and identically distributed*, or i.i.d., sequence of random variables.

Independent random variables play an important role in the theory and application of statistics. In most applications, it is difficult if not impossible to verify independence directly by the definition. Instead, by interpreting $\sigma$-fields as collections of pieces of information about events of interest, we check independence by examining whether such pieces of information are *independent* across the different $\sigma$-fields or random variables.

**Example 10.** In Example 7, $X_1$ and $X_2$ are generally dependent as the numbers of surgeries performed by the same hospital should not be too different in two consecutive years unless there is some disease outbreak or a major population change in the area served by the hospital. If $X_1$ and $X_2$ represent the surgeries performed by two hospitals that serve different populations from say distinct localities in the same year, then they are independent.

If the components of a random vector $\mathbf{X} = (X_1, \ldots, X_k)^\top$ are independent, then $\{X_i \leq a_i; 1 \leq i \leq k\}$ is a collection of independent sets for any $a_i \in R$ $(1 \leq i \leq k)$. It follows that

$$F(\mathbf{a}) = \Pr\left(\bigcap_{i=1}^{k} \{X_i \leq a_i\}\right) = \prod_{i=1}^{k} \Pr(X_i \leq a_i) = \prod_{i=1}^{k} F_i(a_i)$$

So, the distribution function of such a random vector is determined by the CDFs of its individual components. It is readily checked that the PDF or probability distribution function (for discrete $\mathbf{X}$) is the product of the density or probability distribution functions of the component random variables (see exercise). As a special case, if $X_i$ have the same distribution, then the joint PDF or probability mass function of $\mathbf{X}$ is given by

$$f(\mathbf{a}) = \prod_{i=1}^{k} f(a_i), \quad f_{i_1,\ldots i_k} = \prod_{j=1}^{k} f_{i_j}$$

where $f$ and $f_{i_j}$ denote the PDF and probability mass function of $X_i$ depending on whether $\mathbf{X}$ is continuous or discrete. Such a joint PDF or probability mass function is known as the *likelihood function*, which is the premise for maximum likelihood and related inferences in statistics.

### 1.5.3 Expectation

Let $\mathcal{P} = (\Omega, \mathcal{F}, P)$ be a probability space. Let $X$ be a random variable defined on $(\Omega, \mathcal{F}, P)$. Then, we define the expectation of $X$ as:

$$E(X) = \int_{\Omega} X dP \tag{1.26}$$

provided that at least one of $E(X^+) = \int_{\Omega} X^+ dP$ and $E(X^-) = \int_{\Omega} X^- dP$ is finite, where $X^+ = \max(X, 0)$ and $X^- = \max(-X, 0)$. Since $P(\Omega) = 1$, $E(X)$ is an average of $X$.

**Example 11**. Let $X = c$, a constant random variable. Then, $E(X) = \int_{\Omega} c dP = c$.

**Example 12**. Let $B \in \mathcal{B}$ and $X = I_B$. Then,

$$E(X) = \int_{\Omega} I_B dP = \Pr(X = 1)$$

the average of the values of $X$ weighted by the probability of $X = 1$.

**Example 13**. Consider a probability space $\mathcal{P} = (\Omega, \mathcal{F}_2, P)$, with $\Omega$ and $\mathcal{F}_2$ defined in Example 2 of Section 1.2 and $P$ being some probability measure. A random variable $X$ with respect to $\mathcal{F}_2$ must be constant over each of the points in $\{1, 3, 5\}$ and $\{2, 4, 6\}$. Let $a$ and $b$ be the two constants of $X$ on $\{1, 3, 5\}$ and $\{2, 4, 6\}$, respectively. Then,

$$E(X) = a \Pr(\{X = k; k = 1, 3, 5\}) + b \Pr(\{X = m; m = 2, 4, 6\})$$

the average of $a$ and $b$ weighted by the probabilities when $X$ equals the respective constants. In general, if $X$ is discrete, then its expectation is the weighted average of the values of $X$ according to its probability distribution function.

Suppose that $E|X| < \infty$ and $E|Y| < \infty$. Then, the expectation has the following properties:

1. $E(aX + b) = aEX + b$.
2. If $X \geq Y$, then $EX \geq EY$.
3. (*Cauchy–Schwartz inequality*) $E(UV) \leq \sqrt{EU^2 EV^2}$, where $U = X - E(X)$ and $V = Y - E(Y)$.
4. (*Jensen's inequality*) Suppose $\varphi$ is convex (concave), then $E(\varphi(X)) \geq \varphi(E(X))$ $(E(\varphi(X)) \leq \varphi(E(X)))$, provided that both expectations exist.
5. (*Chebyshev's inequality*) Suppose $\varphi : R \to R$, with $\varphi \geq 0$ and nondecreasing. Then, for any $\alpha > 0$, $P(X \geq \alpha) \leq \frac{E(\varphi(X))}{\varphi(\alpha)}$. When $\varphi(x) = x^2$, we have for $X > 0$, $P(X \geq \alpha) \leq \frac{E(X^2)}{\alpha^2}$.

Proof. (1) and (2) follow from similar properties for integration with respect to general measures discussed in Section 1.3. The Cauchy–Schwartz, Jensen's and Chebyshev's inequalities are also readily proved (see exercise).

Let $X$ be a random variable defined on $\mathcal{P} = (\Omega, F, P)$ and $\mu_X$ the induced probability measure. Then, by the change of variable formula, we have

$$E(X) = \int_\Omega X dP = \int_R x d\mu_X$$

Thus, integration with respect to $P$ can be carried out by integration with respect to $\mu_X$. If $X$ is continuous with PDF $f$, then we can compute $E(X)$ by integration with respect to the Lebesgue measure,

$$E(X) = \int_\Omega X dP = \int_R x d\mu_X = \int_R x f(x)\, dx \tag{1.27}$$

More generally, if $g$ is a real measurable function from $\mathcal{R} = (R, B)$ to itself, then $Y = g(X)$ is a random variable. Let $\mu_Y$ denoted the induced measure. Then, it follows from the change of variable formula that

$$E(g(X)) = \int_\Omega g(X)\, dP = \int_R g(x)\, d\mu_X = \int_R y d\mu_Y \tag{1.28}$$

As in the case of $X$, the above integration can be carried out using the Lebesgue measure if $Y$ is continuous.

**Example 14.** Let $X$ be a random variable following a normal distribution, $N(\mu, \sigma^2)$. Then, it follows from (1.27) that

$$
\begin{aligned}
E(X) &= \int_R x f(x)\, dx = \int_R x \frac{1}{\sqrt{2\pi}\sigma} \exp\left(-\frac{(x-\mu)^2}{2\sigma^2}\right) dx \\
&= \int_R \frac{(x-\mu)}{\sqrt{2\pi}\sigma} \exp\left(-\frac{(x-\mu)^2}{2\sigma^2}\right) dx + \int_R \frac{\mu}{\sqrt{2\pi}\sigma} \exp\left(-\frac{(x-\mu)^2}{2\sigma^2}\right) dx \\
&= \int_R x \frac{1}{\sqrt{2\pi}\sigma} \exp\left(-\frac{x^2}{2\sigma^2}\right) dx + \mu \\
&= \mu
\end{aligned}
$$

Let $Y = g(X) = (X-\mu)^2$. Then, by (1.28),

$$
\begin{aligned}
E(Y) &= E\left((X-\mu)^2\right) = \int_R (x-\mu)^2 f(x)\, dx = \frac{1}{\sqrt{2\pi}\sigma} \exp\left(-\frac{(x-\mu)^2}{2\sigma^2}\right) dx \\
&= \sigma^2
\end{aligned}
$$

The *mean* $\mu$ and *variance* $\sigma^2$ completely characterize the normal distribution.

More generally, for each positive integer $k$, $E(X^k)$ $(E(X - E(X))^k)$ is called the $k$th (centered) moment of $X$. Thus, the mean is the first moment and variance the second centered moment of $X$. The first two moments play an important role in distribution-free modeling, which we will discuss in detail in later chapters.

### 1.5.4 Conditional Expectation

Let $\mathcal{P} = (\Omega, \mathcal{F}, P)$ be a probability space and let $X$ and $Y$ be two random variables with respect to $\mathcal{P}$. Let $\sigma(Y)$ denote the $\sigma$-field generated by $Y$. Then, the conditional expectation of $X$ given $Y$, $E(X \mid Y)$, is a $\sigma(Y)$-measurable random variable satisfying:

$$
\int_G E(X \mid Y)\, dP = \int_G X dP, \quad \text{for any } G \in \sigma(Y) \tag{1.29}
$$

The conditional expectation has the interpretation of being a *smoothed* version of $X$ with respect to $Y$.

Note that $E(X \mid Y)$ is unique up to a set of probability 0, that is, if there is another random variable, $g(Y)$, satisfying (1.29), then, $E(X \mid Y) = g(Y)$ a.e., or, $\Pr(E(X \mid Y) \neq g(Y)) = 0$.

Conditional expectation of a random variable $X$ may also be defined with respect to a $\sigma$-field, $\mathcal{G}$. In this case, $E(X \mid \mathcal{G})$ is defined by substituting $\mathcal{G}$ for $\sigma(Y)$ in (1.29).

**Example 15**. Let $Y = c$ be a constant. Then, $\sigma(Y) = \{\Phi, \Omega\}$. Since $E(X \mid Y)$ is $\sigma(Y)$-measurable, it must be a constant on $\Omega$. Thus, it follows that

$$E(X) = \int_\Omega X dP = \int_\Omega E(X \mid Y) dP = E(X \mid Y) \int_\Omega dP = E(X \mid Y)$$

The conditional expectation in this case has smoothed $X$ to a single value.

This example shows that $E(X)$ can be viewed as a conditional expectation corresponding to the trivial $\sigma$-field $\{\Phi, \Omega\}$.

**Example 16**. Let $Y$ be a binary random variable with the value 0 or 1. Then, $\sigma(Y) = \{\Phi, \Omega, Y^{-1}(0), Y^{-1}(1)\}$, where $Y^{-1}(a)$ denotes the preimage of $Y = a$ in $\mathcal{F}$ of $\mathcal{P}$. Since $E(X \mid Y)$ must be a constant on $Y^{-1}(0)$ and $Y^{-1}(1)$, it follows that

$$\begin{aligned} E(X \mid Y = 0) \Pr(Y = 0) &= \int_{Y=0} E(X \mid Y = 0) dP \\ &= \int_{Y=0} X dP = E\left(X I_{\{Y=0\}}\right) \\ E(X \mid Y = 1) \Pr(Y = 1) &= \int_{Y=1} E(X \mid Y = 1) dP \\ &= \int_{Y=1} X dP = E\left(X I_{\{Y=1\}}\right) \end{aligned}$$

Thus, the conditional expectation $E(X \mid Y)$ is given by

$$E(X \mid Y) = \begin{cases} \frac{E\left(X I_{\{Y=0\}}\right)}{\Pr(Y=0)} & \text{if } Y = 0 \\ \frac{E\left(X I_{\{Y=1\}}\right)}{\Pr(Y=1)} & \text{if } Y = 1 \end{cases} \tag{1.30}$$

In this example, $Y$ has been smoothed into a random variable $X$ with two values, $\frac{E\left(X I_{\{Y=0\}}\right)}{\Pr(Y=0)}$ and $\frac{E\left(X I_{\{Y=1\}}\right)}{\Pr(Y=1)}$, which are the averages of $X$ over the sets $\{Y = 0\}$ and $\{Y = 1\}$ defined by the respective levels of the binary $Y$.

**Example 17**. Let $A, B \in \mathcal{F}$, $X = I_A$ and $Y = I_B$. It follows from (1.30) that

$$E(X \mid Y = 1) = \frac{E\left(X\ I_{\{Y=1\}}\right)}{\Pr(Y = 1)} = \frac{E(I_A I_B)}{\Pr(Y = 1)} = \frac{\Pr(A \cap B)}{\Pr(B)}$$

The last quantity is used to define the *conditional* probability of $A$ given $B$, $\Pr(A \mid B)$. The formula $\Pr(A \mid B) = \frac{\Pr(A \cap B)}{\Pr(B)}$ is called the Bayes' theorem.

Note that $E(X \mid Y)$ is a random variable while $\Pr(A \mid B)$ is a value equal to $E(X \mid Y = 1)$.

**Example 18.** Let $X$ and $Y$ be two continuous random variables with a joint PDF $f_{XY}(x, y)$. Denote by $f_X(x)$ and $f_Y(y)$ the PDFs of the marginal $X$ and $Y$. Consider $G = \{Y \leq y\} \in \sigma(Y)$. Then, by the change of variable formula, we have:

$$\begin{aligned}
\int_G X dP &= \int_\Omega I_G X dP = \int_\Omega I_{\{Y \leq y\}} X dP = \int_R \int_R I_{(-\infty,y)} x \cdot f_{XY}(x, y)\, dx dy \\
&= \int_{-\infty}^{y} \int_R x \cdot f_{XY}(x, y)\, dx dy = \int_{-\infty}^{y} f_Y(y)\, dy \int_R x \cdot \frac{f_{XY}(x, y)}{f_Y(y)} dx \\
&= \int_R I_{(-\infty,y)} \mu_Y(dy) \int_R x \cdot\ f_{X|Y}(x) dx \\
&= \int_G \int_R x \cdot\ f_{X|Y}(x) dx dP
\end{aligned}$$

On the other hand, since $\int_G E(X \mid Y)\, dP = \int_G X dP$ by definition, it follows that

$$\int_G E(X \mid Y)\, dP = \int_G \int_R x \cdot\ f_{X|Y}(x) dx dP$$

Since $G$ generates $\sigma(Y)$, it follows from the discussion in Section 1.3 that the above holds for any subset $Q \in \sigma(Y)$. Thus,

$$E(X \mid Y)\, dP = \int_R x \cdot\ f_{X|Y}(x) dx dP, \quad \text{a.e.}$$

The function $f_{X|Y}(x)$ is called the conditional density of $X$ given $Y$. This example shows that the conditional expectation can be evaluated by the conditional density for continuous $X$ and $Y$.

**Theorem 1.** Let $X$ and $Y$ be some random variables, and $\mathcal{G}_1, \mathcal{G}_2 \subseteq F$ some $\sigma$-fields.

1. (*Iterated conditional expectation*) If $\mathcal{G}_1 \subseteq \mathcal{G}_2$, $E[E(X \mid \mathcal{G}_2) \mid \mathcal{G}_1] = E(X \mid \mathcal{G}_1)$.
2. (*Linearity*) $E(aX + bY \mid \mathcal{G}) = aE(X \mid \mathcal{G}) + bE(Y \mid \mathcal{G})$.
3. If $X$ and $Y$ are independent, then $E(X \mid Y) = E(X)$.
4. If $X$ is measurable with respect to $\mathcal{G}$, then $E(X \mid \mathcal{G}) = X$.
5. (*Jensen's inequality*) If $g$ is convex, then $E(g(X) \mid \mathcal{G}) \geq g(E(X \mid \mathcal{G}))$.
6. $E[E(X \mid \mathcal{G}_1) \mid \mathcal{G}_2] = E[E(X \mid \mathcal{G}_2) \mid \mathcal{G}_1] = E(X \mid \mathcal{G}_1)$.

Proof. Here we will prove (1). The others can be shown in a similar fashion (see exercise).

First, note that the left side is $\mathcal{G}_1$-measurable. Now, by definition of $E[E(X \mid \mathcal{G}_2) \mid \mathcal{G}_1]$, we have

$$\int_G E[E(X \mid \mathcal{G}_2) \mid \mathcal{G}_1]\, dP = \int_G E(X \mid \mathcal{G}_2)\, dP, \quad G \in \mathcal{G}_1 \tag{1.31}$$

However, if $G \in \mathcal{G}_1$, then $G \in \mathcal{G}_2$. Applying the definition of $E(X \mid \mathcal{G}_2)$ yields:

$$\int_G E(X \mid \mathcal{G}_2)\, dP = \int_G X dP$$

Thus,

$$\int_G E[E(X \mid \mathcal{G}_2) \mid \mathcal{G}_1]\, dP = \int_G X dP, \quad G \in \mathcal{G}_1 \tag{1.32}$$

It follows from (1.31) and (1.32) that

$$E[E(X \mid \mathcal{G}_2) \mid \mathcal{G}_1] = E(X \mid \mathcal{G}_1)$$

**Example 19**. Let $\mathcal{G}_1 = \{\Phi, \Omega\}$ and $\mathcal{G}_2 = \sigma(Y)$. Then, clearly $\mathcal{G}_1 \subset \mathcal{G}_2$ and it follows from Theorem 1/1 that $E[E(X \mid Y)] = E(X)$. This identity has wide applications in the theory and application of statistics.

## 1.6 Convergence of Random Variables and Vectors

Convergence of sequence of random variables (vectors) plays a key role in the study of behavior of statistics in large samples and is a concept fundamental to the theory of statistics asymptotics. In this section, we review the notion of convergence of random sequence and important results concerning the convergence of a particular class of random sequences consisting of independently and identically distributed random variables (vectors).

### 1.6.1 Modes of Convergence

Let $X_n$ $(n \geq 1)$ be a sequence of random variables defined on some probability space, $\mathcal{P} = (\Omega, \mathcal{F}, P)$. As random variables are real measurable functions, all results discussed in Section 1.3 for the convergence of such a sequence apply to the sequence of random variables. For example, if $X_n \to X$ for all $\omega \in \Omega$ except for a subset with probability measure 0, then $X_n \to X$

a.e. (almost everywhere). Since this definition of convergence relies on the original probability space, it is not widely used in statistics. The definitions most relevant to statistical theory and application are *convergence in probability* and *convergence in distribution*. We discuss convergence in probability below as well as in Section 1.6.2. Convergence in distribution will be reviewed in Section 1.6.3.

Convergence in probability is convergence in measure for real measurable functions when rephrased with respect to the probability measure $P$. Thus, $X_n$ converges to a random variable $X$ in probability, $X_n \rightarrow_p X$, if for any $\delta > 0$, the deterministic sequence, $d_{n,\delta} = \Pr[|X_n - X| > \delta]$ or $c_{n,\delta} = \Pr[|X_n - X| \leq \delta]$ converges to 0 or 1 as $n \rightarrow \infty$, that is, for any $\epsilon > 0$, we can find some integer $N_{\epsilon,\delta}$ such that for all $n \geq N_{\epsilon,\delta}$,

$$d_{n,\delta} = \Pr[|X_n - X| > \delta] < \epsilon, \quad c_{n,\delta} = \Pr[|X_n - X| \leq \delta] \geq 1 - \epsilon \tag{1.33}$$

Thus, under (1.33) the variable or random nature of $X_n$ is captured by a single deterministic sequence $d_{n,\delta}$ or $c_{n,\delta}$ so that its convergence can be evaluated by the criteria for deterministic sequence.

In most statistical applications, $X$ is often an unknown constant $\mu$ that describes some features of a population of interest such as age, disease prevalence or number of hospitalizations, while $X_n$ is an estimate of $\mu$. If $X_n \rightarrow_p \mu$, $X_n$ is a *consistent estimate* of $\mu$. In this case, if $\upsilon_n$ denotes the $X_n$-induced measure on $\mathcal{S} = (R, \mathcal{B})$, $X_n \rightarrow_p \mu$ can be stated as

$$\upsilon_n((\mu - \delta, \mu + \delta)^c) \rightarrow 0 \quad \text{or} \quad \upsilon_n([\mu - \delta, \mu + \delta]) \rightarrow 1 \quad n \rightarrow \infty$$

In this sense, the notion of convergence in probability or consistency of parameter estimate is *independent* of the underlying probability space.

**Example 1**. Let $f(x)$ be a continuous function on $R$. If $X_n \rightarrow_p \mu$, then $f(X_n) \rightarrow_p f(\mu)$.

Let $K > |\mu| > 0$. Then, for any $\epsilon > 0$,

$$\Pr[|f(X_n) - f(\mu)| > \epsilon] \leq \Pr[|f(X_n) - f(\mu)| > \epsilon, |X_n| \leq K] + \\ + \Pr[|X_n| > K] \tag{1.34}$$

Since $f$ is uniformly continuous on $[-K, K]$, there exists $\delta_{\epsilon,K} > 0$ such that

$$\{|f(X_n) - f(\mu)| > \epsilon, |X_n| \leq K\} \subseteq \{|X_n - \mu| > \delta_{\epsilon,K}\} \tag{1.35}$$

We first choose $K_\epsilon$ so that $\Pr[|X_n| > K_\epsilon] < \frac{\epsilon}{2}$. Given such a $K_\epsilon$, we then select $\delta_{\epsilon,K}$ to ensure $\Pr[|X_n - \mu| > \delta_{\epsilon,K}] < \frac{\epsilon}{2}$. Since $X_n \rightarrow_p \mu$, we can find

$N_{\epsilon,K}$ so that for all $n > N_{\epsilon,K}$

$$\begin{aligned}\Pr\left[\left|f\left(X_n\right) - f\left(\mu\right)\right| > \epsilon\right] &\le \Pr\left[\left|X_n - \mu\right| > \delta_{\epsilon,K}\right] + \Pr\left[\left|X_n\right| > K_\epsilon\right] \\ &< \frac{\epsilon}{2} + \frac{\epsilon}{2} = \epsilon\end{aligned} \tag{1.36}$$

The conclusion follows from (1.34)–(1.36).

**Example 2.** Let $X_n \sim N\left(\mu, \frac{\sigma^2}{n}\right)$, that is, $X_n$ follows a normal distribution with mean $\mu$ and variance $\frac{\sigma^2}{n}$ $(n \ge 1)$. Then, $X_n$ is a random sequence. For any $\delta > 0$,

$$d_{n,\delta} = \Pr\left[\left|\frac{X_n - \mu}{\frac{\sigma}{\sqrt{n}}}\right| > \frac{\delta\sqrt{n}}{\sigma}\right] = 2\Phi\left(-\frac{\delta\sqrt{n}}{\sigma}\right) \to \Phi\left(-\infty\right) = 0$$

where $\Phi\left(\cdot\right)$ denotes the CDF of standard normal $N\left(0,1\right)$. It follows that $X_n \to_p \mu$, a consistent estimate of $\mu$.

In this example, if $X_n$ does not follow the normal distribution, it may be difficult to evaluate $\Pr\left[\left|X_n - \mu\right| > \delta\right]$. Fortunately, in many statistical applications, $X_n$ is typically the average of i.i.d. random variables and in this case, it is possible to determine the behavior of $\Pr\left[\left|X_n - \mu\right| > \delta\right]$ for large $n$ as we discuss next.

### 1.6.2 Convergence of Sequence of I.I.D. Random Variables

Statistics and model estimates of interest arising in many applications are often in the form of an average of i.i.d. random variables. For such random sequences, we can determine whether they converge (in probability) as well as the converged limits by the *law of large numbers* (LLN).

**Theorem 1** (*Law of large numbers*). Let $X_i$ be i.i.d. random variables, with finite mean $\mu = E\left(X_i\right)$ and variance $\sigma^2 = Var\left(X_i\right)$. Let $\overline{X}_n = \frac{1}{n}\sum_{i=1}^{n} X_i$ be the average of the first $n$ random variables. Then, the random sequence $\overline{X}_n$ converges to the mean $\mu = E\left(X_i\right)$ in probability, $\overline{X}_n \to_p \mu$, as $n \to \infty$.

Proof. For any $\delta > 0$, it follows from the assumptions that

$$\begin{aligned}E\left(\overline{X}_n - \mu\right)^2 &\ge E\left[\left(\overline{X}_n - \mu\right)^2 I_{\{|\overline{X}_n - \mu \ge \delta|\}}\right] \ge \delta^2 E\left[I_{\{|\overline{X}_n - \mu \ge \delta|\}}\right] \\ &= \delta^2 \Pr\left[\left|\overline{X}_n - \mu \ge \delta\right|\right]\end{aligned}$$

From the above, we immediately obtain the Chebyshev's inequality:

$$\Pr\left[\left|\overline{X}_n - \mu \ge \delta\right|\right] \le \frac{1}{\delta^2} E\left(\overline{X}_n - \mu\right)^2 = \frac{1}{n}\sigma^2 \tag{1.37}$$

For each $\epsilon$, by setting $N_{\epsilon,\delta} = \left\lceil \frac{1}{n}\sigma^2 \right\rceil + 1$, it follows from (1.37) that

$$\Pr\left[\left|\overline{X}_n - \mu \geq \delta\right|\right] \leq \epsilon, \quad \text{for all } n \geq N_{\epsilon,\delta}$$

The above LLN is known as the *weak* law of large numbers. A similar version for the almost everywhere convergence of an i.i.d. random sequence is called the *strong* law of large numbers. In addition, the assumptions of Theorem 1 can also be relaxed so that it can be applied to a broader class of i.i.d. random sequences. But, the version of LLN stated in Theorem 1 suffices for the development of later chapters in this book.

The law of large numbers is widely applied to facilitate the process of determining the convergence of $\overline{X}_n$ when it has an unknown distribution. Random variables arising from almost all applications have a limited range, (the variables have the value 0 outside this range) and thus not only the first two moments , but all higher order moments are finite. For this reason, we apply Theorem 1 without citing the assumption throughout the book unless otherwise noted.

**Example 3** (***Monte Carlo integration***). Let $f(x)$ be a continuous function on $[a, b]$. Let $I(f) = \int_a^b f(x)\,dx$ be the integral of $f$ over $[a, b]$. Let $X_i$ be i.i.d. continuous random variables with PDF $g(x) >$ for all $a \leq x \leq b$. Then, by LLN,

$$\widehat{I}_n(f) = \frac{1}{n}\sum_{i=1}^{n} \frac{f(X_i)}{g(X_i)} = \overline{Y}_n \to_p E(Y_i) = E\left(\int_a^b \frac{f(x)}{g(x)} g(x)\,dx\right) = I(f)$$

where $Y_i = \frac{f(X_i)}{g(X_i)}$. Thus, if $\int_a^b f(x)\,dx$ has no closed form solution, we can approximate $I(f)$ by a random sum $\widehat{I}_n(f) = \frac{1}{n}\sum_{i=1}^{n} \frac{f(X_i)}{g(X_i)}$. If both $a$ and $b$ are finite, the simplest way to compute $\widehat{I}_n(f)$ is to sample $X_i$ from a uniform $U[a, b]$. This Monte Carlo integration plays an important role in Bayesian analysis.

**Example 4.** Let $X_i \sim \text{i.i.d.}(\mu, \sigma^2)$ and $\overline{X}_n = \sum_{i=1}^{n} X_i$, where $(\mu, \sigma^2)$ denotes a distribution with mean $\mu$ and variance $\sigma^2$. For $r > 0$, let $f(x) = x^r$. Then, by applying LLN and Example 2, $\overline{X}_n^r = f(\overline{X}_n) \to_p f(\mu) = \mu^r$. As a special case, by letting $r = k$ $(= 1, 2, \ldots)$, it follows that sample moments $\overline{X}_n^k$ converge to the corresponding population moments $\mu^k$ of all orders.

### 1.6.3 Rate of Convergence of Random Sequence

In the theory and application of statistics, we need to know not only whether an estimate is consistent, but also its sampling variability. To study the latter, we must *normalize* the estimate using the *rate of convergence* of the estimate since convergence in probability does not distinguish random sequences with different rates of convergence as shown by the next example.

**Example 5**. Let $X_i \sim$ i.i.d. $N\left(\mu, \sigma^2\right)$. Then, by LLN, $\overline{X}_n \rightarrow_p \mu$ or equivalently, $\overline{X}_n - \mu \rightarrow_p 0$.

Now, let $Y_n = n^{\frac{1}{4}}\left(\overline{X}_n - \mu\right)$. By applying Chebshev's inequality, we have:

$$\begin{aligned} \Pr\left[\left|n^{\frac{1}{4}}\left(\overline{X}_n - \mu\right)\right| \geq \delta\right] &\leq \frac{E\left[n^{\frac{1}{4}}\left(\overline{X}_n - \mu\right)\right]^2}{\delta^2} = \frac{\sqrt{n}}{\delta^2} Var\left(\overline{X}_n\right) \\ &= \frac{\sigma^2}{\delta^2}\frac{1}{\sqrt{n}} \rightarrow 0 \end{aligned}$$

So, $Y_n \rightarrow_p 0$. However, since $n^{\frac{1}{4}} \uparrow \infty$, $Y_n = n^{\frac{1}{4}}\left(\overline{X}_n - \mu\right)$ has a slower rate of convergence than $\overline{X}_n - \mu$.

The above example shows that we need a new concept of convergence to delineate the different rates of convergence for sequences that converge to the same limit in probability.

Let $X_n$ be a random sequence with CDF $F_n$ $(n \geq 1)$ and $X$ a random variable with CDF $F$. The sequence $X_n$ *converges in distribution to* $X$, $X_n \rightarrow_d X$, if for every continuity point $c$ of $F$,

$$\lim_{n\rightarrow\infty} F_n(c) = F(c)$$

Since $F$ is right continuous, continuity points are those $c$ at which $F(c) = \lim_{x\uparrow c} F(x) = F(c^-)$, the limit of $F$ as $x$ approaches $c$ from left, that is, $x < c$.

Convergence in probability is stronger than convergence in distribution. However, when $X$ is a constant, the two become equivalent as the following example shows.

**Example 6**. $X_n \rightarrow_p \mu$ iff $X_n \rightarrow_d \mu$.

We only establish the first part, that is, $X_n \rightarrow_p \mu$ implies that $X_n \rightarrow_d \mu$. The reverse is similarly proved (see exercise).

By viewing the constant $\mu$ as a special random variable, the CDF of $\mu$ is $F_\mu(c) = I_{\{c\geq\mu\}}$. Such an $F$ is known as a degenerate CDF. Thus, any $c \neq \mu$ is a continuity point of $F_\mu(\cdot)$. To establish the conclusion, we need to show that for any $c \neq \mu$,

$$F_n(c) = \Pr\left(X_n \leq c\right) \rightarrow F_\mu(c)$$

First, consider $c < \mu$. Since $X_n \rightarrow_p \mu$ and $\mu - c \neq 0$, it follows that

$$F_n(c) = \Pr(X_n \leq c) = \Pr[X_n - \mu \leq c - \mu] \leq \Pr[|X_n - \mu| \geq \mu - c] \rightarrow 0$$

Likewise, we can readily show that for $c > \mu$, $F_n(c) \rightarrow 1$. Thus, $F_n(c) \rightarrow F_\mu(c)$ for all continuity points $c\ (\mu)$ of $F_\mu$.

However, the most important use of this new notion of convergence in statistics is to define the rate of convergence of a consistent estimate so that we can study its sampling variability. A random sequence $X_n$ $(n \geq 1)$ has a rate of convergence $n^p$ if

$$\frac{X_n}{n^p} \rightarrow_d X, \quad \text{as } n \rightarrow \infty$$

where $X$ has a nondegenerate distribution.

**Example 7**. In Example 5, since $\sqrt{n}\left(\overline{X}_n - \mu\right) \sim N\left(0, \sigma^2\right)$, $\overline{X}_n - \mu$ converges to 0 at the rate of $\left(\frac{1}{n}\right)^{\frac{1}{2}}$. Similarly, the rate of convergence for $n^{\frac{1}{4}}\left(\overline{X}_n - \mu\right)$ is $\left(\frac{1}{n}\right)^{\frac{1}{4}}$, slower than $\overline{X}_n - \mu$.

If $\overline{X}_n$ is non-normal, it is generally difficult to find its rate of convergence. The *central limit theorem* (CLT) is applied to facilitate the evaluation of $\overline{X}_n$ in general situations.

**Theorem 2** (*Univariate central limit theorem*). Let $X_i$ be i.i.d. random variables, with finite mean $\mu = E(X_i)$ and variance $\sigma^2 = Var(X_i)$. Then, $\sqrt{n}\left(\overline{X}_n - \mu\right) \rightarrow_d N\left(0, \sigma^2\right)$. Since $N\left(0, \sigma^2\right)$ has continuity points on $R^1$ (the real line), this implies that for all $x \in R^1$,

$$F_n(x) = \Pr\left[\sqrt{n}\left(\overline{X}_n - \mu\right) \leq x\right] \rightarrow \Phi\left(\frac{x}{\sigma}\right)$$

where $\Phi(\cdot)$ denotes the CDF of standard normal $N(0, 1)$. As in the case of LLN, the moment assumption of Theorem 2 is satisfied in most applications.

For the mean $\overline{X}_n$ of i.i.d. random variables, the CLT refines the conclusion of LLN by providing both the rate of convergence for the centered mean $\overline{X}_n - \mu$ and the limiting distribution for the normalized sample mean $\sqrt{n}\left(\overline{X}_n - \mu\right)$. The latter, known as *asymptotic normality*, has wide applications in the theory and application of statistics asymptotics. It asserts that for large $n$, $\overline{X}_n$ follows approximately a normal distribution, $N\left(\mu, \frac{1}{n}\sigma^2\right)$, which provides the basis for inference for $\mu$. Throughout the book, we state such an approximation by $\overline{X}_n \sim AN\left(\mu, \frac{1}{n}\sigma^2\right)$ and refer to $AN\left(\mu, \frac{1}{n}\sigma^2\right)$ as the *asymptotic distribution* and $\sigma^2$ as the *asymptotic variance* of $\overline{X}_n$.

**Example 8**. Let $X_i$ be i.i.d. random variables following a Bernoulli distribution with mean $p$, that is,

$$\Pr[X_i = 1] = p, \quad \Pr[X_i = 0] = 1 - p = q$$

Let $\widehat{p}_n = \frac{1}{n}\sum_{i=1}^{n} X_i$. Then, by CLT, $\widehat{p}_n \sim AN\left(p, \frac{1}{n}pq\right)$. Thus, we can use this asymptotic distribution to compute probabilities for binomial random variables. For example, let $r = \sum_{i=1}^{n} X_i$, the number of times when $X_i = 1$. Then,

$$\Pr\left[r \le k\right] = \Pr\left[\frac{\frac{r}{n} - p}{\sqrt{\frac{pq}{n}}} \le \frac{\frac{k}{n} - p}{\sqrt{\frac{pq}{n}}}\right] \simeq \Phi\left(\frac{\frac{k}{n} - p}{\sqrt{\frac{pq}{n}}}\right) = \Phi\left(\frac{k - np}{\sqrt{npq}}\right)$$

The above approximation works well if $p$ is too small. We can also use the asymptotic distribution to approximate confidence intervals for $\widehat{p}_n$:

$$\Pr\left[-z_{\frac{\alpha}{2}} \le z_n \le z_{\frac{\alpha}{2}}\right] = \Pr\left[\widehat{p}_n - z_{\frac{\alpha}{2}}\sqrt{\frac{pq}{n}} \le p \le \widehat{p}_n + z_{\frac{\alpha}{2}}\sqrt{\frac{pq}{n}}\right] \approx 1 - \alpha$$

where $z_{\frac{\alpha}{2}}$ denotes the $\frac{\alpha}{2}$ percentile of the standard normal $N(0,1)$, that is, $\Phi\left(z_{\frac{\alpha}{2}}\right) = \frac{\alpha}{2}$.

We conclude this section by generalizing the concepts of convergence to random vectors. Let $\mathbf{X}_n \in R^k$ be a sequence of $k$-dimensional random vectors. Then, $\mathbf{X}_n$ converges to $\mathbf{X}$ in probability, $\mathbf{X}_n \to_p \mathbf{X}$, if $\|\mathbf{X}_n - \mathbf{X}\| \to_p 0$, where $\|\cdot\|$ denotes the Euclidean distance in $R^k$. Let $F_n$ $(n \ge 1)$ be the CDF of $\mathbf{X}_n$ and $F$ the CDF of $\mathbf{X}$. The sequence $\mathbf{X}_n$ converges in distribution to $\mathbf{X}$, $\mathbf{X}_n \to_d \mathbf{X}$, if for every continuity point $\mathbf{c}$ of $F$, $\lim_{n\to\infty} F_n(\mathbf{c}) = F(\mathbf{c})$. Again, convergence in probability implies convergence in distribution (see exercise). Also, Theorem 2 is similarly generalized to sequences of random vectors.

**Theorem 3** (*Multivariate central limit theorem*)**.** Let $\mathbf{X}_i \in R^k$ be i.i.d. random vectors, with mean $\boldsymbol{\mu} = E(\mathbf{X}_i)$ and variance $\Sigma = Var(\mathbf{X}_i)$. If $\boldsymbol{\mu}$ and $\Sigma$ are finite and $\Sigma$ is full rank, then $\sqrt{n}\left(\overline{\mathbf{X}}_n - \boldsymbol{\mu}\right) \to N(\mathbf{0}, \Sigma)$, that is, for all $\mathbf{y} \in R^k$,

$$F_n(\mathbf{x}) = \Pr\left[\sqrt{n}\left(\overline{\mathbf{X}}_n - \boldsymbol{\mu}\right) \le \mathbf{y}\right] \to \Phi\left(\Sigma^{\frac{1}{2}}\mathbf{y}\right)$$

where $N(\boldsymbol{\mu}, \Sigma)$ denotes a $k$-dimensional normal distribution with mean $\boldsymbol{\mu}$ and variance $\Sigma$, and $\Phi(\cdot)$ the CDF of such a distribution with $\boldsymbol{\mu} = \mathbf{0}$ and $\Sigma = I_k$.

In many applications, we can often verify whether a sequence of random vectors $\mathbf{X}_n$ converges in probability by examining its moments. More specifically, let $r$ $(> 0)$. The sequence $\mathbf{X}_n$ convergence in $r$th mean, $\mathbf{X}_n \to_r \mathbf{X}$, if

$$E\|\mathbf{X}_n - \mathbf{X}\|^r \to 0, \quad n \to \infty$$

It is readily shown that $\mathbf{X}_n \to_r \mathbf{X}$ implies that $\mathbf{X}_n \to_p \mathbf{X}$ (see exercise).

### 1.6.4 Stochastic $o_p(\cdot)$ and $O_p(\cdot)$

As in the study of convergence of deterministic sequence, we often just want to know the rate of convergence of a random sequence rather than its exact limit. To this end, we use the stochastic $o_p(\cdot)$ and $O_p(\cdot)$. First, we review the stochastic $o_p(\cdot)$, which, like its counterpart $o(\cdot)$ for deterministic sequence, is used to indicate a faster rate of convergence for a random sequence to converge to 0 in probability.

For a random sequence $X_n$, $X_n = o_p(1)$ if $X_n \to_p 0$. More generally, if $r_n (> 0) \to 0$, $X_n = o_p(r_n)$ is defined as $r_n^{-1} X_n = o_p(1)$.

**Example 9**. Let $X_i \sim$ i.i.d. $(\mu, \sigma^2)$. Then, it follows from LLN that $\overline{X}_n - \mu \to_p 0$. Thus, $\overline{X}_n - \mu = o_p(1)$.

**Example 10**. Let $X_i$ and $\overline{X}_n$ be defined as in Example 9. Consider another random sequence, $Y_n = \frac{1}{n}(\overline{X}_n - \mu)$. Then, $nY_n = (\overline{X}_n - \mu) \to_p 0$. Thus, $Y_n = o_p\left(\frac{1}{n}\right)$ and $Y_n$ converges to 0 in probability at a rate faster than $\frac{1}{n}$.

Unlike its counterpart $O(\cdot)$ for deterministic sequence, the stochastic $O_p(\cdot)$ is used to indicate *stochastic boundedness* rather than equivalent rate of convergence. We first review the notion of stochastic boundedness.

**Example 11**. Let $X_n = n$ or 0 with $\Pr[X_n = 0] = \frac{1}{n}$ and $\Pr[X_n = n] = 1 - \frac{1}{n}$. Then, $X_n$ has a large mass centered at $n$ and the probability of $X_n$ at this mass point increases to 1 as $n \to \infty$. For any $M > 0$,

$$\Pr[-M \le X_n \le M] = 1 - \Pr[X_n > M] = \frac{1}{n} \to 0$$

Thus, $X_n$ is not bounded in the sense that the probability that $X_n$ is confined to any finite interval $[-M, M]$ decreases to 0 as $n \to \infty$. In other words, we cannot find a sufficiently large interval $[-M, M]$ to trap a desired amount of mass of $X_n$, as a nonzero mass escapes to infinity.

A random sequence $X_n$ is *stochastically bounded* or *bounded in probability*, denoted by $O_p(1)$, if for any $\epsilon > 0$, we can find $M_\epsilon$ and $N_\epsilon$ such that

$$\begin{aligned} &\Pr[|X_n| \le M_\epsilon] \ge 1 - \epsilon, \quad \text{all } n \ge N_\epsilon \quad \text{or equivalently} \\ &\Pr[|X_n| \ge M_\epsilon] \le \epsilon, \quad \text{all } n \ge N_\epsilon \end{aligned}$$

For a bounded random sequence, we can always find a sufficiently large interval so that the probability that the sequence lies outside the interval can be made arbitrarily small.

Note that in many applications, $\Pr[-M \le X_n \le M]$ has a limit for each fixed $M$ as $n \to \infty$. To check stochastic boundedness in this case, we only need to show that for any $\epsilon > 0$, we can find $M_\epsilon$ such that

$$\Pr[|X_n| \le M_\epsilon] \to a \ge 1-\epsilon \quad \text{— or equivalently —} \quad \Pr[|X_n| \ge M_\epsilon] \to b \le \epsilon$$

**Example 12**. Let $X$ be some random variable and $X_n = X$. Then, it is readily shown that $X_n$ is bounded (see exercise).

Let $\mathbf{X}_n = (X_{n1}, \ldots, X_{nk})^\top \in R^k$. We write $\mathbf{X}_n = \mathbf{o}_p(1)$ if $\mathbf{X}_n \to_p \mathbf{0}$. We define $\mathbf{X}_n = \mathbf{O}_p(1)$ if for any $\epsilon > 0$ we can find $M_\epsilon$ and $N_\epsilon$ such that

$$\begin{aligned} &\Pr[\|\mathbf{X}_n\| \le M_\epsilon] \ge 1-\epsilon, \quad \text{for all } n \ge N_\epsilon, \quad \text{or equivalently} \\ &\Pr[\|\mathbf{X}_n\| \ge M_\epsilon] \le \epsilon, \quad \text{for all } n \ge N_\epsilon \end{aligned} \tag{1.38}$$

**Example 13**. If $\mathbf{X}_n \to_d \mathbf{X}$, then $\mathbf{X}_n = \mathbf{O}_p(1)$.

Let $\mathbf{m}_1$ and $\mathbf{m}_2$ denote some continuity points of the CDF $F(\mathbf{x})$ of $\mathbf{X}$. For any $\epsilon > 0$, we can find continuity points $\mathbf{m}_{1\epsilon}$ and $\mathbf{m}_{2\epsilon}$ such that

$$F(\mathbf{m}_{1\epsilon}) - F(\mathbf{m}_{2\epsilon}) \ge 1-\epsilon$$

It follows from $\mathbf{X}_n \to_d \mathbf{X}$ and the above that

$$\Pr[\mathbf{m}_{1\epsilon} \le \mathbf{X}_n \le \mathbf{m}_{2\epsilon}] = F_n(\mathbf{m}_{2\epsilon}) - F_n(\mathbf{m}_{1\epsilon}) \to F(\mathbf{m}_{2\epsilon}) - F(\mathbf{m}_{1\epsilon}) \ge 1-\epsilon.$$

Let $M_\epsilon = \max\{\|m_{1\epsilon}\|, \|m_{2\epsilon}\|\}$. Then,

$$\Pr[\|\mathbf{X}_n\| \le M_\epsilon] \ge \Pr[\mathbf{m}_{1\epsilon} \le \mathbf{X}_n \le \mathbf{m}_{2\epsilon}] \to 1-\epsilon$$

and the conclusion follows.

**Example 14**. Let $\mathbf{X}_i \sim$ i.i.d.$(\boldsymbol{\mu}, \Sigma)$, with $(\boldsymbol{\mu}, \Sigma)$ denoting a multivariate distribution with mean $\boldsymbol{\mu}$ and variance $\Sigma$. Then, by CLT, $\sqrt{n}\left(\overline{\mathbf{X}}_n - \boldsymbol{\mu}\right) \to_d N(\boldsymbol{\mu}, \Sigma)$. Thus, $\sqrt{n}\left(\overline{\mathbf{X}}_n - \boldsymbol{\mu}\right) = \mathbf{O}_p(1)$.

Let $r_n\ (>0) \to 0$. If $r_n^{-1}\mathbf{X}_n = \mathbf{O}_p(1)$, we write $\mathbf{X}_n = \mathbf{O}_p(r_n)$.

**Example 15**. Consider again $Y_n = n^{\frac{1}{4}}\left(\overline{X}_n - \mu\right)$ in Example 7. Since $n^{\frac{1}{4}}Y_n = \sqrt{n}\left(\overline{X}_n - \mu\right) = O_p(1)$, $Y_n = O_p\left(\left(\frac{1}{n}\right)^{\frac{1}{4}}\right)$.

**Example 16**. Let $r_n\ (>0) \to 0$ as $n \to \infty$. Then, $\mathbf{O}_p(r_n) = \mathbf{o}_p(1)$.

For any $\epsilon > 0$, since $\frac{\epsilon}{r_n} \to \infty$ as $n \to \infty$, it follows that

$$\Pr[\|\mathbf{O}_p(r_n)\| \ge \epsilon] = \Pr\left[r_n \left\|\frac{\mathbf{O}_p(r_n)}{r_n}\right\| \ge \epsilon\right] = \Pr\left[\left\|\frac{\mathbf{O}_p(r_n)}{r_n}\right\| \ge \frac{\epsilon}{r_n}\right] \to 0$$

Thus, $\mathbf{O}_p(r_n) \to_p 0$ and the conclusion follows.

Let $r_n, q_n\ (>0) \to 0$ and $\mathbf{X}_n, \mathbf{Y}_n \in R^k$. Listed below are some additional properties regarding $\mathbf{o}_p(\cdot)$ and $\mathbf{O}_p(\cdot)$ (see exercise).

1. $\mathbf{O}_p(1) + \mathbf{O}_p(1) = \mathbf{O}_p(1)$.
2. $\mathbf{O}_p(1) + \mathbf{o}_p(1) = \mathbf{O}_p(1)$.
3. $\mathbf{o}_p(1) + \mathbf{o}_p(1) = \mathbf{o}_p(1)$.
4. $\mathbf{O}_p(1)\,\mathbf{o}_p(1) = \mathbf{o}_p(1)$.
5. $\mathbf{O}_p(r_n) = \mathbf{o}_p(1)$.
6. $\mathbf{X}_n = \mathbf{o}_p(1)$ iff $X_{nj} = o_p(1)$ for all $1 \le j \le k$.
7. $\mathbf{X}_n = \mathbf{O}_p(1)$ iff $X_{nj} = O_p(1)$ for all $1 \le j \le k$.
8. If $\mathbf{X}_n = \mathbf{o}_p(r_n)$ and $\mathbf{Y}_n = \mathbf{o}_p(q_n)$, then $\mathbf{X}_n + \mathbf{Y}_n = \mathbf{o}_p\{\max(r_n, q_n)\}$ and $\mathbf{X}_n^\top \mathbf{Y}_n = \sum_{j=1}^k X_n Y_n = o_p(r_n q_n)$.

**Example 17**. Let $X_i \sim$ i.i.d.$(\mu, \sigma^2)$ and $s_n^2 = \frac{1}{n}\sum_{i=1}^n \left(X_i - \overline{X}_n\right)^2$. Then, $s_n^2 \to_p \sigma^2$.

First, we have

$$s_n^2 - \sigma^2 = \frac{1}{n}\sum_{i=1}^n \left[(X_i - \mu)^2 - \sigma^2\right] + \left(\overline{X}_n - \mu\right)^2$$

By LLN, $\frac{1}{n}\sum_{i=1}^n \left[(X_i - \mu)^2 - \sigma^2\right] = o_p(1)$. In addition, by Property 5, we have:

$$\left(\overline{X}_n - \mu\right)^2 = \left[\frac{1}{\sqrt{n}}\sqrt{n}\left(\overline{X}_n - \mu\right)\right]^2 = \frac{1}{n}O_p^2(1) = o_p(1)$$

Thus,

$$s_n^2 - \sigma^2 = o_p(1) + o_p(1) = o_p(1)$$

# 1.7 Convergence of Functions of Random Vectors

## 1.7.1 Convergence of Functions of Random Variables

In many inference problems in statistics, we often need to determine the distribution of a function of a sequence of random variables. For relatively simple functions, such as a linear combination of random variables, we can apply Slutsky's theorem to determine the limiting distribution of the sequence. For more complex functions, we typically first linearize the function using a stochastic Taylor series expansion and then apply CLT and/or Slutsky's theorem to find the asymptotic distribution of the sequence. The latter approach is also known as the Delta method.

**Theorem 1** (*Slutsky's theorem*). Let $X_n \to_d X$ and $Y_n \to_p c$, where $c$ is a constant. Then,

1. $X_n + Y_n \to_d X + c$
2. $X_n Y_n \to_d cX$.

3. If $c \neq 0$, $X_n/Y_n \rightarrow_d X/c$.

Proof. We only prove (1), with (2) and (3) following a similar argument.

First, note that the continuity points of the CDFs of $X$ and $X + c$ are the same. For a continuity point $a$ of the CDF of $X$, the following is true for any $b > 0$:

$$\{X_n + b \leq a\} \cap \{|Y_n - c| \leq b\} \subseteq \{X_n + Y_n \leq a + c\} \\ \subseteq \{X_n - b \leq a\} \cup \{|Y_n - c| \geq b\}$$

Let $n \rightarrow \infty$, we get

$$\lim \{X_n + b \leq a\} \subseteq \lim \{X_n + Y_n \leq a\} \subseteq \lim \{X_n - b \leq a\}$$

Hence,

$$\lim \Pr(X_n \leq a - b) \leq \lim \Pr(X_n + Y_n \leq a) \leq \lim \Pr\{X_n \leq a + b\}$$

Since $a$ is a continuity point of $X$, we obtain the identity by letting $b \rightarrow 0$ and the conclusion follows.

**Example 1**. Let $X_i \sim \text{i.i.d.}\left(\mu, \sigma^2\right)$ and $\overline{X}_n = \frac{1}{n}\sum_{i=1}^{n} X_i$. Let us find the limit and the asymptotic distribution of the sequence of the second order moments $\overline{X}_n^2$.

By LLN, $\overline{X}_n \rightarrow_p \mu$. We have shown in Section 1.6 that all moments of a convergent random sequence converge to the corresponding moments of the distribution. Thus, $\overline{X}_n^2 \rightarrow_p \mu^2$ and $\overline{X}_n^2$ is a consistent estimate of $\mu^2$.

By CLT and LLN, we have

$$\sqrt{n}\left(\overline{X}_n - \mu\right) \rightarrow_d N\left(0, \sigma^2\right), \quad \overline{X}_n + \mu \rightarrow_p 2\mu$$

It follows from Slutsky's theorem that

$$\sqrt{n}\left(\overline{X}_n^2 - \mu^2\right) = \sqrt{n}\left(\overline{X}_n - \mu\right)\left(\overline{X}_n + \mu\right) \rightarrow_d N\left(0, 4\mu^2\sigma^2\right)$$

Thus, $\sqrt{n}\left(\overline{X}_n^2 - \mu^2\right)$ also has a limiting normal with asymptotic variance $4\mu^2\sigma^2$.

**Example 2**. In Example 1, consider the limit and the asymptotic distribution of the sample variance, $s_n^2 = \frac{1}{n}\sum_{i=1}^{n}\left(X_i - \overline{X}_n\right)^2$.

It follows from LLN and Example 1 that

$$s_n^2 = \frac{1}{n}\sum_{i=1}^{n}\left(X_i - \overline{X}_n\right)^2 = \frac{1}{n}\sum_{i=1}^{n} X_i^2 - \overline{X}_n^2 \rightarrow E\left(X_i^2\right) - \mu^2 = \sigma^2$$

Thus, $s_n^2$ is a consistent estimate of $\sigma^2$.

To find the asymptotic distribution of $s_n^2$, first, note that

$$\sqrt{n}\left(s_n^2-\sigma^2\right)=\frac{\sqrt{n}}{n}\sum_{i=1}^{n}\left[\left(X_i-\mu\right)^2-\sigma^2\right]+\sqrt{n}\left(\overline{X}_n-\mu\right)^2$$

By CLT,

$$\frac{\sqrt{n}}{n}\sum_{i=1}^{n}\left[\left(X_i-\mu\right)^2-\sigma^2\right]\to_d N\left(0,Var\left(\left(X_i-\mu\right)^2\right)\right)$$

In addition,

$$\sqrt{n}\left(\overline{X}_n-\mu\right)^2=\frac{\sqrt{n}}{n}O_p^2\left(1\right)=\frac{1}{\sqrt{n}}O_p\left(1\right)=o_p\left(1\right)$$

It then follows from Slutsky's theorem that

$$\sqrt{n}\left(s_n^2-\sigma^2\right)\to_d N\left(0,Var\left(\left(X_i-\mu\right)^2\right)\right)$$

In Theorem 1, if $X_n\to_p a$, a constant, then the results also hold when convergence in distribution "$\to_d$" is replaced by convergence in probability "$\to_p$" (see exercise). For example, if $X_n\to_p a$ and $Y_n\to_p c$ $(\neq 0)$, then $\frac{X_n}{Y_n}\to_p\frac{a}{c}$.

For complex $f$ such as $\sqrt{\cdot}$, we cannot apply Theorem 1 to find the limiting distribution. As in the study of deterministic function, Taylor series expansion is often used to facilitate the computation. The following is an analogue of such an expansion for random functions.

**Theorem 2** (*Stochastic Taylor series expansion*). Let $X_n$ be a sequence of random variables such that $X_n=a+O_p\left(r_n\right)$, where $r_n\to_p 0$ as $n\to\infty$. If $f$ has continuous derivatives up to order $s+1$ in a neighborhood of $a$, then

$$f\left(X_n\right)=\sum_{k=0}^{s}\frac{f^{(k)}\left(a\right)}{k!}\left(X_n-a\right)^k+o_p\left(r_n\right)$$

where $f^{(k)}\left(a\right)$ denotes the $k$th order derivative. Note that $f\left(X_n\right)$ above is a random sequence, since both $\left(X_n-a\right)^k$ and $o_p\left(r_n\right)$ are random sequences.

Proof. First, note that since $r_n\to 0$, $X_n=a+O_p\left(r_n\right)=a+o_p\left(1\right)$.

For $x\neq a$, consider the deterministic Taylor series expansion:

$$f\left(x\right)=\sum_{k=0}^{s}\frac{f^{(k)}\left(a\right)}{k!}\left(x-a\right)^k+\frac{f^{(s+1)}\left(\xi\right)}{(s+1)!}\left(x-a\right)^{s+1}$$

where $\xi \in (a, x)$. Let

$$h(x) = \left[ f(x) - \sum_{k=0}^{s} \frac{f^{(k)}(a)}{k!} (x-a)^k \right] \div \left[ \frac{(x-a)^s}{s!} \right] \tag{1.39}$$
$$= \frac{s!}{(s+1)!} f^{(s+1)}(\xi)(x-a)$$

Then, $h(a) = 0$. In addition, $h$ is continuous at $a$. By Example 1 of Section 1.6, we have

$$h(X_n) = h(a) + o_p(1) = o_p(1) \tag{1.40}$$

It follows from (1.39)–(1.40) that

$$f(X_n) - \sum_{k=0}^{s} \frac{f^{(k)}(a)}{k!} (X_n - a)^k = s! h(X_n)(X_n - a)^s = o_p(1) O_p^s(r_n)$$
$$= o_p(r_n)$$

The last equality above follows from the property that $o_p(1) O_p^s(r_n) = o_p(r_n)$.

**Example 3**. In Example 2, consider the limit of $s_n = \sqrt{s_n^2}$.

From Example 2, $s_n^2 \to_p \sigma^2$ and thus $s_n^2 = \sigma^2 + o_p(1)$. Let $f(x) = \sqrt{x}$. By Theorem 2, we have

$$s_n = \sigma + \frac{1}{2\sqrt{\sigma}} \left( s_n^2 - \sigma^2 \right) + o_p(1) = \sigma + o_p(1) + o_p(1) = \sigma + o_p(1)$$

The last equality follows from the properties of $o_p(1)$. Thus, $s_n$ is a consistent estimate of $\sigma$.

Note that in this simple example, we may find the limit of $s_n$ without using the Taylor series expansion. Since $f(x) = \sqrt{x}$ $(x > 0)$ is continuous and $s_n^2 \to_p \sigma^2$, it follows from Example 1 of Section 1.6 that $f\left(s_n^2\right) \to f\left(\sigma^2\right) = \sigma$.

**Example 4**. In Example 1, consider the sequence $Y_n = \sqrt{n}\frac{\overline{X}_n - \mu}{s_n}$. By CLT, $\sqrt{n}\left(\overline{X}_n - \mu\right) \to_d N\left(0, \sigma^2\right)$ and from Example 3, $s_n \to_p \sigma$. Thus, it follows from Slutsky's Theorem that

$$\sqrt{n}\frac{\overline{X}_n - \mu}{s_n} = \frac{\sigma}{s_n}\sqrt{n}\frac{\left(\overline{X}_n - \mu\right)}{\sigma} \to_d N(0, 1)$$

In Example 4, if $X_i \sim$ i.i.d.$N\left(\mu, \sigma^2\right)$, then it is well known that $Y_n = \sqrt{n}\frac{\overline{X}_n - \mu}{s_n}$ has a $t$ distribution with $n-1$ degrees of freedom. For non-normal data, $Y_n$ no longer follows the $t$ distribution. However, the above example shows that $Y_n$ has an approximate normal distribution. Since $t$ converges to the standard normal as $n \to \infty$, the difference between $t$ and normal diminishes as sample size increases.

**Example 5**. In Example 3, consider the asymptotic distribution of $s_n$. Since

$$\sqrt{n}\left(s_n^2 - \sigma^2\right) \to_d N\left(0, Var\left(\left(X_i - \mu\right)^2\right)\right)$$

it follows that $s_n^2 - \sigma^2 = O_p\left(\frac{1}{\sqrt{n}}\right)$. By Taylor series expansion,

$$s_n = \sigma + \frac{1}{2\sqrt{\sigma}}\left(s_n^2 - \sigma^2\right) + o_p\left(\frac{1}{\sqrt{n}}\right)$$

Thus,

$$\begin{aligned}\sqrt{n}\left(s_n - \sigma\right) &= \frac{1}{2\sqrt{\sigma}}\sqrt{n}\left(s_n^2 - \sigma^2\right) + \sqrt{n}o_p\left(\frac{1}{\sqrt{n}}\right) \\ &= \frac{1}{2\sqrt{\sigma}}\sqrt{n}\left(s_n^2 - \sigma^2\right) + o_p\left(1\right) \\ &\to_d N\left(0, \frac{1}{4\sigma}Var\left(\left(X_i - \mu\right)^2\right)\right)\end{aligned}$$

Note that although $s_n^2$ is an unbiased estimate of $\sigma^2$, $s_n$ is a biased estimate of $\sigma$. Since $f\left(x\right) = x^{\frac{1}{2}}$ is a concave function, it follows from the Jensen's inequality (see Section 1.5) that

$$E\left(s_n\right) = E\left[f\left(s_n^2\right)\right] \leq f\left[E\left(s_n^2\right)\right] = f\left(\sigma^2\right) = \sigma$$

Thus, $s_n$ always underestimates $\sigma$ in finite samples.

If $X \sim N\left(\mu, \sigma^2\right)$, then it follows from the properties of the normal distribution that any linear combination of $X$ also has a normal distribution, that is, $aX + b \sim N\left(a\mu + b, a^2\sigma^2\right)$, where $a$ and $b$ are constants. This property also carries over to random sequences with asymptotic normal distributions as the next example shows.

**Example 6**. If $X_n \sim AN\left(\mu, \frac{\sigma^2}{n}\right)$, then for any $a, b \in R$,

$$\sqrt{n}\left[\left(aX_n + b\right) - \left(a\mu + b\right)\right] = a\sqrt{n}\left(X_n - \mu\right) \to_d N\left(0, a^2\sigma^2\right)$$

Thus, $aX_n + b \sim AN\left(a\mu + b, \frac{a^2\sigma^2}{n}\right)$.

### 1.7.2 Convergence of Functions of Random Vectors

In Section 1.6, we considered convergence of a sequence of random vectors $\mathbf{X}_n$. By applying a vector-valued function $\mathbf{f}(\mathbf{x})$ to such a sequence, we get a sequence of vector-valued functions of random vectors $\mathbf{f}(\mathbf{X}_n)$. As in the scalar case, the following theorem is quite useful to determine the convergence of such a sequence.

**Theorem 3**. Let $\mathbf{X}_n \in \mathbf{R}^k$ be a sequence of random vectors. If $\mathbf{X}_n - \mathbf{X} = \mathbf{o}_p(1)$ and $\mathbf{f}(\mathbf{x})$ is a continuous mapping from $\mathbf{R}^k$ to $\mathbf{R}^m$, then $\mathbf{f}(\mathbf{X}_n) - \mathbf{f}(\mathbf{X}) = \mathbf{o}_p(1)$, that is, $\mathbf{f}(\mathbf{X}_n) \rightarrow_p \mathbf{f}(\mathbf{X})$.

Likewise, for such a sequence of vector-valued functions of random vectors, we are also interested in its asymptotic distribution. We first discuss a useful result that enables the study of convergence of random vectors through random variables.

**Theorem 4** (*Cramér–Wold device*). Let $\mathbf{X}_n \in \mathbf{R}^k$ be a sequence of random vectors. Then, $\mathbf{X}_n \rightarrow_d \mathbf{X}$ iff $\boldsymbol{\lambda}^\top \mathbf{X}_n \rightarrow_d \boldsymbol{\lambda}^\top \mathbf{X}$ for any constant vector $\boldsymbol{\lambda} \in \mathbf{R}^k$.

The Cramér–Wold device is a useful tool for extending univariate asymptotic results to the multivariate setting. One important application is in defining the asymptotic normal distribution for sequence of random vectors. The random sequence $\mathbf{X}_n$ has an asymptotic multivariate normal distribution with mean $\boldsymbol{\mu}$ and variance $\Sigma$ if for all $\boldsymbol{\lambda} \in \mathbf{R}^k$,

$$\sqrt{n}\boldsymbol{\lambda}^\top (\mathbf{X}_n - \boldsymbol{\mu}) \rightarrow_d N\left(\mathbf{0}, \boldsymbol{\lambda}^\top \Sigma \boldsymbol{\lambda}\right), \quad \text{or} \quad \boldsymbol{\lambda}^\top \mathbf{X}_n \sim AN\left(\boldsymbol{\lambda}^\top \boldsymbol{\mu}, \frac{1}{n}\boldsymbol{\lambda}^\top \Sigma \boldsymbol{\lambda}\right)$$

In Theorem 3 of Section 1.6, we stated a version of asymptotic multivariate normal distribution when the asymptotic variance $\Sigma$ is full rank so that the PDF exists. This alternative definition require no such assumption and thus applies to *degenerate multivariate normal distribution* as well. In most applications, however, $\Sigma$ is full rank and two versions become identical.

By applying Theorem 4, we can immediately show that all linear vector-valued functions of $\mathbf{X}_n$ are asymptotically normal.

**Example 7**. Let $\mathbf{X}_n \sim AN\left(\boldsymbol{\mu}, \frac{1}{n}\Sigma\right)$. Then, the linear function $A\mathbf{X}_n + \mathbf{b}$ is also asymptotically normal, $A\mathbf{X}_n + \mathbf{b} \sim AN\left(A\boldsymbol{\mu} + \mathbf{b}, \frac{1}{n}A\Sigma A^\top\right)$, where $A$ is some $m \times k$ constant matrix and $\mathbf{b}$ a $m \times 1$ constant vector.

For any $\boldsymbol{\lambda} \in \mathbf{R}^m$,

$$\boldsymbol{\lambda}^\top [(A\mathbf{X}_n + \mathbf{b}) - (A\boldsymbol{\mu} + \mathbf{b})] = \boldsymbol{\lambda}^\top A (\mathbf{X}_n - \boldsymbol{\mu}) = \widetilde{\boldsymbol{\lambda}}_n^\top (\mathbf{X}_n - \boldsymbol{\mu})$$

Since $\widetilde{\boldsymbol{\lambda}} = \boldsymbol{\lambda}^\top A \in R$ and $\widetilde{\mathbf{b}} = \boldsymbol{\lambda}^\top \mathbf{b} \in R$, it follows from the assumptions and

Theorem 4 that

$$\widetilde{\boldsymbol{\lambda}}^{\top}(\mathbf{X}_n-\boldsymbol{\mu})\sim AN\left(0,\frac{1}{n}\widetilde{\boldsymbol{\lambda}}^{\top}\Sigma\widetilde{\boldsymbol{\lambda}}\right)=AN\left(0,\frac{1}{n}\boldsymbol{\lambda}^{\top}(A\Sigma A)\boldsymbol{\lambda}\right)$$

Thus, $\boldsymbol{\lambda}^{\top}(A\mathbf{X}_n+\mathbf{b})\sim AN\left(\boldsymbol{\lambda}^{\top}(A\boldsymbol{\mu}+\mathbf{b}),\frac{1}{n}\boldsymbol{\lambda}^{\top}(A\Sigma A)\boldsymbol{\lambda}\right)$ and the conclusion follows.

For general non-linear functions of $\mathbf{X}_n$, the asymptotic distributions are determined by the *Delta method.*

**Theorem 5** (*Delta method*). Let $\mathbf{X}_n\sim AN\left(\boldsymbol{\mu},\frac{1}{n}\Sigma\right)$ and $\mathbf{g}(\mathbf{x})=(g_1(\mathbf{x}),\ldots,g_m(\mathbf{x}))^{\top}$ be a continuous vector-valued function from $\mathbf{R}^k$ to $\mathbf{R}^m$. If $\mathbf{g}(\mathbf{x})$ is differentiable at $\boldsymbol{\mu}$, then $\mathbf{g}(\mathbf{X}_n)\sim AN\left(\mathbf{g}(\boldsymbol{\mu}),\frac{1}{n}D^{\top}\Sigma D\right)$, where $D=\frac{\partial}{\partial\mathbf{x}}\mathbf{g}\mid_{\mathbf{x}=\boldsymbol{\mu}}$ is an $m\times k$ derivative matrix with $\frac{\partial}{\partial\mathbf{x}}\mathbf{g}$ defined as:

$$\frac{\partial}{\partial\mathbf{x}}\mathbf{g}=\begin{pmatrix}\frac{\partial g_1}{\partial x_1} & \cdots & \frac{\partial g_m}{\partial x_1}\\ \vdots & \ddots & \vdots\\ \frac{\partial g_1}{\partial x_k} & \cdots & \frac{\partial g_m}{\partial x_k}\end{pmatrix}_{k\times m}$$

Proof. Let $\mathbf{X}_n=(X_{n1},\ldots,X_{nk})^{\top}$. It follows from Theorem 4 that

$$\sqrt{n}\left(X_{nj}-\mu_j\right)\to_d N\left(0,\sigma_{jj}^2\right),\quad 1\le j\le k$$

Thus, $X_{nj}=\mu_j+O_p\left(\frac{1}{\sqrt{n}}\right)$ $(1\le j\le k)$. By a Taylor series expansion for each $g_l(\mathbf{X}_n)$ $(1\le l\le m)$, we have:

$$g_l(\mathbf{X}_n)=g_l(\boldsymbol{\mu})+\sum_{j=1}^{m}\frac{\partial}{\partial x_j}g_l(\boldsymbol{\mu})\left(X_{nj}-\mu_j\right)+o_p\left(\frac{1}{\sqrt{n}}\right)$$

When expressed in the vector form, the above becomes:

$$\mathbf{g}(\mathbf{X}_n)=\mathbf{g}(\boldsymbol{\mu})+D^{\top}(\mathbf{X}_n-\boldsymbol{\mu})+\mathbf{o}_p\left(\frac{1}{\sqrt{n}}\right)$$

where $\mathbf{o}_p(\cdot)=(o_{p1}(\cdot),\ldots,o_{pm}(\cdot))^{\top}$. It follows from Example 7 that $D^{\top}(\mathbf{X}_n-\boldsymbol{\mu})\sim AN\left(\mathbf{0},\frac{1}{n}D^{\top}\Sigma D\right)$. Thus, for any $\boldsymbol{\lambda}\in\mathbf{R}^m$,

$$\begin{aligned}\sqrt{n}\left[\boldsymbol{\lambda}^{\top}\mathbf{g}(\mathbf{X}_n)-\boldsymbol{\lambda}^{\top}\mathbf{g}(\boldsymbol{\mu})\right]&=\boldsymbol{\lambda}^{\top}\sqrt{n}D^{\top}(\mathbf{X}_n-\boldsymbol{\mu})+\mathbf{o}_p(1)\\&\to_d N\left(0,\boldsymbol{\lambda}^{\top}D^{\top}\Sigma D\boldsymbol{\lambda}\right)\end{aligned}$$

The conclusion follows from Theorem 4.

**Example 8**. Let $X_i \sim$ i.i.d.$\left(\mu, \sigma^2\right)$. Then, the statistic, $t_x = \frac{\overline{X}_n - \mu}{s_n}$, is invariant under linear transformation. In other words, if $Y_i = aX_i + b$ for any $a$ $(> 0)$, $b \in R$, then $Y_i \sim$ i.i.d.$\left(\mu_y, \sigma_y^2\right)$ and $t_y = \frac{\overline{Y}_n - \mu_y}{s_{yn}} = t_x$, where $s_{yn}^2$ denotes the sample variance of $Y_i$. However, the measure of variability does not have such a nice property. For example, if $Y_i = aX_i$, then $\sigma_y^2 = a^2\sigma^2$ and the scale transformation has increased the variance of $X_i$ by a multiplicative factor $a^2$. A scale-invariant measure of variability is the *coefficient of variation*, $\tau = \frac{\sigma}{\mu}$.

We can estimate $\tau$ by $\widehat{\tau}_n = \frac{s_n}{\overline{X}_n}$. It follows from Slutsky's theorem that $\widehat{\tau}_n$ is consistent, that is,

$$\widehat{\tau}_n = \frac{s_n}{\overline{X}_n} \to \frac{\sigma}{\mu} = \tau$$

To find the asymptotic distribution of $\widehat{\tau}_n$, let $f(x, y) = \frac{y}{x}$. Then, $\widehat{\tau}_n = f\left(\overline{X}_n, s_n\right)$ and by the Delta method, $\widehat{\tau}_n \sim AN\left(\tau, \frac{1}{n} D^\top \Sigma D\right)$, where $\Sigma$ is the asymptotic variance of $\left(\overline{X}_n, s_n\right)$ and $D$ is the derivative of $f(x, y)$ with respect to $(x, y)$:

$$D^\top = \left(\frac{\partial f}{\partial x}, \frac{\partial f}{\partial y}\right) = \left(-\frac{\sigma}{\mu^2}, \frac{1}{\mu}\right)$$

In the above approach, it may not be straightforward to find the asymptotic variance $\Sigma$ (see exercise). An alternative is to express the estimate $\widehat{\tau}_n$ differently. To this end, let

$$g(x, y) = x^{-1}\left(y - x^2\right)^{\frac{1}{2}}, \quad U_n = \overline{X}_n, \quad V_n = \frac{1}{n}\sum_{i=1}^{n} X_i^2, \quad \mathbf{Z}_n = \left(U_n, V_n\right)^\top$$

Then, it is readily checked that $\widehat{\tau}_n = g\left(U_n, V_n\right)$. In addition, by CLT,

$$\mathbf{Z}_n \sim AN\left(\begin{pmatrix} \mu \\ \mu^2 + \sigma^2 \end{pmatrix}, \frac{1}{n}\begin{pmatrix} \sigma^2 & E\left(X_1^3\right) - \mu_1\mu_2 \\ & E\left(X_1^4\right) - \mu_2^2 \end{pmatrix}\right) \tag{1.41}$$

By the Delta method, $\widehat{\tau}_n \sim AN\left(\tau, \frac{1}{n} D^\top \Sigma D\right)$, where $\Sigma$ is given by the above and $D$ by

$$D^\top = \left(\frac{\partial f}{\partial x}, \frac{\partial f}{\partial y}\right) = \left(-\frac{\mu_2}{\sigma\mu_1}, \frac{1}{2\sigma\mu_1}\right) \tag{1.42}$$

**Example 9**. Let $\mathbf{Z}_i = \left(X_i, Y_i\right)^\top$ be i.i.d. random vectors $(1 \leq i \leq n)$. Let

$$\gamma = Cov\left(X_1, Y_1\right), \quad \widehat{\gamma}_n = \frac{1}{n}\sum_{i=1}^{n}\left(X_i - \overline{X}_n\right)\left(Y_i - \overline{Y}_n\right)$$

Then, $\sqrt{n}(\widehat{\gamma}_n - \gamma)$ has an asymptotic normal distribution.

Let $\mathbf{W}_i = (X_i, Y_i, X_iY_i)^\top$ and $f(\mathbf{w}) = w_3 - w_1w_2$. Then, $\widehat{\gamma}_n = f(\overline{\mathbf{W}}_n) = f(\overline{X}_n, \overline{Y}_n, \overline{XY}_n)$, where $\overline{XY}_n = \frac{1}{n}\sum_{i=1}^n X_iY_i$. It follows from CLT that

$$\sqrt{n}\left(\overline{\mathbf{W}}_n - E(\mathbf{W}_i)\right) \to_d N(0, \Sigma)$$

where $\Sigma = Var(\mathbf{W}_i)$ is readily computed and estimated. By the Delta method, $\widehat{\gamma}_n \sim AN\left(\gamma, \frac{1}{n}D^\top \Sigma D\right)$, where

$$D^\top = \left(\frac{\partial f}{\partial w_1}, \frac{\partial f}{\partial w_2}, \frac{\partial f}{\partial w_3}\right) = (-w_2, -w_1, 1)$$

Similarly, we can show that the Pearson correlation coefficient

$$\widehat{\rho}_n = \frac{\sum_{i=1}^n \left(X_i - \overline{X}_n\right)\left(Y_i - \overline{Y}_n\right)}{\sqrt{\sum_{i=1}^n \left(X_i - \overline{X}_n\right)^2}\sqrt{\sum_{i=1}^n \left(Y_i - \overline{Y}_n\right)^2}} \tag{1.43}$$

is a consistent estimate of $\rho = Corr(X_1, Y_1) = \frac{Cov(X_1, Y_1)}{\sqrt{Vov(X_1)Var(Y_1)}}$ and asymptotically normal (see exercise).

## 1.8 Exercises

### Section 1.1

1. For the linear regression model discussed in Section 1.1.1 for the comparison of two treatment conditions,

(a) show that the MLE of $\boldsymbol{\beta}$ is $\widehat{\boldsymbol{\beta}} = \left(\widehat{\beta}_1, \widehat{\beta}_2\right)^\top = (\overline{y}_{1\cdot}, \overline{y}_{2\cdot})^\top$, where $\overline{y}_{k\cdot} = \frac{1}{n_k}\sum_{i=1}^n x_{ik}y_i$;

(b) verify (1.4) by showing

$$\left(X^\top X\right)^{-1} = \begin{pmatrix} \frac{n}{n_1} & 0 \\ 0 & \frac{n}{n_2} \end{pmatrix}$$

2. Let $R_{ki}$ denote the rank scores for $y_{ki}$ $(k = 1, 2)$ as defined in Section 1.1.3. Let $W_n = \sum_{i=1}^{n_1} R_{1i}$. Show

(a) $W_n + \sum_{j=1}^{n_2} R_j^2 = \frac{n(n+1)}{2}$, where $n = n_1 + n_2$.

(b) $W_n = U_n + \frac{n_1(n_1+1)}{2}$.

3. In Problem 2 above, show

(a) $E(W_n) = \frac{n_1(n_1+n_2+1)}{2}$.

(b) $Var\left(W_n\right) = \frac{n_1 n_2 (n_1+n_2+1)}{12}$.

**Section 1.2**

1. Let $\Omega$ be some sample space. A power set $\mathcal{P}$ of $\Omega$ is a collection of all subsets of $\Omega$ including $\phi$ and $\Omega$. Show that $\mathcal{P}$ is a $\sigma$-field.

2. Prove the proposition.

3. For any $a \in R$, the real line, show that $(-\infty, a]$ can be expressed as a countable union of the intervals of the form:

(a) $(b, c]$ with $b, c \in R$.

(b) $(a, b)$ with $a, b \in R$.

Thus, intervals of the form $(-\infty, a]$, $(a, b]$ and $(a, b)$ all generate the same Borel $\sigma$-field $\mathcal{B}$. Since $a \in R$, the complement of $(-\infty, a]$ is $(a, \infty)$, $\mathcal{B}$ is also generated by intervals of the form $(a, \infty)$.

4. In problem 2 above, show that $[a, \infty)$ can be expressed as a countable union of $(a, \infty)$ for any $a \in R$. Thus, in light of problem 2 above, the class of intervals $[a, \infty)$ generates $\mathcal{B}$. Since $(-\infty, a)$ is the complement of $[a, \infty)$, $\mathcal{B}$ is also generated by the intervals of the form $(-\infty, a)$.

5. Prove Theorem 1.

**Section 1.3**

1. Let $\mathcal{S} = (\Omega, \mathcal{F})$ and $f_n$ be a real measurable function defined on $\mathcal{S}$ $(n = 1, 2, \ldots)$. Let $g_n = \max_{1 \le k \le n} f_k$ and $h_n = \min_{1 \le k \le n} f_k$. Show that for each $\omega \in \Omega$, $\lim_{n \to \infty} g_n$ and $\lim_{n \to \infty} h_n$ exist if $\infty$ is allowed to be a "value" for the limit.

2. In Problem 1, let $g_n = \sup_{k \ge n} f_k$ and $h_n = \inf_{k \ge n} f_k$. Show that both are monotone sequences and thus their limits exist.

3. Prove Theorems 2 and 3.

4. If $f = \sum_{i=1}^{n} c_i I_{A_i} = \sum_{i=1}^{n} d_i I_{B_i}$, show $\sum_{i=1}^{n} c_i \mu\left(A_i\right) = \sum_{i=1}^{n} d_i \mu\left(B_i\right)$. Thus, the integral $\int_\Omega f d\mu = \sum_{i=1}^{n} c_i \mu\left(A_i\right)$ is well defined.

5. For $f \ge 0$, show

(a) $f_n\left(x\right)$ defined in (1.11) is a monotone increasing sequence of real measurable functions.

(b) $\lim_{n \to \infty} f_n\left(\omega\right) = f\left(\omega\right)$ for each $\omega \in \Omega$.

(c) if $g_n\left(\omega\right)$ is also a monotone increasing sequence of real measurable functions and $f = \lim_{n \to \infty} g_n$, then $\lim_{n \to \infty} \int_\Omega f_n d\mu = \lim_{n \to \infty} \int_\Omega g_n d\mu$.

6. Let $\mathcal{M} = (\Omega, \mathcal{F}, \mu)$ be a measure space and $f$ some real measurable function define on $\mathcal{M}$. Let $v\left(E\right) = \int_E f d\mu$ for any $E \in \mathcal{F}$. Show that $v\left(\cdot\right)$ is also a measure.

7. Let $F\left(x\right)$ be a nondecreasing, right continuous, step function. Let $F_d\left(x\right) = F(x) - F\left(x^+\right)$. Show

(a) the number of points $x$ such that $F_d\left(x\right) > 0$ is countable.

(b) $F(x) - F_d(x)$ is nondecreasing and right continuous.

8. Let $\mathcal{S} = (R, \mathcal{B}, \mu)$ with $\mu$ denoting the Lebsgue measure. Show that for any absolutely continuous measure $\upsilon$ with respect to $\mu$, any single point $a \in R$ has a measure 0, that is, $\upsilon(\{a\}) = 0$.

**Section 1.4**

1. If a sequence of non-negative $p_k$ satisfies (1.12), show $\lim_{k\to\infty} p_k = 0$.

2. For the Poisson probability law defined by (1.13) in Example 2, show

(a) For any $A, B \in \digamma$ with $A \cap B = \phi$, $P(A \cup B) = P(A) + P(B)$.

(b) For a sequence of disjoint sets $A_n$ $(n = 1, 2, \ldots)$, $P\left(\bigcup_{n=1}^{\infty} A_n\right) = \sum_{n=1}^{\infty} P(A_n)$.

3. To show $\int_{-\infty}^{\infty} \frac{1}{\sqrt{2\pi}} \exp\left(-\frac{x^2}{2}\right) dx = 1$, let

$$\int_{-\infty}^{\infty} \frac{1}{\sqrt{2\pi}} \exp\left(-\frac{x^2}{2}\right) dx = a$$

Then,

$$\begin{aligned} a^2 &= \int_{-\infty}^{\infty} \frac{1}{\sqrt{2\pi}} \exp\left(-\frac{x^2}{2}\right) dx \int_{-\infty}^{\infty} \frac{1}{\sqrt{2\pi}} \exp\left(-\frac{y^2}{2}\right) dy \\ &= \int\int_{R^2} \frac{1}{2\pi} \exp\left(-\frac{x^2+y^2}{2}\right) dxdy \end{aligned}$$

Now, prove that $\int\int_{R^2} \frac{1}{2\pi} \exp\left(-\frac{x^2+y^2}{2}\right) dxdy = 1$ by using the polar coordinate transformation, that is, $x = r\cos\theta$ and $y = r\sin\theta$.

4. Verify (1.17), (1.19), and (1.20).

**Section 1.5**

1. Let $\mathcal{P} = (\Omega, \digamma, P)$ be a probability space, where $\Omega = \{0, 1, 2, \ldots\}$, $\digamma$ is the power set of $\Omega$ and $P$ satisfies (1.12). Let $X = k$ be a real function from $\Omega$ to $R$. Show that $X$ is a random variable from $(\Omega, \digamma)$ to $(R, \mathcal{B})$.

2. Let $F(x)$ be the CDF of a random variable $X$ and $F^{-1}(x)$ be defined by (1.24) in Example 4. Then, $Y = F(X)$ is a random variable. Show that $Z = F^{-1}(Y)$ is also a random variable.

3. If the $X_i$ of $\mathbf{X} = (X_1, \ldots, X_k)^\top$ are independent, show that the PDF or probability mass function is the product of the density or probability mass functions of $X_i$ $(1 \le i \le k)$.

4. Prove the Cauchy–Schwartz, Jensen's and Chebyshev's inequalities.

5. Show part (2)–(6) of Theorem 1.

**Section 1.6**

1. Show If $X_n \to_d \mu$, then $X_n \to_p \mu$.

2. Let $\mathbf{X}_n \in R^k$ be a sequence of random vectors. Show

(a) $\mathbf{X}_n \to_p \mathbf{X} \in R^k$ implies that $\mathbf{X}_n \to_d \mathbf{X}$.

(b) if $\mathbf{X} = \boldsymbol{\mu}$, a constant, $\mathbf{X}_n \to_d \mathbf{X}$ implies that $\mathbf{X}_n \to_p \mathbf{X}$.

3. In Example 2, show that $\mathbf{X}_n \to_r \mathbf{X}$ implies that $\mathbf{X}_n \to_p \mathbf{X}$.

4. Let $X$ be some random variable and $X_n = X$ $(n \geq 1)$. Show that the random sequence $X_n$ is stochastically bounded.

5. Show that an equivalent definition of stochastic boundedness in (1.38) is that for any $\epsilon > 0$, we can find $M_\epsilon$ such that

$$\begin{aligned} &\Pr\left[\|\mathbf{X}\|_n \leq M_\epsilon\right] \geq 1-\epsilon, \quad \text{all } n, \quad \text{or equivalently} \\ &\Pr\left[\|\mathbf{X}\|_n \geq M_\epsilon\right] \leq \epsilon, \qquad \text{all } n \end{aligned}$$

6. Let $r_n, q_n\,(>0) \to 0$, as $n \to \infty$. Show the properties regarding $o_p(\cdot)$ and $O_p(\cdot)$ in Section 1.6.4.

7. Let $\mathbf{X}_n, \mathbf{X} \in R^k$. Show that if $\mathbf{X}_n \to_r \mathbf{X}$, then $\mathbf{X}_n \to_p \mathbf{X}$.

**Section 1.7**

1. Prove part (b) and (c) of Slutsky's theorem.

2. In Theorem 1, if $X_n \to_p a$, a constant, show that (1)–(3) still hold true when "$\to_d$" is replaced by "$\to_p$".

3. Let $X_i \sim$i.i.d.$(\mu, \sigma^2)$. Let $X_n = \overline{X}_n$, $Y_n = \frac{1}{n}\sum_{i=1}^n X_i^2$ and $\mathbf{Z}_n = (X_n, Y_n)^\top$.

(a) Show that $\mathbf{Z}_n$ has a bivariate asymptotic normal distribution and find the asymptotic variance.

(b) Find an appropriate functional $f(x,y)$ so that $f(X_n, Y_n) = s_n^2 = \frac{1}{n}\sum_{i=1}^n \left(X_i - \overline{X}_n\right)^2$.

(c) Use the Delta method to find the asymptotic distribution of $s_n^2$.

4. In Example 8,

(a) find the asymptotic distribution of $\mathbf{Z}_n = \left(\overline{X}_n, s_n\right)^\top$.

(b) use the result in (a) in conjunction with the Delta method to find the asymptotic distribution of $\widehat{\tau} = \frac{s_n}{\overline{X}_n}$.

(c) show that the asymptotic variance from (b) is $D^\top \Sigma D$ with $\Sigma$ given in (1.41) and $D$ in (1.42).

5. Prove Theorem 3.

6. Let $\mathbf{X}_n, \mathbf{X} \in R^k$, and $F_n(\mathbf{x})$ and $F(\mathbf{x})$ be the CDF of $\mathbf{X}_n$ and $\mathbf{X}$. Let

$$\phi_n(\mathbf{t}) = \int_{\mathbf{R}^k} \exp\left(i\mathbf{t}^\top \mathbf{x}\right) dF_n(\mathbf{x}), \quad \phi(\mathbf{t}) = \int_{\mathbf{R}^k} \exp\left(i\mathbf{t}^\top \mathbf{x}\right) dF(\mathbf{x})$$

denote the *characteristic functions* of $\mathbf{X}_n$ and $\mathbf{X}$, respectively. Show

(a) $\mathbf{X}_n \to_d \mathbf{X}$ is equivalent to convergence of $\phi_n(\mathbf{t})$ to $\phi(\mathbf{t})$ for every $\mathbf{t}$ $(\in \mathbf{R}^m)$.

(b) use (a) to prove Theorem 4 (Cramér–Wold Device).

7. In Problem 9, show

(a) $\widehat{\rho}_n$ is a consistent and asymptotically normal estimate of $\rho$.

(b) if $\mathbf{Z}_i \sim$ i.i.d.$N(\boldsymbol{\mu}, \boldsymbol{\Sigma})$, the asymptotic variance of $\widehat{\rho}_n$ is $\left(1 - \rho^2\right)^2$.

# Chapter 2

# Models for Cross-Sectional Data

In this chapter, we discuss regression models for cross-sectional study data analysis. Cross-sectional study designs employ a single assessment time to obtain a snapshot of a study population with respect to the outcomes of interest. Cross-sectional data can also arise from longitudinal and other related study designs. For example, in longitudinal study designs, a cohort of study subjects is followed up for a period of time and assessed repeatedly during the study period. If we look only at the data collected from a single assessment, such as the end of the study, the data is cross-sectional. Longitudinal study data captures both between-individual differences and within-subject dynamics, offering the opportunity to study more complicated biological, psychological and behavioral hypotheses, especially those involving changes in outcomes of interest over time such as causal effects and disease progression. In contrast, cross-sectional data can only be used to examine relationships among outcomes at a single time point. Thus, any relationship or association among outcomes can be seriously confounded by individual differences among the study subjects and must be interpreted with caution. In particular, cross-sectional study data in general does not permit causal inference about the outcomes of interest.

Despite such limitations, however, analysis of cross-sectional data is still quite popular in many areas of research. First, cross-sectional designs are much less costly than their longitudinal counterparts and logistically are much easier to carry out. Although inappropriate for causal inference, analyses of cross-sectional data still provide useful information for assessing and generating hypotheses concerning association among outcomes of interest that serve as a precursor for formal investigation using longitudinal studies. Second, cross-sectional data can also arise from longitudinal study designs. For example, many longitudinal studies employ a pre- and post-treatment design in which each subject is assessed first at baseline before some treatment of interest is given and then again at some point post-treatment. For such a special longitudinal design, treatment effect can be analyzed using methods for cross-sectional data such as the analysis of covariance model (see Section 2.1 for more details). Such analyses even provide valid inference when there are missing data or subject dropout at the post-treatment assessment (see Chapter 4 for more discussion on the topic of missing data). Third, when studying causal or risk factors for some disease of interest, the standard of randomized controlled clinical trial may not be an option; in these situations, the case-control retrospective study can be a useful alternative. For example, it is impossible to carry out a randomized controlled trial to study the causal effect of smoking on lung cancer because of the inability to control exposure status by randomly assigning subjects to the smoking and no smoking condition. Finally, for a rare disease, controlled longitudinal or observational cohort study designs may not yield a sufficiently large event rate to ensure adequate power. Although increasing sample size is an option, logistic considerations and prohibitively high costs often argue against such a prospective design in favor of a case-control retrospective alternative.

Linear regression is probably the most popular approach for modeling a response or dependent variable as a function of a set of independent variables. However, a major limitation is that it only applies to continuous response. In addition, the classic normal distribution assumption further limits the applicability of such models to real study data, as data distributions in many studies are often at odds with such a mathematical model. Nonetheless, it remains popular among practitioners and serves a good starting point to discuss the concepts of regression and parametric inference. In Section 2.1, we first discuss the linear regression model and then its extension to the generalized linear model to address the first major limitation. In Section 2.2, we address the second distribution concern by studying a class of distribution-free models.

## 2.1 Parametric Regression Models

Parametric models are widely used for data analysis in studies in the biomedical, behavioral and social sciences. For regression analysis, one of the study outcomes is designated as the *dependent* or *response* variable and is modeled as a function of a set of other outcomes or *independent* variables. For parametric regression models, some parametric distribution is assumed for the conditional distribution of the response given the independent variables. Such an assumption provides the premise for statistical inference of model parameters of interest. Although a variety of parametric models have been developed and used in statistical applications, linear regression is probably the most popular choice for modeling continuous response. In this section, we discuss this class of models along with inference for model parameters.

### 2.1.1 Linear Regression Model

Consider a study with $n$ subjects and let $y_i$ be a continuous dependent variable of interest and $\mathbf{x}_i = (x_{i1}, \ldots, x_{ip})^\top$ a $p \times 1$ vector of independent variables from the $i$th subject $(1 \leq i \leq n)$. Note that unlike Chapter 1, we will use lowercase letters to denote random variables and vectors throughout the rest of the book unless otherwise noted. Interest lies in modeling $y_i$ as a function of $\mathbf{x}_i$. The linear regression for modeling such a relationship is defined by the following model statement:

$$y_i = \mathbf{x}_i^\top \boldsymbol{\beta} + \epsilon_i, \quad \epsilon_i \sim \text{i.i.d. } N\left(0, \sigma^2\right), \quad 1 \leq i \leq n \tag{2.1}$$

where $\boldsymbol{\beta} = \left(\beta_1, \beta_2, \ldots, \beta_p\right)^\top$ is a $p \times 1$ vector of parameters, $\epsilon_i$ denotes the error term in the model, $N\left(\mu, \sigma^2\right)$ denotes a normal distribution with mean $\mu$ and variance $\sigma^2$, and i.i.d. stands for independently and identically distributed. In (2.1), $\epsilon_i \sim$i.i.d. $N\left(0, \sigma^2\right)$ means that $\epsilon_i$ are independently and identically distributed with the same normal distribution $N\left(0, \sigma^2\right)$. The linear function $\eta_i = \mathbf{x}_i^\top \boldsymbol{\beta}$ is often called the *linear predictor*.

In the linear regression model (2.1), we are interested in estimating $\boldsymbol{\beta}$ and making inference about this parameter vector. By viewing (2.1) as a relationship in which $y_i$ depends on $\mathbf{x}_i$, a known $\boldsymbol{\beta}$ will enable us to predict the mean response $y_i$ for a given $\mathbf{x}_i$. This is the reason why $y_i$ and $\mathbf{x}_i$ are often called the *dependent* (or *response)* and the *independent* (or *predictor*) variables, respectively.

In many applications, especially in epidemiologic studies, a subset of $\mathbf{x}$ is of primary interest and the variables within such a subset are often called *explanatory* variables. The remaining variables in $\mathbf{x}$ are used to control for heterogeneity of study sample such as differences in demographic and clinical outcomes, and for this reason, they are often called *covariates* or *confounding* variables. The presence of such variables in the model (2.1) is to control for the additional variability in $y_i$ that reflects the heterogeneity of the subjects in the study sample not explained by the explanatory variables or predictors. In this book, we will not carefully distinguish the different types of independent variables, although such a distinction plays a critical role in causal inference.

Because of scale differences between $y_i$ and the components of $\mathbf{x}_i$, it is typically necessary to include an *intercept* term in the model (2.1) to calibrate such differences. For this reason, one of the components of $\mathbf{x}_i$, say $x_{i1}$, is usually set to 1 so that the corresponding parameter $\beta_1$ serves as this calibration coefficient and the linear predictor of the model has the form:

$$\eta_i = \beta_1 + x_{i2}\beta_2 + \ldots + x_{ip}\beta_p \tag{2.2}$$

**Example 1 (*ANOVA*).** Consider a study comparing $g$ different treatment conditions, each of which has $n_k$ subjects $(1 \leq j \leq n_k,\ 1 \leq k \leq g)$. Let $y_i$ denote the continuous response of interest. We assume that subject $i$ is in the $k$th treated group if $\sum_{l=0}^{k-1} n_l + 1 \leq i \leq \sum_{l=0}^{k} n_l$ $(n_0 \equiv 0, 1 \leq k \leq g)$. Thus, $y_1,\ldots,y_{n_1}$ represent data from the first group, $y_{n_1+1},\ldots,y_{n_1+n_2}$ denote the data from the second group, etc.

Define $g$ binary indicators for each of the treatment conditions as follows:

$$x_{ik} = \begin{cases} 1 & \text{if subject } i \text{ receives treatment } k \\ 0 & \text{if otherwise} \end{cases} \tag{2.3}$$

Then, the linear model in (2.1) with the linear predictor $\eta_i = \beta_1 x_{i1} + \cdots + x_{ig}\beta_g$ is equivalently expressed as:

$$y_i = \beta_k + \epsilon_i, \quad \epsilon_i \sim \text{i.i.d. } N\left(0, \sigma^2\right) \tag{2.4}$$
$$\sum_{l=0}^{k-1} n_l + 1 \leq i \leq \sum_{l=0}^{k} n_l, \quad n_0 \equiv 0, \quad 1 \leq k \leq g$$

In the above, $\beta_k$ is the mean of the response $y_i$ across the subjects within the $k$th treatment group. By rewriting $y_i$ as $y_{kj}$ when $\sum_{l=0}^{k-1} n_l + 1 \leq i \leq \sum_{l=0}^{k} n_l$ and changing $\beta_k$ to $\mu_k$, (2.4) becomes the familiar analysis of variance (ANOVA) model:

$$y_{kj} = \mu_k + \epsilon_{kj}, \quad \epsilon_{kj} \sim \text{i.i.d. } N\left(0, \sigma^2\right) \quad 1 \leq j \leq n_k, \quad 1 \leq k \leq g \tag{2.5}$$

The most popular approach for studying treatment difference is to compare the mean response $\mu_k$ across the different treatment groups. Thus, under this classic treatment comparison paradigm, we are interested in testing the following hypothesis:

$$H_0 : \mu_k = \mu \quad \text{for all } k \quad \text{versus} \quad H_a : \mu_k \neq \mu_l \quad \text{for some } k \neq l \tag{2.6}$$

For example, if $g = 2$, we have: $H_0 : \mu_1 = \mu_2$ versus $H_0 : \mu_1 \neq \mu_2$. For $g > 2$, we are not only interested in whether the group means $\mu_k$ are the same, but also how the difference arises when the means do differ across the groups. Thus, if $H_0$ is rejected, we then follow with post-hoc comparisons among $\mu_k$ to identify the groups that differ from each other.

In Example 1, if we use the linear predictor with an intercept, $\eta_i = \beta_0 + \beta_1 x_{i1} + \ldots + x_{ig}\beta_g$, the ANOVA becomes:

$$y_i = \beta_0 + \beta_k + \epsilon_i, \quad \epsilon_i \sim \text{i.i.d. } N\left(0, \sigma^2\right) \tag{2.7}$$
$$\sum_{l=0}^{k-1} n_l + 1 \leq i \leq \sum_{l=0}^{k} n_l, \quad n_0 \equiv 0, \quad 1 \leq k \leq g$$

In this case, $\beta_k$ are not identifiable since the model above involves more than $g$ number of parameters. To address identifiability, some constraint must be imposed. One popular approach is to set one of $\beta_k$ to 0. For example, by setting $\beta_g = 0$, we can identify the remaining $g$ parameters, $\beta_0$, $\beta_1$,...,$\beta_{g-1}$. This *cell reference* coding method is equivalent to including only the first $g - 1$ binary indicators $x_{i1}$, $x_{i2}$, ..., $x_{i(g-1)}$. Under this parameterization scheme, $\beta_0$ is interpreted as the mean response for the $g$th treatment and $\beta_k$ represents the difference in mean response between the $k$th and $g$th treatment conditions. Another common method for identifiability is to impose the constraint, $\sum_{k=1}^{g} \beta_k = 0$, and then solve for $\beta_g$. This *effect* coding method is also widely used by popular software packages, such as SAS.

The hypothesis for testing treatment difference in mean response takes different forms depending on the coding schemes used. For example, under the reference cell coding method, (2.6) is equivalent to

$$\begin{aligned} &H_0 : \beta_k = 0 \quad \text{for all } 1 \leq k \leq g-1 \quad \text{versus} \\ &H_a : \beta_k \neq 0 \quad \text{for some } 1 \leq k \leq g-1 \end{aligned} \tag{2.8}$$

For example, if $g = 2$, the above yields $H_0 : \beta_1 = 0$ versus $H_a : \beta_1 \neq 0$.

In many applications, differences in patients with respect to their disease history and comorbid conditions often alter the effect of treatment. Understanding such a relationship not only enables us to obtain unbiased treatment effect, but also affords us an opportunity to develop individualized treatment therapy. The next example generalizes ANOVA to model the impact of patients differences on treatment effect.

**Example 2 (ANCOVA).** Within the context of Example 1, let $z_{ki}$ denote the vector of covariates from the $i$th subject within the $k$th treatment condition. Consider the following linear model:

$$\begin{aligned} y_{kj} &= \mu_k + \boldsymbol{\beta}_k^\top \mathbf{z}_{ki} + \epsilon_{kj}, \quad \epsilon_{kj} \sim \text{i.i.d. } N\left(0, \sigma^2\right) \\ &1 \leq j \leq n_k, \quad 1 \leq k \leq g \end{aligned} \tag{2.9}$$

The above is similar to the ANOVA model in (2.5) except for the extra term $\beta_k^\top z_{ki}$ to account for the effect of covariates $z_{ki}$ on treatment differences. If $\beta_k = \beta$, that is, $z_{ki}$ does not interact with the treatment conditions, (2.9) is known as the analysis of covariance model (ANCOVA). In this case, ANCOVA allows us to derive treatment effect by controlling for $z_{ki}$. Otherwise, there exists covariate by treatment interaction. In treatment studies, it is important to identify such interactions to study the differential effect of a treatment when applied to subjects from different study populations, such as different ethnicity, gender, medical and psychiatric history and co-morbid conditions so that effective treatment interventions can be develop to account for such patients' differences.

### 2.1.2 Inference for Linear Models

For parametric models, the method of maximum likelihood is the most popular approach for inference about model parameters. We first discuss computations and properties of parameter estimates for general parametric models obtained under this approach and then specialize the results to the linear regression model.

Consider a sample of $n$ subjects and let $y_i$ denote some outcome of interest $(1 \leq i \leq n)$. We assume that the distribution of $y_i$ is prescribed by a family of probability density or distribution function (PDF) $f(y_i \mid \boldsymbol{\theta})$ indexed by a $q \times 1$ vector of parameters $\boldsymbol{\theta}$ depending on whether $y_i$ is continuous or discrete. Note that since the theory of maximum likelihood applies to both continuous and discrete outcomes, we will refer to either the probability distribution or density function as PDF throughout the rest of the book for the sake of convenience. When the distinction between the two is important, we will be specific about which type of distribution is being considered. The likelihood function is defined as the product of the PDFs from the $n$ responses: $L_n(\boldsymbol{\theta}) = \prod_{i=1}^n f(y_i \mid \boldsymbol{\theta})$. For a fixed $\boldsymbol{\theta}$, as noted in

Chapter 1, $f(\cdot \mid \boldsymbol{\theta})$ is a real measurable function with respect to the Borel $\sigma$-field and as such the composite function $f(y_i \mid \boldsymbol{\theta})$ is a random variable. Thus, as the product of these random variables, the likelihood $L_n(\boldsymbol{\theta})$ is also a random variable. When applied to data in a real study, the likelihood becomes a function of $\boldsymbol{\theta}$ conditional on the observed $y_i$. The *maximum likelihood estimate* (MLE) is defined as the value of $\boldsymbol{\theta}$ that maximizes the likelihood function $L_n(\boldsymbol{\theta})$ when viewed as a function of $\boldsymbol{\theta}$.

As will become clear shortly, the log of the likelihood defined below plays a much more critical role for inference about $\boldsymbol{\theta}$ than the likelihood function itself:

$$l_n(\boldsymbol{\theta}) = \log(L_n(\boldsymbol{\theta})) = \sum_{i=1}^{n} \log f(y_i \mid \boldsymbol{\theta})$$

Since $\log(\cdot)$ is a monotone increasing function, the MLE of $\boldsymbol{\theta}$ can be equivalently defined as the value that maximizes the log-likelihood, $l_n(\boldsymbol{\theta})$. To discuss computations, as well as properties of MLE, we will be taking first and second order derivatives of $l_n(\boldsymbol{\theta})$. It is helpful to review the definition of vector-valued function and some basic properties. Let $\mathbf{f}(\boldsymbol{\theta})$ be a $n \times 1$ and $\mathbf{h}(\boldsymbol{\theta})$ a $1 \times m$ vector-valued function of $\boldsymbol{\theta}$. The derivatives of $\frac{\partial}{\partial \boldsymbol{\theta}}\mathbf{f}$ and $\frac{\partial}{\partial \boldsymbol{\theta}}\mathbf{g}$ are defined as follows:

$$\frac{\partial}{\partial \boldsymbol{\theta}}\mathbf{f} = \begin{pmatrix} \frac{\partial f_1}{\partial \theta_1} & \cdots & \frac{\partial f_n}{\partial \theta_1} \\ \vdots & \ddots & \vdots \\ \frac{\partial f_1}{\partial \theta_q} & \cdots & \frac{\partial f_n}{\partial \theta_q} \end{pmatrix}_{q \times n}, \quad \frac{\partial}{\partial \boldsymbol{\theta}}\mathbf{g} = \begin{pmatrix} \frac{\partial g_1}{\partial \theta_1} & \cdots & \frac{\partial g_1}{\partial \theta_q} \\ \vdots & \ddots & \vdots \\ \frac{\partial g_m}{\partial \theta_1} & \cdots & \frac{\partial g_m}{\partial \theta_q} \end{pmatrix}_{m \times q} \tag{2.10}$$

Thus, $\frac{\partial}{\partial \boldsymbol{\theta}}\mathbf{g} = \left(\frac{\partial}{\partial \boldsymbol{\theta}}\mathbf{g}^\top\right)^\top$. As a special case, if $f(\boldsymbol{\theta})$ is a scalar function, it follows from (2.10) that $\frac{\partial f}{\partial \boldsymbol{\theta}} = \left(\frac{\partial}{\partial \theta_1} f, \ldots, \frac{\partial}{\partial \theta_1} f\right)^\top$ is a $q \times 1$ column vector. Let $A$ be a $m \times n$ matrix of constants, $\mathbf{g}(\boldsymbol{\theta})$ a $m \times 1$ vector-valued function of $\boldsymbol{\theta}$ and $h(\boldsymbol{\theta})$ a function of $\boldsymbol{\theta}$. Then, we have (see exercise):

1. $\frac{\partial}{\partial \boldsymbol{\theta}}(A\mathbf{f}) = \left(\frac{\partial}{\partial \boldsymbol{\theta}}\mathbf{f}\right) A^\top$.
2. $\frac{\partial}{\partial \boldsymbol{\theta}}(h\mathbf{f}) = \left(\frac{\partial}{\partial \boldsymbol{\theta}}h\right) \mathbf{f}^\top + h\frac{\partial}{\partial \boldsymbol{\theta}}\mathbf{f}$.
3. $\frac{\partial}{\partial \boldsymbol{\theta}}\left(\mathbf{g}^\top A\mathbf{f}\right) = \left(\frac{\partial}{\partial \boldsymbol{\theta}}\mathbf{g}\right) A\mathbf{f} + \left(\frac{\partial}{\partial \boldsymbol{\theta}}\mathbf{f}\right) A^\top \mathbf{g}$.

If the maximum of $l_n(\boldsymbol{\theta})$ is achieved at an internal point $\widehat{\boldsymbol{\theta}}$, then this value should be a critical point of $l_n(\boldsymbol{\theta})$, that is, $\widehat{\boldsymbol{\theta}}$ must satisfy the following condition:

$$\mathbf{u}_n(\boldsymbol{\theta}) = \frac{\partial l_n(\boldsymbol{\theta})}{\partial \boldsymbol{\theta}} = \sum_{i=1}^{n} \frac{\partial}{\partial \boldsymbol{\theta}} \log f(y_i \mid \boldsymbol{\theta}) = \mathbf{0} \tag{2.11}$$

As noted earlier, the derivative $\mathbf{u}_n(\boldsymbol{\theta})$ of $l_n(\boldsymbol{\theta})$ is a $q \times 1$ column vector and is known as the *score statistic vector* (when viewed as a function of the random variables $y_i$). The equation in (2.11) is known as the *score equation.* Like $l_n(\boldsymbol{\theta})$, the score statistic vector $\mathbf{u}_n(\boldsymbol{\theta})$ plays an important role in the study of MLE.

**Example 3 (*One-group ANOVA with known variance*).** If the linear model contains only the intercept, that is, $\mathbf{x}_i = x_{i1} \equiv 1$, then (2.1) reduces to the following simpler model:

$$y_i = \beta_1 + \epsilon_i, \quad \epsilon_i \sim \text{i.i.d. } N\left(0, \sigma^2\right), \quad 1 \leq i \leq n$$

In this special case, the intercept $\beta_1 = E(y_i)$ is the mean of $y_i$ and the above linear regression models the mean of $y_i$ from a single study population.

Assume that $\sigma^2$ is known so that $\beta_1$ is the only parameter of the model. The log-likelihood function is given by

$$l_n(\beta_1) = -\frac{n}{2}\log(2\pi) - \frac{n}{2}\log\left(\sigma^2\right) - \frac{1}{2\sigma^2}\sum_{i=1}^{n}(y_i - \beta_1)^2 \tag{2.12}$$

The score statistic $u_n(\beta_1)$ is given by

$$u_n(\beta_1) = \frac{\partial}{\partial \mu} l_n(\beta_1) = \frac{1}{\sigma^2}\sum_{i=1}^{n}(y_i - \beta_1)$$

By setting $u_n(\beta_1) = 0$ and solving for $\beta_1$, we obtain the MLE: $\widehat{\beta}_1 = \overline{y}_n = \frac{1}{n}\sum_{i=1}^{n} y_i$. It follows from the law of large numbers (LLN) that the MLE $\widehat{\beta}_1$ is a consistent estimate of $\beta_1$ (see Chapter 1 for LLN and definition of consistency). Further, as $\widehat{\beta}_1$ is a linear combination of i.i.d. normal variates $y_i$, it follows from the properties of normal distribution that $\widehat{\beta}_1 \sim N\left(\beta_1, \frac{\sigma^2}{n}\right)$. This sampling distribution allows us to make inference about $\beta_1$. For example, consider testing the hypothesis: $H_0 : \beta_1 = 0$ versus $H_a : \beta_1 \neq 0$. Under $H_0$, we have: $\widehat{\beta}_1 \sim N\left(0, \frac{\sigma^2}{n}\right)$. Thus, the test statistic, $t = \frac{\sqrt{n}\widehat{\beta}_1}{\sigma} \sim N(0,1)$ and the p-value is given by: $p = 2(1 - \Phi(|t|))$, where $\Phi(\cdot)$ denotes the cumulative distribution function (CDF) of the standard normal $N(0,1)$.

**Example 4 (*ANOVA with unknown variance*).** Now, assume that $\sigma^2$ is unknown in Example 3. Then, the normal sampling distribution of $\widehat{\beta}_1$ derived in the above example cannot be used for inference about $\beta_1$. We must consider a different statistic.

Let $\boldsymbol{\theta} = \left(\beta_1, \sigma^2\right)^\top$. The log-likelihood function is the same as the one given in (2.1), but the score statistic vector $\mathbf{u}(\boldsymbol{\theta})$ now contains the additional component for the derivative of the log-likelihood with respect to $\sigma^2$:

$$\frac{\partial}{\partial \mu} l_n(\boldsymbol{\theta}) = -\frac{1}{\sigma^2}\sum_{i=1}^n (y_i - \mu), \quad \frac{\partial}{\partial \sigma^2} l_n(\boldsymbol{\theta}) = -\frac{n}{2\sigma^2} + \frac{1}{2\sigma^4}\sum_{i=1}^n (y_i - \beta_1)^2$$

By setting $\mathbf{u}(\boldsymbol{\theta}) = \mathbf{0}$ and solving for $\boldsymbol{\theta}$, we obtain the MLE $\widehat{\boldsymbol{\theta}} = \left(\widehat{\beta}_1, \widehat{\sigma}^2\right)^\top$:

$$\widehat{\beta}_1 = \overline{y}_n = \frac{1}{n}\sum_{i=1}^n y_i, \quad \widehat{\sigma}^2 = \frac{1}{n}\sum_{i=1}^n (y_i - \overline{y}_n)^2 = \frac{n-1}{n} s_n^2 \tag{2.13}$$

where $s_n^2 = \frac{1}{n-1}\sum_{i=1}^n (y_i - \overline{y}_n)^2$ is the usual sample variance. It is a well-known fact that $\widehat{\beta}_1$ and $s_n^2$ are independent and $\frac{(n-1)s_n^2}{\sigma^2} \sim \chi^2_{n-1}$, a $\chi^2$ distribution with $n-1$ degrees of freedom (see exercise).

Thus, the MLE of $\widehat{\beta}_1$ is still the sample mean as in Example 3. Unlike Example 3, however, we cannot perform inference about $\beta_1$ based on the normal sampling distribution $N\left(\beta_1, \frac{\sigma^2}{n}\right)$, as $\sigma^2$ is unknown. Fortunately, by the independence between $\widehat{\beta}_1$ and $s_n^2$ and the fact that $\frac{(n-1)s_n^2}{\sigma^2} \sim \chi^2_{n-1}$, we have: $\frac{\sqrt{n}(\widehat{\beta}_1 - \beta_1)}{s_n} \sim t_{n-1}$, a univariate $t$ distribution with $n-1$ degrees of freedom. This $t$ sampling distribution is widely used for inference about $\beta_1$.

Note that the MLE of $\widehat{\sigma}^2$ is not the same as the sample variance $s_n^2$. However, as we have shown in Chapter 1, $\widehat{\sigma}^2$ and $s_n^2$ are both consistent estimates of $\sigma^2$ with the same asymptotic distribution. Thus, they are asymptotically equivalent.

**Example 5** (***Linear regression***). Consider the general linear regression model (2.1) with a general $\mathbf{x}_i$ and unknown $\sigma^2$. Let $\boldsymbol{\theta} = \left(\boldsymbol{\beta}^\top, \sigma^2\right)^\top$ denote the vector of model parameters. The log-likelihood function is given by

$$l_n(\boldsymbol{\theta}) = -\frac{n}{2}\log\left(2\pi\sigma^2\right) - \frac{1}{2\sigma^2}\sum_{i=1}^n \left(y_i - \mathbf{x}_i^\top \boldsymbol{\beta}\right)^2$$

Using the properties of differentiation of vector-valued function, the score vector equation is readily calculated

$$\begin{aligned}\frac{\partial}{\partial \boldsymbol{\beta}} l_n &= \frac{1}{\sigma^2}\sum_{i=1}^{n}\left(\mathbf{x}_i y_i - \mathbf{x}_i \mathbf{x}_i^\top \boldsymbol{\beta}\right) = \mathbf{0}, \\ \frac{\partial}{\partial \sigma^2} l_n &= -\frac{n}{2\sigma^2} + \frac{1}{2\sigma^4}\sum_{i=1}^{n}\left(y_i - \mathbf{x}_i^\top \boldsymbol{\beta}\right)^2\end{aligned} \tag{2.14}$$

By solving for $\boldsymbol{\theta}$, we obtain the MLE $\widehat{\boldsymbol{\theta}} = \left(\widehat{\boldsymbol{\beta}}^\top, \widehat{\sigma}^2\right)^\top$:

$$\widehat{\boldsymbol{\beta}} = \left(\sum_{i=1}^{n} \mathbf{x}_i \mathbf{x}_i^\top\right)^{-1}\left(\sum_{i=1}^{n} \mathbf{x}_i y_i\right), \quad \widehat{\sigma}^2 = \frac{1}{n}\sum_{i=1}^{n}\left(y_i - \mathbf{x}_i^\top \widehat{\boldsymbol{\beta}}\right)^2 \tag{2.15}$$

If $\mathbf{x}_i$ is regard as fixed, then it is readily shown (see exercise) that

$$\begin{aligned}&\frac{\sqrt{n}}{\sqrt{MSE}}\left(\widehat{\boldsymbol{\beta}} - \boldsymbol{\beta}\right) \mid MSE \sim N\left(\mathbf{0}, \frac{\sigma^2}{MSE}\left(\frac{1}{n}\sum_{i=1}^{n} \mathbf{x}_i \mathbf{x}_i^\top\right)^{-1}\right) \\ &\frac{(n-p)\,MSE}{\sigma^2} = \frac{n\widehat{\sigma}^2}{\sigma^2} \sim \chi^2_{n-p}\end{aligned} \tag{2.16}$$

where $MSE = \frac{n\widehat{\sigma}^2}{n-p}$ is known as the error mean square. Again, as in Example 4, $\widehat{\boldsymbol{\beta}}$ and $\widehat{\sigma}^2$ are independent (see exercise).

As $\sigma^2$ is unknown, we cannot use the multivariate normal distribution in (2.16) for inference about $\boldsymbol{\beta}$. However, using the distribution result for $\widehat{\sigma}^2$ in (2.16), it can be shown that $\widehat{\boldsymbol{\beta}}$ follows a multivariate $t$ distribution (see exercise):

$$\frac{\sqrt{n}}{\sqrt{MSE}}\left(\widehat{\boldsymbol{\beta}} - \boldsymbol{\beta}\right) \sim t\left(\mathbf{0}, \left(\frac{1}{n}\sum_{i=1}^{n} \mathbf{x}_i \mathbf{x}_i^\top\right)^{-1}, n-p\right) \tag{2.17}$$

where $t(\boldsymbol{\mu}, \Sigma, v)$ denotes a multivariate $t$ with mean $\boldsymbol{\mu}$, variance $\Sigma$ and degrees of freedom (or shape parameter) $v$, with the following PDF:

$$\frac{\Gamma\left(\frac{v+p}{2}\right)}{\Gamma\left(\frac{v}{2}\right)(\pi v)^{\frac{p}{2}}}\,|\Sigma|^{-\frac{1}{2}}\left(1 + \frac{(\mathbf{y} - \boldsymbol{\mu})^\top \Sigma\,(\mathbf{y} - \boldsymbol{\mu})}{v}\right)^{-\frac{v+p}{2}} \tag{2.18}$$

The multivariate $t$ distribution has nice properties similar to those of the multivariate normal. In particular, it is readily shown that any sub-vector also has a $t$ distribution (see exercise). Thus, we can make inference for each individual $\beta_k$ using the marginal univariate $t$ distribution of $\widehat{\beta}_k$.

When using the distribution result in (2.17), $\sum_{i=1}^{n} \mathbf{x}_i \mathbf{x}_i^\top$ is presumed to be a constant matrix. As noted in Section 1.1 of Chapter 1, the response $y_i$ and the set of independent variables $\mathbf{x}_i$ are often concurrently observed and as such $\sum_{i=1}^{n} \mathbf{x}_i \mathbf{x}_i^\top$ also varies from sample to sample. Thus, unless for the special case of a constant $\mathbf{x}_i$, the sampling distribution of the MLE $\widehat{\boldsymbol{\beta}}$ in general does not follow the $t$ distribution as given in (2.17). In such applications, we must rely on the asymptotic distribution of MLE for valid inference of model parameters.

To discuss the asymptotic distribution of the MLE $\widehat{\boldsymbol{\beta}}$ for linear regression, we first present results concerning maximum likelihood estimates for general parametric models as summarized by the following theorem. To avoid a lengthy list of assumptions for the legitimacy of performing certain mathematical operations, such as exchanging the order of differentiation and integration, we follow the convention of subsuming all such operations under the phrase, "under mild regularity conditions," throughout the book.

**Theorem 1**. Let $y_i$ be i.i.d. random variables with the PDF $f(y_i \mid \boldsymbol{\theta})$ $(1 \leq i \leq n)$. Let $\widehat{\boldsymbol{\theta}}$ denote the maximum likelihood estimate of $\boldsymbol{\theta}$. Then, under mild regularity conditions, we have

1. The maximum likelihood estimate $\widehat{\boldsymbol{\theta}}$ is consistent and asymptotically normal.

2. The asymptotic variance of $\widehat{\boldsymbol{\theta}}$ is given by

$$\begin{aligned} \Sigma_\theta &= E^{-1}\left[-\frac{\partial^2}{\partial \boldsymbol{\theta} \partial \boldsymbol{\theta}^\top} \log f(y_i \mid \boldsymbol{\theta})\right] \qquad (2.19) \\ &= E^{-1}\left[\left(\frac{\partial}{\partial \boldsymbol{\theta}} \log f(y_i \mid \boldsymbol{\theta})\right)\left(\frac{\partial}{\partial \boldsymbol{\theta}} \log f(y_i \mid \boldsymbol{\theta})\right)^\top\right] \end{aligned}$$

where $\frac{\partial^2}{\partial \boldsymbol{\theta} \partial \boldsymbol{\theta}^\top} \log f(y_i \mid \boldsymbol{\theta}) = \frac{\partial}{\partial \boldsymbol{\theta}}\left(\frac{\partial}{\partial \boldsymbol{\theta}} \log f(y_i \mid \boldsymbol{\theta})\right)^\top$.

Proof. First, we show that the normalized score statistic vector, $\mathbf{u}_n(\boldsymbol{\theta}) = \frac{1}{n}\frac{\partial}{\partial \boldsymbol{\theta}} l_n(\boldsymbol{\theta})$, is asymptotically normal. Note that the normalized $\mathbf{u}_n(\boldsymbol{\theta})$ is slightly different from the score statistic vector $\mathbf{u}_n(\boldsymbol{\theta})$ used in the above examples with the normalizing factor $\frac{1}{n}$ used to control the rate at which $\mathbf{u}_n(\boldsymbol{\theta})$ converges to 0 in probability.

For notational brevity, let $\mathbf{w}_i(\boldsymbol{\theta}) = \frac{\partial}{\partial \boldsymbol{\theta}} \log f(y_i \mid \boldsymbol{\theta})$. Since

$$\mathbf{u}_n(\boldsymbol{\theta}) = \frac{1}{n}\sum_{i=1}^{n} \mathbf{w}_i(\boldsymbol{\theta}) = \frac{1}{n}\sum_{i=1}^{n} \frac{\partial}{\partial \boldsymbol{\theta}} \log f(y_i \mid \boldsymbol{\theta}) \qquad (2.20)$$

it follows from LLN that

$$\begin{aligned}\mathbf{u}_n(\boldsymbol{\theta}) \to_p E[\mathbf{w}_i(\boldsymbol{\theta})] &= E\left[\frac{\partial}{\partial\boldsymbol{\theta}}\log f(y_i \mid \boldsymbol{\theta})\right] \\ &= \int \frac{1}{f(y \mid \boldsymbol{\theta})}\left[\frac{\partial}{\partial\boldsymbol{\theta}} f(y \mid \boldsymbol{\theta})\right] f(y \mid \boldsymbol{\theta})\, dy = \int \frac{\partial}{\partial\boldsymbol{\theta}} f(y \mid \boldsymbol{\theta})\, dy \\ &= \frac{\partial}{\partial\boldsymbol{\theta}}\int f(y \mid \boldsymbol{\theta})\, dy = \mathbf{0}\end{aligned}$$

By CLT,

$$\sqrt{n}\mathbf{u}_n(\boldsymbol{\theta}) \to_d N\left(0, \Sigma_u = Var\left(\frac{\partial}{\partial\boldsymbol{\theta}}\log f(y_i \mid \boldsymbol{\theta})\right)\right)$$

Further, since

$$\frac{\partial^2 \log f(y \mid \boldsymbol{\theta})}{\partial\boldsymbol{\theta}\partial\boldsymbol{\theta}^\top} = \frac{-1}{f^2(y \mid \boldsymbol{\theta})}\frac{\partial}{\partial\boldsymbol{\theta}} f(y \mid \boldsymbol{\theta})\left[\frac{\partial}{\partial\boldsymbol{\theta}} f(y \mid \boldsymbol{\theta})\right]^\top + \frac{1}{f(y \mid \boldsymbol{\theta})}\frac{\partial^2 f(y \mid \boldsymbol{\theta})}{\partial\boldsymbol{\theta}\partial\boldsymbol{\theta}^\top}$$

it follows that

$$\begin{aligned}\Sigma_u = Var(\mathbf{w}_i(\boldsymbol{\theta})) &= Var\left(\frac{\partial}{\partial\boldsymbol{\theta}}\log f(y_i \mid \boldsymbol{\theta})\right) \qquad (2.21)\\ &= E\left[\left(\frac{\partial}{\partial\boldsymbol{\theta}}\log f(y_i \mid \boldsymbol{\theta})\right)\left(\frac{\partial}{\partial\boldsymbol{\theta}}\log f(y_i \mid \boldsymbol{\theta})\right)^\top\right] \\ &= E\left[\frac{-\partial^2 \log f(y_i \mid \boldsymbol{\theta})}{\partial\boldsymbol{\theta}\partial\boldsymbol{\theta}^\top}\right] + E\left[\frac{1}{f(y_i \mid \boldsymbol{\theta})}\frac{\partial^2 f(y_i \mid \boldsymbol{\theta})}{\partial\boldsymbol{\theta}\partial\boldsymbol{\theta}^\top}\right] \\ &= -E\left[\frac{\partial^2}{\partial\boldsymbol{\theta}\partial\boldsymbol{\theta}^\top}\log[f(y_i \mid \boldsymbol{\theta})]\right]\end{aligned}$$

Now, consider the distribution of MLE $\widehat{\boldsymbol{\theta}}$. For convenience and without the loss of generality, assume a scalar $\theta$ in which case, $u_n(\theta)$ is also a scalar.

First, we show consistency of $\widehat{\theta}$. Consider a neighborhood $N(\theta)$ of $\theta$. Assume that

$$\left|\frac{\partial^3 f(y \mid \lambda)}{\partial\lambda^3}\right| \le h(y), \quad \text{all } \lambda \in N(\theta)$$

Then,

$$\left|\frac{\partial^2}{\partial\lambda^2}u_n(\lambda)\right| \le \frac{1}{n}\sum_{i=1}^n\left|\frac{\partial^2 w_i(\lambda)}{\partial\lambda^2}\right| \le \frac{1}{n}\sum_{i=1}^n h(y_i) = h_n$$

It then follows from a Taylor series expansion (Chapter 1) that

$$\begin{aligned} u_n(\lambda) &= u_n(\theta) + \frac{\partial}{\partial\lambda} u_n(\theta)(\lambda - \theta) + \frac{1}{2}\frac{\partial^2}{\partial\lambda^2} u_n(\xi)(\lambda - \theta)^2 \\ &= u_n(\theta) + \frac{\partial}{\partial\lambda} u_n(\theta)(\lambda - \theta) + \frac{1}{2}\zeta h_n(\lambda - \theta)^2 \end{aligned}$$

where $\xi \in N(\theta)$ and $|\zeta| < 1$. It follows from LLN that

$$\begin{aligned} u_n(\lambda) &= o_p(1), \quad h_n = E[h(y_i)] + o_p(1), \\ \frac{\partial}{\partial\lambda} u_n(\theta) &= E\left[\frac{\partial}{\partial\lambda} w_i(\theta)\right] + o_p(1) = E\left[\frac{\partial^2}{\partial\lambda^2} \log f(y_i \mid \theta)\right] + o_p(1) \\ &= -E\left[\frac{\partial}{\partial\lambda} \log f(y_i \mid \theta)\right]^2 + o_p(1) \\ &= -v_\theta + o_p(1) \end{aligned}$$

Thus,

$$\begin{aligned} u_n(\lambda) &= o_p(1) + [-v(\theta) + o_p(1)](\lambda - \theta) + [E(h(y_i)) + o_p(1)](\lambda - \theta)^2 \\ &= -v(\theta)(\lambda - \theta) + E(h(y_i))(\lambda - \theta)^2 + z_n \end{aligned}$$

where $z_n = o_p(1)$.

For each integer $k$, let

$$\lambda_1 = \theta - \frac{1}{k}, \quad \lambda_2 = \theta + \frac{1}{k}, \quad \Omega_{n,\frac{1}{k}} = \left\{|z_n| < \frac{1}{2k} v(\theta)\right\}, \quad \text{all } n \geq 1.$$

Since $z_n = o_p(1)$, we can find $N_{\frac{1}{k}}$ such that

$$\Pr\left[\Omega_{n,\frac{1}{k}}\right] \geq 1 - \frac{1}{k}, \quad \text{all } n \geq N_{\frac{1}{k}}$$

For such $n$, we have:

$$u_n(\lambda_2) = -v(\theta)\frac{1}{k} + z_n < 0, \quad u_n(\lambda_1) = v(\theta)\frac{1}{k} + z_n > 0$$

Since $u_n(\lambda)$ is continuous in $\lambda$, the interval $[\lambda_1, \lambda_2]$ contains a solution of the likelihood, that is, $u_n(\eta) = 0$ for some $\eta \in [\lambda_1, \lambda_2]$. Choose some $n > N_{\frac{1}{k}}$ and let

$$\widehat{\theta}_k = \inf\left\{\lambda : \left\{\theta - \frac{1}{k} \leq \lambda \leq \theta + \frac{1}{k}\right\} \bigcap \{u_n(\lambda) = 0\}\right\} \tag{2.22}$$

Then, $\widehat{\theta}_k$ is a random sequence (see exercise). Further,

$$\Pr\left[\left|\widehat{\theta}_k - \theta\right| \geq \frac{1}{k}\right] \leq \Pr\left[\Omega^c_{n,\frac{1}{k}} \text{ for some } n \geq N_{\frac{1}{k}}\right] \leq \frac{1}{k} \to 0$$

Thus, $\widehat{\theta}_k$ is a consistent estimate of $\theta$.

Next, we show the asymptotic normality of $\widehat{\theta}$. It follows from a stochastic Taylor series expansion that

$$u_n\left(\widehat{\theta}\right) = u_n(\theta) + \frac{\partial}{\partial \boldsymbol{\theta}} u_n(\theta)\left(\widehat{\theta} - \theta\right) + \frac{1}{2}\zeta h_n\left(\widehat{\theta} - \theta\right)^2 \tag{2.23}$$

where $|\zeta| < 1$. Since $|\zeta h_n| \leq |h_n|$, $\zeta h_n = O_p(1)$. Solving the above for $\left(\widehat{\theta} - \theta\right)$ yields

$$\sqrt{n}\left(\widehat{\theta} - \theta\right) = -\left[\frac{\partial u_n(\theta)}{\partial \theta} + o_p(1)\right]^{-1} \sqrt{n} u_n(\theta)$$

Since

$$\frac{\partial}{\partial \theta} u_n(\theta) + o_p(1) \to_p -v_\theta, \quad \sqrt{n} u_n(\theta) \to_d N(0, v_\theta)$$

it follows from Slutsky's theorem that

$$\sqrt{n}\left(\widehat{\theta} - \theta\right) \to_d N\left(0, \sigma^2_{\widehat{\theta}} = v_\theta^{-1} v_\theta v_\theta^{-1} = v_\theta^{-1}\right) \tag{2.24}$$

By generalizing (2.24) to the case of vector $\boldsymbol{\theta}$, we have:

$$\sqrt{n}\left(\widehat{\boldsymbol{\theta}} - \boldsymbol{\theta}\right) \to_d N(0, \Sigma_\theta)$$

with the asymptotic variance $\Sigma_\theta$ given in (2.19).

The results in this theorem provide the basis for inference about $\boldsymbol{\theta}$ for parametric models $f(y \mid \boldsymbol{\theta})$ when the sampling distribution of MLE $\widehat{\boldsymbol{\theta}}$ is not in the form of an available analytic distribution such as the normal or $t$ distribution. In particular, we can use Theorem 1 to address inference about $\boldsymbol{\beta}$ for the linear regression model in Example 5 when $\sum_{i=1}^n \mathbf{x}_i \mathbf{x}_i^\top$ is not a constant matrix. Note that the asymptotic variance $\Sigma_\theta$ of $\widehat{\boldsymbol{\theta}}$ in Theorem 1 is not the variance of $\widehat{\boldsymbol{\theta}}$, but rather the variance of the limiting normal of $\sqrt{n}\left(\widehat{\boldsymbol{\theta}} - \boldsymbol{\theta}\right)$. In fact, $\widehat{\boldsymbol{\theta}}$ follows approximately a normal distribution $N\left(\boldsymbol{\theta}, \frac{1}{n}\Sigma_\theta\right)$ with mean $\boldsymbol{\theta}$ and variance $\frac{1}{n}\Sigma_\theta$ for large samples.

**Example 6**. For the linear regression model in Example 5, it follows from (2.14) that the second derivative of the log density function $\log f(y_i \mid \boldsymbol{\theta})$ with respect to $\boldsymbol{\theta} = \left(\boldsymbol{\beta}^\top, \sigma^2\right)^\top$ is

$$-\frac{\partial^2}{\partial \boldsymbol{\beta} \partial \boldsymbol{\beta}^\top} \log f(y_i \mid \boldsymbol{\theta}) = \frac{-1}{\sigma^2} \mathbf{x}_i \mathbf{x}_i^\top, \quad \frac{\partial^2 l_n}{\partial \boldsymbol{\beta} \partial \sigma^2} = \frac{-1}{\sigma^4} \left(\mathbf{x}_i y_i - \mathbf{x}_i \mathbf{x}_i^\top \boldsymbol{\beta}\right) \tag{2.25}$$

$$\frac{\partial^2 l_n}{\partial \left(\sigma^2\right)^2} = -\frac{1}{2\sigma^4} - \frac{1}{\sigma^6} \left(y_i - \mathbf{x}_i^\top \boldsymbol{\beta}\right)^2$$

By taking the expectation of the above (see Chapter 1 for review of expectation and conditional expectation), we have

$$E\left[-\frac{\partial^2}{\partial \boldsymbol{\beta} \partial \boldsymbol{\beta}^\top} \log f(y_i \mid \boldsymbol{\theta})\right] = \frac{-1}{\sigma^2} E\left(\mathbf{x}_i \mathbf{x}_i^\top\right) \tag{2.26}$$

$$E\left(\frac{\partial^2 l_n}{\partial \boldsymbol{\beta} \partial \sigma^2}\right) = \frac{-1}{\sigma^4} E\left(\mathbf{x}_i y_i - \mathbf{x}_i \mathbf{x}_i^\top \boldsymbol{\beta}\right)$$

$$E\left(\frac{\partial^2 l_n}{\partial \left(\sigma^2\right)^2}\right) = \frac{1}{2\sigma^4} - \frac{1}{\sigma^6} E\left(y_i - \mathbf{x}_i^\top \boldsymbol{\beta}\right)^2 = -\frac{1}{2\sigma^4}$$

Since

$$\begin{aligned} E\left(\mathbf{x}_i y_i - \mathbf{x}_i \mathbf{x}_i^\top \boldsymbol{\beta}\right) &= E\left(\mathbf{x}_i y_i\right) - E\left(\mathbf{x}_i \mathbf{x}_i^\top\right) \boldsymbol{\beta} \\ &= E\left(\mathbf{x}_i E\left(y_i \mid \mathbf{x}_i\right)\right) - E\left(\mathbf{x}_i \mathbf{x}_i^\top\right) \boldsymbol{\beta} \\ &= E\left(\mathbf{x}_i \mathbf{x}_i^\top \boldsymbol{\beta}\right) - E\left(\mathbf{x}_i \mathbf{x}_i^\top\right) \boldsymbol{\beta} \\ &= \mathbf{0} \end{aligned}$$

it follows that

$$\Sigma_\theta = \begin{pmatrix} \frac{1}{\sigma^2} E\left(\mathbf{x}_i \mathbf{x}_i^\top\right) & 0 \\ & \frac{1}{2\sigma^4} \end{pmatrix}^{-1} = \begin{pmatrix} \sigma^2 E^{-1}\left(\mathbf{x}_i \mathbf{x}_i^\top\right) & 0 \\ & 2\sigma^4 \end{pmatrix} \tag{2.27}$$

Thus, $\Sigma_\theta$ is the asymptotic variance of the MLE $\widehat{\boldsymbol{\theta}} = \left(\widehat{\boldsymbol{\beta}}^\top, \widehat{\sigma}^2\right)^\top$. Further, it follows from (2.27) and properties of multivariate normal distribution that the marginal vector $\widehat{\boldsymbol{\beta}}$ is also asymptotically normal with the asymptotic variance $\Sigma_\beta = \sigma^2 E^{-1}\left(\mathbf{x}_i \mathbf{x}_i^\top\right)$. By Slutsky's theorem, a consistent estimate of $\Sigma_\beta$ is given by $\widehat{\Sigma}_\beta = \widehat{\sigma}^2 \left(\frac{1}{n} \sum_{i=1}^n \mathbf{x}_i \mathbf{x}_i^\top\right)^{-1}$. Thus, for large $n$, $\widehat{\boldsymbol{\beta}}$ follows approximately the following normal distribution:

$$\widehat{\boldsymbol{\beta}} \sim AN\left(\boldsymbol{\beta}, \frac{1}{n}\widehat{\Sigma}_{\beta}\right) = AN\left(\boldsymbol{\beta}, \frac{\widehat{\sigma}^2}{n}\left(\frac{1}{n}\sum_{i=1}^{n}\mathbf{x}_i\mathbf{x}_i^{\top}\right)^{-1}\right) \tag{2.28}$$

where $AN(\boldsymbol{\theta},\Sigma)$ denotes the *asymptotic or approximate* normal distribution with mean $\boldsymbol{\theta}$ and variance $\Sigma$. Note that the variance estimate $\frac{1}{n}\widehat{\Sigma}_{\beta}$ in the asymptotic distribution of $\widehat{\boldsymbol{\beta}}$ in (2.28) is the estimated asymptotic variance $\widehat{\Sigma}_{\beta}$ divided by the sample size $n$. By comparing the above with (2.16), it is clear that even when $\mathbf{x}_i$ varies from sample to sample, the normal sampling distribution of $\widehat{\boldsymbol{\beta}}$ in (2.16) still provides valid inference of $\boldsymbol{\beta}$ for large sample $n$.

It is seen from (2.27) that $\widehat{\boldsymbol{\beta}}^{\top}$ and $\widehat{\sigma}^2$ are uncorrelated in the asymptotic distribution. Since uncorrelatedness implies independence for variables following the multivariate normal distribution, $\widehat{\boldsymbol{\beta}}^{\top}$ and $\widehat{\sigma}^2$ are asymptotically independent. This independence makes it possible to use $\Sigma_{\beta}$ as the asymptotic variance for inference about $\boldsymbol{\beta}$ since otherwise the inverse of $\Sigma_{\theta}$ would also involve the covariance between $\widehat{\boldsymbol{\beta}}$ and $\widehat{\sigma}^2$. Note that the asymptotic variance of the marginal $\widehat{\sigma}^2$ in (2.27) suggests that we can also make inference about the nuisance scale parameter $\sigma^2$ using the asymptotic normal distribution: $\widehat{\sigma}^2 \sim AN\left(\sigma^2, \frac{2\sigma^4}{n}\right)$ rather than the $\chi^2$ distribution in (2.16) for large samples. For fixed $\mathbf{x}_i$, this asymptotic distribution can also be obtained from the $\chi^2_{n-p}$ distribution of $\widehat{\sigma}^2$ given in (2.16) (see exercise).

One of the primary reasons for the popularity of maximum likelihood inference is the optimal property that maximum likelihood estimates are asymptotically most efficient. This is best discussed by introducing the notion of *information matrix.*

As before, let $l_n(\boldsymbol{\theta})$ denote the log-likelihood and let

$$I_n^o(\boldsymbol{\theta}) = -\frac{\partial^2}{\partial\boldsymbol{\theta}\partial\boldsymbol{\theta}^{\top}} l_n(\boldsymbol{\theta}) = \sum_{i=1}^{n} -\frac{\partial^2}{\partial\boldsymbol{\theta}\partial\boldsymbol{\theta}^{\top}} \log f(y_i \mid \boldsymbol{\theta})$$

The $I_n^o(\boldsymbol{\theta})$ above is known as the *observed information matrix.* By taking expectation, we obtain the *expected* or *Fisher information matrix* as follows:

$$I_n^e(\boldsymbol{\theta}) = E\left[I_n^o(\boldsymbol{\theta})\right] = nE\left[-\frac{\partial^2}{\partial\boldsymbol{\theta}\partial\boldsymbol{\theta}^{\top}} \log f(y_i \mid \boldsymbol{\theta})\right] = nI_1^e(\boldsymbol{\theta}) \tag{2.29}$$

Thus, it follows from Theorem 1 that the asymptotic variance of $\widehat{\boldsymbol{\theta}}$ can be expressed as the inverse of the expected information matrix, that is,

$\Sigma_\theta = \left[I_1^e(\boldsymbol{\theta})\right]^{-1}$. The identity (2.29) shows that the expected information for a sample of size $n$ is the sum of the individual expected information from each individual $I_1^e(\boldsymbol{\theta})$. It is readily checked that $I_n^e(\boldsymbol{\theta}) = Var(\mathbf{u}_n(\boldsymbol{\theta}))$, where $\mathbf{u}_n(\boldsymbol{\theta}) = \frac{\partial}{\partial \boldsymbol{\theta}} l_n(\boldsymbol{\theta})$ is the score statistic vector (see exercise).

**Example 7**. In Example 6, we computed the second derivatives of the log-likelihood function for the linear regression model (2.1). It follows from (2.25) that

$$I_n^o(\boldsymbol{\theta}) = \sum_{i=1}^n \begin{pmatrix} \frac{1}{\sigma^2}\mathbf{x}_i\mathbf{x}_i^\top & \frac{1}{\sigma^4}\left(\mathbf{x}_i y_i - \mathbf{x}_i\mathbf{x}_i^\top \boldsymbol{\beta}\right) \\ & \frac{1}{2\sigma^4} + \frac{1}{\sigma^6}\left(y_i - \mathbf{x}_i^\top \widehat{\boldsymbol{\beta}}\right)^2 \end{pmatrix} \tag{2.30}$$

$$I_1^e(\boldsymbol{\theta}) = \begin{pmatrix} \frac{1}{\sigma^2}E\left(\mathbf{x}_i\mathbf{x}_i^\top\right) & 0 \\ & \frac{1}{2\sigma^4} \end{pmatrix} = \begin{pmatrix} I_1^e(\boldsymbol{\beta}) & 0 \\ & I_1^e\left(\sigma^2\right) \end{pmatrix}, \quad I_n^e(\boldsymbol{\theta}) = nI_1^e(\boldsymbol{\theta})$$

As noted earlier in Example 6, the off-diagonal block of $I_n^e(\boldsymbol{\theta})$ is zero. Thus, we can simply use the inverse of the marginal information for $\boldsymbol{\beta}$, $\left[I_1^e(\boldsymbol{\beta})\right]^{-1}$, as the asymptotic variance for inference about the parameter vector of interest $\boldsymbol{\beta}$. Note that the off-diagonal block of the observed information matrix $I_n^o(\boldsymbol{\theta})$ also vanishes when substituting the MLE $\widehat{\boldsymbol{\theta}}$ in place of $\boldsymbol{\theta}$.

For a scalar variable $\theta$, $l_n(\theta)$ is a function of $\theta$ conditional on the observed $y_i$ $(1 \leq i \leq n)$. It follows from calculus that for a concave (down) function $g$, the second derivative of $g$ provides the degree of curvature for measuring sharpness of the function around the point at which $g$ achieves its maximum. Thus, $I_n^o(\theta)$ measures the curvature of the log-likelihood function $l_n(\theta)$ around the MLE $\widehat{\theta}$, while $I_1^e(\theta)$ measures the same curvature averaged over the joint distribution of $y_i$. A highly peaked $l_n(\theta)$ around $\widehat{\theta}$ yields a large second order derivative (in absolute value) which in turn leads to a smaller asymptotic variance $\sigma_\theta^2$ or more accurate estimate $\widehat{\theta}$ of $\theta$, while a more blunt curve from $l_n(\theta)$ corresponds to a lower second derivative and a larger asymptotic variance or less accurate estimate. For vector $\boldsymbol{\theta}$, the interpretation is similar except that $I_n^o(\boldsymbol{\theta})$ and $I_1^e(\boldsymbol{\theta})$ measure the curvature of a surface defined by $l_n(\boldsymbol{\theta})$ around the MLE $\widehat{\boldsymbol{\theta}}$.

The Fisher information matrix $I_1^e(\boldsymbol{\theta})$ not only provides the asymptotic variance of the MLE, and but also furnishes an asymptotic bound for all estimates satisfying certain regularity conditions. To illustrate, consider a parametric model for continuous response defined by the density function $f(y; \boldsymbol{\theta})$. For convenience, assume a scalar $\theta$. Given an i.i.d. sample $y_i$ from this model $(1 \leq i \leq n)$, let $\widehat{\theta}$ be the MLE of $\theta$ and $\sigma_\theta^2$ the asymptotic variance of $\widehat{\theta}$. If $T_n = T_n(y_1, \ldots, y_n)$ is any unbiased estimate of $\theta$, then

$$E(T_n) = \int T_n f_n(y_1, \ldots, y_n; \theta) \, dy_1 \cdots dy_n = \theta$$

where $f_n(y_1, \ldots, y_n; \theta) = \prod_{i=1}^{n} f(y_i; \theta)$ is the joint density or likelihood function. It follows that

$$\begin{aligned} Cov(T_n, u_n) = E(T_n u_n) &= \int T_n \frac{\partial l_n(\theta)}{\partial \theta} f_n(y_1, \ldots, y_n; \theta) \, dy_1 \cdots dy_n \qquad (2.31) \\ &= \int T_n \left( \frac{\partial}{\partial \theta} f_n(y_1, \ldots, y_n; \theta) \right) dy_1 \cdots dy_n \\ &= \frac{\partial}{\partial \theta} \left( \int T_n f_n(y_1, \ldots, y_n; \theta) \, dy_1 \cdots dy_n \right) = \frac{\partial}{\partial \theta} E(T_n) \\ &= 1 \end{aligned}$$

As in Theorem 1, we assume in (2.31) the regularity conditions to ensure the validity of the change of order of integration and differentiation. By Cauchy–Schwartz inequality, $Cov(T_n, u_n) \leq \sqrt{Var(T_n) Var(u_n)}$ (see Chapter 1). Thus, it follows from (2.31) and Theorem 1 that

$$Var(T_n) \geq \frac{1}{Var(u_n)} = [I_n^e(\theta)]^{-1} = [n I_1^e(\theta)]^{-1} = \frac{\sigma_\theta^2}{n} \qquad (2.32)$$

The above inequality, known as the Cramér-Rao bound, shows that the variance of any unbiased estimate $T_n$ is as large as the asymptotic variance of the MLE $\widehat{\theta}$.

The Cramér–Rao bound in (2.32) also holds for vector $\boldsymbol{\theta}$. In this case, $\mathbf{T}_n$ is a vector, $\mathbf{I}_n^e(\boldsymbol{\theta})$ and $\mathbf{I}_1^e(\boldsymbol{\theta})$ are matrices, $\Sigma_\theta$ is the asymptotic variance of the MLE $\widehat{\boldsymbol{\theta}}$ (or MLE-based estimate $\boldsymbol{\psi}(\widehat{\boldsymbol{\theta}})$) and $\geq$ in (2.32) means that the symmetric matrix $Var(\mathbf{T}_n) - [\mathbf{I}_n^e(\boldsymbol{\theta})]^{-1}$ is non-negative. More generally, for any vector-valued, differentiable function $\boldsymbol{\psi}(\boldsymbol{\theta})$, it follows from the Delta method (see Chapter 1),

$$\sqrt{n} \left( \boldsymbol{\psi}\left(\widehat{\boldsymbol{\theta}}\right) - \boldsymbol{\psi}(\boldsymbol{\theta}) \right) \to_d N\left(0, \Sigma_\psi = D^\top(\boldsymbol{\theta}) \Sigma_\theta D(\boldsymbol{\theta})\right) \qquad (2.33)$$

where $D = \frac{\partial \boldsymbol{\psi}}{\partial \boldsymbol{\theta}}$. If $\mathbf{T}_n$ is any unbiased estimate of $\boldsymbol{\psi}(\boldsymbol{\theta})$, then it is readily shown that (2.32) is also true with $\sigma_\theta^2$ replaced by the asymptotic variance $\Sigma_\psi$ of the estimate $\boldsymbol{\psi}\left(\widehat{\boldsymbol{\theta}}\right)$ given in (2.33) (see exercise).

For consistent estimates, the Cramér–Rao bound does not apply directly. However, it is natural to anticipate that there may be a similar asymptotic bound for comparing the asymptotic variance of consistent estimate with that of the MLE. This turns out to be true and there is an asymptotic version for the Cramér–Rao bound for consistent estimates that are regular and asymptotically linear. For this reason, the Fisher information can be used to compare and define asymptotically efficient estimates in large sample theory. An estimate is asymptotically efficient at $\boldsymbol{\theta}$ for estimating a differentiable $\psi(\boldsymbol{\theta})$ if it is regular at $\boldsymbol{\theta}$ and asymptotically normal with the asymptotic variance given by the Cramér–Rao bound. Thus, the MLE is asymptotically efficient for parametric models.

### 2.1.3 General Linear Hypothesis

When making inference about a single model parameter, such as regression coefficient $\beta$ for a continuous or binary independent variable, hypotheses can be expressed as:

$$H_0 : \beta = a, \quad \text{versus} \quad H_a : \beta \neq a \tag{2.34}$$

where $a$ is a known constant. For example, if there are two treatment conditions in Example 4 of Section 2.1.1, $g = 2$ and under the cell reference coding method, $\beta_1$ represents the difference in mean response between the two treatment conditions. The null $H_0$ in (2.34) with $a = 0$ tests the hypothesis of no between-treatment difference. Testing for hypotheses of the form (2.34) can be carried out using the distribution of the parameter estimate.

**Example 8 (*Two-group ANOVA*).** Consider comparing two treatment conditions using the linear model in Example 4 of Section 2.1.1. Under the cell reference coding method,

$$y_i = \beta_0 + \beta_1 x_{i1} + \epsilon_i, \quad \epsilon_i \sim \text{i.i.d. } N\left(0, \sigma^2\right)$$

where $y_i$ is from the first treatment if $1 \leq i \leq n_1$ and from the second if $n_1+1 \leq i \leq n_2$ $(n = n_1 + n_2)$. Let $\boldsymbol{\beta} = (\beta_0, \beta_1)^\top$. It follows from Example 5 of Section 2.1.2 that the MLE $\frac{\sqrt{n}(\widehat{\boldsymbol{\beta}}-\boldsymbol{\beta})}{MSE}$ has a bivariate $t$ distribution, $t\left(0, \boldsymbol{\Sigma}_\beta = \left(\frac{1}{n}\sum_{i=1}^n \mathbf{x}_i \mathbf{x}_i^\top\right)^{-1}, n-2\right)$. Since the marginal $\widehat{\beta}_1$ also has a $t$ distribution, we can test the null $H_0 : \beta_1 = 0$ using the statistic $\frac{\sqrt{n}\widehat{\beta}_1}{MSE} \sim t\left(0, \sigma^2_{\beta_1}, n-2\right)$, where $\sigma^2_{\beta_1}$ is the diagonal element of $\boldsymbol{\Sigma}_\beta$ corresponding to $\beta_1$.

Many hypotheses of interest, however, cannot be expressed in this simple form. For example, if testing the hypothesis in Example 8 under the cell mean or ANOVA model (2.5) in Example 3 of Section 2.1.1, the hypothesis is given by:

$$H_0 : \mu_1 - \mu_2 = 0 \quad \text{versus} \quad \mu_1 - \mu_2 \neq 0 \tag{2.35}$$

which is not in the form of (2.34) as it involves more than a single parameter.

The general linear hypothesis for a model parameterized by a $p \times 1$ vector $\boldsymbol{\theta}$ has the following form:

$$H_0 : K\boldsymbol{\theta} = \mathbf{b} \quad \text{versus.} \quad H_a : K\boldsymbol{\theta} \neq \mathbf{b} \tag{2.36}$$

where $K$ is some full rank matrix of dimension $l \times p$ and $\mathbf{b}$ some $l \times 1$ column vector, all consisting of known constants ($l \leq p$). When $\mathbf{b} = \mathbf{0}$, the null in (2.36) is called a *linear contrast.*

**Example 9 (*Three-group ANOVA*).** In Example 8 above, if comparing three treatment conditions, the linear model under the cell reference coding is given by

$$y_i = \beta_0 + \beta_1 x_{i1} + \beta_2 x_{i2} + \epsilon_i, \quad \epsilon_i \sim \text{i.i.d. } N\left(0, \sigma^2\right)$$

where $y_i$ is from the $k$th treatment if $\sum_{j=1}^{k-1} n_j + 1 \leq i \leq \sum_{j=1}^{k} n_j$, with $n_0 \equiv 0$ and $n_k$ denoting the sample size of the $k$th treatment group ($1 \leq k \leq 3$). The null of no difference across treatments can be expressed as a linear contrast:

$$H_0 : K\boldsymbol{\beta} = \mathbf{0}, \quad K = \begin{pmatrix} 0 & 1 & 0 \\ 0 & 0 & 1 \end{pmatrix}, \quad \boldsymbol{\beta} = (\beta_0, \beta_1, \beta_2)^\top$$

For the linear regression in (2.1), if $\mathbf{x}_i$ are fixed such as by design in Example 7, $\widehat{\boldsymbol{\beta}}$ follows a multivariate normal given in (2.16). From the definition of $\chi^2$ distribution, we have

$$Q_w^2 = n\left(K\widehat{\boldsymbol{\beta}} - \mathbf{b}\right)^\top \left(K\Sigma_\beta K^\top\right)^{-1} \left(K\widehat{\boldsymbol{\beta}} - \mathbf{b}\right) \sim \chi_l^2 \tag{2.37}$$

Since $\widehat{\boldsymbol{\beta}}$ and MSE are independent, so are $Q_w^2$ and MSE. Thereby, it follows from (2.16) and (2.37) that

$$\frac{Q_w^2/l}{\text{MSE}/\sigma^2} = \frac{\frac{1}{l}\left(K\widehat{\boldsymbol{\beta}} - \mathbf{b}\right)^\top \left[K\left(\sum_{i=1}^n \mathbf{x}_i\mathbf{x}_i^\top\right)^{-1} K^\top\right]^{-1} \left(K\widehat{\boldsymbol{\beta}} - \mathbf{b}\right)^\top}{\text{MSE}} \sim F_{l,n-p}$$

where $F_{u,v}$ denotes an $F$ distribution with the degrees of freedom $u$ for the numerator and $v$ for the denominator. This $F$ distribution is widely used for inference about the general linear hypotheses.

When $\mathbf{x}_i$ are concurrently sampled with $y_i$, we use the asymptotic distribution of the MLE for inference about the general linear hypothesis. The two most popular approaches for testing the null of linear hypothesis are the *Wald* and *likelihood ratio* tests.

If $\widehat{\boldsymbol{\theta}} \sim AN\left(\boldsymbol{\theta}, \frac{1}{n}\Sigma_\theta\right)$, then it follows from the properties of multivariate normal distribution that $K\widehat{\boldsymbol{\theta}} \sim N\left(K\boldsymbol{\theta}, \frac{1}{n}K\Sigma_\theta K^\top\right)$. Under the null, $K\widehat{\boldsymbol{\theta}} \sim AN\left(\mathbf{b}, \frac{1}{n}K\Sigma_\theta K^\top\right)$. By a slight abuse of notation, let $Q_w^2$ denote the *Wald* statistic obtained from (2.37) by replacing $\Sigma_\beta$ with $\widehat{\Sigma}_\beta$. It follows from Slutsky's theorem that $Q_w^2$ has an asymptotic $\chi_l^2$ distribution for large $n$. Alternatively, we can use the *likelihood ratio test* (LRT) to examine the general linear hypothesis.

Let $L_n(\boldsymbol{\theta})$ denote the log-likelihood function and let $\widehat{\boldsymbol{\theta}}$ the MLE of $\boldsymbol{\theta}$. Let $\widehat{\boldsymbol{\theta}}_0$ be the MLE of $\boldsymbol{\theta}$ under the null $H_0$ in (2.36). Then, the *likelihood ratio* based statistic, $-2\ln R\left(\widehat{\boldsymbol{\theta}}_0\right) = -2\ln\left(\frac{L_n(\widehat{\boldsymbol{\theta}}_0)}{L_n(\widehat{\boldsymbol{\theta}})}\right)$, has an asymptotic $\chi_l^2$ distribution under $H_0$ as asserted by the following theorem.

**Theorem 2**. Under mild regularity assumptions,

$$-2\ln R\left(\widehat{\boldsymbol{\theta}}_0\right) = -2\ln\left(\frac{L_n\left(\widehat{\boldsymbol{\theta}}_0\right)}{L_n\left(\widehat{\boldsymbol{\theta}}\right)}\right) \rightarrow_d \chi_l^2, \quad \text{as } n \rightarrow \infty \tag{2.38}$$

Proof. Let $\mathbf{u}_n(\boldsymbol{\theta}) = \frac{1}{n}\frac{\partial l_n}{\partial \boldsymbol{\theta}}$. We prove (2.38) in three steps.

Case 1. First, assume $K = I_p$. In this case, the null in (2.36) reduces to $H_0 : \boldsymbol{\theta} = \mathbf{b}$. For notational brevity, we continue to use $\boldsymbol{\theta}$, but with the understanding that $\boldsymbol{\theta}$ is a known vector under $H_0$.

By (2.23), we have

$$\mathbf{0} = \mathbf{u}_n\left(\widehat{\boldsymbol{\theta}}\right) = \mathbf{u}_n(\boldsymbol{\theta}) + \frac{\partial}{\partial \boldsymbol{\theta}^\top}\mathbf{u}_n(\boldsymbol{\theta})\left(\widehat{\boldsymbol{\theta}} - \boldsymbol{\theta}\right) + o_p\left(n^{-\frac{1}{2}}\right)$$

It follows that

$$\left(\widehat{\boldsymbol{\theta}} - \boldsymbol{\theta}\right)^\top \mathbf{u}_n(\boldsymbol{\theta}) = -\left(\widehat{\boldsymbol{\theta}} - \boldsymbol{\theta}\right)^\top \frac{\partial}{\partial \boldsymbol{\theta}^\top}\mathbf{u}_n(\boldsymbol{\theta})\left(\widehat{\boldsymbol{\theta}} - \boldsymbol{\theta}\right) + \left(\widehat{\boldsymbol{\theta}} - \boldsymbol{\theta}\right)^\top o_p\left(n^{-\frac{1}{2}}\right)$$

By substituting $\frac{1}{n}\frac{\partial l_n}{\partial \boldsymbol{\theta}}$ for $\mathbf{u}_n(\boldsymbol{\theta})$, we have

$$\begin{aligned}\left(\widehat{\boldsymbol{\theta}}-\boldsymbol{\theta}\right)^\top \frac{\partial l_n(\boldsymbol{\theta})}{\partial \boldsymbol{\theta}} &= -\left(\widehat{\boldsymbol{\theta}}-\boldsymbol{\theta}\right)^\top \frac{\partial^2 l_n(\boldsymbol{\theta})}{\partial \boldsymbol{\theta}\partial \boldsymbol{\theta}^\top}\left(\widehat{\boldsymbol{\theta}}-\boldsymbol{\theta}\right)+ \\ &\quad + n\left(\widehat{\boldsymbol{\theta}}-\boldsymbol{\theta}\right)^\top o_p\left(n^{-\frac{1}{2}}\right) \\ &= \left(\widehat{\boldsymbol{\theta}}-\boldsymbol{\theta}\right)^\top I_n^o(\boldsymbol{\theta})\left(\widehat{\boldsymbol{\theta}}-\boldsymbol{\theta}\right)+n\left(\widehat{\boldsymbol{\theta}}-\boldsymbol{\theta}\right)^\top o_p\left(n^{-\frac{1}{2}}\right)\end{aligned} \tag{2.39}$$

Since

$$\begin{aligned}n\left(\widehat{\boldsymbol{\theta}}-\boldsymbol{\theta}\right)^\top o_p\left(n^{-\frac{1}{2}}\right) &= \sqrt{n}\left(\widehat{\boldsymbol{\theta}}-\boldsymbol{\theta}\right)^\top \sqrt{n}o_p\left(n^{-\frac{1}{2}}\right) \\ &= O_p(1)\, o_p(1) = o_p(1)\end{aligned} \tag{2.40}$$

it follows from (2.39) and (2.40) that

$$\left(\widehat{\boldsymbol{\theta}}-\boldsymbol{\theta}\right)^\top \frac{\partial l_n(\boldsymbol{\theta})}{\partial \boldsymbol{\theta}} = \left(\widehat{\boldsymbol{\theta}}-\boldsymbol{\theta}\right)^\top I_n^o(\boldsymbol{\theta}) . \left(\widehat{\boldsymbol{\theta}}-\boldsymbol{\theta}\right)+o_p(1) \tag{2.41}$$

Now, expand $l_n\left(\widehat{\boldsymbol{\theta}}\right)$ around $l_n(\boldsymbol{\theta})$ to obtain

$$\begin{aligned}l_n\left(\widehat{\boldsymbol{\theta}}\right)-l_n(\boldsymbol{\theta}) &= \left(\widehat{\boldsymbol{\theta}}-\boldsymbol{\theta}\right)^\top \frac{\partial l_n(\boldsymbol{\theta})}{\partial \boldsymbol{\theta}}+ \\ &\quad +\frac{1}{2}\left(\widehat{\boldsymbol{\theta}}-\boldsymbol{\theta}\right)^\top \frac{\partial^2 l_n(\boldsymbol{\theta})}{\partial \boldsymbol{\theta}\partial \boldsymbol{\theta}^\top}\left(\widehat{\boldsymbol{\theta}}-\boldsymbol{\theta}\right)+o_p(1) \\ &= \left(\widehat{\boldsymbol{\theta}}-\boldsymbol{\theta}\right)^\top \frac{\partial l_n(\boldsymbol{\theta})}{\partial \boldsymbol{\theta}}-\frac{1}{2}\left(\widehat{\boldsymbol{\theta}}-\boldsymbol{\theta}\right)^\top I_n^o(\boldsymbol{\theta})\left(\widehat{\boldsymbol{\theta}}-\boldsymbol{\theta}\right)+o_p(1)\end{aligned} \tag{2.42}$$

By combining (2.41) and (2.42), we have

$$\begin{aligned}-2\ln R(\boldsymbol{\theta}_0) &= 2\left(l_n\left(\widehat{\boldsymbol{\theta}}\right)-l_n(\boldsymbol{\theta})\right) \\ &= \left(\widehat{\boldsymbol{\theta}}-\boldsymbol{\theta}\right)^\top I_n^o(\boldsymbol{\theta})\left(\widehat{\boldsymbol{\theta}}-\boldsymbol{\theta}\right)+o_p(1)\end{aligned} \tag{2.43}$$

Since

$$\sqrt{n}\left(\widehat{\boldsymbol{\theta}}-\boldsymbol{\theta}\right) \to_d N\left(0, \left(I_1^e(\boldsymbol{\theta})\right)^{-1}\right), \quad \frac{1}{n}I_n^o(\boldsymbol{\theta}) \to_p I_1^e(\boldsymbol{\theta}) \tag{2.44}$$

the conclusion follows from (**??**), (2.44), and an application of Slutsky's theorem.

Case 2. Assume $K = (I_l, \mathbf{0})$ with $I_l$ denoting the $l \times l$ identity matrix. Then, the null becomes $H_0 : \boldsymbol{\theta}_1 = \mathbf{b}$. In this case, the first $l \times 1$ subvector of $\boldsymbol{\theta} = \left(\boldsymbol{\theta}_1^\top, \boldsymbol{\theta}_2^\top\right)^\top$ is known under $H_0$ and the MLE of $\boldsymbol{\theta}$ under $H_0$ is simply $\widetilde{\boldsymbol{\theta}}_2(\mathbf{b})$, which may depend on $\mathbf{b}$. Note that we use different notation for the MLE under $H_0$ since $\widetilde{\boldsymbol{\theta}}_2$ is generally different from the corresponding $(p-l)$ subvector $\widehat{\boldsymbol{\theta}}_2$ of the unconstrained MLE $\widehat{\boldsymbol{\theta}} = \left(\widehat{\boldsymbol{\theta}}_1^\top, \widehat{\boldsymbol{\theta}}_2^\top\right)^\top$. For notational convenience, we use $\boldsymbol{\theta}_1$ for $\mathbf{b}$ and denote $\widetilde{\boldsymbol{\theta}}_2(\mathbf{b})$ simply by $\widetilde{\boldsymbol{\theta}}_2$.

Let $I_n^o = \begin{pmatrix} I_{n11}^o & I_{n12}^o \\ & I_{n22}^o \end{pmatrix}$. By a similar argument as in the proof of Case 1, we have

$$-2\left[l_n(\boldsymbol{\theta}) - l_n\left(\widehat{\boldsymbol{\theta}}\right)\right] = \left(\widehat{\boldsymbol{\theta}} - \boldsymbol{\theta}\right)^\top I_n^o \left(\widehat{\boldsymbol{\theta}} - \boldsymbol{\theta}\right) + o_p(1) \tag{2.45}$$

$$-2\left[l_n(\boldsymbol{\theta}) - l_n\left(\boldsymbol{\theta}_1, \widetilde{\boldsymbol{\theta}}_2\right)\right] = \left(\widetilde{\boldsymbol{\theta}}_2 - \boldsymbol{\theta}_2\right)^\top I_{n22}^o \left(\widetilde{\boldsymbol{\theta}}_2 - \boldsymbol{\theta}_2\right) + o_p(1)$$

By the consistency of $\widehat{\boldsymbol{\theta}}$ and $\widetilde{\boldsymbol{\theta}}_2$ under $H_0$, it can be shown (see exercise) that

$$\begin{aligned}\left(\widehat{\boldsymbol{\theta}} - \boldsymbol{\theta}\right)^\top I_n^o \left(\widehat{\boldsymbol{\theta}} - \boldsymbol{\theta}\right) &= \left(\widehat{\boldsymbol{\theta}}_1 - \boldsymbol{\theta}_1\right)^\top I_{n11}^o \left(\widehat{\boldsymbol{\theta}}_1 - \boldsymbol{\theta}_1\right) + \\ &\quad + \left(\widehat{\boldsymbol{\theta}}_2 - \boldsymbol{\theta}_2\right)^\top I_{n22}^o \left(\widehat{\boldsymbol{\theta}}_2 - \boldsymbol{\theta}_2\right) + o_p(1) \\ \left(\widetilde{\boldsymbol{\theta}}_2 - \boldsymbol{\theta}_2\right)^\top I_{n22}^o \left(\widetilde{\boldsymbol{\theta}}_2 - \boldsymbol{\theta}_2\right) &= \left(\widehat{\boldsymbol{\theta}}_2 - \boldsymbol{\theta}_2\right)^\top I_{n22}^o \left(\widehat{\boldsymbol{\theta}}_2 - \boldsymbol{\theta}_2\right)^\top + o_p(1)\end{aligned} \tag{2.46}$$

It follows from (2.45) and (2.46) that

$$\begin{aligned}-2\frac{L\left(\widetilde{\boldsymbol{\theta}}_2\right)}{L\left(\widehat{\boldsymbol{\theta}}\right)} &= -2\left(l_n\left(\boldsymbol{\theta}_1, \widetilde{\boldsymbol{\theta}}_2\right) - l_n\left(\widehat{\boldsymbol{\theta}}\right)\right) \\ &= 2\left(l_n(\boldsymbol{\theta}) - l_n\left(\boldsymbol{\theta}_1, \widetilde{\boldsymbol{\theta}}_2\right)\right) - 2\left(l_n(\boldsymbol{\theta}) - l_n\left(\widehat{\boldsymbol{\theta}}\right)\right) \\ &= \left(\widehat{\boldsymbol{\theta}}_1 - \boldsymbol{\theta}_1\right)^\top I_{n11}^o \left(\widehat{\boldsymbol{\theta}}_1 - \boldsymbol{\theta}_1\right) + o_p(1) \\ &\to \chi_l^2\end{aligned}$$

Case 3. Consider $H_0$ in the general form given in (2.36) with $l < p$. By viewing the rows of $K$ as $l$ linearly independent vectors in $R^p$, it follows from the theory of linear algebra that there exists $p - l$ linearly independent vectors such that together with the rows of $K$ they form a basis for $R^p$. Let $G$ be a matrix whose rows are formed by these $p - l$ vectors. Then,

$\boldsymbol{\xi} = \left(K^\top, G^\top\right)^\top \boldsymbol{\theta} = \left(\boldsymbol{\xi}_1^\top, \boldsymbol{\xi}_2^\top\right)^\top$ is a linear transformation of $\boldsymbol{\theta}$. Under $H_0$, $\boldsymbol{\xi}_1 = K\boldsymbol{\theta} = \mathbf{b}$ and the MLE of $\boldsymbol{\xi}_2$ is $\widehat{\boldsymbol{\xi}}_2(\mathbf{b})$. It follows from Case 2 that $-2\frac{L(\widehat{\boldsymbol{\xi}}_2(\boldsymbol{\xi}_1))}{L(\widehat{\boldsymbol{\xi}})} \to_d \chi_l^2$. The conclusion follows from expressing $\widehat{\boldsymbol{\xi}}$ and $\widehat{\boldsymbol{\xi}}_2(\boldsymbol{\xi}_1)$ in terms of $\boldsymbol{\theta}$ as $\widehat{\boldsymbol{\theta}} = \left(K^\top, G^\top\right)^{-\top} \widehat{\boldsymbol{\xi}}$ and $\widehat{\boldsymbol{\theta}}_0 = \left(K^\top, G^\top\right)^{-\top} \left(\mathbf{b}^\top, \widehat{\boldsymbol{\xi}}_2^\top\right)^\top$ and the invariance of the likelihood function under such linear transformations.

**Example 10**. In Example 9, let $\mathbf{x}_i = (1, x_{i1}, x_{i2})^\top$ and consider the linear hypothesis:

$$H_0 : \beta_1 - \beta_2 = 0 \quad \text{versus} \quad H_a : \beta_1 - \beta_2 \neq 0$$

Let

$$\xi_0 = \beta_0, \quad \xi_1 = \beta_1 - \beta_2, \quad \xi_2 = \beta_2 \tag{2.47}$$

Then, when expressed in terms of $\boldsymbol{\xi} = (\xi_0, \xi_1, \xi_2)^\top$, the linear model is

$$y_i = \xi_0 + (\xi_1 + \xi_2) x_{i1} + \xi_2 x_{i2} + \epsilon_i, \quad \epsilon_i \sim \text{i.i.d. } N\left(0, \sigma^2\right) \tag{2.48}$$

Under $H_0$, $\xi_1 = 0$ and (2.48) reduces to

$$y_i = \xi_0 + \xi_2 (x_{i1} + x_{i2}) + \epsilon_i, \quad \epsilon_i \sim \text{i.i.d. } N\left(0, \sigma^2\right) \tag{2.49}$$

Let $\widehat{\boldsymbol{\zeta}} = \left(\widehat{\xi}_0, \widehat{\xi}_1, \widehat{\xi}_2, \widehat{\sigma}^2\right)^\top$ and $\widetilde{\boldsymbol{\zeta}}_0 = \left(\widetilde{\xi}_0, \widetilde{\xi}_2, \widetilde{\sigma}^2\right)^\top$ be the MLE of $\widetilde{\boldsymbol{\zeta}} = \left(\boldsymbol{\xi}^\top, \sigma^2\right)^\top$ for the *full* and *reduced* models in (2.48) and (2.49), respectively. Then, the MLEs of $\boldsymbol{\theta} = \left(\boldsymbol{\beta}^\top, \sigma^2\right)^\top$ for the respective full and reduced models are readily obtained by transforming $\widehat{\boldsymbol{\zeta}}$ and $\widetilde{\boldsymbol{\zeta}}_0$ to the scale of $\boldsymbol{\theta}$ by (2.47). For example, the MLE $\widehat{\boldsymbol{\theta}}$ for the full model is given by

$$\widehat{\beta}_0 = \widehat{\xi}_0, \quad \widehat{\beta}_1 = \widehat{\xi}_1 + \widehat{\xi}_2, \quad \widehat{\beta}_2 = \widehat{\xi}_2$$

It follows from Theorem 2 that $-2\log R\left(\widetilde{\boldsymbol{\theta}}_0\right) = -2\ln\left(\frac{L_n(\widetilde{\boldsymbol{\theta}}_0)}{L_n(\widehat{\boldsymbol{\theta}})}\right)$ has an approximate $\chi_1^2$ distribution for large $n$, where $L_n\left(\widetilde{\boldsymbol{\theta}}_0\right)$ and $L_n\left(\widehat{\boldsymbol{\theta}}\right)$ are the likelihood functions corresponding to the full and reduced models in (2.48) and (2.49).

Some software packages may only implement tests for linear contrasts. When using such software, we have to rewrite the general linear hypothesis in terms of linear contrast:

$$H_0 : K\boldsymbol{\beta} - \mathbf{b} = \mathbf{0} \quad \text{versus} \quad H_a : K\boldsymbol{\beta} - \mathbf{b} \neq \mathbf{0}$$

By performing the transform $\boldsymbol{\gamma} = \boldsymbol{\beta} - K^\top \left(K^\top K\right)^{-1} \mathbf{b}$ and reparameterizing the linear model in terms of $\boldsymbol{\gamma}$, it follows from (2.1) that

$$y_i - \mathbf{x}_i^\top K^\top \left(K^\top K\right)^{-1} \mathbf{b} = \mathbf{x}_i^\top \boldsymbol{\gamma} + \epsilon_i, \quad \epsilon_i \sim \text{i.i.d. } N\left(0, \sigma^2\right)$$

Thus, by redefining the response, $z_i = y_i - \mathbf{x}_i^\top K^\top \left(K^\top K\right)^{-1} \mathbf{b}$, we can test the original null through a linear contrast $H_0 : K\boldsymbol{\gamma} = \mathbf{0}$. Alternatively, we can use an *offset* term without performing the linear transformation. We discuss this alternative approach in Section 2.1.5.

### 2.1.4 Generalized Linear Models

The linear regression model discussed in the preceding sections assumes a continuous response $y$. In applications, other types of dependent variables also arise. For example, binary response is quite popular in many areas of research such as depression diagnosis in mental health studies and infection of the human immunodeficiency virus (HIV), in investigation of infectious disease. Since such a variable only takes on two values and the linear predictor $\eta$ of the linear regression model is unconstrained and can in theory range from $-\infty$ to $\infty$, linear regression is generally inappropriate for modeling such a type of response.

The *generalized linear model* (GLM) is a generalization of linear regression to accommodate different types of response, including binary response. This general class of models frames a wide range of seemingly disparate problems of statistical modeling and inference within an elegant unifying framework. This new integrated modeling approach provides great power and flexibility to model a wide variety of response variables. Examples of generalized linear models include linear regression, logistic regression and Poisson log-linear models.

To appreciate the limitation of linear regression and set the stage for GLM, let $\mu_i = E\left(y_i \mid \mathbf{x}_i\right)$, the conditional mean of $y_i$ given $\mathbf{x}_i$, and express the linear regression model in (2.1) equivalently as:

$$\begin{aligned} & y_i \mid \mathbf{x}_i \sim \text{i.d. } N\left(\mu_i, \sigma^2\right) \\ & \mu_i = E\left(y_i \mid \mathbf{x}_i\right) = \eta_i = \mathbf{x}_i^\top \boldsymbol{\beta}, \quad 1 \leq i \leq n \end{aligned} \tag{2.50}$$

where i.d. denotes *independently distributed*. It is clear from (2.50) that $\mu_i$ is a linear function of $\mathbf{x}_i$. In addition, the right side of the model, $\eta_i = \mathbf{x}_i^\top \boldsymbol{\beta}$, has range in the real line $R$, which concurs with the range of $\mu_i$ on the left side. If $y_i$ is binary, however,

$$\mu_i = E\left(y_i \mid \mathbf{x}_i\right) = \Pr\left[y_i = 1 \mid \mathbf{x}_i\right] \tag{2.51}$$

The conditional mean $\mu_i$ in (2.51) is a value between 0 and 1. In this case, it does not make sense to equate $\mu_i$ with the linear predictor $\eta_i$ as for the linear regression model in (2.50). In addition, the normal assumption does not apply to binary response.

The generalization of linear regression to accommodate other types of response is made possible by modifying the following two components of the linear regression in (2.50):

1. *Random component.* This part specifies the conditional distribution of the response $y$ given the dependent variables $\mathbf{x}$. In the linear regression, a normal distribution is assumed.

2. *Systematic component.* This part links the conditional mean of $y$ given $\mathbf{x}$ to the linear predictor $\mathbf{x}$ by a one-to-one function $g$:

$$g(\mu) = \eta = \beta_1 x_1 + \cdots + \beta_p x_p = \mathbf{x}^\top \boldsymbol{\beta} \tag{2.52}$$

In the linear regression model, $g(\mu) = \mu$, an identity function.

In the nomenclature of generalized linear models, $g(\cdot)$ is called a *link* function. By varying the distribution function for the random component and the link function in the systematic part, we can use the class of GLM to model many different types of response.

**Example 11 (*Models for binary response*).** Consider a sample of $n$ subjects. Let $y_i$ be a binary response of interest and $\mathbf{x}_i$ a set of independent variables. Logistic regression is the most popular choice to model the relationship of $y_i$ as a function of $\mathbf{x}_i$. As a member of GLM, the random component specifies the conditional distribution of $y_i$ given $\mathbf{x}_i$ as a Bernoulli or binomial with sample size 1, with the probability of success given by $E(y_i \mid \mathbf{x}_i) = \pi(\mathbf{x}_i) = \pi_i$. For the systematic part, the conditional mean $\mu_i$ is linked to the linear predictor by the logit function: $g(\pi) = \log(\pi/(1-\pi))$. Thus, this logistic model in terms of a GLM is expressed as:

$$y_i \mid \mathbf{x}_i \sim \text{i.d. BI}(\pi(\mathbf{x}_i); 1), \quad g(\pi_i) = \log(\pi_i/(1-\pi_i)), \quad 1 \leq i \leq n$$

where BI$(\pi; 1)$ denotes a binomial distribution with sample size 1 and probability of success $\pi$. The log transform of the odds $\pi/(1-\pi)$ is also known as the logit transform of $\pi$.

In addition to the logit link, other popular functions used in practice for modeling binary response include

1. The probit link: $g(\pi) = \Phi^{-1}(\pi)$, where $\Phi(\cdot)$ is the cumulative distribution function of the standard normal.

2. The complementary log–log function: $g(\pi) = \log(-\log(1-\pi))$.

**Example 12**. In Example 11, consider a special case with a single binary predictor $x_i$ such as gender with $x_i = 1$ for male and $x_i = 0$ for female subjects in the study. Then, $\pi(x) = \Pr(Y = 1 \mid x)$ is the conditional probability of response $y = 1$ given $x$. The systematic component of the logistic model is

$$\text{logit}\left[\pi(x)\right] = \text{logit}\left[\Pr(Y = 1 \mid X = x)\right] = \log\left(\frac{\pi(x)}{1 - \pi(x)}\right) = \beta_0 + \beta_1 x$$

It follows from the above that the log odds ratio $OR$(odds of response of male to that of female) is

$$\begin{aligned}\log(OR) &= \log\left(\frac{\pi(1)/(1-\pi(1))}{\pi(0)/(1-\pi(0))}\right) = \text{logit}(\pi(1)) - \text{logit}(\pi(0)) \\ &= (\beta_0 + \beta_1) - \beta_0 = \beta_1\end{aligned}$$

Thus, the parameter $\beta_1$ is the log odds ratio or, equivalently, $\exp(\beta_1)$ is the odds ratio for comparing the response between the male and female subjects.

As in linear regression, we can develop ANOVA like GLMs for comparing multiple treatment groups using the different coding methods discussed in Section 2.1.1. For example, within the context of Example 11, suppose there are three treatment groups and interest lies in comparing the mean response of $y_i$ across three treatment conditions. Let $x_{ik}$ be a binary indicator for the $k$th treatment condition defined in (2.3) with $1 \leq k \leq 3$. Then, under cell reference coding, the linear predictor is $\eta_i = \mathbf{x}_i^\top \boldsymbol{\beta} = \beta_0 + \beta_1 x_{i1} + \beta_2 x_{i2}$ and the systematic component of the logistic model is

$$\text{logit}\left[\pi(\mathbf{x}_i)\right] = \log\left(\frac{\pi(\mathbf{x}_i)}{1 - \pi(\mathbf{x}_i)}\right) = \beta_0 + \beta_1 x_{i1} + \beta_2 x_{i2}$$

The log odds for each treatment condition is given by

$$\begin{aligned}\text{For group } k \ (= 1, 2) : \log\left(\frac{\pi_k}{1 - \pi_k}\right) &= \beta_0 + \beta_k \\ \text{For group } 3 : \log\left(\frac{\pi_3}{1 - \pi_3}\right) &= \beta_0\end{aligned} \tag{2.53}$$

where $\pi_k$ is the mean response of $y$ for group $k$. From (2.53), we obtain the log odds ratios for comparing groups 1 and 2 to group 3:

$$\log\left(\frac{\pi_1}{1-\pi_1}/\frac{\pi_3}{1-\pi_3}\right) = \beta_1, \quad \log\left(\frac{\pi_2}{1-\pi_2}/\frac{\pi_3}{1-\pi_3}\right) = \beta_2$$

Thus, under this coding method, the intercept $\beta_0$ is identified as the log odds for group 3, while $\beta_1$ and $\beta_2$ represent the log odds ratios for comparing each of the remaining groups to this group. Note that there is no particular reason to use group 3 as the reference level and any treatment condition can be designated as the reference group.

In a case-control study on examining the relationship between some exposure variable and disease of interest, we first select a random sample from a population of diseased subjects or cases. Such a population is usually retrospectively identified by reviews of patients' medical history and records. We then randomly select a matched sample of disease-free individuals or controls from a nondiseased population based on similar demographic and clinical variables. Thus, unlike most prospective studies, the number of diseased and nondiseased subjects are fixed *a priori*, while the number of exposed and nonexposed subjects are random under such a retrospective study design.

**Example 13 (*Logistic model for case-control study*).** In example 11, suppose that $y_i$ and $x_i$ represent the status of disease and exposure of interest. Let $z$ be a dummy variable denoting whether a subject is sampled or not from a population of interest. Since the sampling process is independent of the exposure status, we have

$$p_1(x) = \Pr[z = 1 \mid y = 1, x] = \Pr[z = 1 \mid y = 1] = p_1$$
$$p_0(x) = \Pr[z = 1 \mid y = 0, x] = \Pr[z = 1 \mid y = 0] = p_0$$

It follows from the above and the Bayes' theorem that the disease probability among sampled individuals with exposure status $x$ is

$$\Pr[y = 1 \mid z = 1, x] = \frac{p_1(x)\Pr[y = 1 \mid x]}{p_1(x)\Pr[y = 1 \mid x] + p_0(x)\Pr[y = 0 \mid x]}$$
$$= \frac{p_1 \exp(\beta_0 + \beta_1 x)}{p_0 + p_1 \exp(\beta_0 + \beta_1 x)} = \frac{\exp(\beta_0^* + \beta_1 x)}{1 + \exp(\beta_0^* + \beta_1 x)}$$

where $\beta_0^* = \beta_0 + \log(p_1/p_0)$. We can see that the logistic model for the retrospective case-control study sample has the same coefficient $\beta_1$ for the exposure variable $x_i$ as the model for the prospective study sample, except for a different intercept. For this reason, the logistic model is also widely used in case control studies to assess the relationship between exposure and disease variables, provided that the intercept is interpreted with caution.

All three link functions are similar (McCullagh and Nelder, 1989). In particular, the logit and probit functions are almost identical over the interval $0.1 \leq \pi \leq 0.9$. Thus, it is usually difficult to discriminate between these two link functions on the grounds of model parameter estimates. Although any of the three link functions above [and other functions that map $(0,1)$ onto to $R$] can be applied to link the mean of response to the linear predictor, the logit link is the most popular. One reason for its popularity is its simple interpretation as the logarithm of the odds of response and the fact that odds ratios are functions of model parameters. Second, the logit link is the only link function that can be applied to both prospective and respective case-control study designs without altering the interpretation of model parameters.

The logit link is also known as the *canonical* link for modeling binary response. This term is derived from the exponential family of distributions. If $y$ is from the exponential family of distributions, the PDF of $y$ can be expressed in the form:

$$f(y \mid \theta) = \exp\left[(y\theta - b(\theta))/a(\phi) + c(y,\phi)\right] \tag{2.54}$$

where $\theta$ is the *canonical parameter* (if $\phi$ is known) and $\phi$ is the dispersion or scale parameter. The exponential family includes many distributions, such as normal, binomial, and exponential. For example, in the normal distribution case $N\left(\mu,\sigma^2\right)$, $\theta = \mu$, $\phi = \sigma^2$ and $a\left(\sigma^2\right) = \sigma^2$. However, $\theta$ is not the mean (or probability of success) for the Bernoulli or binomial $\mathrm{BI}(\pi,1)$ with sample size 1. The *canonical* link is defined as a one-to-one function $g(\cdot)$ that maps the mean $\mu = E(y)$ of $y$ to the parameter $\theta$, that is, $g(\mu) = \theta$. It is readily shown that for the exponential family defined above, $\mu = \frac{d}{d\theta}b(\theta) = h(\theta)$, and hence the canonical link is $g(\mu) = h(\mu)^{-1}$ (see exercise).

When modeling binary response, $y$ follows a binomial $\mathrm{BI}(\pi,1)$ with the following distribution function:

$$f(y \mid \pi) = \pi^y (1-\pi)^{1-y} = \exp\left[y\log\left(\pi/(1-\pi)\right) + \log(1-\pi)\right]$$

where $\pi = E(y)$. It follows from (2.54) that for this particular model,

$$\theta = \log\frac{\pi}{1-\pi}, \qquad b(\theta) = -\log(1-\pi) = \log(1+e^{\theta})$$

By definition, the canonical link is $g(\pi) = \log\frac{\pi}{1-\pi}$. Alternatively, since $\frac{\partial}{\partial\theta}b(\theta) = \frac{e^{\theta}}{1+e^{\theta}}$, the canonical link is also given by the inverse of this derivative, $g(\pi) = \log\left(\frac{\pi}{1-\pi}\right)$. Thus, the logit function is the canonical link for modeling binary response. Similarly, it is easily shown that the identity function is the canonical link for modelling a continuous random variable following the normal distribution.

Except for these key differences, however, all three link functions have similar interpretations. For example, as they are all monotone increasing functions, the corresponding model parameters $\boldsymbol{\beta}$ indicate the same direction of association (signs) even though their values are generally different across the three different models.

In many applications, we often encounter multi-level response variables which cannot be collapse into two-level binary response. For example, classification of blood types is a four-level response with qualitative categories O, A, B, AB. In cancer research, cancer patients are classified into four stages from I to IV to indicate disease severity and prognosis. In mental health research, subjects may be diagnosed as one of the three categories, no depression, subsyndromal depression (SSD) and major depression, with patients diagnosed with major depression being more paralyzed mentally than those with SSD, which typically show some symptoms of depression.

If we study such *polytomous* responses further, we can identify them as one of the following types:

1. *Nominal scale.* The categories of such a response variable are regarded as exchangeable and totally devoid of structure. For example, the color of cars and the type of disciplines for graduate studies belong to such a response type.

2. *Ordinal scale.* The categories of this type of response are ordered in terms of preference or severity. For example, cancer staging and depression diagnoses produce responses that are ordered based on the severity of disease of interest.

For nominal response, the *generalized logit* or *multinomial response model* is the most popular approach. Suppose that a nominal response $y$ has $J$ categories. This model designates one category as a reference level and then pairs each other response category to this reference category. Usually the last category is chosen as the reference level. Of course, for nominal response, the "last" category is not well defined, as the response categories are exchangeable. Thus, the selection of the reference level is arbitrary and is usually a matter of convenience.

To define the generalized logit model under GLM, let $\mathbf{y}_i = (y_{i1}, \ldots, y_{iJ})^\top$ be a vector of binary responses $y_{ij}$ from the $i$th subject with $y_{ij} = 1$ if the response is in the $j$th category and $y_{ij} = 0$ if otherwise $(1 \leq j \leq J)$. Let

$$\pi_{ij} = \pi_j(\mathbf{x}_i) = \Pr(y_{ij} = 1 \mid \mathbf{x}_i), \quad j = 1, \ldots, J$$

the conditional probability of response in the $j$th category given $\mathbf{x}_i$. The generalized logit model is defined as follows:

$$\mathbf{y}_i \mid \mathbf{x}_i \sim \text{i.d. MN}\left(\boldsymbol{\pi}_i; 1\right), \quad \boldsymbol{\pi}_i = \left(\boldsymbol{\pi}_{i1}, \ldots, \boldsymbol{\pi}_{iJ}\right) \tag{2.55}$$
$$\log\left(\frac{\pi_j\left(\mathbf{x}_i\right)}{\pi_J\left(\mathbf{x}_i\right)}\right) = \alpha_j + \boldsymbol{\beta}_j^\top \mathbf{x}_i = \eta_{ij}, \quad j = 1, \ldots, J-1$$

where $\text{MN}(\boldsymbol{\pi}; 1)$ denotes the multinomial distribution with sample size 1 for the random component and $\boldsymbol{\alpha} = \left(\alpha_1, \ldots, \alpha_{J-1}\right)^\top$ is a vector of parameters of the *baseline* generalized logits in the absence of $\mathbf{x}$. When $J = 2$, (2.55) reduces to the logistic regression for binary response. These $J-1$ logits determine the parameters for any other pairs of the response categories:

$$\log\frac{\pi_j\left(\mathbf{x}\right)}{\pi_k\left(\mathbf{x}\right)} = \log\frac{\pi_j\left(\mathbf{x}\right)}{\pi_J\left(\mathbf{x}\right)} - \log\frac{\pi_k\left(\mathbf{x}\right)}{\pi_J\left(\mathbf{x}\right)}$$

From the defining equation of $\eta_j$, we obtain the probability of response of the $j$th category:

$$\pi_j\left(\mathbf{x}_i\right) = \frac{\exp\left(\alpha_j + \boldsymbol{\beta}_j^\top \mathbf{x}_i\right)}{1 + \sum_{i=1}^{J-1}\exp\left(\alpha_j + \boldsymbol{\beta}_j^\top \mathbf{x}_i\right)} = \frac{\exp\left(\eta_{ij}\right)}{1 + \sum_{i=1}^{J-1}\exp\left(\eta_{ij}\right)}$$
$$j = 1, \ldots, J-1$$

By setting $\alpha_J = 0$ and $\boldsymbol{\beta}_J = \mathbf{0}$ and including $j = J$ in the modeling of generalized logit, we can express the probability of response for all $J$ categories in a uniform fashion as:

$$\pi_j\left(\mathbf{x}\right) = \frac{\exp\left(\alpha_j + \boldsymbol{\beta}_j^\top \mathbf{x}\right)}{1 + \sum_{i=1}^{J-1}\exp\left(\alpha_j + \boldsymbol{\beta}_j^\top \mathbf{x}\right)} = \frac{\exp(\eta_j)}{\sum_{i=1}^{J}\exp(\eta_i)}, \quad j = 1, \ldots, J$$

**Example 14** (***Generalized logit model for nominal response***). Consider a three level depression diagnosis outcome $y$ with 0 for no depression, 1 for SSD and 2 for major depression. Although these categories are ordinal, for illustration purposes, we treat them as nominal and model the response using the generalized logit model. For convenience, consider a single binary predictor, gender denoted by $x$, with $x = 1$ for male and $x = 0$ for female subjects.

If $y = 0$ is selected as the reference level, then the generalized logit model is given by

$$\log\frac{\pi_j\left(x\right)}{\pi_0(x)} = \alpha_j + \beta_j x, \quad j = 1, 2$$

It then follows that

$$\beta_j = \log \frac{\pi_j(1)}{\pi_0(1)} - \log \frac{\pi_j(0)}{\pi_0(0)} = \log \left( \frac{\frac{\pi_j(1)}{\pi_j(0)}}{\frac{\pi_0(1)}{\pi_0(0)}} \right) = \log \left( \frac{\frac{\pi_j(1)}{\pi_j(1)+\pi_0(1)} / \frac{\pi_0(1)}{\pi_j(1)+\pi_0(1)}}{\frac{\pi_j(0)}{\pi_j(0)+\pi_0(0)} / \frac{\pi_0(0)}{\pi_j(0)+\pi_0(0)}} \right)$$

By viewing $\frac{\pi_1(x)}{\pi_1(x)+\pi_0(x)} / \frac{\pi_0(x)}{\pi_1(x)+\pi_0(x)}$ as the conditional odds of SSD for the subgroup of SSD and no depression subjects, we may interpret $\beta_1$ as the log odds ratio of SSD for comparing the male and female subjects within this subgroup. Similarly, $\beta_2$ may be interpreted as the log odds ratio of major depression for comparing the male and female subjects within the subgroup of major depression and no depression subjects.

For ordinal response, it is natural to model the cumulative response probabilities. Consider a $J$-level ordinal response $y$. Again, let $\mathbf{y}_i = (y_{i1}, \ldots, y_{iJ})^\top$ be a vector of binary responses with $y_{ij} = 1$ if the response is in category $j$ and $y_{ij} = 0$ if otherwise $(1 \le j \le J)$. The popular GLM for ordinal response is the proportional odds model:

$$\mathbf{y}_i \mid \mathbf{x}_i \sim \text{i.d. MN}(\boldsymbol{\pi}_i; 1), \quad \boldsymbol{\pi}_i = (\boldsymbol{\pi}_{i1}, \ldots, \boldsymbol{\pi}_{iJ}) \tag{2.56}$$

$$\log\left(\frac{\gamma_j(\mathbf{x}_i)}{1-\gamma_j(\mathbf{x}_i)}\right) = \alpha_j + \boldsymbol{\beta}^\top \mathbf{x}_i, \quad \alpha_1 \le \alpha_2 \le \ldots \le \alpha_{J-1}, \quad 1 \le j \le J-1$$

where $\gamma_j(\mathbf{x}_i) = \sum_{k=1}^{j} \pi_{ik}(\mathbf{x}_i)$ is the cumulative probability of response up to and including category $j$ conditional on the vector of independent variables $\mathbf{x}_i$. Unlike the generalized logit model, logit transform is applied to the cumulative probabilities $\gamma_j(\mathbf{x})$. Since $\gamma_j(\mathbf{x})$ increases as a function of $j$, the logit transform of $\gamma_j(\mathbf{x})$ is also a monotone increasing function of $j$ and as a result the $\alpha_j$'s must satisfy the order constraint in (2.56). Of course, these two sets of probabilities $\gamma_j(\mathbf{x})$ and $\pi_j(\mathbf{x})$ are equivalent, that is, one completely determines the other. However, models based on cumulative probabilities are easier to interpret for ordinal response than similar models based on the categorical probabilities of individual responses.

**Example 15 (*Proportional odds model for ordinal response*).** Consider Example 14, but now model depression diagnosis using the proportional odds model. The systematic part of the model is given by:

$$\log\left(\frac{\gamma_j(x)}{1-\gamma_j(x)}\right) = \alpha_j + \beta x, \quad j = 0, 1$$

It then follows that

$$\beta = \log\left(\frac{\gamma_j(1)}{1-\gamma_j(1)}\right) - \log\left(\frac{\gamma_j(0)}{1-\gamma_j(0)}\right) = \log\left(\frac{\gamma_j(1)/(1-\gamma_j(1))}{\gamma_j(0)/(1-\gamma_j(0))}\right)$$

For $j = 0$, the right side above is the log odds ratio of no depression (versus SSD or depression) for comparing the male and female subjects, while for $j = 1$, it becomes the log odds ratio of no depression or SSD (versus depression) for comparing the male and female subjects. Under the proportional odds model, the two log odds ratios (or odds ratios) are assumed to be the same. This is in stark contrast with the generalized logit model where the generalized log odds ratios are allowed to change as a function of the response categories. In applications, we can check the proportionality by first assuming different $\boldsymbol{\beta}_j$ and then testing the null hypothesis of a common $\boldsymbol{\beta}$.

With the general specification of GLM, not only can we readily model other types of response, such as continuous and binary, but also the same type of response with different link functions, such as logistic and probit. The next example shows that we can also change the random component when modeling the same response variable.

**Example 16 (*Linear regression with t distribution*).** Consider the linear regression in (2.50). The random component assumes a normal distribution for the continuous response $y_i$. A major limitation of the normal distribution is that its tails decay to 0 too rapidly at an exponential rate, making it ill-suited for modeling data distributions with relatively thicker tails. A popular alternative is to use a distribution that does not decay at such a fast rate, such as the $t$ distribution.

Let

$$\begin{aligned} y_i \mid \mathbf{x}_i &\sim \text{ i.d. } t\left(\mu_i, \sigma^2, v\right) \\ \mu_i &= E\left(y_i \mid \mathbf{x}_i\right) = \mathbf{x}_i^\top \boldsymbol{\beta}, \quad 1 \leq i \leq n \end{aligned} \tag{2.57}$$

where $t\left(\mu, \sigma^2, v\right)$ denotes a $t$ distribution with mean $\mu$, scale parameter $\sigma^2$ and degrees of freedom $v$. The above GLM has the same systematic part as the normal-based model in (2.50), except for a different distribution assumption in the random component. In comparison to the normal-based counterpart, the model in (2.57) accounts for thicker tails in the conditional distribution of $y_i$ given $\mathbf{x}_i$ through an appropriate choice of the degree of freedom $v$. Since $t\left(\mu, \sigma^2, v\right)$ converges to $N\left(\mu, \sigma^2\right)$ as $v \to \infty$, the $t$-based linear regression (2.57) is expected to provide estimates similar to those of its normal-based counterpart (2.50) for large $v$.

Another popular type of outcome in many applications is count response. Count response such as the number of heart attacks, suicide attempts, abortions and birth defects arises quite often in studies in the biomedical, behavioral and social sciences. Since count response only takes on non-negative integers, linear regression is inappropriate for modeling this type of response.

On the other hand, unlike the other discrete responses considered above, count response cannot be expressed in the form of proportions as the range of such a response is theoretically unbounded.

**Example 17** (***Poisson log-linear model for count response***). Consider a study of $n$ subjects and let $y_i$ be a count response. The most popular approach for such a response is the Poisson log-linear model with the random and systematic components specified as follows:

$$y_i \mid \mathbf{x}_i \sim \text{i.d. Poisson}(\mu_i) \tag{2.58}$$
$$\log(\mu_i) = \log(E(y_i \mid \mathbf{x}_i)) = \eta_i = \mathbf{x}_i^\top \boldsymbol{\beta}, \quad 1 \leq i \leq n$$

where Poisson($\mu$) denotes a Poisson distribution with the following probability distribution function:

$$f_P(y_i \mid \mu) = \frac{\mu^{y_i}}{y_i!} \exp(-\mu), \quad \mu > 0, \quad y_i = 0, 1, \ldots \tag{2.59}$$

Since $\mu \in (0, \infty)$, $\log(\cdot)$ is an appropriate link as it maps $(0, \infty)$ to $(-\infty, \infty)$, the range of the linear predictor $\eta$.

One major problem with the Poisson log-linear model is *overdispersion.* Under the model assumption, it is readily checked that the conditional mean $E(y_i \mid \mathbf{x}_i)$ and variance $Var(y_i \mid \mathbf{x}_i)$ of $y_i$ given $\mathbf{x}_i$ satisfy (exercise):

$$E(y_i \mid \mathbf{x}_i) = Var(y_i \mid \mathbf{x}_i) = \mu_i \tag{2.60}$$

In many applications, the conditional variance $Var(y_i \mid \mathbf{x}_i)$ often exceeds the conditional mean $\mu_i$, causing overdispersion and making the Poisson model inappropriate for modeling such data. When overdispersion occurs, the standard errors of the parameter estimates of the Poisson model are artificially deflated, giving rise to exaggerated effect size estimates and false significant findings.

One common cause of overdispersion is data clustering. Within the context of cross-sectional studies, clustered data often arise when subjects sampled from within a common habitat (cluster), such as families, schools and communities are more similar than those sampled across different habitats, leading to correlated responses within a cluster. The existence of data clusters invalidates the usual independent sampling assumption, rendering statistical methods developed based on independence of observations inapplicable to such data.

Overdispersion can often be empirically detected by goodness of fit statistics for generalized linear models. When deemed present, overdispersion may be corrected post hoc by using robust estimates of the asymptotic variance rather than the one based on the MLE in Theorem 1. We discuss this option in Section 2.2 after introducing the distribution-free generalized linear models. A better alternative is to use models that directly address data clustering, the cause of overdispersion. One popular approach is the mixed-effects model, which employs *latent variables* or *random effects* to model the effect of data clustering. We discuss this general approach for modeling clustered data in Chapter 4. Below, we consider one particular example of this class of models for count response.

**Example 18 (*Negative binomial log-linear model for count response*).** In Example 17, if overdispersion occurs, we can no longer model the count response using the Poisson log-linear model. A popular approach for modeling such overdispersed data is the negative binomial model:

$$y_i \mid \mathbf{x}_i \sim \text{ i.d. NB}\left(\mu_i, \alpha\right) \tag{2.61}$$
$$\log\left(\mu_i\right) = \log\left(E\left(y_i \mid \mathbf{x}_i\right)\right) = \mathbf{x}_i^\top \boldsymbol{\beta}, \quad 1 \leq i \leq n$$

where NB$(\mu, \alpha)$ denotes the negative binomial distribution with the following probability distribution function:

$$f_{NB}\left(y \mid \mu, \alpha\right) = \frac{\Gamma\left(y + \frac{1}{\alpha}\right)}{y!\Gamma\left(\frac{1}{\alpha}\right)} \left(\frac{1}{1+\alpha\mu}\right)^{\frac{1}{\alpha}} \left(\frac{\mu}{\frac{1}{\alpha}+\mu}\right)^{y_i} \tag{2.62}$$
$$\mu > 0, \quad \alpha > 0, \quad y = 0, 1, \ldots$$

It is readily shown that the mean and variance of the NB distribution are given by (see exercise):

$$E\left(y \mid \mu, \alpha\right) = \mu, \quad Var\left(y \mid \mu, \alpha\right) = \mu(1 + \alpha\mu) \tag{2.63}$$

Thus, unless $\alpha = 0$, the variance is always larger than the mean $\mu$, addressing the issue of overdispersion. Note that $\alpha$ indicates the degree of overdispersion and as $\alpha \downarrow 0$ (decreases to 0) this dispersion tends to 0 and thereby $f_{NB}(y) \to f_P(y)$ (see exercise).

Data clustering is not the only source of overdispersion for count response. Too often, the distribution of count response arising in biomedical and psychosocial research is dominated by a preponderance of zeros that exceeds the expected frequency of the Poisson or negative binomial model. In this book, we use the term *structural zeros* to refer to excess zeros that fall either above or below the number of zeros expected by a statistical model such as the Poisson. For example, when modeling the frequency of condom protected vaginal sex over a period of time as in a sexual health study among heterosexual, adolescent females, the presence of structural zeros may reflect a proportion of subjects who either practice abstinence or engage in unprotected vaginal sex, thereby inflating the proportion of sampling zeros under the Poisson law. As in the case of data clustering, the presence of structural zeros also causes overdispersion. However, an even more serious problem with applications of the Poisson model to such data is that structural zeros not only impact the conditional variance, but also the conditional mean, leading to biased estimates of model parameters. Thus, unlike the situation with data clustering, we can no longer fix the problem by only considering the conditional variance. Rather, we must tackle this issue on both fronts. The zero-inflated Poisson (ZIP) model is a popular approach to address the twin effects of structural zeros on both the mean and variance. Since ZIP is developed based on *finite mixture*, we first illustrate the notion of mixture through a simple example.

**Example 19 (*Mixture of two normals*).** All models considered so far are unimodal, that is, there is one unique mode in the conditional distribution of the response $y$ given $\mathbf{x}$. When mixing two or more such unimodal distributions, we obtain a mixture of distributions which typically has more than one mode.

Let $y$ be sampled from $N\left(\mu_1, \sigma_1^2\right)$ or $N\left(\mu_2, \sigma_2^2\right)$ according to some probability $p\ (> 0)$ and $(1-p)$, respectively. Let $z_i = 1$ if $y_i$ is sampled from $N\left(\mu_1, \sigma_1^2\right)$ and 0 if otherwise. It is readily shown that $y$ is distributed with the following density function (see exercise):

$$y \sim f\left(y \mid p, \mu_1, \sigma_1^2, \mu_2, \sigma_2^2\right) = p\phi_1\left(y \mid \mu_1, \sigma_1^2\right) + (1-p)\,\phi_2\left(y \mid \mu_2, \sigma_2^2\right) \quad (2.64)$$

where $\phi_k\left(y \mid \mu_k, \sigma_k^2\right)$ denotes the PDF of $N\left(\mu_k, \sigma_k^2\right)$ $(k = 1, 2)$. Shown in Figure 2.1 is an equal mixture $(p = 0.5)$ based on two normals: $N(3, 0.3)$ and $N(1, 0.2)$. As seen from the plot, there are two modes in this distribution.

It follows from (2.64) that the mean of $y$ is $E(y) = p\mu_1 + (1-p)\,\mu_2$, the weighted average of the means of the two component normals (see exercise).

Distributions with multiple modes arise in practice all the time. For example, consider a study to compare efficacy of two treatment conditions. At the end of the study, if treatment difference does exist, then the distribution of the entire study sample (subjects from both treatments) will be bimodal with the models corresponding to the respective treatment conditions. In general, with the codes for treatment assignment, we can identify the two groups of subjects and therefore model each component of the bimodal distribution using, say a normal distribution, $N\left(\mu_k, \sigma_k^2\right)$ $(k = 1, 2)$.

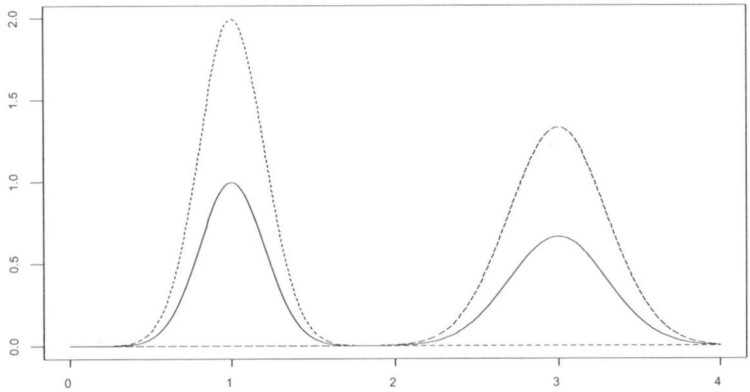


Figure 2.1. Probability density function of an equal mixture of two normals (solid line), along with the probability density function of each normal component, $N(1, 0.2)$ (dotted line) and $N(3, 0.3)$ (dashed line).

The parameter vector in this case is $\boldsymbol{\theta} = \left(\mu_1, \mu_2, \sigma_1^2, \sigma_2^2\right)^\top$. We can then assess treatment effect by comparing the means of the two normals $\mu_k$ using the ANOVA.

Now, suppose that the treatment assignment codes are unavailable. In this case, we have no information to identify the subjects assigned to the two treatment conditions. As a result, we cannot tease out the two individual components of the bimodal distribution by modeling the distribution of $y$ for each group of subjects. For example, if $y$ follows $N\left(\mu_k, \sigma_k^2\right)$ for the $k$th group, we do not know whether $y_i$ is from $N\left(\mu_1, \sigma_1^2\right)$ or $N\left(\mu_2, \sigma_2^2\right)$. Thus, we have to model the combined groups as a whole using a mixture of two normals. The parameter vector for this two-component normal mixture, $\boldsymbol{\theta} = \left(p, \mu_1, \mu_2, \sigma_1^2, \sigma_2^2\right)^\top$, now also includes the mixing proportion $p$.

Thus, when differences exist among a number of subgroups within a study sample, we can explicitly model such between-group differences using ANOVA or ANOVA like GLMs (for discrete response) if the subgroups are identifiable based on information other than the response such as treatment conditions as in an intervention study. When such subgroups cannot be identified, we can implicitly account for their differences using finite mixtures. By viewing structural zeros as the result of a subgroup of subjects who, unlike the rest of the subjects, are not susceptible to experiencing the event being modeled by the count response, we can employ finite mixtures to account for their presence.

**Example 20 (*The zero-inflated Poisson model*).** In Example 17, suppose that there are structural zeros in the distribution of $y_i$ when modeled by the Poisson. Consider a two-component *mixture model* consisting of a Poisson and a degenerate distribution centered at 0 defined by

$$f_M(y) = \rho f_0(y) + (1-\rho) f_P(y), \quad y = 0, 1, \ldots \tag{2.65}$$

where $f_P(y)$ is the Poisson distribution given in (2.59) and $f_0(y) = I_{\{y=0\}}$ denotes the distribution of the constant 0 By reexpressing this ZIP model, we obtain

$$f_M(y) = \begin{cases} \rho f_0(0) + (1-\rho) f_P(0) & \text{if } y = 0 \\ (1-\rho) f_P(y_i) & \text{if } y > 0 \end{cases} \tag{2.66}$$

Thus, the Poisson probability at 0, $f_P(0)$, is modified by: $f_M(0) = \rho + (1-\rho) f_P(0)$, with the degenerate distribution component $f_0(y)$ to account for structural zeros.

Since $0 \leq f_M(0) \leq 1$, it follows that $\frac{1}{1-\exp(\mu)} \leq \rho \leq 1$. Thus, $\rho$ can be negative. When $0 < \rho < 1$, $\rho$ represents the amount of *positive structural* zeros above and beyond the sampling zeros expected by the Poisson distribution $f_P(y)$. A negative $\rho$ implies that the amount of *structural* zeros falls below the expected sampling zeros of the Poisson. Although not as popular as the former, the latter *zero-inflated* or *negative structural zero* case may also arise in practice. For example, in the sexual health example discussed earlier, the distribution of condom protected vaginal sex may initially exhibit positive structural zeros at the beginning study because of a large number of people who never consider using condoms. If study contains some intervention for promoting safe sex, then behavioral changes may take place and many of these initial non-condom users may start practicing safe sex during or after the intervention. Such changes will reduce or remove the positive structural zeros, and may even result in negative structural zeros in the distribution if a substantial number of these non-condom users change their sexual behaviors at the end of the study.

The mean and variance of ZIP defined in (2.65) are given by (see exercise)

$$E(y) = (1-\rho)\mu, \quad Var(y_i) = \mu(1-\rho)(1+\mu\rho) \tag{2.67}$$

If $0 < \rho < 1$, $E(y) < Var(y)$ and thus the presence of positive structural zeros leads to overdispersion. By comparing (2.67) to (2.60) and (2.63), it is seen that the positive structural zeros also affect the mean response $E(y)$.

For regression analysis, we need to model both the Poisson mean $\mu$ and the amount of structural zeros $\rho$. Let $\mathbf{u}_i$ and $\mathbf{v}_i$ be two subsets of $\mathbf{x}_i$. Note that $\mathbf{u}_i$ and $\mathbf{v}_i$ may overlap and thus generally do not form a partition of $\mathbf{x}_i$. The ZIP regression model is given by

$$y_i \mid \mathbf{x}_i \sim \text{i.d. ZIP}\left(\mu_i, \rho_i\right) \tag{2.68}$$
$$h_1\left(\rho_i\right) = \operatorname{logit}\left(\rho_i\right) = \mathbf{u}_i^\top \boldsymbol{\beta}_u, \quad h_2\left(\mu_i\right) = \log\left(\mu_i\right) = \mathbf{v}_i^\top \boldsymbol{\beta}_v$$

Thus, the Poisson mean $\mu_i$ and structural zeros $\rho_i$ are linked to the two linear predictors $\mathbf{u}_i^\top \boldsymbol{\beta}_u$ and $\mathbf{v}_i^\top \boldsymbol{\beta}_v$ by the log and logit functions, respectively.

Note that mixture-based models such as the normal mixture and ZIP do not fall under the rubric of generalized linear models. For example, for the ZIP regression defined in (2.68), the mean response, $E\left(y_i \mid \mathbf{x}_i^\top\right) = \left(1-\rho_i\right)\mu_i$, is not directly linked to a linear predictor as in the standard GLM such as the NB model in (2.61). This distinction will become clearer when we discuss the distribution-free generalized linear model in Section 2.2.

### 2.1.5 Inference for Generalized Linear Models

Like linear regression, inference for generalized linear models as well as the mixture-based models is carried out by the method of maximum likelihood. Thus, the results in Theorem 1 apply to this general class of models as well. However, unlike linear regression, closed-form solutions to the score equation are usually not available for maximum likelihood estimates and numerical methods must be employed to find the MLE. The *Newton–Raphson* method is the most popular approach for computing the MLE numerically.

For a given GLM, let $\boldsymbol{\theta}$ be the vector of model parameters and $l_n\left(\boldsymbol{\theta}\right)$ the log-likelihood function of the model. As discussed in Section 2.1.3, the maximum likelihood estimate is a solution to the score equation:

$$\mathbf{u}_n\left(\boldsymbol{\theta}\right) = \frac{\partial}{\partial \boldsymbol{\theta}} l_n\left(\boldsymbol{\theta}\right) = \mathbf{0}$$

When the above is not solvable in closed form, the *Newton–Raphson* method is used to obtain numerical solutions. This method is based on a Taylor series expansion of the score statistic vector $\mathbf{u}_n\left(\boldsymbol{\theta}\right)$ around some point $\boldsymbol{\theta}^{(0)}$ near the MLE:

$$\mathbf{u}_n\left(\boldsymbol{\theta}\right) = \mathbf{u}_n\left(\boldsymbol{\theta}^{(0)}\right) + \left(\frac{\partial}{\partial \boldsymbol{\theta}} \mathbf{u}_n\left(\boldsymbol{\theta}^{(0)}\right)\right)^\top \left(\boldsymbol{\theta} - \boldsymbol{\theta}^{(0)}\right) + \mathbf{o}\left(\left\|\boldsymbol{\theta} - \boldsymbol{\theta}^{(0)}\right\|\right) \tag{2.69}$$

where $\|\cdot\|$ denotes the Euclidean norm and $\mathbf{o}(\delta)$ a $p \times 1$ vector with its length approaching 0 faster than $\delta \downarrow 0$. By ignoring the higher order term $\mathbf{o}\left(\left\|\boldsymbol{\theta} - \boldsymbol{\theta}^{(0)}\right\|\right)$ in (2.69) and solving for $\boldsymbol{\theta}$, we obtain:

$$\begin{aligned} \boldsymbol{\theta}^{(1)} &= \boldsymbol{\theta}^{(0)} + \left[-\frac{\partial}{\partial \boldsymbol{\theta}} \mathbf{u}_n\left(\boldsymbol{\theta}^{(0)}\right)\right]^{-1} \mathbf{u}_n\left(\boldsymbol{\theta}^{(0)}\right) \\ &= \boldsymbol{\theta}^{(0)} + \left(I_n^o\left(\boldsymbol{\theta}^{(0)}\right)\right)^{-1} \mathbf{u}_n\left(\boldsymbol{\theta}^{(0)}\right) \end{aligned} \tag{2.70}$$

where $I_n^o(\boldsymbol{\theta}) = -\frac{\partial^2}{\partial \boldsymbol{\theta} \partial \boldsymbol{\theta}^\top} l_n(\boldsymbol{\theta})$ is the observed information matrix. Substituting $\boldsymbol{\theta}^{(1)}$ for $\boldsymbol{\theta}^{(0)}$ in (2.70) and solving the resulting equation yields a new $\boldsymbol{\theta}^{(2)}$. By repeating this process, we obtain a sequence $\boldsymbol{\theta}^{(k)}$, which converges to the MLE $\widehat{\boldsymbol{\theta}}$ as $n \to \infty$, that is, $\widehat{\boldsymbol{\theta}} = \lim_{k \to \infty} \boldsymbol{\theta}^{(k)}$. In practice, the process ends after a finite number of iterations once some convergence criterion is achieved. The Newton–Raphson algorithm converges very rapidly. For most models, convergence is usually reached after 10–30 iterations. However, starting values are critical for the iterations to converge. Figure 2.2 depicts what could happen with two different starting values for the case of a scalar parameter $\theta$. In this case, the algorithm would converge, when initiated with some value $\theta^{(0)}$ as depicted on the left part of the diagram. A different starting value as shown on the right part of the diagram would drive $\theta^{(k)}$ to $\infty$. When convergence is not reached in about 30 iterations, it is recommended that the algorithm be stopped and restarted with a new initial value.

General linear hypothesis can be similarly tested using the Wald or likelihood ratio tests. However, unlike linear regression, it is not possible to test nonlinear contrasts by simply reparameterizing the model. For example, consider the general linear hypothesis (2.36) with $\mathbf{b} \neq \mathbf{0}$. As in the linear model case, we can perform the transformation $\boldsymbol{\gamma} = \boldsymbol{\beta} - K^\top \left(K^\top K\right)^{-1} \mathbf{b}$ so that the null hypothesis can be expressed as a linear contrast in terms of $\boldsymbol{\gamma}$, $H_0 : K\boldsymbol{\gamma} = \mathbf{0}$. For GLMs, the linear predictor has the form:

$$g(\mu_i) = \mathbf{x}_i^\top \boldsymbol{\beta} = \mathbf{x}_i^\top K^\top \left(K^\top K\right)^{-1} \mathbf{b} + \mathbf{x}_i^\top \boldsymbol{\gamma} = c_i + \mathbf{x}_i^\top \boldsymbol{\gamma} \tag{2.71}$$

For mixture-based models such as ZIP, the linear predictors can be expressed in the above form. Unlike linear regression, however, we can no longer transform the response $y_i$ to absorb the extra *offset* term $c_i$ in (2.71). Most software packages, such as SAS allow for the specification of such *offset* terms when fitting GLMs. Note that $c_i$ cannot be treated as a covariate since it has a known coefficient 1.

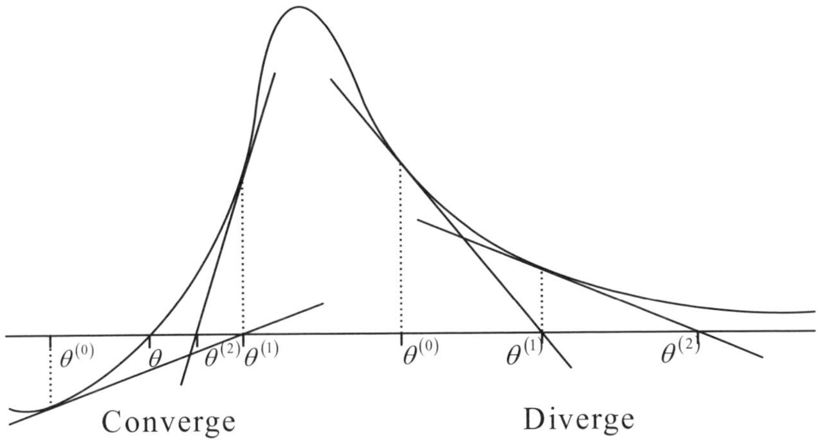


Figure 2.2. Diagram showing the effect of the starting point on the convergence of Newton–Raphson algorithm.

**Example 21**. Consider Example 17 again. The log-likelihood function is

$$l_n(\boldsymbol{\beta}) = \sum_{i=1}^{n} \left[ y_i \mathbf{x}_i^\top \boldsymbol{\beta} - \exp\left(\mathbf{x}_i^\top \boldsymbol{\beta}\right) - \log y_i! \right]$$

Thus, the score function and observed information are given by

$$\mathbf{u}_n(\boldsymbol{\beta}) = \frac{\partial}{\partial \boldsymbol{\beta}} l_n(\boldsymbol{\beta}) = \sum_{i=1}^{n} \left[ y_i \mathbf{x}_i - \exp\left(\mathbf{x}_i^\top \boldsymbol{\beta}\right) \mathbf{x}_i \right] \tag{2.72}$$

$$I_n^o(\boldsymbol{\beta}) = -\frac{\partial}{\partial \boldsymbol{\beta}} \mathbf{u}_n(\boldsymbol{\beta}) = \sum_{i=1}^{n} \exp\left(\mathbf{x}_i^\top \boldsymbol{\beta}\right) \mathbf{x}_i \mathbf{x}_i^\top, \; I_1^e(\boldsymbol{\beta}) = E\left[\exp\left(\mathbf{x}_i^\top \boldsymbol{\beta}\right) \mathbf{x}_i \mathbf{x}_i^\top\right]$$

By applying (2.70) to $\mathbf{u}_n(\boldsymbol{\beta})$ and $I_n^o(\boldsymbol{\beta})$ above, we can numerically obtain the MLE $\widehat{\boldsymbol{\beta}}$. It follows from Theorem 1 that the asymptotic distribution of $\widehat{\boldsymbol{\beta}}$ is $AN\left(\boldsymbol{\beta}, \boldsymbol{\Sigma}_\beta = \frac{1}{n}\left[I_1^e(\boldsymbol{\beta})\right]^{-1}\right)$. Since $I_1^e(\boldsymbol{\beta})$ cannot be expressed in closed form, we estimate it by: $\frac{1}{n} I_n^o(\boldsymbol{\beta})$. It follows from Slutsky's theorem that a consistent estimate of $\boldsymbol{\Sigma}_\beta$ is given by

$$\widehat{\boldsymbol{\Sigma}}_\beta = \frac{1}{n}\left[\frac{1}{n} I_n^o\left(\widehat{\boldsymbol{\beta}}\right)\right]^{-1} = \left[I_n^o\left(\widehat{\boldsymbol{\beta}}\right)\right]^{-1} \tag{2.73}$$

**Example 22**. Consider the ZIP regression model in Example 20. The conditional PDF of $y_i$ given $\mathbf{x}_i$ is

$$f_M(y_i \mid \mathbf{x}_i, \boldsymbol{\theta}) = h_1^{-1}\left(\mathbf{u}_i^\top \boldsymbol{\beta}_u\right) f_0(y_i) + \left[1 - h_2^{-1}\left(\mathbf{v}_i^\top \boldsymbol{\beta}_u\right)\right] f_P(y_i \mid \mu_i)$$

where $\boldsymbol{\theta}^\top = \left(\boldsymbol{\beta}_u^\top, \boldsymbol{\beta}_v^\top\right)^\top$ and

$$f_P(y_i \mid \mu_i) = \frac{\mu_i^{y_i}}{y_i!}\exp(-\mu_i) = \frac{\left(h_2^{-1}\left(\mathbf{v}_i^\top \boldsymbol{\beta}_u\right)\right)^{y_i}}{y_i!}\exp\left(-h_2^{-1}\left(\mathbf{v}_i^\top \boldsymbol{\beta}_u\right)\right)$$

The log-likelihood function is given by:

$$\begin{aligned}
l_n(\boldsymbol{\theta}) &= \sum_{i=1}^n \log f_M(y_i \mid \mathbf{x}_i, \boldsymbol{\theta}) \\
&= \sum_{i=1}^n I_{\{y_i=0\}} \log\left[h_1^{-1}\left(\mathbf{u}_i^\top \boldsymbol{\beta}_u\right) + \left(1 - h^{-1}\left(\mathbf{x}_{2i}^\top \gamma\right)\right) f_P(0 \mid \mu_i)\right] + \\
&+ \sum_{i=1}^n I_{\{y_i>0\}} \log\left[\left(1 - h_2^{-1}\left(\mathbf{v}_i^\top \boldsymbol{\beta}_u\right)\right) f_P(y_i \mid \mu_i)\right]
\end{aligned}$$

where $I_{\{\cdot\}}$ denotes a set indicator. The score $\mathbf{u}_n(\boldsymbol{\theta})$ and observed information matrix $I_n^o(\boldsymbol{\theta})$ are also readily calculated, albeit with more complex expressions than those in Example 16.

## 2.2 Distribution-Free (Semiparametric) Models

In Section 2.1, we discussed the classic linear regression model for continuous response and its extension, the generalized linear model (GLM), to accommodate other types of outcome variables such as binary and count response. The key step in the generalization of linear regression to the class of GLM is the extension of the link function in the systematic part of the linear model so that the mean response can be meaningfully related to the linear predictor for other types of response.

In this section, we consider generalizing the other, random, component of the GLM. More specifically, we will completely remove this component from the specification of GLM so that no parametric distribution is imposed on the response variable. The resulting models are *distribution-free* and are applicable to a wider class of data distributions. This distribution-free property is especially desirable for power analysis. Without real data, it would be impossible to verify a parametric distribution model, and the robust property of the distribution-free models will ensure reliable power and sample size estimates regardless of the data distribution.

The removal of the random component of GLM, however, entails serious ramifications for inference of model parameters: without a distribution model specified in this part of GLM, it is not possible to use the method of maximum likelihood for parameter estimation and inference. Thus, we must introduce and develop a new and alternative inference paradigm.

### 2.2.1 Distribution-Free Generalized Linear Models

Recall that the parametric GLM has the following components:

1. *Random component.* This part specifies the conditional distribution of the response $y$ given the independent variables $\mathbf{x}$.

2. *Systematic component.* This part links the conditional mean of $y$ given $\mathbf{x}$ to the linear predictor $\eta$ by a link function:

$$g(\mu) = \eta = \beta_0 + \beta_1 x_1 + \ldots + \beta_p x_p = \mathbf{x}^\top \boldsymbol{\beta} \tag{2.74}$$

By removing the random component, we obtain the distribution-free GLM with only the systematic part specified in (2.74).

**Example 1 (*Linear regression*).** Consider a sample of $n$ subjects. Let $y_i$ be a continuous response and $\mathbf{x}_i$ a vector of independent variables of interest. The normal-based linear regression has the following form:

$$y_i \mid \mathbf{x}_i \sim \text{i.d. } N\left(\mu_i, \sigma^2\right), \quad g(\mu_i) = \mu_i = \eta_i = \mathbf{x}_i^\top \boldsymbol{\beta}, \quad 1 \leq i \leq n \tag{2.75}$$

By excluding the distribution specification $y_i \sim N\left(\mu_i, \sigma^2\right)$ from (2.75), we obtain the distribution-free version of the normal based linear regression for modeling the conditional mean of $y_i$ given $\mathbf{x}_i$ as a linear function of $\mathbf{x}_i$.

In the special case when $y_i$ represent individual responses from a study comparing $g$ treatment conditions and $\mathbf{x}_i$ are defined by the $g$ binary indicators as in Example 1 of Section 2.1.1, we obtain the distribution-free ANOVA:

$$E(y_i) = \beta_k, \quad \sum_{l=0}^{k-1} n_l + 1 \leq i \leq \sum_{l=0}^{k} n_l, \quad n_0 \equiv 0, \quad 1 \leq k \leq g$$

or equivalently,

$$E(y_{kj}) = \mu_k, \quad 1 \leq j \leq n_k, \quad 1 \leq k \leq g$$

where $y_{kj}$ denotes the response from the $j$th subject within the $k$th treatment group.

**Example 2** (***Linear regression with*** $t$ ***distribution***). In Example 1, suppose that we want to model $y_i$ using a $t$ distribution to accommodate thicker tails in the data distribution under the parametric formulation, that is,

$$y_i \mid \mathbf{x}_i \sim \text{i.d. } t\left(\mu_i, \sigma^2, v\right), \quad g\left(\mu_i\right) = \mu_i = \eta_i = \mathbf{x}_i^\top \boldsymbol{\beta}, \quad 1 \leq i \leq n \tag{2.76}$$

where $t\left(\mu, \sigma^2, v\right)$ denotes a $t$ distribution with mean $\mu$, scale parameter $\sigma^2$ and shape parameter $v$. This alternative model differs from the normal based linear regression in (2.75) only in the random component. Thus, by eliminating this component, we obtain from (2.76) the same distribution-free linear model, $\mu_i = \mathbf{x}_i^\top \boldsymbol{\beta}$, as in Example 1.

The normal-based linear model in Example 1 may yield biased inference when the response $y_i$ follows a $t$ distribution. Likewise, the $t$-based model in Example 2 may produce biased estimates of $\boldsymbol{\beta}$ if the assumption of $t$ distribution is violated (e.g., skewness in the data distribution). Being independent of such parametric assumptions, the distribution-free linear model yields valid inference regardless of the shape or form of the distribution of the response so long as the model in the systematic part is specified correctly.

**Example 3** (***Log-linear model***). In Example 1, now suppose that $y_i$ is a count response. As discussed in Section 2.1.4, the most popular parametric GLM for such a response variable is the Poisson log-linear model:

$$y_i \mid \mathbf{x}_i \sim \text{i.d. Poisson}\left(\mu_i\right), \quad \log\left(\mu_i\right) = \eta_i = \mathbf{x}_i^\top \boldsymbol{\beta}, \quad 1 \leq i \leq n \tag{2.77}$$

By removing the distribution specification $y_i \sim$Poisson$(\mu_i)$, (2.75) yields the distribution-free log-linear model, $\log\left(\mu_i\right) = \eta_i = \mathbf{x}_i^\top \boldsymbol{\beta}$.

**Example 4** (***Log-linear model with negative binomial distribution***). In Example 3, suppose that data are clustered. As discussed in Section 2.1.4, the Poisson-based log-linear model generally yields biased inference when applied to such data. A popular remedial alternative is the following negative binomial model:

$$y_i \mid \mathbf{x}_i \sim \text{i.d. NB}\left(\mu_i, \alpha\right), \quad \log\left(\mu_i\right) = \mathbf{x}_i^\top \boldsymbol{\beta}, \quad 1 \leq i \leq n \tag{2.78}$$

By comparing the two models in (2.77) and (2.78), we see that the systematic components of Poisson- and NB-based log-linear models are identical. Thus, with the random component excluded, (2.77) and (2.78) define the same distribution-free log-linear model.

The distribution-free version of the model in all the examples above can be expressed in a general form:

$$g(\mu_i) = g(E(y_i \mid \mathbf{x}_i)) = \eta_i = \mathbf{x}_i^\top \boldsymbol{\beta}, \quad 1 \le i \le n \tag{2.79}$$

where $g(\cdot)$ is a link function appropriate for the type of response variable in the regression model. The link function is specified in the same way as in the systematic component of parametric GLM. For example, for binary response, only functions that map $(0, 1)$ to $(-\infty, \infty)$ such as logit and probit are appropriate. Although no parametric model is specified for the distribution of the response, the systematic component in (2.79) still involves parametric assumptions for the link function and the linear predictor. For this reason, distribution-free GLMs are often called *semi-parametric* models.

### 2.2.2 Inference for Generalized Linear Models

Inference for parametric GLM is carried out by the method of maximum likelihood, which is discussed in Section 2.1.5. As the log-likelihood function is required for deriving model estimates and their inference, the method of maximum likelihood does not apply to the distribution-free GLM. The most popular approach for inference for this new class of models is the method of *estimating equation.* We first illustrate the ideas behind this approach with examples and then discuss the properties of estimates obtained under this alternative inference paradigm.

**Example 5.** Consider the normal-based linear model in Example 1 of Section 2.2.1:

$$y_i \mid \mathbf{x}_i \sim \text{i.d. } N\left(\mu_i, \sigma^2\right), \quad \mu_i = \mathbf{x}_i^\top \boldsymbol{\beta}, \quad 1 \le i \le n \tag{2.80}$$

The log-likelihood function and score statistic vector with respect to the parameters of interest $\boldsymbol{\beta}$ are given by

$$l_n\left(\boldsymbol{\beta},\sigma^2\right) = -\frac{n}{2}\log\left(2\pi\sigma^2\right) - \frac{1}{2\sigma^2}\sum_{i=1}^n (y_i - \mu_i)^2 \tag{2.81}$$

$$\mathbf{u}_n\left(\boldsymbol{\beta},\sigma^2\right) = \frac{\partial l_n\left(\boldsymbol{\beta},\sigma^2\right)}{\partial \boldsymbol{\beta}} = \sum_{i=1}^n \frac{1}{\sigma^2}(y_i - \mu_i)\,\mathbf{x}_i = \sum_{i=1}^n \frac{1}{\sigma^2}(y_i - \mu_i)\frac{\partial \mu_i}{\partial \boldsymbol{\beta}}$$

Let

$$D_i = \frac{\partial \mu_i}{\partial \boldsymbol{\beta}}, \quad V_i = 1, \quad S_i = y_i - \mu_i \tag{2.82}$$

Then, the score equation defined by $\mathbf{u}_n$ in (2.81) can be expressed as:

$$\mathbf{u}_n(\boldsymbol{\beta}) = \sum_{i=1}^{n} D_i V_i^{-1} S_i = \mathbf{0} \qquad (2.83)$$

Unlike the score vector in (2.81), the above equation only involves $\boldsymbol{\beta}$, which suffices to define the MLE of this parameter vector of interest.

**Example 6**. Consider the Poisson log-linear model in Example 3. The likelihood and score statistic are given by

$$l_n(\boldsymbol{\beta}) = \sum_{i=1}^{n} \left[-\mu_i + y_i \log(\mu_i)\right], \quad \mathbf{u}_n(\boldsymbol{\beta}) = \sum_{i=1}^{n} \frac{1}{\mu_i}(y_i - \mu_i)\frac{\partial}{\partial \boldsymbol{\beta}}\mu_i$$

Let $V_i = \mu_i$, but define $D_i$ and $S_i$ the same way as in (2.82). Then, the score equation can again be expressed in the form of (2.83).

**Example 7** (***Logistic regression***). Consider a sample of $n$ subjects. Let $y_i$ be a binary response and $\mathbf{x}_i$ a vector of independent variables of interest. The parametric GLM for modeling $y_i$ as a function of $\mathbf{x}_i$ is given by

$$y_i \mid \mathbf{x}_i \sim \text{i.d. BI}(\mu_i, 1), \quad g(\mu_i) = \mathbf{x}_i^\top \boldsymbol{\beta}, \quad 1 \leq i \leq n$$

where $g(\mu_i)$ is some appropriate link function, such as the logit and probit. The log-likelihood and score vector are given by

$$l_n(\boldsymbol{\beta}) = \sum_{i=1}^{n} \left[y_i \log(\mu_i) + (1 - y_i)\log(1 - \mu_i)\right]$$

$$\mathbf{u}_n(\boldsymbol{\beta}) = \sum_{i=1}^{n} \frac{1}{\mu_i(1 - \mu_i)}(y_i - \mu_i)\frac{\partial}{\partial \boldsymbol{\beta}}\mu_i$$

By letting $V_i = \mu_i(1 - \mu_i)$, but leaving $D_i$ and $S_i$ in (2.82) unchanged, we can again express the score equation in the form of (2.83).

In the three examples above, the score equation follow a common, general form given in (2.83). This is not a coincidence and in fact this expression holds true more generally if the conditional distribution of $y_i$ given $\mathbf{x}_i$ in the random component of parametric GLM is from the exponential family of distributions defined in (2.54). For such a distribution, the log-likelihood function and score statistic vector with respect to $\boldsymbol{\beta}$ are given by

$$l_n = \sum_{i=1}^{n} \frac{1}{a(\phi)}(y_i\theta_i - b(\theta_i)) + \sum_{i=1}^{n} c(y_i, \phi) \qquad (2.84)$$

$$\mathbf{u}_n = \frac{\partial l_n}{\partial \boldsymbol{\beta}} = \sum_{i=1}^{n} \frac{1}{a(\phi)}\left(y_i - \frac{d}{d\theta_i}b(\theta_i)\right)\frac{\partial \theta_i}{\partial \boldsymbol{\beta}}$$

Further, it is readily checked that (see exercise),

$$E(y_i \mid \mathbf{x}_i) = \mu_i = \frac{d}{d\theta_i} b(\theta_i), \quad Var(y \mid \mathbf{x}_i) = a(\phi) \frac{d^2}{d\theta_i^2} b(\theta_i) \qquad (2.85)$$

$$\frac{d\theta_i}{d\mu_i} = \left( \frac{d^2}{d\theta_i^2} b(\theta) \right)^{-1} = v(\mu_i) = V_i$$

It follows from (2.84) and (2.85) that

$$\mathbf{u}_n = \sum_{i=1}^{n} \frac{1}{a(\phi)} (y_i - \mu_i) \frac{d\theta_i}{d\mu_i} \frac{\partial \mu_i}{\partial \boldsymbol{\beta}} = \sum_{i=1}^{n} \frac{1}{a(\phi)} V_i (y_i - \mu_i) \frac{\partial \mu_i}{\partial \boldsymbol{\beta}}$$

Thus, by setting $\mathbf{u}_n$ to 0 and eliminating $a(\phi)$, we obtain a score equation with respect to $\boldsymbol{\beta}$ that has the same form as the estimating equation in (2.83).

We know from the discussion in Section 2.1.4 that inference for parametric GLM by the method of maximum likelihood is really driven by the score equation. The likelihood function provides the basis for constructing the score statistic vector, but computations as well as asymptotic distributions of maximum likelihood estimates can all be derived from the score statistic vector. The method of estimating equation is premised on a score-like statistic vector.

Consider the class of distribution-free GLM defined by (2.79). Although no parametric distribution is assumed for the response $y_i$, $D_i = \frac{\partial \mu_i}{\partial \boldsymbol{\beta}}$ and $S_i = y_i - \mu_i$ are still well defined given the model specification in (2.79). Thus, given some choice of $V_i = v(\mu_i)$, we can still define a score-like vector $\mathbf{w}_n(\boldsymbol{\beta})$ and associated *estimating equation*:

$$\mathbf{w}_n(\boldsymbol{\beta}) = \sum_{i=1}^{n} D_i V_i^{-1} S_i = \mathbf{0}, \quad D_i = \frac{\partial}{\partial \boldsymbol{\beta}} \mu_i, \quad V_i = v(\mu_i) \qquad (2.86)$$

$$S_i = y_i - \mu_i$$

In the above, $S_i$ is called the *theoretical residual* (to differentiate it from the *observed residual* with estimated $\mu_i$) and is the only quantity that contains the response $y_i$. The quantity $V_i$ is assumed to a function of $\mu_i$. As discussed earlier, with the right selection of $V_i$, the estimating equation in (2.86) yields the MLE of $\boldsymbol{\beta}$ for the parametric GLM when $y_i$ is modeled by the exponential family of distributions.

*Estimating equation estimates* are defined in the same way as the MLE — by solving the equation in (2.86). As in the case of MLE, a closed-form solution typically does not exist and numerical estimates can be obtained by applying the Newton-Raphson algorithm. For example, by expanding $\mathbf{w}_n(\boldsymbol{\beta})$ around the estimating equation estimate using a stochastic Taylor series expansion and ignoring the higher order terms, we have

$$\mathbf{w}_n\left(\boldsymbol{\beta}^{(k+1)}\right) = \mathbf{w}_n\left(\boldsymbol{\beta}^{(k)}\right) + \left(\frac{\partial}{\partial \boldsymbol{\beta}} \mathbf{w}_n\left(\boldsymbol{\beta}^{(k)}\right)\right)^\top \left(\boldsymbol{\beta}^{(k+1)} - \boldsymbol{\beta}^{(k)}\right)$$

By setting $\mathbf{w}_n\left(\boldsymbol{\beta}^{(k+1)}\right) = 0$ and solving for $\boldsymbol{\theta}^{(k+1)}$, we obtain

$$\boldsymbol{\beta}^{(k+1)} = \boldsymbol{\beta}^{(k)} + \left[-\frac{\partial}{\partial \boldsymbol{\beta}} \mathbf{w}_n\left(\boldsymbol{\beta}^{(k)}\right)\right]^{-\top} \mathbf{w}_n\left(\boldsymbol{\beta}^{(k)}\right) \tag{2.87}$$

where $A^{-\top} = \left(A^{-1}\right)^\top$ for a matrix $A$. Starting with some initial $\boldsymbol{\beta}^{(0)}$, we use (2.87) to iterate until some convergence criterion is reached and the limit $\widehat{\boldsymbol{\beta}}$ is the estimating equation estimate of $\boldsymbol{\beta}$. The calculation of $\frac{\partial}{\partial \boldsymbol{\beta}} \mathbf{w}_n(\boldsymbol{\beta})$ may be complicated in some cases. An alternative is to compute the estimate $\widehat{\boldsymbol{\beta}}$ using the following algorithm:

$$\boldsymbol{\beta}^{(k+1)} = \boldsymbol{\beta}^{(k)} + \left(D_i V_i^{-1} D_i^\top\right)^{-1} \mathbf{w}_n\left(\boldsymbol{\beta}^{(k)}\right) \tag{2.88}$$

While this algorithm may converge at a slower rate, it avoids the calculation of $\frac{\partial}{\partial \boldsymbol{\beta}} \mathbf{w}_n$.

**Example 8**. Consider the distribution-free log-linear model specified by only the systematic component in Example 3. Let $V_i = \mu_i$. The estimating equation is given by

$$\mathbf{w}_n(\boldsymbol{\beta}) = \sum_{i=1}^{n} D_i V_i^{-1} S_i = \sum_{i=1}^{n} (y_i - \mu_i)\, \mathbf{x}_i = \mathbf{0}$$

It is readily checked (see exercise) that

$$\frac{\partial}{\partial \boldsymbol{\beta}} \mathbf{w}_n(\boldsymbol{\beta}) = -\sum_{i=1}^{n} \mathbf{x}_i \left(\frac{\partial \mu_i}{\partial \boldsymbol{\beta}}\right)^\top = \sum_{i=1}^{n} \mu_i \mathbf{x}_i \mathbf{x}_i^\top \tag{2.89}$$

$$D_i V_i^{-1} D_i^\top = \mu_i \frac{\partial \mu_i}{\partial \boldsymbol{\beta}} \left(\frac{\partial \mu_i}{\partial \boldsymbol{\beta}}\right)^\top = \mu_i \mathbf{x}_i \mathbf{x}_i^\top$$

Thus, for this particular example, the two updating schemes (2.87) and (2.88) are identical.

When using the estimating equation in (2.86) to find estimates of $\boldsymbol{\beta}$ for the distribution-free GLM, we must select some function for $V_i$. We know from the above discussion that if the conditional distribution of $y_i$ given $\mathbf{x}_i$ is a member of the exponential family, $V_i$ can be determined to enable the estimating equation to yield the MLE of $\boldsymbol{\beta}$. In most applications, however, this distribution is unknown and the selection of $V_i$ becomes arbitrary. Fortunately, the estimating equation still yields consistent estimates of $\boldsymbol{\beta}$ regardless of the choice of $V_i$. Theorem 1 below summarizes the asymptotic properties of the estimating equation estimates.

**Theorem 1** (*Properties of estimating equation estimates*). Let $\widehat{\boldsymbol{\beta}}$ denote the estimate obtained by the method of estimating equation. Let $B = E\left(D_i V_i^{-1} D_i^{\top}\right)$ and $V_i = v\left(\mu_i\right)$, a known function of $\mu_i$. Then,

1. $\widehat{\boldsymbol{\beta}} \rightarrow_p \boldsymbol{\beta}$
2. $\sqrt{n}\left(\widehat{\boldsymbol{\beta}} - \boldsymbol{\beta}\right) \rightarrow_p N\left(0, \boldsymbol{\Sigma}_\beta\right)$, where $\boldsymbol{\Sigma}_\beta$ is given by

$$\boldsymbol{\Sigma}_\beta = B^{-1} E\left(D_i V_i^{-2} Var\left(y_i \mid \mathbf{x}_i\right) D_i^{\top}\right) B^{-1} \tag{2.90}$$

Proof. The proof follows from an argument similar to the one used to prove Theorem 1 of Section 2.1.2 for the asymptotic properties of MLE.

First, note that the score-like vector, $\mathbf{w}_n(\boldsymbol{\beta}) = \frac{1}{n}\sum_{i=1}^n D_i V_i^{-1} S_i$, is asymptotically normal,

$$\sqrt{n}\mathbf{w}_n(\boldsymbol{\beta}) \rightarrow_d N\left(\mathbf{0}, \Sigma_w = E\left[D_i V_i^{-2} Var\left(y_i \mid \mathbf{x}_i\right) D_i^{\top}\right]\right) \tag{2.91}$$

This follows immediately from an application of CLT to $\mathbf{w}_n(\boldsymbol{\beta})$ and the following identities:

$$E\left(D_i V_i^{-1} S_i\right) = E\left[D_i V_i^{-1} E\left(S_i \mid \mathbf{x}_i\right)\right] = \mathbf{0} \tag{2.92}$$

$$Var\left(D_i V_i^{-1} S_i\right) = E\left[D_i V_i^{-2} E\left(S_i^2 \mid \mathbf{x}_i\right) D_i^{\top}\right] = E\left[D_i V_i^{-2} Var\left(y_i \mid \mathbf{x}_i\right) D_i^{\top}\right]$$

Now, consider the estimating equation estimate $\widehat{\boldsymbol{\beta}}$. By an argument similar to the proof of consistency for MLE, we can show that $\widehat{\boldsymbol{\beta}}$ is consistent regardless of the choice of $V_i$ (see exercise).

To prove asymptotic normality, consider a Taylor series expansion of $\mathbf{w}_n\left(\widehat{\boldsymbol{\beta}}\right)$ around $\mathbf{w}_n(\boldsymbol{\beta})$:

$$\mathbf{w}_n\left(\widehat{\boldsymbol{\beta}}\right) = \mathbf{w}_n(\boldsymbol{\beta}) + \left(\frac{\partial}{\partial \boldsymbol{\beta}} \mathbf{w}_n\left(\widehat{\boldsymbol{\beta}}\right)\right)^{\top}\left(\widehat{\boldsymbol{\beta}} - \boldsymbol{\beta}\right) + o_p(1)\left(\widehat{\boldsymbol{\beta}} - \boldsymbol{\beta}\right)$$

from which it follows that

$$\sqrt{n}\left(\widehat{\boldsymbol{\beta}}-\boldsymbol{\beta}\right)=\left(-\frac{\partial \mathbf{w}_n(\boldsymbol{\beta})}{\partial \boldsymbol{\beta}}+o_p(1)\right)^{-\top}\sqrt{n}\mathbf{w}_n(\boldsymbol{\beta}) \qquad (2.93)$$

By LLN, $\frac{\partial}{\partial \boldsymbol{\beta}}\mathbf{w}_n(\boldsymbol{\beta}) \rightarrow_p E\left[\frac{\partial}{\partial \boldsymbol{\beta}}\left(D_i V_i^{-1} S_i\right)\right]$. Further, it is readily checked that (see exercise)

$$\begin{aligned} E\left[\frac{\partial}{\partial \boldsymbol{\beta}}\left(D_i V_i^{-1} S_i\right)\right] &= E\left[S_i \frac{\partial}{\partial \boldsymbol{\beta}}\left(D_i V_i^{-1}\right)\right]+ \\ &\quad + E\left[\left(\frac{\partial}{\partial \boldsymbol{\beta}} S_i\right) V_i^{-1} D_i^{\top}\right] \\ &= E\left(E\left(S_i \mid \mathbf{x}_i^{\top}\right)\frac{\partial}{\partial \boldsymbol{\beta}}\left(D_i V_i^{-1}\right)\right)-E\left(D_i V_i^{-1} D_i^{\top}\right) \\ &= -E\left(D_i V_i^{-1} D_i^{\top}\right)=-B \end{aligned} \qquad (2.94)$$

Thus, by applying the Delta method and Slutsky's theorem, it follows from (2.91) and (2.93) and (2.94) that $\sqrt{n}\left(\widehat{\boldsymbol{\beta}}-\boldsymbol{\beta}\right) \rightarrow_d N\left(\mathbf{0}, \boldsymbol{\Sigma}_\beta\right)$ with the asymptotic variance $\boldsymbol{\Sigma}_\beta$ given in (2.90).

There are two ways to estimate the asymptotic variance $\boldsymbol{\Sigma}_\beta$ in (2.90) depending on the choice of $V_i$. If $Var(y_i \mid \mathbf{x}_i)=\sigma^2 V_i$, (2.90) simplifies to

$$\Sigma_\beta^{MB}=B^{-1}E\left[D_i V_i^{-2} Var(y_i \mid \mathbf{x}_i) D_i^{\top}\right]B^{-1}=\sigma^2 B^{-1} \qquad (2.95)$$

If $\sigma^2$ is known, it follows from an application of LLN that a consistent estimate of $\boldsymbol{\Sigma}_\beta$ is given by:

$$\widehat{\boldsymbol{\Sigma}}_\beta^{MB}=\sigma^2\widehat{B}^{-1}=\sigma^2\left(\frac{1}{n}\sum_{i=1}^{n}\widehat{D}_i\widehat{V}_i^{-1}\widehat{D}_i^{\top}\right)^{-1} \qquad (2.96)$$

where $\widehat{D}_i$, $\widehat{V}_i$ and $\widehat{B}$ denote the estimates of $D_i$, $V_i$ and $B$ obtained by substituting $\widehat{\boldsymbol{\beta}}$ in place of $\boldsymbol{\beta}$. If $y_i$ is modeled by the exponential family of distributions, $\sigma^2=\phi$ and $\Sigma_\beta^{MB}$ in (2.95) is the asymptotic variance of the MLE of $\boldsymbol{\beta}$ and $\widehat{\boldsymbol{\Sigma}}_\beta^{MB}$ above is a consistent estimate of $\Sigma_\beta^{MB}$ (see exercise). For example, for the Poisson log-linear model in Example 6, $\sigma^2=1$ and $\widehat{\boldsymbol{\Sigma}}_\beta^{MB}$ is a consistent estimate of the asymptotic variance of the MLE of $\boldsymbol{\beta}$. If $\sigma^2$ is unknown, we can substitute the following estimate based on the Pearson residuals in place of $\sigma^2$ in (2.96):

$$\widehat{\sigma}^2=\frac{1}{n}Q_P^2, \quad Q_P^2=\sum_{i=1}^{n} v^{-\frac{1}{2}}\left(\widehat{\mu}_i\right)\left(y_i-\widehat{\mu}_i\right)^2, \quad \widehat{\mu}_i=g^{-1}\left(\mathbf{x}_i^{\top}\widehat{\boldsymbol{\beta}}\right) \qquad (2.97)$$

Note that when $\sigma^2$ is unknown, $\boldsymbol{\Sigma}_\beta$ in (2.90) is generally different from the asymptotic variance of the MLE of $\boldsymbol{\beta}$ under the parametric GLM even when $y_i$ is modeled by the exponential family of distributions. This is because the MLEs of $\boldsymbol{\beta}$ and other (nuisance) parameters of the distribution may be asymptotically correlated and as such the off-diagonal block of the Fisher information matrix that represents such asymptotic correlations may not be zero. However, if $y_i$ given $\mathbf{x}_i$ follows a normal distribution, $\boldsymbol{\Sigma}_\beta$ in (2.90) yields the asymptotic variance of the MLE of $\boldsymbol{\beta}$ (see Example 9 below).

If $Var(y_i \mid \mathbf{x}_i) \neq \sigma^2 V_i$ or if $y_i$ conditional on $\mathbf{x}_i$ does not follow the exponential family of distributions, the model-based $\widehat{\boldsymbol{\Sigma}}_\beta^{MB}$ no longer estimate $\Sigma_\beta$ in (2.90). Since

$$\Sigma_\beta = B^{-1} E\left[D_i V_i^{-2} Var(y_i \mid \mathbf{x}_i) D_i^\top\right] B^{-1} = B^{-1} E\left(D_i V_i^{-2} S_i^2 D_i^\top\right) B^{-1}$$

it follows from Slutsky's theorem that we can estimate $\boldsymbol{\Sigma}_\beta$ using the following *sandwich estimate*:

$$\widehat{\boldsymbol{\Sigma}}_\beta^{SW} = \widehat{B}_n^{-1}\left(\frac{1}{n}\sum_{i=1}^n \widehat{D}_i \widehat{V}_i^{-2} \widehat{S}_i^2 \widehat{D}_i^\top\right)\widehat{B}_n^{-1} \tag{2.98}$$

where $\widehat{D}_i$, $\widehat{V}_i$, $\widehat{S}_i$ and $\widehat{B}$ denote the estimates of $D_i$, $V_i$, $S_i$ and $B$ obtained by substituting $\widehat{\boldsymbol{\beta}}$ in place of $\boldsymbol{\beta}$.

**Example 9**. Consider the distribution-free linear regression model in Example 1 without the random component. Let $V_i = 1$. Then, we have

$$B = E\left(D_i V_i^{-1} D_i^\top\right) = E\left(\mathbf{x}_i \mathbf{x}_i^\top\right) \tag{2.99}$$

$$\Sigma_\beta = E^{-1}\left(\mathbf{x}_i \mathbf{x}_i^\top\right) E\left[\mathbf{x}_i\left(y_i - \mathbf{x}_i^\top \boldsymbol{\beta}\right)^2 \mathbf{x}_i^\top\right] E^{-1}\left(\mathbf{x}_i \mathbf{x}_i^\top\right)$$

Thus, a consistent estimate of the asymptotic variance of the estimating equation estimate is given by

$$\widehat{\boldsymbol{\Sigma}}_\beta^{SW} = \left(\sum_{i=1}^n \mathbf{x}_i \mathbf{x}_i^\top\right)^{-1}\left(n\sum_{i=1}^n \mathbf{x}_i\left(y_i - \mathbf{x}_i^\top \widehat{\boldsymbol{\beta}}\right)^2 \mathbf{x}_i^\top\right)\left(\sum_{i=1}^n \mathbf{x}_i \mathbf{x}_i^\top\right)^{-1}$$

If $Var(y_i \mid \mathbf{x}_i) = \sigma^2$, we can use the model-based asymptotic variance, $\Sigma_\beta^{MB} = \sigma^2 E^{-1}\left(\mathbf{x}_i \mathbf{x}_i^\top\right)$, for more efficient inference about $\boldsymbol{\beta}$. Further, if $y_i$ given $\mathbf{x}_i$ follows a normal distribution, $\Sigma_\beta^{MB}$ actually becomes the asymptotic variance of the MLE of $\boldsymbol{\beta}$. This follows from comparing $\Sigma_\beta^{MB}$ with the asymptotic variance of the MLE of $\boldsymbol{\beta}$ derived in Example 5 of Section 2.1.2.

**Example 10**. Consider the distribution-free log-linear model specified by the systematic component in Example 3. Let $V_i = \mu_i$. Then,

$$B = E\left(\mu_i \mathbf{x}_i \mu_i \mathbf{x}_i^\top\right) = E\left(\mu_i \mathbf{x}_i \mathbf{x}_i^\top\right) \tag{2.100}$$
$$\Sigma_\beta = E^{-1}\left(\mu_i \mathbf{x}_i \mathbf{x}_i^\top\right) E\left[\left(y_i - \mu_i\right)^2 \mathbf{x}_i \mathbf{x}_i^\top\right] E^{-1}\left(\mu_i \mathbf{x}_i \mathbf{x}_i^\top\right)$$

The estimating equation in (2.86) yields a consistent estimate of $\boldsymbol{\beta}$ with the asymptotic variance $\Sigma_\beta$ regardless of the distribution of $y_i$. As in Example 9, if $Var\left(y_i \mid \mathbf{x}_i\right) = \mu_i$ or if $y_i$ given $\mathbf{x}_i$ follows a Poisson, the model-based $\Sigma_\beta^{MB} = E^{-1}\left(\mu_i \mathbf{x}_i \mathbf{x}_i^\top\right)$ should be used to provide more efficient inference about $\boldsymbol{\beta}$.

As discussed in Section 2.1.4, overdispersion may occur when there is data clustering. The MLE of $\boldsymbol{\beta}$ from the Poisson log-linear model is still consistent. However, its asymptotic variance $\Sigma_\beta^{MB} = E^{-1}\left(\mu_i \mathbf{x}_i \mathbf{x}_i^\top\right)$ underestimates the sampling variability. We can detect overdispersion by examining some goodness of fit statistics. For example, under the Poisson model, $v^{-\frac{1}{2}}\left(\widehat{\mu}_i\right)\left(y_i - \widehat{\mu}_i\right) = \widehat{\mu}_i^{-\frac{1}{2}}\left(y_i - \widehat{\mu}_i\right) \approx 1$ so that the Pearson goodness of fit statistic $Q_P^2$ in (2.97) is close to $n$ for large $n$. Thus, if $\frac{Q_P^2}{n}$ is significantly larger than 1, overdispersion is indicated and $\widehat{\boldsymbol{\Sigma}}_\beta^{SW}$ should be used for inference. Alternatively, we may simply use $\widehat{\boldsymbol{\Sigma}}_\beta = \widehat{\sigma}^2\left(\frac{1}{n}\sum_{i=1}^n \mu_i \mathbf{x}_i \mathbf{x}_i^\top\right)^{-1}$ to correct for overdispersion. This simpler estimate is based on the particular form of $V_i$ and may be biased if $Var\left(y_i \mid \mathbf{x}_i\right) \neq \sigma^2 \mu_i$. For example, if $y_i$ given $\mathbf{x}_i$ follows a negative binomial, then $Var\left(y_i \mid \mathbf{x}_i\right) = \mu_i(1 + \alpha\mu_i)$ for some known constant $\alpha$. In this case, only the sandwich estimate is consistent and valid for inference.

When $y_i$ is modeled by a distribution from the exponential family, the estimating equation can yield the MLE with the right choice of $V_i$. As shown by the next example, this is generally not true when the distribution does not belong to the exponential family.

**Example 11**. Consider the linear regression model with $y_i$ conditional on $\mathbf{x}_i$ given following the $t$ distribution in Example 2. The log-likelihood function is given by (see exercise):

$$l_n = n \log \frac{\Gamma\left(\frac{v+1}{2}\right)}{\Gamma\left(\frac{v}{2}\right)\left(\sigma^2 \pi v\right)^{\frac{1}{2}}} - \frac{v+1}{2} \sum_{i=1}^n \log\left[1 + \frac{\sigma^2}{v}\left(y_i - \mu_i\right)^2\right] \tag{2.101}$$

Suppose that $v$ and $\sigma^2$ are both known. Then, the MLE of $\boldsymbol{\beta}$ is defined by the score equation (see exercise):

$$\mathbf{u}_n = \sum_{i=1}^{n} \mathbf{x}_i \left[1 + \frac{\sigma^2}{v} (y_i - \mu_i)^2\right]^{-1} (y_i - \mu_i) = \mathbf{0} \qquad (2.102)$$

Although similar, the above is not an estimating equation since $V_i = 1 + \frac{\sigma^2}{v} (y_i - \mu_i)^2$ depends on the response $y_i$.

Inference for general linear hypotheses proceeds in the same fashion as for parametric models except that the likelihood ratio test no longer applies as it relies on parametric distribution assumptions. For example, for testing the null of the form, $H_0 : K\boldsymbol{\beta} = \mathbf{b}$, we can use the Wald statistic,

$$Q_w^2 = n \left(K\widehat{\boldsymbol{\beta}} - \mathbf{b}\right)^{\top} \left(K\widehat{\Sigma}_{\beta} K^{\top}\right)^{-1} \left(K\widehat{\boldsymbol{\beta}} - \mathbf{b}\right) \sim \chi_l^2$$

where $\widehat{\boldsymbol{\beta}}$ is the estimating equation estimate of $\boldsymbol{\beta}$ and $\widehat{\Sigma}_{\beta}$ some consistent estimate of the asymptotic variance $\Sigma_{\beta}$ of $\widehat{\boldsymbol{\beta}}$.

For parametric models, the MLE is optimal in that it is asymptotically most efficient. Similar optimal properties exist for estimating equation estimates from distribution-free GLMs. However, as the distribution of response is totally unspecified, that is, no analytic form is posited as in the case of parametric models, the vector of parameters for distribution-free models is of infinite dimension. As a result, the Fisher information cannot be computed directly to provide the optimal lower bound for the variance of such parameter estimates. Asymptotic efficiency of estimates for a distribution-free model is defined as the infimum of the information over all parametric submodels embedded in such a model (e.g., Bickel et al.1993; Tsiatis,2006).

For estimates defined by the estimating equation, their efficiency depends on the choice of $V$. If specified correctly, that is if $V = Var(y \mid \mathbf{x})$, the estimating equation estimate is asymptotically efficient. Thereby, in theory, there exists an asymptotically efficient estimate in the classes of estimating equation estimates. In most applications, however, this correct $V$ is unknown. Various approaches have been considered for the selection of $V$ which we will not delve into in this book. We emphasis that regardless of the choice of $V$, the estimating equation estimates are always consistent. This robustness property is what we value most.

## 2.3 Exercises

### Section 2.1

1. Verify the two basic properties regarding differentiation of vector-valued function.

2. Show that $\widehat{\beta}_1$ and $\widehat{\sigma}^2$ defined in (2.15) are stochastically independent. As a special case, it follows that $\widehat{\beta}_1$ and $\widehat{\sigma}^2$ defined in (2.13) are independent.

3. Let $s_n^2$ be defined in (2.13). Show that $\frac{(n-1)s_n^2}{\sigma^2} \sim \chi_{n-1}^2$.

4. Let $\mathbf{y}$ be a $p \times 1$ random vector and $\lambda$ a random variable. Suppose that $\mathbf{y}$ and $\lambda$ are independent and satisfy the following:

$$\mathbf{y} \mid \lambda \sim N\left(\boldsymbol{\mu}, \lambda\Sigma\right), \quad \frac{v}{\lambda} \sim \chi_v^2, \quad v > 0, \quad \Sigma > 0$$

where $\Sigma > 0$ means that the symmetric $p \times p$ matrix $\Sigma$ is positive definite.

(a) Show that the unconditional distribution of $\mathbf{y}$ has a multivariate $t$ distribution, $t\left(\boldsymbol{\mu}, \Sigma, v\right)$, with the density function given by (2.17).

(b) Show that any subvector of $\mathbf{y}$ also has a multivariate $t$ distribution.

(c) Show that for any vector $\mathbf{a}$ of known elements any linear combination $\mathbf{a}^\top \mathbf{y}$ follows a univariate $t$ distribution, $t\left(\mathbf{a}^\top \boldsymbol{\mu}, \mathbf{a}^\top \Sigma \mathbf{a}, v\right)$.

5. Consider the MLE $\widehat{\boldsymbol{\beta}}$ and $\widehat{\sigma}^2$ in the linear regression model given in (2.15).

(a) Show that $\widehat{\boldsymbol{\beta}}$ and $\widehat{\sigma}^2$ are independent.

(b) Show that $\frac{(n-p)MSE}{\sigma^2} \sim \chi_{n-p}^2$ with MSE defined in (2.16).

(c) Show that $\widehat{\boldsymbol{\beta}}$ follows a multivariate $t$ distribution given in (2.17).

6. For fixed $\mathbf{x}_i$, $\widehat{\sigma}^2$ is a $\chi_{n-p}^2$ variable as shown in (2.16), that is,

$$\sqrt{n}\left(\widehat{\sigma}^2 - \sigma^2\right) = \sqrt{n}\sigma^2 \left(\frac{\chi_{n-p}^2}{n} - 1\right)$$

Show that $\sqrt{n}\left(\widehat{\sigma}^2 - \sigma^2\right) \rightarrow_d N\left(0, 2\sigma^4\right)$.

7. Show that $\widehat{\theta}_k$ $(k \geq 1)$ defined in (2.22) in the proof of Theorem 1 is a sequence of random variables.

8. Use (2.20) and (2.21) to show that $I_n^e(\boldsymbol{\theta}) = Var\left(\mathbf{u}_n(\boldsymbol{\theta})\right)$, where $\mathbf{u}_n(\boldsymbol{\theta}) = \frac{\partial}{\partial \boldsymbol{\theta}} l_n(\boldsymbol{\theta})$ is the score statistic vector.

9. Consider a parametric model $f(y; \boldsymbol{\theta})$. Let $\boldsymbol{\psi}(\boldsymbol{\theta})$ be some smooth function of $\boldsymbol{\theta}$ and $\mathbf{T}_n$ an unbiased estimate of $\boldsymbol{\psi}(\boldsymbol{\theta})$. Let $\widehat{\boldsymbol{\theta}}$ be the MLE of $\boldsymbol{\theta}$, $\Sigma_\theta$ the asymptotic variance of $\widehat{\boldsymbol{\theta}}$ and $\Sigma_\psi$ the asymptotic variance of $\boldsymbol{\psi}\left(\widehat{\boldsymbol{\theta}}\right)$ given in (2.33). Show $Var\left(\mathbf{T}_n\right) \geq \frac{1}{n}\Sigma_\psi$.

10. Verify (2.46) in the proof of Theorem 2.

11. For the exponential family of distributions defined in (2.54), show

(a) $E(y) = \frac{d}{d\theta} b(\theta)$.

(b) $Var(y) = a(\phi) \frac{d^2}{d\theta^2} b(\theta)$.

12. Show that for the Poisson log-linear model defined in (2.58), the conditional mean $E(y_i \mid \mathbf{x}_i)$ and variance $Var(y_i \mid \mathbf{x}_i)$ are identical and given by (2.60).

13. Let $y$ be a continuous response following the negative binomial distribution defined in (2.62). Verify that the mean and variance of $y$ are given by (2.63).

14. Show that the probability distribution function $f_{NB}$ for the negative binomial model defined in (2.62) converges to the Poisson distribution function $f_P$ in (2.59) when $\alpha$ decreases to 0.

15. Let $y$ be sampled from a mixture of $f_k(y \mid \boldsymbol{\theta}_k)$, where $f_k(y \mid \boldsymbol{\theta}_k)$ denotes two PDFs ($k = 1, 2$). Let $z = 1$ if $y$ is sampled from $f_1$ and $z = 0$ if otherwise. Let $p = E(z) > 0$.

(a) Show that the PDF of $y$ is given by

$$f(y \mid p, \boldsymbol{\theta}_1, \boldsymbol{\theta}_2) = pf_1(y \mid \boldsymbol{\theta}_1) + (1-p) f_2(y \mid \boldsymbol{\theta}_2)$$

(b) If $f_k(y \mid \boldsymbol{\theta}_k) = N(\mu_k, \sigma_k^2)$, find the mean and variance of $y$.

(c) In (b), find the maximum likelihood estimates of $p$, $\mu_k$ and $\sigma_k^2$ $(k = 1, 2)$.

16. Verify (2.67).

**Section 2.2**

1. Let $y_i$ and $\mathbf{x}_i$ be some continuous response and vector of independent variables of interest $(1 \le i \le n)$. Consider a GLM for relating $y_i$ to $\mathbf{x}_i$ with the distribution of $y_i$ from the exponential family defined in (2.54). Verify (2.84) and 2.85).

2. Show that at convergence the estimates of $\boldsymbol{\beta}$ obtained from (2.87) and (2.88) are identical.

3. Verify (2.89) and (2.92).

4. Show that the estimating equation estimate $\widehat{\boldsymbol{\beta}}$ is consistent regardless of the choice of $V_i$.

5. Verify (2.94) by applying the properties regarding differentiation of vector-valued function in Section 2.1.2.

6. Within the context of Example 1, suppose that $\phi$ is known.

(a) Find the Fisher information matrix $I_n^e = E\left(-\frac{\partial^2}{\partial \boldsymbol{\beta} \partial \boldsymbol{\beta}^\top} l_n\right)$, where $l_n$ is the log-likelihood function in (2.84).

(b) Show that the asymptotic variance $\left(\frac{1}{n} I_n^e\right)^{-1}$ of the MLE of $\boldsymbol{\beta}$ is the same as the model-based asymptotic variance of the estimating equation estimate given in (2.95).

7. Show that if $\alpha$ is known, the negative binomial distribution of $y_i$ given $\mathbf{x}_i$ in Example 2 is a member of the exponential family of distributions.

8. Use the density function for the $t$ distribution in (2.18) to verify (2.101) and (2.102).

# Chapter 3

# Univariate U-Statistics

In Chapter 2, we discussed various regression models for modeling the relationship between some response of interest and a function of other independent variables. We started with the normal based linear regression and then generalized this class of classic models to address different response types and distribution assumptions, culminating in the development of the distribution-free, or semiparametric, generalized linear models (GLM). This general class of regression models not only applies to all popular types of response but also imposes no parametric distribution assumption on the response, yielding a powerful and flexible platform for regression analyses of cross-sectional data arising in practical studies.

In the distribution-free GLM, a parametric structure is assumed to link the mean of a designated dependent or response variable to the linear predictor (a combination of other independent variables) in order to characterize the change in the dependent variable in response to a change in the linear predictor. This systematic component is at the core of GLM as it provides the essential bridge to connect the dependent and the set of independent variables. Thus, the distribution-free GLM places minimal assumptions on modeling the relationship between the mean of the response and the independent variables. Although such minimal specifications yield more robust inference when applied to the wide array of data distributions arising in real study data, important features of the data distribution of the response may get lost. For example, as discussed in Chapter 2, the Poisson and negative binomial become indistinguishable when implemented under distribution-free GLM, resulting in an inability to detect data clustering and a lack of efficiency. Since the difference between the Poisson and negative binomial is reflected in the variance, it is necessary to include models the variance to obtain the required information in order to distinguish the two models to address data clustering and efficiency.

In addition to the negative binomial model, many other statistical models are not amenable to treatment by GLM. Two such examples, the Pearson correlation and the Mann–Whitney–Wilcoxon test, are discussed in Chapter 1. Because the Pearson correlation models the concurrent changes between two variables rather than the mean of one variable as a function of the other, GLM is not applicable. The difference between these two modeling approaches is also amply reflected in their analytic forms; the Pearson correlation involves second order moments, while the distribution-free GLM models only the mean response, or first order moment. The form of the rank-based Mann–Whitney–Wilcoxon test deviates even more from that of GLM as it models the difference between two data distributions without using any of the moments of the outcome variables.

In this chapter, we introduce univariate U-statistics and discuss their applications to a wide array of statistical problems some of which address the limitations of the distribution-free GLM including modeling higher order moments, while others require models that cannot be subsumed under the rubric of GLM or treated by the conventional asymptotic methods discussed in Chapter 1. We limit our attention to cross-sectional data without data clustering. In Chapters 5 and 6, we take up multivariate U-statistics and discuss their applications to repeated, correlated responses within a longitudinal or clustered data setting along with addressing missing data within such a setting.

## 3.1 U-Statistics and Associated Models

Since Hoeffding's (1948) foundational work, U-statistics have been widely used in both theoretical and applied statistical research. What is a U-Statistic and how is it different from the statistics that we have seen in Chapters 1 and 2? From a technical point of view, U-statistics are different both in their appearance and in the methods used to study their asymptotic behaviors. U-statistics are not in the usual form of a sum of i.i.d. random variables either expressed exactly or asymptotically through a stochastic Taylor series expansion as are all the statistics and model estimates discussed in Chapters 1 and 2.

Take for example an i.i.d. sample $y_i$ with mean $\mu$ and variance $\sigma^2$. The sample mean $\overline{y}_n$ and variance $s_n^2$ given below are unbiased and consistent estimates of $\mu$ and variance $\sigma^2$:

$$\overline{y}_n = \frac{1}{n}\sum_{i=1}^{n} y_i, \quad s_n^2 = \frac{1}{n-1}\sum_{i=1}^{n}(y_i - \overline{y}_n)^2 \tag{3.1}$$

The sample mean is a sum of i.i.d. random variables $y_i$. Although the sample variance is not in such a form, it can be expressed as an i.i.d. sum by applying a Taylor series expansion around $\mu$ (see Chapter 1):

$$s_n^2 = \frac{1}{n-1}\sum_{i=1}^{n}(y_i - \mu)^2 + o_p(1) \tag{3.2}$$

For statistics and/or estimates that are in the form of an i.i.d. sum such as $\overline{y}_n$ and $s_n^2$, we can find their asymptotic distributions by applying the various methods and techniques discussed in Chapter 1. As will be seen shortly, the distinctive appearance of U-statistics makes it impossible to express them as an i.i.d. sum in either closed form or asymptotic approximation using a Taylor series expansion.

From an applications standpoint, statistics and estimates that can be expressed as an i.i.d. sum either exactly or through a Taylor series expansion typically arise from parameters of interest defined by a single-subject response. For example, the mean $\mu = E(y_i)$ and variance $\sigma^2 = E(y_i^2) - E^2(y_i)$, all defined by a single-subject $y_i$. More generally, the class of distribution-free GLM is also defined in such a fashion since the random component in the defining model, $g(E(y_i \mid \mathbf{x}_i)) = \mathbf{x}_i^\top \boldsymbol{\beta}$, involves only a single-subject response $y_i$ as discussed in Chapter 2. Many parameters and statistics of interest, such as the Mann–Whitney–Wilcoxon test, cannot be defined by such single-response based statistical models. Further, even for those that can be modeled by a single-subject response, the study of the asymptotic behavior of their associated statistics and estimates can be greatly facilitated when formulated within the U-statistics setting.

In this Section, we first introduce the one-sample U-statistic and then generalize it to two and more samples. We then discuss some classic properties of U-statistics. We develop inference theories for univariate U-statistics in Section 3.2.

### 3.1.1 One Sample U-Statistics

Consider an i.i.d. sample of $n$ subjects. Let $y_i$ be some response of interest with mean $\mu$ and variance $\sigma^2$ $(1 \leq i \leq n)$. In most applications, we are interested in estimating the population mean $\mu$. As discussed in Chapter 2, we can model this parameter of interest using either a parametric or non-parametric GLM. For example, we may assume that $y_i$ follows a normal distribution, in which case we obtain the maximum likelihood estimate (MLE) of $\mu$ and can use this estimate along with its sampling distribution for inference about $\mu$. If we are unwilling to posit any parametric model for the distribution of $y_i$, we can still make inference about $\mu$ by modeling $y_i$ using the distribution-free GLM.

In many other applications, we may be interested in inference about the variance $\sigma^2$. For example, when comparing two effective interventions for treating some medical and/or psychological condition of interest, we may be interested in examining difference in variances if both treatments yield similar mean responses. If one treatment has significantly smaller variability in subjects' responses, it is likely to be the preferred treatment choice for the disease, as it is more likely to give patients the expected treatment effect. Such considerations may be especially important in effectiveness research, where the potency of an effective treatment is often diluted when given to patients with comorbid health conditions. A treatment with smaller variability across patients with different comorbid conditions compromises less than its competing alternative in terms of treatment efficacy when prescribed to patients with diverse comorbid conditions.

As in the case of inference for $\mu$, we can use the MLE of $\sigma^2$ and its sampling distribution to make inference about $\sigma^2$ under a parametric GLM. However, if we are unwilling to use a parametric model for the data distribution of $y_i$, we will not be able to model this variance parameter using a distribution-free GLM. Fortunately, in this simple problem, we know that we can estimate $\sigma^2$ using the sample variance $s_n^2$. Further, by using the asymptotic distribution of this estimate that we developed in Chapter 1, we can also make inference about $\sigma^2$.

This example shows the limitation of GLM and points to the need to develop a new approach for systematically modeling higher order moments such as the variance of $y_i$ in this example, as it may be difficult to find estimates and their asymptotic distributions when dealing with other more complex problems. The power of U-statistics lies exactly in addressing this type of problem.

Let us consider again the estimate $s_n^2$. As shown below, we can rewrite it in a *U-statistic,* or *symmetric*, form involving pairs of responses:

$$s_n^2 = \frac{1}{n(n-1)}\left((n-1)\sum_{i=1}^{n} y_i^2 - \sum_{i\neq j} y_i y_j\right) \tag{3.3}$$
$$= \frac{1}{n(n-1)}\sum_{i\neq j}\frac{1}{2}(y_i - y_j)^2 = \frac{2}{n(n-1)}\sum_{(i,j)\in C_2^n}\frac{1}{2}(y_i - y_j)^2$$

where $C_2^n = \{(i,j); 1 \leq i < j \leq n\}$ denotes the set of all distinct combinations of pairs $(i,j)$ from the integer set $\{1, 2, \ldots, n\}$. Note that two pairs $(i,j)$ and $(k,l)$ are not distinct if one can be obtained from the other through a permutation. For example, $(1,2)$ and $(2,1)$ are not distinct, but $(1,2)$ and $(1,3)$ are two distinct pairs. By comparing the above expression to the one in (3.1), it is seen that the U-statistic expression employs pairs of responses to express $s_n^2$ as a function of between-subject variability rather than deviations of individual response from the mean $\overline{y}_n$ of the distribution $y_i$. As will be seen shortly, this shift of modelling paradigm is especially critical for studying statistics defined by responses from multiple subjects.

Let $h(y_i, y_j) = \frac{1}{2}(y_i - y_j)^2$. This *kernel* function is symmetric with respect to the pair of responses $y_i$ and $y_j$, that is, $h(y_i, y_j) = h(y_j, y_i)$. The alternative U-statistic expression in (3.3) shows that $s_n^2$ is the sample mean of this kernel function over all pairs of subjects in the sample:

$$s_n^2 = \binom{n}{2}^{-1}\sum_{(i,j)\in C_2^n} h(y_i, y_j) = \frac{1}{2}(y_i - y_j)^2 \tag{3.4}$$

Further,

$$E(h(y_i, y_j)) = \frac{1}{2}\left[E\left(y_i^2\right) - 2E(y_i)E(y_j) + E\left(y_j^2\right)\right] \tag{3.5}$$
$$= \sigma^2$$

Thus, it follows immediately from (3.4) and (3.5) that the sample variance $s_n^2$ is an unbiased estimate of $\sigma^2$. In comparison, it is more involved to show unbiasedness of $s_n^2$ using the traditional expression in (3.1).

Not only can we express the variance, or centered second order moment, estimate $s_n^2$, but also other order moments, such as the sample mean $\overline{x}_n$ in a U-statistic form similar to (3.4) involving a symmetric kernel $h$.

**Example 1 (*U-statistic for sample mean*).** Let $y_i$ be an i.i.d. sample with mean $\theta$ and variance $\sigma^2$ $(1 \leq i \leq n)$. Now consider estimating the $k$th order moment $\theta = E\left(y^k\right)$ (assuming it exists). Let $h(y) = y^k$. In this case, the *one-argument* kernel function $h(y)$ is, of course, symmetric. Further,

$$\widehat{\theta} = \binom{n}{1}^{-1} \sum_{i=1}^{n} h(y_i) = \frac{1}{n} \sum_{i=1}^{n} y_i^k$$

As $E(h(y)) = \theta$, it immediately follows that $\widehat{\theta}$ is an unbiased estimate of $\theta$. The above is the usual estimate for the $k$th-order moment.

**Example 2 (*U-statistic for proportion*).** In Example 1, let $z_i = I_{\{y_i > 0\}}$ $(1 \leq i \leq n)$. Let $h(z) = z_i$. Then, $\theta = E(h(z)) = \Pr(y_i > 0)$ and

$$\widehat{\theta} = \binom{n}{1}^{-1} \sum_{i=1}^{n} h(y_i) = \frac{1}{n} \sum_{i=1}^{n} I_{\{y_i > 0\}}$$

where $I_{\{\cdot\}}$ denotes a set indicator. Thus, the U-statistic $\widehat{\theta}$ is the usual proportion of $y_i > 0$.

The next example shows that it is also easy to construct U-statistics to estimate powers of the mean $\mu^k$ $(k = 1, 2, \ldots)$.

**Example 3 (*U-statistic for powers of mean*).** In Example 1, consider estimating the square of the mean $\mu^2$. Let $h(y_1, y_2) = y_1 y_2$. Then,

$$E[h(y_1, y_2)] = E(y_1 y_2) = E(y_1) E(y_2) = \mu^2$$

the parameter of interest. The U-statistic based on this kernel is:

$$U_n = \binom{n}{2}^{-1} \sum_{(i,j) \in C_2^n} h(y_i, y_j) = \binom{n}{2}^{-1} \sum_{(i,j) \in C_2^n} y_i y_j$$

Again, $U_n$ is an unbiased estimate of $\mu^2$.

In Chapter 1, we indicated that we could use the theory of U-statistics to facilitate the study of the asymptotic behavior of the Pearson correlation estimate. In the example below, let us first express a related quantity in the form of a U-statistic.

**Example 4 (*U-statistic for covariance*).** Let $\mathbf{z}_i = (x_i, y_i)^\top$ be i.i.d. pairs of continuous random variables $(1 \leq i \leq n)$. Consider estimating the covariance between $x_i$ and $y_i$, $\theta = Cov(x_i, y_i)$. Let

$$h(\mathbf{z}_i, \mathbf{z}_j) = \frac{1}{2}(x_i - x_j)(y_i - y_j)$$

Then,

$$\begin{aligned} E[h(\mathbf{z}_i, \mathbf{z}_j)] &= \frac{1}{2}[E(x_i y_i) - E(x_i) E(y_i)] + \frac{1}{2}[E(x_j y_j) - E(x_j) E(y_j)] \\ &= \frac{1}{2} Cov(x_i, y_i) + \frac{1}{2} Cov(x_j, y_j) = \theta \end{aligned}$$

Thus, the U-statistic below:

$$U_n = \binom{n}{2}^{-1} \sum_{(i,j)\in C_2^n} h(\mathbf{z}_i, \mathbf{z}_j) \tag{3.6}$$

is an unbiased estimate of $\theta = Cov(x, y)$.

Note that as the study of the Pearson correlation estimate requires multivariate U-statistics, we will defer the discussion of this important statistic to Chapter 5 after we introduce multivariate U-statistics.

In all the examples above, the U-statistics based estimates for these popular statistical parameters of interest can also be expressed as traditional sample averages over single subjects or functions of such sample averages. For example, $U_n$ in (3.6) can be reexpressed in the familiar form as (see exercise):

$$\begin{aligned} U_n &= \binom{n}{2}^{-1} \sum_{(i,j)\in C_2^n} \frac{1}{2}(x_i - x_j)(y_i - y_j) \\ &= \frac{1}{n-1}\sum_{i=1}^{n}(x_i - \overline{x}_n)(y_i - \overline{y}_n) \end{aligned} \tag{3.7}$$

However, as will be seen shortly, many statistics and estimates of interest can only be expressed in the U-statistic form by using multiple subjects' responses. However, the power of U-statistics not only lies in studying such *intrinsic* or *endogenous* multisubject based statistics and estimates, but also in facilitating asymptotic inference for parameters amenable to the traditional treatment discussed in Chapters 1 and 2. Before proceeding with more examples, let us give a formal definition of U-statistics.

Consider an i.i.d. sample of $p \times 1$ random vectors $\mathbf{y}_i$ $(1 \le i \le n)$. Let $h(\mathbf{y}_1, \ldots, \mathbf{y}_m)$ be a symmetric function with $m$ arguments or input vectors, that is, $h(\mathbf{y}_1, \ldots, \mathbf{y}_m) = h(\mathbf{y}_{i_1}, \ldots, \mathbf{y}_{i_m})$ for any permutation $(i_1, \ldots, i_m)$ of $(1, 2, \ldots, m)$. We define a univariate one-sample, $m$-argument $U$-statistic as follows:

$$U_n = \binom{n}{m}^{-1} \sum_{(i_1,\ldots,i_m)\in C_m^n} h(\mathbf{y}_{i_1}, \ldots, \mathbf{y}_{i_m}) \tag{3.8}$$

where $C_m^n = \{(i_1, \ldots, i_m); 1 \le i_1 < \ldots < i_m \le n\}$ denotes the set of all distinct combinations of $m$ indices $(i_1, \ldots, i_m)$ from the integer set $\{1, 2, \ldots, n\}$. Under this formal definition, all U-statistic based estimates in the above examples have two arguments, except for the ones in Example 1 and 2 where the estimates are expressed as one-argument U-statistics. Let $\theta = E[h(\mathbf{y}_1, \ldots, \mathbf{y}_m)]$, then,

$$\begin{aligned} E(U_n) &= \binom{n}{m}^{-1} \sum_{(i_1,\ldots,i_m)\in C_m^n} E[h(\mathbf{y}_{i_1}, \ldots, \mathbf{y}_{i_m})] \qquad (3.9) \\ &= \binom{n}{m}^{-1} \sum_{(i_1,\ldots,i_m)\in C_m^n} E[h(\mathbf{y}_1, \ldots, \mathbf{y}_m)] \\ &= \theta \end{aligned}$$

Thus, $U_n$ is an unbiased estimate of $\theta$.

Note that *univariate* refers to the dimension of $U_n$. Thus, the i.i.d. sample $\mathbf{y}_i$ in the definition of a univariate U-statistic can be multivariate such as in Example 4. In each of the examples above, the kernel function $h$ is symmetric. For cases in which a kernel is not symmetric, it is easy to construct a symmetric version.

**Example 5**. Consider again the problem of estimating $\sigma^2$ in Example 1. Let

$$h(y_i, y_j) = y_i^2 - y_i y_j, \quad 1 \le i, j \le n, \ i \ne j$$

Then,

$$E[h(y_i, y_j)] = E(y_i^2) - E(y_i) E(y_j) = \sigma^2$$

However, $h(y_i, y_j)$ is not symmetric, that is, $h(y_i, y_j) \ne h(y_j, y_i)$. We can readily construct a symmetric version of $h(x_i, x_j)$ by

$$\begin{aligned} g(x_i, x_j) &= \frac{1}{2}[h(x_i, x_j) + h(x_j, x_i)] = \frac{1}{2}(x_i^2 + x_j^2 - 2x_i x_j) \\ &= \frac{1}{2}(x_i - x_j)^2 \end{aligned}$$

which gives rise to the same kernel as in (3.5). This example shows that kernels for a $U$-statistic are not unique.

More generally, if $h(\mathbf{y}_1, \ldots, \mathbf{y}_m)$ is not symmetric, we can readily construct a symmetric version by

$$g(\mathbf{y}_1, \ldots, \mathbf{y}_m) = \frac{1}{m!} \sum_{(i_1,\ldots,i_m)\in P^m} h(\mathbf{y}_{i_1}, \ldots, \mathbf{y}_{i_m}) \qquad (3.10)$$

where $P^m$ denotes the set of all permutations of the integer set $\{1, , \ldots, m\}$.

One major limitation of the product–moment correlation for two continuous outcomes is its dependence on a linear relationship to serve as a valid measure of association between the two outcomes. For example, if two outcomes follow a curved relationship, the product–moment correlation may be low even though the two outcomes are closely related, giving rise to incorrect indication of association between the two variables. One popular alternative to address this limitation is Kendall's $\tau$.

**Example 6 (*Kendall's* $\tau$).** Let $\mathbf{z}_i = (x_i, y_i)^\top$ be i.i.d. continuous bivariate random variables $(1 \leq i \leq n)$ as defined in Example 4. Unlike the Pearson estimate of the product–moment correlation, which relies on moments to capture the relationship between $x$ and $y$, Kendall's $\tau$ defines association between $x_i$ and $y_i$ by considering pairs of $\mathbf{z}_i$ based on the notion of concordance and discordance.

More specifically, two pairs $\mathbf{z}_i = (x_i, y_i)^\top$ and $\mathbf{z}_j = (x_j, y_j)^\top$ are concordant if

$$x_i < x_j \quad \text{and} \quad y_i < y_j, \quad \text{or} \quad x_i > x_j \quad \text{and} \quad y_i > y_j$$

and discordant if

$$x_i < x_j \quad \text{and} \quad y_i > y_j, \quad \text{or} \quad x_i > x_j \quad \text{and} \quad y_i < y_j$$

If a pair of subjects share the same value in either $x$ or $y$ or both, that is, the two subjects have a tie in either $x$ $(x_i = x_j)$ or $y$ $(y_i = y_j)$ or both, then it is neither a concordant nor a discordant pair. For continuous outcomes, the probability of having such tied observations is 0, though in practice it is likely to have a few tied observations because of limited precision in measurement. If there are more concordant pairs in the sample, $x$ and $y$ are positively associated, that is, larger $x$ leads to larger $y$ and vice versa. Likewise, more discordant pairs imply negative association. A similar number of concordant and discordant pairs in the sample indicates weak association between the two variables.

We can express concordant and discordant pairs in a compact form as:

$$\begin{aligned} \text{Concordant pairs: } & (x_i - x_j)(y_i - y_j) > 0 \\ \text{Discordant pairs: } & (x_i - x_j)(y_i - y_j) < 0 \end{aligned}$$

Kendall's $\tau$ is defined based on the probability of concordant pairs:

$$\text{Kendall's } \tau : p_c = \Pr\left[(x_i - x_j)(y_i - y_j) > 0\right] \tag{3.11}$$

A value of $p_c$ closer to 0 (1) indicates a higher degree of negative (positive) association. In practice, we often normalize $p_c$ so that it ranges from $-1$ to 1 to have a similar interpretation as the product–moment correlation:

$$\text{Normalized Kendall's } \tau : \tau = 2p_c - 1 \tag{3.12}$$

Now, define a kernel for the normalized Kendall's $\tau$ in (3.12) as follows:

$$h(\mathbf{z}_i, \mathbf{z}_j) = 2I_{\{(x_i - x_j)(y_i - y_j) > 0\}} - 1$$

Clearly, $h(\mathbf{z}_i, \mathbf{z}_j)$ is symmetric and $E[h(\mathbf{z}_i, \mathbf{z}_j)] = \tau$. Thus, the following U-statistic based on this kernel is an unbiased estimate of $\tau$:

$$\begin{aligned} U_n &= \binom{n}{2}^{-1} \sum_{(i,j) \in C_2^n} \left(2I_{\{(x_i - x_j)(y_i - y_j) > 0\}} - 1\right) \qquad (3.13) \\ &= \binom{n}{2}^{-1} (\text{number of concordant pairs}) - 1 \end{aligned}$$

Unlike all the previous examples, the U-statistic in (3.13) is intrinsically or endogenously multisubject based, as it is impossible to express it as a sum of i.i.d. random variables. Another classic statistic that is not subjected to the treatment of traditional asymptotic theory is the one-sample Wilcoxon signed rank statistic. This statistic also employs ranks, but unlike the two-sample Mann–Whitney–Wilcoxon rank sum test discussed in Section 1.1 of Chapter 1, it tests whether an i.i.d. sample from a population of interest is distributed symmetrically about some point $\theta$.

More specifically, let $y_i$ be an i.i.d. sample from a continuous distribution $(1 \leq i \leq n)$. Let $R_i$ be the rank score of $y_i$ based on its absolute value $|y_i|$ (see Section 1.1 in Chapter 1 for details on creation of rank scores). We are interested in testing whether the point of symmetry $\mu$ is equal to some hypothesized known value $\mu_0$, that is, $H_0 : \mu = \mu_0$. Since in most applications, $\mu_0 = 0$ and the consideration for the general situation with $\mu_0 \neq 0$ is similar (by a transformation $y_i - \mu_0$), we focus on the special case $\mu_0 = 0$ without the loss of generality.

**Example 7** (***One-sample Wilcoxon signed rank test***). The Wilcoxon signed rank statistic, $W_n^+$, for testing $H_0$ is defined by

$$\text{Wilcoxon signed rank test} : W_n^+ = \sum_{i=1}^n I_{\{y_i > 0\}} R_i \qquad (3.14)$$

This statistic ranges from 0 to $\frac{n(n+1)}{2}$ (see exercise). Under $H_0$, about one-half of the $y_i$'s are negative and their rankings are comparable to those of positive $y_i$'s because of the symmetry. It follows that $W_n^+$ has mean $\frac{n(n+1)}{4}$, half of the range $\frac{n(n+1)}{2}$ (see exercise). Further, we show in Section 3.1.3 that $W_n^+$ can be expressed as:

$$W_n^+ = \sum_{i=1}^n I_{\{y_i > 0\}} + \sum_{1 \leq i < j \leq n} I_{\{y_i + y_j > 0\}} \qquad (3.15)$$

It then follows from (3.15) that (see exercise)

$$\begin{aligned}\binom{n}{2}^{-1} W_n^+ &= \binom{n}{2}^{-1} \sum_{i=1}^{n} I_{\{y_i>0\}} + \binom{n}{2}^{-1} \sum_{1\leq i<j\leq n} I_{\{y_i+y_j>0\}} \qquad (3.16)\\ &= \frac{2}{n-1}\frac{1}{n}\sum_{i=1}^{n} h_1(y_i) + \binom{n}{2}^{-1} \sum_{1\leq i<j\leq n} h_2(y_i, y_j)\\ &= \frac{2}{n-1} U_{n1} + U_{n2}\end{aligned}$$

In the above, $U_{n1}$ is a U-statistic in Example 1 and $U_{n2}$ is also a U-statistic defined by the symmetric kernel $h_2(y_1, y_2) = I_{\{y_i+y_j>0\}}$. Later in Section 3.2, we will show that $\binom{n}{2}^{-1} W_n^+$ and $U_{n2}$ have the same asymptotic distribution and thus inference for $H_0$ can be carried out equivalently by using only the asymptotic distribution of $U_{n2}$ for large samples.

The U-statistic $U_{n2}$ is an unbiased estimate of $\theta = E(h_2(y_i, y_j)) = \Pr(y_i + y_j > 0)$. By taking expectation on both sides of the equality in (3.16) under $H_0 : \mu_0 = 0$ , we have (see exercise)

$$\binom{n}{2}^{-1} E(W_n^+) = \binom{n}{2}^{-1} \frac{n(n+1)}{4} = \frac{1}{2}, \qquad (3.17)$$

$$\begin{aligned}\frac{2}{n-1} E(U_{n1}) + E(U_{n2}) &= \frac{2}{n-1} \Pr(y_i > 0) + \Pr(y_i + y_j > 0)\\ &= \frac{1}{n-1} + \theta\end{aligned}$$

It follows that

$$\theta = \frac{1}{2} + \frac{1}{n-1} \to \frac{1}{2}, \quad \text{as } n \to \infty$$

Thus, we can express the null $H_0 : \mu_0 = 0$ in terms of $\theta$ as:

$$H_0 : \theta = \Pr(y_i + y_j \leq 0) = \frac{1}{2}$$

All the examples above involve continuous outcomes. Many important statistics for discrete or categorical data can also be expressed in U-statistics form.

Consider two categorical ordinal outcomes $u$ and $v$. Suppose $u$ has $K$ and $v$ has $M$ levels indexed by $k$ and $m$, respectively. The notion of concordant and discordant pairs has also been applied to categorical outcomes to create indices of association between two categorical ordinal outcomes such as Goodman and Kruskal's $\gamma$ and Kendall's $\tau_b$, as we discuss in the next example.

**Example 8 (*Kendall's $\tau_b$ and Goodman and Kruskal's $\gamma$*).** Consider an i.i.d. sample of bivariate ordinal outcomes $\mathbf{z}_i = (u_i, v_i)^\top$ $(1 \le i \le n)$ defined above. For a pair of subjects $\mathbf{z}_i$ and $\mathbf{z}_j$, concordance and discordance are defined similarly as for continuous outcomes:

$$(\mathbf{z}_i, \mathbf{z}_j) \equiv \begin{cases} \text{concordant} & \text{if } u_i < (>) u_j, \ v_i < (>) v_j \\ \text{discordant} & \text{if } u_i > (<) u_j, \ v_i < (>) v_j \\ \text{neither} & \text{if otherwise} \end{cases}$$

Note that unlike continuous response, there is usually a sizable number of tied observations. Let $p_{cp}$ and $p_{dp}$ denote the probability of concordant and discordant pairs. Also, let $p_u$ and $p_v$ be the probabilities of having ties in $u$ and $v$, respectively.

Kendall's $\tau_b$ and Goodman and Kruskal's $\gamma$ are defined as:

$$\tau_b = \frac{p_{cp} - p_{dp}}{\sqrt{(1 - p_u)(1 - p_v)}}, \quad \gamma = \frac{p_{cp} - p_{dp}}{p_{cp} + p_{dp}} \tag{3.18}$$

where $p_u = \Pr(u_i = u_j)$ and $p_v = \Pr(v_i = v_j)$. Both coefficients range from $-1$ to 1 (see exercise). Both indices are widely used to measure association between two ordinal outcomes. If $u$ and $v$ are independent, $p_{cp} - p_{dp} = 0$ and thus $\tau_b = \gamma = 0$.

Let $h(\mathbf{z}_i, \mathbf{z}_j) = I_{\{(u_i - u_j)(v_i - v_j) > 0\}} - I_{\{(u_i - u_j)(v_i - v_j) < 0\}}$. Then, the mean of $h(\mathbf{z}_i, \mathbf{z}_j)$ is

$$\begin{aligned} \delta = E(h(\mathbf{z}_i, \mathbf{z}_j)) &= E\left(I_{\{(u_i - u_j)(v_i - v_j) > 0\}}\right) - E\left(I_{\{(u_i - u_j)(v_i - v_j) < 0\}}\right) \\ &= p_{cp} - p_{dp} \end{aligned}$$

In addition, since

$$h(\mathbf{z}_i, \mathbf{z}_j) = \begin{cases} 1 & \text{if } \mathbf{y}_i \text{ and } \mathbf{y}_j \text{ are concordant} \\ -1 & \text{if } \mathbf{y}_i \text{ and } \mathbf{y}_j \text{ are discordant} \\ 0 & \text{if otherwise} \end{cases}$$

it follows that the U-statistic based on $h(\mathbf{z}_i, \mathbf{z}_j)$ has the following form:

$$U_n = \binom{n}{2}^{-1} \sum_{(i,j) \in C_2^n} h(\mathbf{z}_i, \mathbf{z}_j) = \binom{n}{2}^{-1} (C - D)$$

where $C$ and $D$ denote the number of concordant and discordant pairs in the sample, respectively. Under the null of no association between $u$ and $v$, $\delta = p_{cp} - p_{dp} = 0$ and the U-statistic can be used to test this null.

For nominal outcomes, Chi-squares and Fisher's exact tests are widely used to test independence between two such outcomes. We can also develop U-statistics to test independence between two nominal outcomes. In the next example, we consider a relatively simple case with two binary outcomes, as the general setting involving two multilevel nominal outcomes requires multivariate U-statistics and thus will be deferred to Chapter 5.

**Example 9 (*Test for independence between two binary outcomes*).** Within the context of Example 8, assume that $u$ and $v$ are both binary outcomes. For convenience, we recode $u$ and $v$ so that they all have values 0 and 1. Independence between $u$ and $v$ is equivalent to the following condition,

$$p_{uv} = \Pr(uv = 1) = \Pr(u = 1)\Pr(v = 1) = p_u p_v \tag{3.19}$$

Let $h(\mathbf{z}_i, \mathbf{z}_j) = \frac{1}{2}(u_i - u_j)(v_i - v_j)$. Define the following U-statistic:

$$U_n = \binom{n}{2}^{-1} \sum_{(i,j)\in C_2^n} h(\mathbf{z}_i, \mathbf{z}_j) = \binom{n}{2}^{-1} \sum_{(i,j)\in C_2^n} \frac{1}{2}(u_i - u_j)(v_i - v_j)$$

This U-statistic is an unbiased estimate of $\delta = E(h(\mathbf{z}_i, \mathbf{z}_j)) = p_{uv} - p_u p_v$. The condition for independence in (3.19) is equivalent to null $H_0 : \delta = 0$. Thus, we can use the distribution of $U_n$ to test the independence between $u$ and $v$.

### 3.1.2 Two-Sample and General $K$ Sample U-Statistics

Now consider two i.i.d. samples of random vectors, $\mathbf{y}_{1i}$ $(1 \leq i \leq n_1)$ and $\mathbf{y}_{2j}$ $(1 \leq j \leq n_2)$. We define a (univariate) two-sample $U$-statistic with $m_1$ and $m_2$ arguments as follows:

$$U_n = \left[\binom{n_1}{m_1}\binom{n_2}{m_2}\right]^{-1} \sum_{l=1}^{2} \sum_{(i_1,\ldots,i_{m_l})\in C_{m_l}^{n_l}} h_{1i_1,\ldots,1i_{m_1};2i_1,\ldots,2i_{m_2}} \tag{3.20}$$

where $h_{1i_1,\ldots,1i_{m_1};2i_1,\ldots,2i_{m_2}} = h\left(\mathbf{y}_{1i_1},\ldots,\mathbf{y}_{1i_{m_1}};\mathbf{y}_{2i_1},\ldots,\mathbf{y}_{2i_{m_2}}\right)$ and $h$ is some symmetric kernel function. Note that symmetry is defined with respect to arguments within each sample, that is, $h$ is unchanged when permuting $(i_1,\ldots,i_{m_l})$ within each $l$ $(l = 1, 2)$. Note also that although the random vectors $\mathbf{y}_{ki}$ may have varying dimensions, they usually have the same dimension in most applications. As in the one-sample U-statistics case, $U_n$ is an unbiased estimate of the mean of the kernel statistic:

$$\theta = E\left[h\left(\mathbf{y}_{1i_1}, \ldots, \mathbf{y}_{1i_{m_1}}, ; \mathbf{y}_{2i_1}, \ldots, \mathbf{y}_{2i_{m_2}}, \right)\right]$$

As noted in the beginning of Section 3.1, in many applications, we are also interested in comparing variances among different treatment groups. The example below defines a U-statistic for comparing variances between two samples.

**Example 10 (*Model for comparing two group variances*).** Let $y_{ki}$ be two i.i.d. samples from two continuous distributions with mean $\mu_k$ and variance $\sigma_k^2$ ($1 \leq i \leq n_k$, $1 \leq k \leq 2$). Define a symmetric kernel as follows:

$$h\left(y_{1i}, y_{1j}; y_{2m}, y_{2l}\right) = \frac{1}{2}\left(y_{1i} - y_{1j}\right)^2 - \frac{1}{2}\left(y_{2m} - y_{2l}\right)^2 \tag{3.21}$$

Then, it follows that

$$\theta = E\left[h\left(y_{1i}, y_{1j}; y_{2m}, y_{2l}\right)\right] = \sigma_1^2 - \sigma_2^2$$

If $\theta = 0$, then $y_{1i}$ and $y_{2j}$ have the same variance. Thus, the two-sample $U$-statistic given by

$$U_n = \left[\binom{n_1}{2}\binom{n_2}{2}\right]^{-1} \sum_{(i,j)\in C_2^{n_1}} \sum_{(m,l)\in C_2^{n_2}} h\left(y_{1i}, y_{1j}; y_{2m}, y_{2l}\right)$$

can be used to test the null of equal between-sample variance, $H_0 : \sigma_1^2 = \sigma_2^2$.

Note that kernel functions are also not unique for two-sample U-statistics. For example, we can define a different kernel for this problem as follows:

$$g\left(y_{1i}, y_{1j}; y_{2m}, y_{2l}\right) = \left(y_{1i}^2 - y_{1i}y_{1j}\right) - \left(y_{2k}^2 - y_{2m}y_{2l}\right) \tag{3.22}$$

This kernel is not symmetric. As in the one-sample U-statistics case, we can readily create a symmetric version using an argument similar to (3.10) (see exercise). Further, $h\left(y_{1i}, y_{1j}; y_{2m}, y_{2l}\right)$ in (3.21) is a symmetric version of $g\left(y_{1i}, y_{1j}; y_{2m}, y_{2l}\right)$.

Like the one-sample case, the power of U-statistics lies in studying statistics that are intrinsically defined by responses from multiple subjects. For example, in Section 1.1, we discussed the Mann–Whitney–Wilcoxon rank sum statistic to test the equality of two distributions that differ by a location shift. As noted there, this statistic has two representations that were introduced independently by Wilcoxon (1945) and Mann–Whitney (1947) but are equivalent. In the example below, we develop the U-statistic for the Mann–Whitney form of the test. We will show their equivalence in the next section after we introduce order statistics.

**Example 11.** (***Mann–Whitney–Wilcoxon rank sum test***). Let $y_{1i}$ and $y_{2j}$ be two i.i.d. samples from two continuous distributions with CDFs $F_1$ and $F_2$, respectively ($1 \leq i \leq n_1$, $1 \leq j \leq n_2$). We assume that $F_k$ differ by a location shift, that is, $F_1(y) = F_2(y - \mu)$, where $\mu$ is some unknown scalar. Interest lies in testing the null of no location shift, $H_0 : \mu = 0$ or equivalently,

$$H_0 : \Pr(y_{1i} - y_{2j} \leq 0) = \frac{1}{2} \quad \text{versus} \quad H_a : \Pr(y_{1i} - y_{2j} \leq 0) \neq \frac{1}{2} \tag{3.23}$$

Let $h(y_{1i}; y_{2j}) = I_{\{y_{1i} - y_{2j} \leq 0\}}$. Then,

$$\theta = E[h(y_{1i}; y_{2j})] = \Pr(y_{1i} - y_{2j} \leq 0)$$

Following (3.20) and (3.23), the associated U-statistic is given by

$$U_n = \left[\binom{n_1}{1}\binom{n_2}{1}\right]^{-1} \sum_{i=1}^{n_1}\sum_{j=1}^{n_2} h(y_{1i}; y_{2j}) = \frac{1}{n_1 n_2} \sum_{i=1}^{n_1}\sum_{j=1}^{n_2} I_{\{y_{1i} - y_{2j} \leq 0\}}$$

The above $U_n$ is an unbiased estimate of $\theta$. The null in (3.23) can be expressed in terms of $\theta$ as: $H_0 : \theta = \frac{1}{2}$.

Diagnostic tests are widely used in biomedical and psychosocial research for screening diseases and disorders of interest. Many diagnostic tests yield continuous readings. However, to facilitate decision making in practice, such test outcomes are often dichotomized to give binary outcomes. Sensitivity and specificity are used to provide a measure of accuracy of a diagnostic test, determined by some cut-off point in dichotomizing the continuous test outcome.

Consider a diagnostic test for detecting some medical or mental health condition, such as HIV or depression. Consider two i.i.d. samples with one consisting of $n_1$ diseased and the other of $n_2$ disease-free subjects. Let $y_{1i}$ and $y_{2j}$ denote the continuous test outcomes from the diseased and non-disease samples ($1 \leq i \leq n_1$,$1 \leq j \leq n_2$). We assume that larger values of the test outcome indicate the presence of the disease.

Let $F_k(t) = \Pr(y_{ki} \leq t)$ be the CDF of $y_{ki}$ ($k = 1, 2$). Given a cut-point $c$, the test sensitivity and specificity are defined by

$$\eta(c) = \Pr[y_{1i} \geq c] = 1 - F_1(c), \quad \xi(c) = \Pr[y_{2j} \leq c] = F_2(c)$$

We can estimate the sensitivity and specificity by the following statistics:

$$\widehat{\eta}(c) = \frac{1}{n_1}\sum_{i=1}^{n_1} I_{\{y_{1i}\geq c\}}, \quad \widehat{\xi}(c) = \frac{1}{n_1}\sum_{j=1}^{n_1} I_{\{z_j\leq c\}}$$

Following Example 1 of 3.1.1, both estimates can be viewed as one-sample U-statistics.

Since $\eta(c)$ and $\xi(c)$ depend on the cut-off $c$, they vary as $c$ changes. By varying $c$, we can bring either sensitivity or specificity arbitrarily close to 1. Thus, to compare performance between two diagnostic tests, we must examine sensitivity and specificity simultaneously. The Receiver Operating Characteristics (ROC) curve is widely used as an index of performance of test kit by studying $\eta(c)$ and $\xi(c)$ jointly along the continuum $c$.

An ROC curve is the plot of the bivariate $(1-\eta(c), \xi(c))$ as a function of $c$ with a theoretical range $(-\infty, \infty)$. A good test kit should maintain high values of both $\eta(c)$ and $\xi(c)$ across all values of $c$. The area under the ROC curve (AUC) $\theta$ can be expressed as (see exercise):

$$\theta = \int_0^1 \eta(\xi)\, d\xi = \int_{-\infty}^{+\infty} [1 - F_1(t)]\, dF_2(t) = 1 - \int_{-\infty}^{+\infty} F_1(t)\, dF_2(t) \tag{3.24}$$

$$= 1 - E[F_1(y_{2j})] = 1 - E\left[E_{Y_1}\left(I_{\{y_{1i}\leq y_{2j}\}} \mid y_{2j}\right)\right]$$

$$= 1 - E\left(I_{\{y_{1i}\leq y_{2j}\}}\right) = E\left(I_{\{y_{2j}\leq y_{1i}\}}\right)$$

In the next example, we show how to construct a U-statistic for estimating this parameter $\theta$.

**Example 12** (***U-statistic for single ROC curve***). Let $y_{1i}$ and $y_{2j}$ denote the continuous test outcomes from the diseased and non-disease samples as defined above ($1 \leq i \leq n_1$,$1 \leq j \leq n_2$). Define a symmetric kernel as: $h(y_{1i}; y_{2j}) = I_{\{y_{2j}\leq y_{1i}\}}$. Then, the following two-sample U-statistic is an unbiased estimate of the AUC parameter of interest $\theta$:

$$\widehat{\theta}_n = \left[\binom{n_1}{1}\binom{n_2}{1}\right]^{-1} \sum_{i=1}^{n_1}\sum_{j=1}^{n_2} I_{\{y_{2j}\leq y_{1i}\}} \tag{3.25}$$

We can readily extend the two-sample U-statistics to general $K$-sample U-statistics. Consider $K$ i.i.d. samples of random vectors, $\mathbf{y}_{ki}$ ($1 \leq i \leq n_k$, $1 \leq k \leq K$). Let

$$h_{1i_1,\ldots,1i_{m_1};\ldots;Ki_1,\ldots,Ki_{m_K}} = h\left(\mathbf{y}_{1i_1},\ldots,\mathbf{y}_{1i_{m_1}};\ldots;\mathbf{y}_{Ki_1},\ldots,\mathbf{y}_{Ki_{m_K}}\right)$$

be some symmetric function with respect to input arguments within each sample, that is, $h$ is constant when $\mathbf{y}_{ki_1}, \ldots, \mathbf{y}_{ki_{m_k}}$ are permuted within each $k$th sample ($1 \leq k \leq K$). A $K$-sample $U$-statistic with $m_k$ arguments for the $k$th sample has the general form:

$$U_n = \left[\prod_{k=1}^{K} \binom{n_k}{m_k}\right]^{-1} \sum_{k=1}^{K} \sum_{(i_1,\ldots,i_{m_k}) \in C_{m_k}^{n_k}} h_{1i_1,\ldots,1i_{m_1};\ldots;Ki_1,\ldots,Ki_{m_K}} \tag{3.26}$$

As in the one- and two-sample case, $U_n$ is an unbiased estimate of $\theta = E(h)$.

ROC curves are widely used to compare accuracy of multiple diagnostic tests. For example, in biomedical research such analyses are used to compare performance of tests by different manufacturers for detecting infectious diseases such as HIV. In mental health and psychosocial research, ROC curve analyses are often used to compare different instruments for diagnoses of certain mental health conditions and behavioral patterns. Within the context of Example 12, now suppose that we have two test kits and want to compare their AUCs. Either model below can be applied depending on the design of the study.

**Example 13** (***U-statistic for comparing AUCs with independent samples***). If we have two additional i.i.d. samples from the same diseased and disease-free study populations as in Example 12, then we can apply the second test kit to these two samples to obtain outcomes for computing the AUCs between the two test kits. Let the diseased and disease-free sample have $n_3$ and $n_4$ subjects, respectively. Denote the outcomes from the two samples by $y_{3i}$ and $y_{4j}$ ($1 \leq i \leq n_3$, $1 \leq j \leq n_4$).

Then, as in Example 12, we can use the U-statistic obtained by substituting $y_{3i}$ and $y_{4j}$ in place of $y_{1i}$ and $y_{2j}$ and $n_3$ and $n_4$ in place of $n_1$ and $n_2$ in (3.25) to estimate AUC for this second test. If we are interested in the null of no difference in AUC between the two test kits, we can define the following U-statistic to test such a null:

$$U_n = \left[\prod_{k=1}^{4} \binom{n_k}{1}\right]^{-1} \sum_{i=1}^{n_1} \sum_{j=1}^{n_2} \sum_{m=1}^{n_3} \sum_{l=1}^{n_4} \left(I_{\{y_{2j} \leq y_{1i}\}} - I_{\{y_{4l} \leq y_{3m}\}}\right) \tag{3.27}$$

The above is a four-sample U-statistic defined by the kernel function $h = I_{\{y_{21} \leq y_{11}\}} - I_{\{y_{41} \leq y_{31}\}}$. The mean of $h$ is

$$\delta = E\left(h\left(y_{11}; y_{21}; y_{31}; y_{41}\right)\right) = E\left(I_{\{y_{21} \leq y_{11}\}}\right) - E\left(I_{\{y_{41} \leq y_{31}\}}\right) = \theta_1 - \theta_2$$

where $\theta_k$ denotes the AUC of test kit $k$ $(= 1, 2)$. Thus, the null $H_0 : \theta_1 = \theta_2$ is equivalent to the null $H_0 : \delta = 0$. It is interesting to note that the four-sample U-statistic can be expressed as the difference between two two-sample U-statistics:

$$U_n = (n_1 n_2)^{-1} \sum_{i=1}^{n_1} \sum_{j=1}^{n_2} I_{\{y_{2j} \leq y_{1i}\}} - (n_3 n_4)^{-1} \sum_{m=1}^{n_3} \sum_{l=1}^{n_4} I_{\{y_{4l} \leq y_{3m}\}} \tag{3.28}$$

The study design in the above example does not sufficiently control for subjects' sampling variability. If sample sizes are small, differences in AUCs may also reflect heterogeneity between the two samples within the same disease status. In applications, multiple tests are often applied to the same subject to control for heterogeneity across subjects if the order in which the tests are administered has no effect on test outcomes. Unlike the study design in Example 13, test outcomes become correlated since they are obtained from the same subjects. Thus, new statistics must be constructed to test the equality of AUCs between two test kits, as we discuss in the next example.

**Example 14** (***U-statistic for comparing AUCs with correlated outcomes***). Consider applying two tests to each of the two i.i.d. samples of diseased and disease-free subjects as assumed in Example 12. Let $y_{1i}$ and $y_{3i}$ denote the test outcomes from the first and second tests when applied to the diseased, and $y_{2i}$ and $y_{4i}$ the test results from the first and second kits when administered to the disease-free sample. The AUC for each of the test kits is estimated by the U-statistic in (3.25) in Example 12. However, to test the null of no between-kit difference, the U-statistic in (3.27) does not apply, as there only two i.i.d. samples in the design of this study.

To define an appropriate U-statistic, let $\mathbf{z}_{1i} = (y_{1i}, y_{3i})^\top$ and $\mathbf{z}_{2j} = (y_{2i}, y_{4i})^\top$. Then, $h(\mathbf{z}_{1i}; \mathbf{z}_{2j}) = I_{\{y_{2j} \leq y_{1i}\}} - I_{\{y_{4j} \leq y_{3i}\}}$ is a symmetric kernel and this kernel-based U-statistic,

$$U_n = \left[ \binom{n_1}{1} \binom{n_2}{1} \right]^{-1} \sum_{i=1}^{n_1} \sum_{j=1}^{n_2} \left( I_{\{y_{2j} \leq y_{1i}\}} - I_{\{y_{4j} \leq y_{3i}\}} \right)$$

is an unbiased estimate of the difference in AUCs between the two kits, $\gamma = E[h(\mathbf{z}_{1i}; \mathbf{z}_{2j})] = \theta_1 - \theta_2$ with $\theta_k$ again denoting the AUC of test $k$ $(= 1, 2)$. The null of no between-test difference corresponds to the null $H_0 : \gamma = 0$.

### 3.1.3 Representation of U-Statistic by Order Statistic

U-statistics have a close relationship with order statistics, which play an important role in statistical inference. In this section, we first introduce and discuss some properties of order statistics. We then provide a *derivation* of U-statistics from the perspective of order statistics.

Let $y_i$ $(1 \leq i \leq n)$ be an i.i.d. sample from a continuous distribution. Let $y_{(i)}$ denote the order statistic of the sample, that is,

$$y_{(1)} < y_{(2)} < \cdots < y_{(n-1)} < y_{(n)}$$

Let $\mathbf{v} = \left(y_{(1)}, y_{(2)}, \cdots, y_{(n)}\right)^{\top}$ be a vector of the above order statistic. As shown in Chapter 1, $\mathbf{v}$ is a random vector. Although $y_i$ are i.i.d. random variables, the ordered $y_{(i)}$ are not. Clearly, $y_{(i)}$ are not independent. The theorem below carefully characterizes the marginal as well as the joint distribution of the ordered multivariate statistic vector $\mathbf{v}$, which in particular shows that $y_{(i)}$ are not identically distributed.

**Theorem 1**. Let $f(x)$ and $F(x)$ denote the PDF and CDF of $y_i$. Then,

1. The marginal density function of the $j$th component of $\mathbf{v}$ is given by:

$$g_j(v) = \frac{n!}{(n-j)!\,(j-1)!} F^{j-1}(v)\left[1 - F(v)\right]^{n-j} f(v)$$

2. The joint density of the order statistic vector $\mathbf{v}$ is given by:

$$g_{\mathbf{v}}(v_1, \ldots, v_n) = n! I_{\{v_1 < v_2 <, \ldots, < v_n\}} \prod_{i=1}^{n} f(v_i)$$

Proof. 1. Let $G_j(v)$ be the CDF of $v_j = y_{(j)}$. Then, for any $u \in R$, we have

$$\begin{aligned} G_j(x) &= \Pr(v_j \leq u) = \Pr\left(y_{(j)} \leq u\right) = \Pr(\text{at least } j \ y\text{'s } \leq u) \\ &= \sum_{i=j}^{n} \Pr(\text{exactly } i \ y\text{'s } \leq u) = \sum_{i=j}^{n} \binom{n}{i} F^{j-1}(u)\left[1 - F(u)\right]^{n-j} \\ &= \frac{n!}{(n-j)!\,(j-1)!} \int_0^{F(u)} t^{j-1}(1-t)^{n-j}\, dt \end{aligned}$$

The last integral is the incomplete Beta function. Now, let $H(\cdot)$ be the CDF of the Beta distribution, $Beta(\alpha, \beta)$, with $\alpha = j$ and $\beta = n - j + 1$.

Then, we have: $G_j(u) = H(F(u))$. Differentiating $G_j(u)$ with respect to $u$ yields (see exercise):

$$g_j(u) = h(F(u)) f(u) \tag{3.29}$$
$$= \frac{n!}{(n-j)!(j-1)!} F^{j-1}(u) [1 - F(u)]^{n-j} f(u)$$

where $h(\cdot)$ denotes the PDF of $Beta(\alpha = j, \beta = n - j + 1)$.

2. Without the loss of generality, consider the case with $n = 2$ and $\mathbf{v} = (y_{(1)}, y_{(2)})^\top$. In addition, we need the following lemma (see exercise).

**Lemma**. Let $\mathbf{x} \in R^m$ be a random variable with a joint PDF $f(\mathbf{x})$. Let $\mathbf{y} = \mathbf{g}(\mathbf{x}) \in R^m$ be a transformation of $\mathbf{x}$. Suppose that there exists a partition, $D_k \subset R^m$, of the support of $\mathbf{x}$ and associated transformations, $\mathbf{g}_k(\mathbf{x})$, defined on each $D_k$ such that each $\mathbf{g}_k(\mathbf{x})$ is a one-to-one mapping $(1 \leq k \leq K)$. Then, $\mathbf{y}$ has the following PDF:

$$f_y(\mathbf{y}) = \sum_{k=1}^{K} f_x\left(\mathbf{g}_k^{-1}(\mathbf{y})\right) \left| \frac{d}{d\mathbf{y}} \mathbf{g}_k^{-1}(\mathbf{y}) \right|$$

where $\left| \frac{d}{d\mathbf{y}} \mathbf{g}_k^{-1}(\mathbf{y}) \right|$ is the usual Jacobian matrix.

Now let $\mathbf{y} = (y_1, y_2)^\top$ and consider the transformation, $\mathbf{v} = \mathbf{h}(\mathbf{y})$, which is not one-to-one. However, within each $D_k$ $(k = 1, 2)$ defined below,

$$D_1 = \{(y_1, y_2) : y_1 < y_2\}, \quad D_2 = \{(y_1, y_2) : y_2 < y_1\}$$

$\mathbf{h}(\mathbf{y})$ becomes one-to-one. Denote by $\mathbf{h}_k$ the component of $\mathbf{h}(\mathbf{y})$ when restricted to $D_k$ $(k = 1, 2)$. Then, it follows from the lemma that

$$g_{\mathbf{v}}(v_1, v_2) = f(v_1, v_2) \left| \frac{\partial \mathbf{h}_1(y_1, y_2)}{\partial (v_1, v_2)} \right| + f(v_1, v_2) \left| \frac{\partial \mathbf{h}_2(y_1, y_2)}{\partial (v_1, v_2)} \right|$$
$$= f(v_1) f(v_2) (|1| + |-1|)$$
$$= 2f(v_1) f(v_2)$$

The example below illustrates the lemma with a relative simple case of a random variable.

**Example 15**. Let $x \sim N(0, 1)$ and $y = h(x) = x^2$. Then, $y = h(x)$ is not a one-to-one function. By partitioning $R$ into $D_1 \cup D_2 = (-\infty, 0] \cup (0, \infty)$, we can define two components $h_k$ that are one-to-one when restricted to $D_k$, $h_k = h(x)$ if $x \in D_k$. The inverse of $h_k$ exists on each partition $D_k$:

$$x = h_1^{-1}(y) = -\sqrt{y} \quad \text{if } x \in D_1, \quad x = h_2^{-1}(y) = \sqrt{y} \quad \text{if } x \in D_2$$

It follows that the density of $y$ is given by

$$\begin{aligned} f_y(y) &= f_x\left(h_1^{-1}(y)\right)\left|\frac{d}{d(y)}h_1^{-1}(y)\right| + f_x\left(h_2^{-1}(y)\right)\left|\frac{d}{d(y)}h_2^{-1}(y)\right| \\ &= \frac{1}{\sqrt{2\pi}}e^{-\frac{1}{2}y}\left|\frac{-1}{2\sqrt{y}}\right| + \frac{1}{\sqrt{2\pi}}e^{-\frac{1}{2}y}\frac{1}{2\sqrt{y}} \\ &= \frac{1}{\sqrt{2\pi}}e^{-\frac{1}{2}y}y^{-\frac{1}{2}} \end{aligned}$$

Thus, $x^2$ follows $\chi_1^2$, a $\chi^2$ distribution with one degree of freedom.

**Example 16**. Let $y_i$ be an i.i.d. sample from a continuous distribution with CDF $F(y)$. Let $u_i = F(y_i)$ be the probability integral transformation of $y_i$. We have shown in Chapter 1 that $u_i$ has a uniform distribution between 0 and 1, that is, $u_i \sim$ i.i.d. $U[0,1]$.

Now consider a similar transformation of the order statistic $y_{(i)}$, $v_i = F\left(y_{(i)}\right)$. Unlike $u_i$, $v_i$ are not independent. Further, it follows from Theorem 1 that the marginal $v_i$ has a beta distribution, $Beta(\alpha, \beta)$, with $\alpha = j$ and $\beta = n - j + 1$, while the joint $v_i$ has the following density function:

$$f_{\mathbf{v}}(v_1, \ldots, v_n) = \begin{cases} n! \text{ if } & v_1 < v_2 <, \ldots, < v_n \\ 0 & \text{if elsewhere} \end{cases}$$

**Example 17 (*Wilcoxon signed rank test*)**. In Example 7 of Section 3.1.1, we discussed the one-sample Wilcoxon signed rank statistic and claimed the equality in (3.15). We are now in a position to show this identity using the properties of order statistics.

First, it is readily checked that (see exercise):

$$I_{\{y_i>0\}}R_i = \sum_{j=1}^{i} I_{\left\{y_{(i)}+y_{(j)}>0\right\}} \tag{3.30}$$

Then, it follows from the above that

$$\begin{aligned} W_n^+ &= \sum_{i=1}^{n} I_{\{y_i>0\}}R_i = \sum_{i=1}^{n}\sum_{j=1}^{i} I_{\left\{y_{(i)}+y_{(j)}>0\right\}} = \sum_{i=1}^{n}\sum_{j=1}^{i} I_{\{y_i+y_j>0\}} \qquad (3.31) \\ &= \sum_{i=j=1}^{n} I_{\{2y_i>0\}} + \sum_{1\le i<j\le n} I_{\{y_i+y_j>0\}} = nU_{1n} + \binom{n}{2}U_{2n} \end{aligned}$$

where $U_{1n}$ and $U_{2n}$ are defined in (3.16) for Example 7 of Section 3.1.1.

**Example 18 (*Mann–Whitney–Wilcoxon rank sum test*).** In Section 1.1 of Chapter 1, we presented two forms of the two-sample Mann–Whitney–Wilcoxon rank sum test for testing whether there is a location shift between two distributions. More specifically, consider a study with two treatment groups with $n_1$ subjects in the first and $n_2$ in the second group. Let $y_{1i}$ $(1 \leq i \leq n_1)$ and $y_{2j}$ $(1 \leq j \leq n_2)$ denote the continuous response of interest from the first and second groups, respectively. Let $R_i^1$ denote the rank scores assigned to the observations from the first group that are created based on the combined observations. The two statistics are given by:

$$\text{Wilcoxon rank sum statistic: } W_n = \sum_{i=1}^{n_1} R_i^1 \tag{3.32}$$

$$\text{Mann–Whitney statistic: } U_n = \sum_{i=1}^{n_1} \sum_{j=1}^{n_2} I_{\{y_{2j} - y_{1i} \leq 0\}}$$

We have shown in Example 11 of Section 3.1.2 that the normalized Mann–Whitney form, $\frac{1}{n_1 n_2} U_n$, is a U-statistic. As in Example 17 above, we can also express the rank variables $R_i$ as a function of the order statistic $y_{(ki)}$ as follows (see exercise):

$$R_i^1 = \sum_{j=1}^{n_2} I_{\left\{y_{(2j)} - y_{(1i)} \leq 0\right\}} + \sum_{j=1}^{n_1} I_{\left\{y_{(1j)} - y_{(1i)} \leq 0\right\}} \tag{3.33}$$

By using this order statistic based expression, it is readily shown that under the null of identical CDF between the two samples (see exercise),

$$W_n = U_n + \frac{n_1 (n_1 + 1)}{2} \tag{3.34}$$

Further, under the null the mean of $W_n$ is $E(W_n) = \frac{n_1(n_1+n_2+1)}{2}$ (see exercise). By taking expectation of the two sides in (3.34), we obtain $E\left(I_{\left\{y_{(2j)} - y_{(1i)} \leq 0\right\}}\right) = \frac{1}{2}$ (see exercise). This result was also derived in Example 11 based on the symmetry of the distribution of the combined sample of $y_{2j}$ and $y_{1i}$ under the null. Thus, the null of no location shift corresponds to $H_0 : E\left(I_{\left\{y_{(2j)} - y_{(1i)} \leq 0\right\}}\right) = \frac{1}{2}$ for the Mann–Whitney statistic.

Examples 17 and 18 indicate a close relationship between order- and U-statistics. In fact, U-statistics can be *derived* from order statistics. Before proceeding with such an undertaking, let us first derive some useful results for evaluation of conditional expectation.

Let $y_i$ $(1 \leq i \leq n)$ be an i.i.d. random sample and $\mathbf{v} = \left(y_{(1)}, \ldots, y_{(n)}\right)^\top$ the corresponding order statistic vector. Let $\pi$ be a permutation of the integer set $\{1, ..., n\}$. Recall that a permutation of a set is a one-to-one mapping from the set to itself. We use $\pi(k)$ to denote the image of $k$ under the permutation. For example, if $\pi$ maps $(1, 2, 3)$ to $(2, 3, 1)$, $\pi(1) = 2$.

As shown in Chapter 3, $\mathbf{v}$ is a random vector. Since $\mathbf{v}$ is determined by the sample observations $\mathbf{y} = (y_1, \ldots, y_n)^\top$, the joint density of $f_{\mathbf{v},\mathbf{y}}(v_1, \cdots, v_n, y_1, \ldots, y_n)$ equals the marginal density of $f_{\mathbf{y}}(y_1, \ldots, y_n)$. Hence, for any permutation $\pi$ of $\{1, ..., n\}$, it follows from Theorem 1 and the Bayes' theorem that

$$\begin{aligned}
\Pr\left(y_1 = v_{\pi(1)}, \ldots, y_n = v_{\pi(n)} \mid \mathbf{v}\right) &= \frac{f_{\mathbf{v},\mathbf{y}}(v_1, \cdots, v_n, y_1, \ldots, y_n)}{f_{\mathbf{v}}(v_1, \cdots, v_n)} \qquad (3.35) \\
&= \frac{f_{\mathbf{y}}(v_{\pi(1)}) \cdots f\left(v_{\pi(n)}\right)}{n! f(v_1) \cdots f(v_n)} \\
&= \frac{1}{n!}
\end{aligned}$$

From the above joint distribution of $\mathbf{y}$, we can readily derive the marginal distribution of any subset of $\mathbf{y}$ conditional on the order statistic vector $\mathbf{v}$. First, consider the event $\{y_1 = v_i\}$. Since this event consists of $(n-1)!$ events of the form $\left\{y_1 = v_{\pi(1)}, \ldots, y_n = v_{\pi(n)}\right\}$ where the permutations $\pi$ satisfy $\pi(1) = i$, it follows that

$$\Pr(y_1 = v_i \mid \mathbf{v}) = \frac{1}{n!} \sum_{\pi \in P_{1i}^n} \Pr\left(y_1 = v_{\pi(1)}, \ldots, y_n = v_{\pi(n)} \mid \mathbf{v}\right) = \frac{1}{n} \qquad (3.36)$$

where $P_{1i}^n$ denotes the set of all permutations of $\{1, ..., n\}$ satisfying $\pi(1) = i$. Now consider the event $\{y_1 = v_i, y_2 = v_j\}$ $(i \neq j)$. This corresponds to all $(n-2)!$ events $\left\{y_1 = v_{\pi(1)}, \ldots, y_n = v_{\pi(n)}\right\}$ with $\pi(1) = i$ and $\pi(2) = j$. It follows that

$$\Pr(y_1 = v_i, y_2 = v_j \mid \mathbf{v}) = \frac{1}{n(n-1)} \qquad (3.37)$$

and

$$\Pr(\{y_1, y_2\} = \{v_i, v_j\} \mid \mathbf{v}) = \frac{2}{n(n-1)} = \frac{1}{\binom{n}{2}} \qquad (3.38)$$

Note that the equality of the sets $\{y_1, y_2\} = \{v_i, v_j\}$ in (3.38) implies two distinct events: $\{y_1 = v_i, y_2 = v_j\}$ and $\{y_1 = v_j, y_2 = v_i\}$, which have the same probability in the above setting.

More generally, let $\{i_1, ..., i_m\}$ be a subset of $\{1, ..., n\}$. The event $\{y_1 = v_{i_1}, y_2 = v_{i_2}, ..., y_m = v_{i_m}\}$ is the union of $(n-m)!$ events of the form

$\{y_1 = v_{\pi(1)}, \ldots, y_n = v_{\pi(n)}\}$ with $\pi$ satisfying $\pi(k) = i_k$ for $k = 1, ..., m$. Thus, it follows that

$$\Pr\left(y_1 = v_{i_1}, y_2 = v_{i_2}, ..., y_m = v_{i_m} \mid \mathbf{v}\right) = \frac{(n-m)!}{n!} \tag{3.39}$$
$$\Pr\left(\{y_1, ..., y_m\} = \{v_{i_1}, ..., v_{i_m}\} \mid \mathbf{v}\right) = m!\frac{(n-m)!}{n!} = \frac{1}{\binom{n}{m}}$$

Now we are ready to provide an order statistic perspective of the U-statistics. Let $y_i$, $\mathbf{y}$ and $\mathbf{v}$ be defined as before. Let $g(\cdot)$ be a Borel measurable function from $R^m$ to $R$. Then, $g(y_1, \ldots, y_m)$ is a random variable. For $m = n$, it follows from (3.35) that

$$E\left[g(y_1, \ldots, y_n) \mid \mathbf{v}\right] = \frac{1}{n!} \sum_{(i_1, \ldots, i_n) \in P^n} g(y_{i_1}, \ldots, y_{i_n}) \tag{3.40}$$

where $P^n$ denotes the set of permutations of $\{1, \ldots, n\}$. If $g(y_1, \ldots, y_n)$ is symmetric, then the above reduces to:

$$E\left[g(y_1, \ldots, y_n) \mid \mathbf{v}\right] = g\left(y_{(1)}, \ldots, y_{(n)}\right) = g(y_1, \ldots, y_n)$$

Below, we discuss some examples of how the notion of conditional expectation is used to show unbiasedness of symmetric functions as defined in a U-statistic. For the concept of conditional expectation and associated properties, the reader is referred to Chapter 1.

**Example 19**. Let $h(y)$ be a symmetric and measurable function. Then, we have

$$E\left[h(y_1) \mid \mathbf{v}\right] = \frac{1}{n}\sum_{i=1}^{n} h(y_i) \tag{3.41}$$

To justify this, we can use several approaches. First, let $g(y_1, \ldots, y_n) = h(y_1)$. Then, $g(y_1, \ldots, y_n)$ is clearly a random variable. It follows from (3.40) that

$$E\left[h(y_1) \mid \mathbf{v}\right] = \frac{1}{n!} \sum_{(i_1, \ldots, i_n) \in P^n} g(y_{i_1}, \ldots, y_{i_n}) = \frac{(n-1)!}{n!} \sum_{(i_1, \ldots, i_n) \in P^n} h(y_{i_1})$$

from which (3.41) follows.

Alternatively, we can apply (3.35), which immediately yields (3.41).

**Example 20**. Consider a symmetric and measurable function with $m$ arguments $h(y_1, \ldots, y_m)$. Following the first approach in the above example, let $g(y_1, \ldots, y_n) = h(y_1, \ldots, y_m)$. Then, (3.40) yields:

$$E[h(y_1, \ldots, y_m) \mid \mathbf{v}] = \binom{n}{m}^{-1} \sum_{(i_1, \ldots, i_m) \in C_m^n} h(y_{i_1}, \ldots, y_{i_m}) \qquad (3.42)$$

Alternatively, by applying (3.35), we also obtain the above.

The two examples above show that a one-group, $m$-argument U-statistic defined by a symmetric kernel $h$ can be viewed as the conditional expectation of the kernel given the order statistic $y_{(1)}, \ldots, y_{(n)}$. From this perspective, we can readily show some optimal properties of the U-statistic.

Let $T_n(y_1, \ldots, y_n)$ be some unbiased estimate of some parameter $\theta$ of interest. Define a U-statistic by

$$U_n = \frac{1}{n!} \sum_{(i_1, \ldots, i_n) \in P^n} T_n(y_{i_1}, \ldots, y_{i_n}) \qquad (3.43)$$

Then, it follows from the law of iterated conditional expectation that

$$U_n = E[T_n(y_1, \ldots, y_n) \mid \mathbf{v}]$$

Since $g(x) = x^2$ is a convex function, it follows from Jensen's inequality for conditional expectation (see Chapter 1) that

$$E^2(T_n \mid \mathbf{v}) = g(E(T_n \mid \mathbf{v})) \leq E(g(T_n) \mid \mathbf{v}) = E\left(T_n^2 \mid \mathbf{v}\right) \qquad (3.44)$$

By taking expectation again on both sides in (3.44), we have

$$E\left(U_n^2\right) = E\left[E^2(T_n \mid \mathbf{v})\right] \leq E\left[E\left(T_n^2 \mid \mathbf{v}\right)\right] = E\left(T_n^2\right)$$

Since $E(U_n) = E(T_n) = \theta$, it follows that $Var(U_n) \leq Var(T_n)$. Thus, given any unbiased estimate, we can construct a more efficient U-statistic. It follows consequently that the estimate with minimal variance among all unbiased estimates (if it exists) must be a U-statistic. In the above example, if $T_n$ itself is a U-statistic, then $U_n = T_n$ and thus the procedure cannot be continued indefinitely.

**Example 21**. Let $y_i$ be an i.i.d. sample with mean $\mu$ and variance $\sigma^2$. Let $\widehat{\theta}_1 = T_n(y_1, \ldots, y_n) = y_1$. This single-observation based estimate is unbiased since $E\left(\widehat{\theta}_1\right) = \mu$. The variance of $\widehat{\theta}_1$ is $Var\left(\widehat{\theta}_1\right) = \sigma^2$.

Now consider the estimate $U_n$ in (3.43) defined by $T_n$. The U-statistic in this special case yields

$$U_n = \frac{1}{n!} \sum_{(i_1, \ldots, i_n) \in P^n} T_n(X_{i_1}, \ldots, X_{i_n}) = \frac{1}{n} \sum_{i=1}^{n} X_i$$

The above $U_n$ is the well-known sample mean with a much smaller variance $Var(U_n) = \frac{\sigma^2}{n}$.

### 3.1.4 Martingale Structure of U-Statistic

An increasing family of $\sigma$-fields, $\{\mathcal{F}_n; n \geq 1\}$, is called a *filtration*, that is, $\mathcal{F}_n \subseteq \mathcal{F}_m$ for any $n \leq m$. A sequence of random variables $x_n$ is called a *martingale* if for any $n$ $(= 1, 2, \ldots)$,

1. $x_n$ is measurable $\mathcal{F}_n$,
2. $E(x_{n+1} \mid \mathcal{F}_n) = x_n$.

It immediately follows from these defining properties and the law of iterated conditional expectation that for any integers $l \geq k$,

$$\begin{aligned} E(x_l \mid \mathcal{F}_k) &= E[E(x_l \mid \mathcal{F}_{l-1}) \mid \mathcal{F}_k] = E(x_{l-1} \mid \mathcal{F}_k) = \ldots = E(x_{k+1} \mid \mathcal{F}_k) \\ &= x_k \\ E(x_l) &= E[E(x_l \mid \mathcal{F}_k)] \\ &= E(x_k) \end{aligned}$$

Thus, a martingale $x_n$ has a constant mean.

As discussed in Chapter 1, the conditional expectation $E(x_{n+1} \mid \mathcal{F}_n)$ represents the average of $x_{n+1}$ with respect to the $\sigma$-field $\mathcal{F}_n$. If we interpret $n$ as time, $x_{n+1}$ is a future event whose outcome is unknown to $\mathcal{F}_n$, $\sigma$-field only containing information up to and including the present time $n$. Thus, $E(x_{n+1} \mid \mathcal{F}_n)$ is our best guess or prediction of a future outcome using all available information at the present time in $\mathcal{F}_n$. We will exploit such an interpretation in the evaluation of conditional expectation for several examples later. The reader may want to review the definition, properties and examples of conditional expectation in Chapter 1 before proceeding further.

**Example 22**. Let $y_i$ be an i.i.d. random sample with mean $E(y_i) = 0$. Let

$$x_n = \sum_{i=1}^{n} y_i, \quad \mathcal{F}_n = \sigma(x_1, \ldots, x_n) = \sigma(y_1, \ldots, y_n)$$

where $\sigma(x_1, \ldots, x_n)$ denotes the $\sigma$-field generated by the $n$ random variables $x_i$ $(1 \leq i \leq n)$. It is clear that $\mathcal{F}_1 \subseteq \mathcal{F}_2 \ldots \subseteq \mathcal{F}_n \ldots$ and thus $\{\mathcal{F}_n; n \geq 1\}$ is a filtration. Since $x_n$ is measurable $\mathcal{F}_n$, it follows that

$$\begin{aligned} E(x_{n+1} \mid \mathcal{F}_n) &= \sum_{i=1}^{n} y_i + E(y_{n+1} \mid \mathcal{F}_n) = x_n + E(y_{n+1}) \\ &= x_n + 0 = x_n \end{aligned}$$

Thus, $x_n$ is a martingale. Note that in this example, $\mathcal{F}_n$ does not provide information about $y_{n+1}$ as $y_i$ are independent. As a result, $E(y_{n+1} \mid \mathcal{F}_n) =$

$E(y_{n+1})$ is simply the average of $y_{n+1}$ itself without borrowing any information from the $y_i$ variables observed prior to time $n+1$.

**Example 23**. Let $y_i$ be a sequence of random variables with mean $E(y_i) = 0$ $(i \geq 1)$. Instead of assuming independence as Example 22, let $E(y_i \mid y_1, ..., y_{i-1}) = 0$ for $i = 2, 3....$ Let

$$x_n = \sum_{i=1}^{n} y_i, \quad \mathcal{F}_n = \sigma(x_1, \ldots, x_n) = \sigma(y_1, \ldots, y_n)$$

Since $x_n$ is measurable $F_n$, it follows that

$$E(x_{n+1} \mid \mathcal{F}_n) = \sum_{i=1}^{n} y_i + E(y_{n+1} \mid \mathcal{F}_n) = x_n + E(y_{n+1})$$
$$= x_n + 0 = x_n$$

Thus, $x_n$ is again a martingale.

In the above example, if we interpret $y_n$ as the earnings on the $n$th play by a person in a gambling game, $x_n$ represents the gambler's fortune at time $n$. The quantity $E(y_{n+1} \mid \mathcal{F}_n)$ represents the expected earnings on the $(n+1)$th play by the gambler. Since this expected earning is 0, the gambler has a fair chance to win or lose on the $(n+1)$th play. The $\sigma$-field $\mathcal{F}_n$ represents the information available to the gambler right before the $(n+1)$th play and $E(x_{n+1} \mid \mathcal{F}_n)$ the expected fortune after the $(n+1)$ play. The martingale structure in this example represents a fair gambling game.

**Example 24**. In Example 22, let $y_i$ be an independent sample with mean $E(y_i) = \mu \neq 0$. Let $x_n$ and $\mathcal{F}_n$ be defined as in that example. It is readily checked that $\{\mathcal{F}_n; n \geq 1\}$ is again a filtration. However, since

$$E(x_{n+1} \mid \mathcal{F}_n) = \sum_{i=1}^{n} y_i + E(y_{n+1} \mid \mathcal{F}_n) = y_n + E(y_{n+1})$$
$$= x_n + \mu \neq x_n$$

$x_n$ is not a martingale. For $\mu > 0$, $E(x_{n+1} \mid \mathcal{F}_n) > x_n$ and $x_n$ is called a *submartingale*, while for $\mu < 0$, $E(x_{n+1} \mid \mathcal{F}_n) < x_n$ and $x_n$ is called a *supermartingale*.

Within the context of a gambling game, the submartingale structure of $x_n$ implies that the game is not fair and in favor of the gambler, while the supermartingale structure indicates an unfair game against the gambler. Since

$$E(x_{n+1}) = E[E(x_{n+1} \mid \mathcal{F}_n)] = E(x_n) + \mu$$

the gambler is expected to win (lose) $\mu$ amount of earnings on each additional plan with a total fortune (debt) of $n\mu$ after $n$ plays if $\mu$ is positive (negative).

The martingale structure also has applications in other areas of research. For example, for a short period of time, the day-to-day fluctuation of a stock price or investment may follow a martingale structure because of the relatively stable economy. However, the monthly or annual yields of a stock price or investment are likely to follow a submartingale as they are augmented by the infusion of company and investment profits.

Let $h(y_1, \ldots, y_m)$ be a symmetric kernel of a one-group $U$-statistic with $m$ arguments and $\theta = E[h(y_1, \ldots, y_m)]$. Let $\mathcal{F}_0 = \{\Phi, \Omega\}$ be the trivial $\sigma$-field and $\mathcal{F}_k = \sigma(y_1, \ldots, y_k)$ for $1 \le k \le m$. For an integer $k$ $(1 \le k \le m)$, let

$$\widetilde{h} = h - \theta, \quad h_k(y_1, \ldots, y_k) = E(h \mid y_1, \ldots, y_k), \quad \widetilde{h}_k = h_k - \theta$$

Then, $h_k(y_1, \ldots, y_k)$ is a random variable since it is the conditional expectation of $h$ given $y_1, \ldots, y_k$. Since $E(h_k) = 0$, the *centered* version, $\widetilde{h}_k$, has mean $E\left(\widetilde{h}_k\right) = 0$.

Since $\mathcal{F}_k \subseteq \mathcal{F}_l$ for $k \le l$, it follows from the law of iterated conditional expectation that for $1 \le k \le m$

$$\begin{aligned} E(h_1(y_1) \mid \mathcal{F}_0) &= E[E(h \mid y_1) \mid \Omega] = E(h) = \theta \\ E\left(\widetilde{h}_1(y_1) \mid \mathcal{F}_0\right) &= E[(h_1(y_1) - \theta) \mid \mathcal{F}_0] = 0 \\ E(h_k \mid \mathcal{F}_{k-1}) &= E[E(h \mid \mathcal{F}_k) \mid \mathcal{F}_{k-1}] = E(h \mid \mathcal{F}_{k-1}) = h_{k-1} \\ E\left(\widetilde{h}_k \mid \mathcal{F}_{k-1}\right) &= E(h_k \mid \mathcal{F}_{k-1}) - \theta = h_{k-1} - \theta = \widetilde{h}_{k-1} \\ h_m &= E(h \mid y_1, \ldots, y_m) = h \end{aligned} \tag{3.45}$$

Note that for notational brevity, we expressed $E(h_k \mid y_1, \ldots, y_{k-1})$ using $E(h_k \mid \mathcal{F}_{k-1})$ in (3.45).

For $1 \le k \le m$, define

$$g_k(y_1, \ldots, y_k) = \widetilde{h}_k - \sum_{l=1}^{k-1} \sum_{(i_1, \ldots, i_l) \in C_l^k} g_l(y_{i_1}, \ldots, y_{i_l})$$

To appreciate the above, consider three special cases with $k = 1$, 2 and 3 for which the above reduces to

$$g_1(y_1) = \widetilde{h}_1, \quad g_2(y_1, y_2) = \widetilde{h}_2 - g_1(y_1) - g_1(y_2) \tag{3.46}$$

$$g_3(y_1, y_2, y_3) = \widetilde{h}_3 - \sum_{i=1}^{3} g_1(y_i) - \sum_{(i,j)\in C_2^3} g_2(y_i, y_j)$$

By applying (3.45) to the above, we obtain

$$\begin{aligned} E(g_1) &= E\left(\widetilde{h}_1\right) = 0 \\ E[g_2(y_1, y_2) \mid y_1] &= E\left(\widetilde{h}_2 \mid y_1\right) - g_1(y_1) - E[g_1(y_2)] = \widetilde{h}_1 - \widetilde{h}_1 - E\left(\widetilde{h}_1\right) \\ &= 0 \end{aligned}$$

By exchangeability and symmetry, we have

$$E[g_2(y_1, y_3) \mid y_1] = E[g_2(y_2, y_3) \mid y_2] = E[g_2(y_1, y_2) \mid y_1] = 0 \tag{3.47}$$

It follows from (3.45) and (3.47) that

$$\begin{aligned} E(g_3 \mid y_1, y_2) &= E\left(\widetilde{h}_3 \mid y_1, y_2\right) - \sum_{i=1}^{3} E[g_1(y_i) \mid y_1, y_2] \\ &\quad - \sum_{(i,j)\in C_2^3} E[g_2(y_i, y_j) \mid y_1, y_2] \\ &= \widetilde{h}_2(y_1, y_2) - g_1(y_1) - g_1(y_2) - g_2(y_1, y_2) - \\ &\quad - E[g_2(y_1, y_3) \mid y_1, y_2] - E[g_2(y_2, y_3) \mid y_1, y_2] \\ &= g_2(y_1, y_2) - g_2(y_1, y_2) - E[g_2(y_1, y_3) \mid y_1] \\ &\quad - E[g_2(y_2, y_3) \mid y_2] \\ &= 0 \end{aligned}$$

More generally, it is similarly shown that (see exercise)

$$E[g_k(y_1, \ldots, y_k) \mid \mathcal{F}_{k-1}] = 0, \quad 1 \leq k \leq m \tag{3.48}$$

Note that since $y_i$ are an i.i.d. sample, $y_i$ are exchangeable. Thus, (3.48) holds true for any $k$-member subset $y_{i_1}, \ldots, y_{i_k}$ with $\mathcal{F}_{k-1}$ defined by any of the $k-1$ variables of this subset.

The following theorem shows the martingale structure of a one-group $U$-statistic.

**Theorem 2.** Let

$$S_{kn} = \sum_{(i_1,\ldots,i_k)\in C_k^n} g_k\left(y_{i_1},\ldots,y_{i_k}\right), \quad 1 \le k \le m \tag{3.49}$$

Then, we have:

1. For each $k$, when considered as a random sequence of $n$, $S_{kn}$ is a martingale with respect to the filtration, $\{\mathcal{F}_n = \sigma(y_1,\ldots,y_n)\,; n \ge k\}$.

2. The one-sample U-statistic, $U_n$, defined in (3.8) has the following representation:

$$U_n - \theta = \sum_{k=1}^{m} \binom{m}{k} \binom{n}{k}^{-1} S_{kn} \tag{3.50}$$

Proof. We consider the case with $m = 2$ and the more general case is similarly considered. By (3.49), we have

$$\begin{aligned}
S_{1n} &= \sum_{i=1}^{n} g_1(y_i) = \sum_{i=1}^{n} \widetilde{h}_1(y_i), \\
S_{2n} &= \sum_{(i,j)\in C_2^n} g_2(y_i, y_j) = \sum_{(i,j)\in C_2^n} \left[\widetilde{h}_2(y_i, y_j) - g_1(y_i) - g_1(y_j)\right]
\end{aligned}$$

It follows that

$$\begin{aligned}
U_n - \theta &= \binom{n}{2}^{-1} \sum_{(i,j)\in C_2^n} (h - \theta) = \binom{n}{2}^{-1} \sum_{(i,j)\in C_2^n} \widetilde{h}(y_i, y_j) \\
&= \binom{n}{2}^{-1} \sum_{(i,j)\in C_2^n} \left[g_2(y_i, y_j) + g_1(y_i) + g_1(y_j)\right] \\
&= \binom{n}{2}^{-1} \left[S_{2n} + \sum_{(i,j)\in C_2^n} \left(g_1(y_i) + g_1(y_j)\right)\right] \\
&= \binom{n}{2}^{-1} \left[S_{2n} + (n-1) S_{1n}\right] \\
&= \binom{n}{2}^{-1} S_{2n} + \frac{2}{n} S_{1n}
\end{aligned}$$

The last expression is the same as the right side in (3.50) when written out in terms of $S_{2n}$ and $S_{1n}$.

To show that both $S_{1n}$ and $S_{2n}$ are martingales with respect to $\mathcal{F}_n$, first consider $S_{1n}$. For $1 \leq l \leq n$, if $i > l$, then $y_i$ and $\mathcal{F}_l$ are independent. It follows that

$$E\left[g_1\left(y_i\right) \mid \mathcal{F}_l\right] = E\left[g_1\left(y_i\right)\right] = E\left[\widetilde{h}_1\left(y_i\right)\right] = 0$$

Otherwise, for $1 \leq i \leq l$, $g_1\left(y_i\right)$ is measurable $\mathcal{F}_l$ and $E\left[g_1\left(y_i\right) \mid \mathcal{F}_l\right] = g_1\left(y_i\right)$. It follows that

$$E\left(S_{1n} \mid \mathcal{F}_l\right) = \sum_{i=1}^{n} E\left[g_1\left(y_i\right) \mid \mathcal{F}_l\right] = \sum_{i=1}^{l} g_1\left(y_i\right) = S_{1l}$$

Now, consider $S_{2n}$. If $i > l$, then $y_i$ and $\mathcal{F}_l$ are independent. It follows from (3.48) that

$$\begin{aligned} E\left[g_2\left(y_i, y_j\right) \mid \mathcal{F}_l\right] &= E\left[E\left(g_2\left(y_i, y_j\right) \mid y_j, y_1, \ldots, y_l\right) \mid \mathcal{F}_l\right] \\ &= E\left[E\left(g_2\left(y_i, y_j\right) \mid y_j\right) \mid \mathcal{F}_l\right] = E\left(0 \mid \mathcal{F}_l\right) \\ &= 0 \end{aligned} \tag{3.51}$$

Similarly, it is readily checked that the above also holds true if $j > l$ (see exercise). Thus, it follows that

$$E\left(S_{2n} \mid \mathcal{F}_l\right) = \sum_{(i,j) \in C_2^n} E\left[g_2\left(y_i, y_j\right) \mid \mathcal{F}_l\right] = \sum_{(i,j) \in C_2^l} g_2\left(y_i, y_j\right) = S_{2l}$$

**Example 25**. Let $y_i$ be an i.i.d. random sample and consider the sample mean: $U_n = \overline{y}_n$, which is a one-sample $U$-statistic with one argument. Thus, $m = c = 1$ and the kernel $h(y) = y$. It follows from Theorem 2 that

$$g_1(y) = h(y) = y, \quad S_{1n} = \sum_{i=1}^{n} \left(y_i - \theta\right)$$

$$U_n - \theta = \frac{1}{n} \sum_{i=1}^{n} \left(y_i - \theta\right)$$

and $S_{1n}$ is a martingale with respect to the filtration

$$\left\{\mathcal{F}_n = \sigma\left(y_1, \ldots, y_n\right); n \geq 1\right\}$$

We can also verify this directly using the definition of martingale:

$$E\left(S_{1(n+1)} \mid \mathcal{F}_n\right) = S_{1n} + E\left(y_{n+1} - \theta \mid \mathcal{F}_n\right) = S_{1n}$$

Note that it follows from (3.50) that

$$U_n - \theta = \binom{m}{1}\binom{n}{1}^{-1} S_{1n} + \sum_{c=2}^{m} \binom{m}{c}\binom{n}{c}^{-1} S_{cn} \tag{3.52}$$
$$= \frac{m}{n}\sum_{i=1}^{n} \widetilde{h}_1(y_i) + \sum_{c=2}^{m} \binom{m}{c}\binom{n}{c}^{-1} S_{cn}$$

Thus, the centered U-statistic-like quantity $U_n - \theta$ ($U_n - \theta$ is a U-statistic if $\theta$ is known) can be expressed as an i.i.d. random sum $\frac{m}{n}\sum_{i=1}^{n} \widetilde{h}_1(y_i)$ plus an extra term $\sum_{c=2}^{m} \binom{m}{c}\binom{n}{c}^{-1} S_{cn}$. This decomposition plays an important role in the study of asymptotic properties of U-statistics. If the second term in (3.52) is *asymptotically negligible*, that is, $U_n$ and the first term have the same asymptotic distribution, then inference for $\theta$ can be based on the asymptotic distribution of the i.i.d. sum. This is indeed the case as we show in Section 3.2.

## 3.2 Inference for U-Statistics

In this section, we discuss inference for univariate U-statistics. More specifically, we want to investigate the asymptotic properties of such statistics when used as estimates of parameters of interest. In the case of U-statistics, however, it is not possible to directly apply conventional methods [e.g., law of large numbers (LLN), central limit theorem (CLT)] to examine the large same behavior of such statistics, as in the case of the study of regression models in Chapter 2. To illustrate this point, consider again the problem of estimating the variance, $\sigma^2$, of an i.i.d. random sample $y$ introduced in the beginning of Section 3.1. Based on the preceding section, we can represent the usual sample variance $s_n^2$ in the form of a one-group U-statistic as follows:

$$s_n^2 = \binom{n}{2}^{-1} \sum_{(i,j)\in C_2^n} h(y_i, y_j) = \binom{n}{2}^{-1} \sum_{(i,j)\in C_2^n} \frac{1}{2}(y_i - y_j)^2 \tag{3.53}$$

By taking expectation, it immediately follows that $s_n^2$ is an unbiased estimate of $\sigma^2$. However, if we want to study the asymptotic behaviors of this statistic, such as consistency, the U-statistic expression above is not in a form that is conducive to the usual treatment by direct applications of LLN and CLT. The problem lies in the unique definition of U-statistic.

Unlike the statistics we studied in Chapter 2, U-statistics are defined by responses from multiple rather than a single subject. For example, in the case of the sample variance $s_n^2$, the U-statistic in (3.53) expresses the variability of $y_i$ by averaging squared differences over all pairs of responses, yielding a sum of correlated random variables. Although we have seen sums of correlated random variables before, they are fundamentally different from the U-statistic in (3.53). For example, the conventional expression of $s_n^2$ in (3.1) is also a sum of dependent variables. However, by rewriting $s_n^2$ in the following form:

$$s_n^2 = \frac{1}{n} \sum_{i=1}^{n} (y_i - \mu)^2 + (\overline{y}_n - \mu)^2 \tag{3.54}$$

we can apply LLN, CLT, and Slutsky's theorem to obtain the asymptotic properties of $s_n^2$ (see Chapter 1 for details). The key difference between the two approaches is that the quantity that causes the dependence in the conventional expression, $\overline{y}_n$, converges to $\mu$ in probability so that the dependence can be removed *asymptotically* by substituting $\mu$ in place of $\overline{y}_n$ as expressed in (3.54). However, the dependence in the U-statistic expression in (3.53) cannot be *eliminated* in such a *substitution* fashion.

The basic idea behind tackling the dependence of a U-statistic is to create a random quantity, called *projection*, in the usual form of a sum of i.i.d. random variables that approximates the original U-statistic so well that they become indistinguishable asymptotically, that is, they have the same asymptotic distribution. Since we know how to find asymptotic distributions of sums of i.i.d. random variables, the creation of such a projection of random variables will then allow us to find the asymptotic distribution of the underlying U-statistic. The existence of such a sum of i.i.d. random variables was suggested by Theorem 2 of Section 3.1.4. As noted earlier in the discussion of this theorem, we can express a one-sample U-statistic with $m$ arguments as follows:

$$\begin{aligned} U_n - \theta &= \binom{m}{1}\binom{n}{1}^{-1} S_{1n} + \sum_{k=2}^{m} \binom{m}{k}\binom{n}{k}^{-1} S_{kn} \\ &= \frac{m}{n} \sum_{i=1}^{n} \widetilde{h}_1 (y_i) + \sum_{k=2}^{m} \binom{m}{k}\binom{n}{k}^{-1} S_{kn} \end{aligned} \tag{3.55}$$

where $\theta = E(h)$ and $\widetilde{h}_1 (y_1) = E[(h_1 (y_1, \ldots, y_m) - \theta) \mid y_1]$. The first term above is a sum of i.i.d. random quantities. Thus, if the second term in (3.55) is asymptotically negligible, that is, $U_n$ and the first term have the same asymptotic distribution, then $\frac{m}{n} \sum_{i=1}^{n} \widetilde{h}_1 (y_i)$ will be the projection we seek. We now formalize this approach and systematically study such a decomposition of a U-statistic.

### 3.2.1 Projection of U-statistic

Consider again the class of one-sample U-statistics with $m$ arguments:

$$U_n = \binom{n}{m}^{-1} \sum_{(i_1,\ldots,i_m)\in C_m^n} h(\mathbf{y}_{i_1},\ldots,\mathbf{y}_{i_m}) \tag{3.56}$$

where $C_m^n \{(i_1,\ldots,i_m); 1 \le i_1 < \ldots < i_m \le n\}$ denotes the set of all distinct combinations of $m$ indices $(i_1,\ldots,i_m)$ from the integer set $\{1,2,\ldots,n\}$ and $h(y_{i_1},\ldots,y_{i_m})$ a symmetric kernel function defining an i.i.d. sample of $p \times 1$ random vectors $\mathbf{y}_i$ $(1 \le i \le n)$. The projection of $U_n$, denoted by $\widehat{U}_n$, is defined as follows:

$$\widehat{U}_n = \sum_{i=1}^{n} E(U_n \mid \mathbf{y}_i) - (n-1)\theta \tag{3.57}$$

where $\theta = E[h(\mathbf{y}_{i_1},\ldots,\mathbf{y}_{i_m})]$ is the parameter of interest. The appearance of $E(U_n \mid \mathbf{y}_i)$ is a bit deceiving since the fact that $U_n$ is a statistic easily carries the impression that $\sum_{i=1}^n E(U_n \mid \mathbf{y}_i)$ is also a statistic. However, as is seen shortly in Example 1, $E(U_n \mid \mathbf{y}_i)$ is generally a function of $\theta$ and thus is not a statistic. Nonetheless, what is important about $\widehat{U}_n$ is the fact that it is an i.i.d. sum to allow us to readily study its properties using conventional asymptotic techniques. Note that the first term in (3.57) is a sum of i.i.d. random variables, $E(U_n \mid \mathbf{y}_i)$, which is the conditional expectation of $U_n$ given each individual response $\mathbf{y}_i$ $(1 \le i \le n)$, while the second part is a normalizing factor so that $E\left(\widehat{U}_n\right) = \theta$. Thus, $\widehat{U}_n$ can be viewed as *projecting* $U_n$ onto each $\mathbf{y}_i$ of the i.i.d. sample $(1 \le i \le n)$.

**Example 1 (*U-statistic for sample mean*).** Let $y_i$ be an i.i.d. sample with mean $\mu$ and variance $\sigma^2$ $(1 \le i \le n)$. Consider estimating $\mu$ with the sample mean, $U_n = \overline{y}_n$, which is a $U$-statistic with kernel $h(y) = y$. Since $E(y_j \mid y_i) = y_i$ if $j = i$ and $E(y_j \mid y_i) = \mu$ if otherwise, it follows that

$$E(U_n \mid y_i) = E(\overline{y}_n \mid y_i) = \frac{1}{n}\sum_{j=1}^{n} E(y_j \mid y_i)$$
$$= \frac{1}{n} y_i + \frac{(n-1)}{n}\mu$$

Thus, $\sum_{i=1}^{n} E(U_n \mid y_i) = \sum_{i=1}^{n} \frac{1}{n} y_i + \mu(n-1)$ and

$$\widehat{U}_n = \sum_{i=1}^{n} \left[ \frac{1}{n} y_i + \frac{(n-1)}{n} \mu \right] - (n-1)\mu = \overline{y}_n \tag{3.58}$$

In this case, the projection $\widehat{U}_n$ is the same as the original U-statistic. However, it is seen that the first term, $\sum_{i=1}^{n} E(U_n \mid y_i)$, involves the unknown parameter $\mu$ and thus is not a statistic. Nonetheless, the projection $\widehat{U}_n$ in this particular application is a statistic and is the same as the original statistic $\overline{y}_n$.

**Example 2** (***U-statistic for sample variance***). In Example 1, consider the sample variance $s_n^2$ as an estimate of $\sigma^2$. As shown in the preceding section, the U-statistic expression of $s_n^2$ has a symmetric kernel $h(y_j, y_k) = \frac{1}{2}(y_j - y_k)^2$. It follows that

$$E\left(s_n^2 \mid y_i\right) = \binom{n}{2}^{-1} \sum_{(j,k)\in C_2^n} E\left(h(y_j, y_k) \mid y_i\right)$$

To compute $E(h(y_j, y_k) \mid y_i)$, notice that when $(j,k)$ ranges over $C_2^n$, either $j$ or $k$ equals $i$ or neither is equal to $i$. In the first case, that is, $j = i$ or $k = i$, we have by symmetry of $h(y_j, y_k)$:

$$\begin{aligned} E\left(h(y_i, y_k) \mid y_i\right) = E\left(h(y_j, y_i) \mid y_i\right) &= E\left[\frac{1}{2}(y_i - y_k)^2 \mid y_i\right] \\ &= \frac{1}{2}\left(y_i^2 - 2\mu y_i + E\left(y_i^2\right)\right) \end{aligned} \tag{3.59}$$

In the second case, that is, $j \neq i$ and $k \neq i$, it follows that

$$E\left(h(y_j, y_k) \mid y_i\right) = E\left(h(y_j, y_k)\right) = \sigma^2 \tag{3.60}$$

Thus, $E(h(y_j, y_k) \mid y_i)$ can be characterized by whether $i \in \{j,k\}$. Let $h_1(y_i) = E(h(y_j, y_k) \mid y_i)$ when $i \in \{j,k\}$. Then, we can compactly express (3.59) and (3.60) as:

$$E\left(h(y_j, y_k) \mid y_i\right) = \begin{cases} h_1(y_i) & \text{if } i \in \{j,k\} \\ \sigma^2 & \text{if } i \notin \{j,k\} \end{cases} \tag{3.61}$$

Note that because of the exchangeability of $y_i$, (3.61) can be rewritten as: $h_1(y_1) = E(h(y_1, y_2) \mid y_1)$. We introduced $h_1$ as well as its centered version

$\widetilde{h}_1 = h_1 - \theta$ in Section 3.1.4 for a general one-sample U-statistic with $m$ arguments in the form of

$$h_1(y_1) = E[h(y_1, \ldots, y_m) \mid y_1]$$

where $\theta$ is the mean of the kernel $h$. This conditional expectation plays a critical role in the study of asymptotic behavior of U-statistics.

It follows from (3.61) that

$$\begin{aligned}
E\left(s_n^2 \mid y_i\right) &= \binom{n}{2}^{-1}\left[\sum_{j \neq i} E\left(h\left(y_j, y_i\right) \mid y_i\right) + \sum_{(j,k) \in D_{jk}} E\left(h\left(y_j, y_k\right) \mid y_i\right)\right] \\
&= \binom{n}{2}^{-1}\left[\binom{n-1}{1} h_1\left(y_i\right) + \binom{n-1}{2}\sigma^2\right] \\
&= \binom{n}{2}^{-1}\left[(n-1)\frac{1}{2}\left(y_i^2 - 2\mu y_i + E\left(y_i^2\right)\right) + \binom{n-1}{2}\sigma^2\right] \\
&= \frac{1}{n}\left(y_i^2 - 2\mu y_i + E\left(y_i^2\right)\right) + \frac{n-2}{n}\sigma^2
\end{aligned}$$

where $D_{jk} = \{(l,m) : (l,m) \in C_2^n, l \neq i, m \neq i\}$. It follows that the projection of $U_n = s_n^2$ is given by

$$\begin{aligned}
\widehat{U}_n &= \sum_{i=1}^{n}\left[\frac{1}{n}\left(y_i^2 - 2\mu y_i + E\left(y_i^2\right)\right) + \frac{n-2}{n}\sigma^2\right] - (n-1)\sigma^2 \\
&= \frac{2}{n}\sum_{i=1}^{n}\frac{1}{2}\left(y_i^2 - 2\mu y_i + E\left(y_i^2\right)\right) - \sigma^2
\end{aligned}$$

Unlike Example 1, neither the i.i.d. sum $\sum_{i=1}^{n} E(U_n \mid y_i)$ nor the projection itself is a statistic. In fact, the projection bears no resemblance to the original U-statistic in this example at all. Nonetheless, $\widehat{U}_n$ has the desired form of a sum of an i.i.d. random quantities.

For the general $\widehat{U}_n$ in (3.57), by applying an argument similar to the projection in Example 2, we have

$$\begin{aligned}
E\left(U_n \mid y_i\right) &= \binom{n}{m}^{-1}\left[\sum_{(i_1,\ldots,i_m) \in C_m^n}\left(\sum_{i \in G_i} E\left(h \mid y_i\right) + \sum_{i \notin G_i} E\left(h \mid y_i\right)\right)\right] \\
&= \binom{n}{m}^{-1}\left[\binom{n-1}{m-1} h_1\left(y_i\right) + \binom{n-1}{m}\theta\right] \\
&= \frac{m}{n} h_1\left(y_i\right) + \frac{n-m}{n}\theta
\end{aligned}$$

where $G_i = \{i_1, \ldots, i_m\}$. It follows that the projection $\widehat{U}_n$ can be expressed as:

$$\widehat{U}_n = \frac{m}{n} \sum_{i=1}^{n} \widetilde{h}_1(y_i) + \theta \tag{3.62}$$

As $U_n$ is an unbiased estimate of the parameter of interest $\theta$, it is convenient to center $U_n$ as follows when studying its asymptotic behaviors:

$$U_n - \theta = \binom{n}{m}^{-1} \sum_{(i_1,\ldots,i_m) \in C_m^n} \widetilde{h}(y_{i_1}, \ldots, y_{i_m}) \tag{3.63}$$

where $\widetilde{h} = h - \theta$. Analogously, we can define a *centered projection*:

$$\widehat{U}_n - \theta = \sum_{i=1}^{n} E(U_n \mid y_i) - n\theta = \sum_{i=1}^{n} \left[E(U_n \mid y_i) - \theta\right] \tag{3.64}$$

It follows from (3.62) that

$$\widehat{U}_n - \theta = \frac{m}{n} \sum_{i=1}^{n} \widetilde{h}_1(y_i) \tag{3.65}$$

Since $E\left(\widetilde{h}_1(y_i)\right) = E\left[E(h \mid y_i) - \theta\right] = 0$, it immediately follows from (3.65) that $E\left(\widehat{U}_n\right) = \theta$. Thus, both $U_n - \theta$ and $\widehat{U}_n - \theta$ have mean zero.

**Example 3**. Consider Example 1 again. We have $U_n = \overline{y}_n$ and $h(y) = y$. Since

$$h_1(y_i) = E\left[h(y_i) \mid y_i\right] = E(y_i \mid y_i) = y_i$$
$$\widetilde{h}_1(y_i) = h_1(y_i) - \theta = y_i - \theta$$

it follows that

$$\widehat{U}_n - \theta = \frac{1}{n} \sum_{i=1}^{n} \widetilde{h}_1(y_i) = \frac{1}{n} \sum_{i=1}^{n} (y_i - \theta)$$

It immediately follows from LLN and CLT that

$$\widehat{U}_n \to_p \theta, \quad \sqrt{n}\left(\widehat{U}_n - \theta\right) = \sqrt{n}\left(\overline{y}_n - \theta\right) \to_d N\left(0, \sigma^2\right)$$

For this U-statistic, we can also apply CLT directly to $U_n$ to obtain the same asymptotic distribution as above.

**Example 4**. In Example 2, we have

$$\widehat{U}_n - \sigma^2 = \frac{1}{n} \sum_{i=1}^{n} 2 \left[ h_1(y_i) - \sigma^2 \right]$$

By applying LLN and CLT, we have $\widehat{U}_n \to_p \sigma^2$ and

$$\sqrt{n} \left( \widehat{U}_n - \sigma^2 \right) = \frac{\sqrt{n}}{n} \sum_{i=1}^{n} 2 \left[ h_1(y_i) - \theta \right] \to_p N \left( 0, 4Var \left( h_1(y_1) \right) \right) \qquad (3.66)$$

To find $Var(h_1(y_1))$, first note that

$$\begin{aligned}
\widetilde{h}_1(y_1) &= E\left[h(y_1, y_2) \mid y_1\right] - \sigma^2 = E\left[\frac{1}{2}(y_1 - y_2)^2 \mid y_1\right] - \sigma^2 \\
&= \frac{1}{2}\left[y_1^2 - y_1\mu + E\left(y_2^2\right)\right] - \sigma^2 = \frac{1}{2}\left(y_1^2 - y_1\mu + \mu^2 + \sigma^2\right) - \sigma^2 \\
&= \frac{1}{2}\left[(y_1 - \mu)^2 - \sigma^2\right]
\end{aligned}$$

Thus,

$$\begin{aligned}
Var\left(h_1(y_1)\right) &= E\left(\widetilde{h}_1(y_1)\right)^2 = \frac{1}{4} Var\left((y_1 - \mu)^2\right) \\
&= \frac{1}{4}\left(\mu_4 - \sigma^4\right)
\end{aligned} \qquad (3.67)$$

where $\mu_4 = E(y_1 - \mu)^4$ denotes the fourth centered moment of $y_i$. It follows from (3.66) and (3.67) that

$$\sqrt{n}\left(\widehat{U}_n - \sigma^2\right) \to_p N\left(0, \mu_4 - \sigma^4\right) \qquad (3.68)$$

In this simple example, we can also find the asymptotic distribution of $s_n^2$ directly using a Taylor stochastic series expansion as follows:

$$s_n^2 - \sigma^2 = \frac{1}{n}\sum_{i=1}^{n}(y_i - \overline{y}_n)^2 - \sigma^2 = \frac{1}{n}\sum_{i=1}^{n}\left[(y_i - \mu)^2 - \sigma^2\right] + o_p\left(n^{-\frac{1}{2}}\right)$$

By applying CLT and Slutsky's theorem to the above decomposition, we obtain

$$\sqrt{n}\left(s_n^2 - \sigma^2\right) \to_p N\left(0, \mu_4 - \sigma^4\right)$$

Both Examples 3 and 4 show that the projection and the U-statistic have the same asymptotic distribution. This is not a coincidence and is in fact true for the general U-statistics, as we show in the next section. As noted earlier, the projection in general is not a statistic and thus the projection itself can not be used for statistical inference. The role of the projection is to serve as a vehicle from which to derive the asymptotic distribution of the underlying U-statistic.

Note that in (3.68), we must assume $\mu_4 \neq \sigma^4$ since otherwise the asymptotic distribution is degenerate and $s^2$ becomes asymptotically a constant. In practice, the situation with $\mu_4 = \sigma^4$ is unlikely to occur. A notable exception is the Bernoulli random variable. If $y_i$ is a binary variable with the value 0 and 1, then

$$p = E(y_1), \quad \sigma^2 = p(1-p), \quad \mu_4 = (1-p)^4 p + p^4 (1-p)$$

When $p = \frac{1}{2}$, $\mu_4 = \left(\frac{1}{2}\right)^4$, $\sigma^4 = \left(\frac{1}{2}\right)^4$ and thus $\mu_4 - \sigma^4 = 0$. In this special case, the squared individual deviation $\left(y_i - \frac{1}{4}\right)^2$ from the theoretical mean $\frac{1}{2}$ is a constant $\frac{1}{4}$. By taking variability in the sample mean in the definition of $s_n^2$ into consideration, we have

$$\begin{aligned} s^2 &= \frac{1}{n}\sum_{i=1}^{n}(y_i - \overline{y}_n)^2 = \frac{1}{n}\sum_{i=1}^{n}\left(y_i - \frac{1}{2}\right)^2 + o_p\left(\frac{1}{\sqrt{n}}\right) \\ &= \frac{1}{4} + o_p\left(\frac{1}{\sqrt{n}}\right) \end{aligned}$$

Thus, $s^2$ is asymptotically a constant $\frac{1}{4}$.

### 3.2.2 Asymptotic Distribution of One-Group U-Statistic

To show that the centered projection, $\widehat{U}_n - \theta$, has the same asymptotic distribution as the centered U-statistic, $U_n - \theta$, consider the following relationship:

$$\sqrt{n}(U_n - \theta) = \sqrt{n}\left(\widehat{U}_n - \theta\right) + e_n \tag{3.69}$$

where $e_n$ is the *normalized* difference between $U_n$ and $\widehat{U}_n$,

$$e_n = \sqrt{n}(U_n - \theta) - \sqrt{n}\left(\widehat{U}_n - \theta\right) = \sqrt{n}\left(U_n - \widehat{U}_n\right) \tag{3.70}$$

If $e_n = o_p(1)$, then it follows from Slutsky's theorem that $U_n - \theta$ and $\widehat{U}_n - \theta$ have the same asymptotic distribution. This is indeed the case, but requires some effort to prove. We summarize the asymptotic results regarding the one-sample U-statistic in a theorem below.

**Theorem 1** (Univariate one-sample). Let

$$h_1(\mathbf{y}_1) = E[h(\mathbf{y}_1, \ldots, \mathbf{y}_m) \mid \mathbf{y}_1], \quad \widetilde{h}_1(\mathbf{y}_1) = h_1(\mathbf{y}_1) - \theta$$
$$e_n = \sqrt{n}\left(U_n - \widehat{U}_n\right), \quad \sigma_h^2 = Var\left[\widetilde{h}_1(\mathbf{y}_1)\right]$$

Then, $e_n \to_p 0$. Thus,

1. $U_n$ is consistent, that is, $U_n \to_p \theta$.
2. $U_n$ is asymptotically normal, $\sqrt{n}(U_n - \theta) \to_d N\left(0, \sigma_U^2 = m^2\sigma_h^2\right)$.

Proof. In light of the relationship between $U_n$ and $\widehat{U}_n$ as expressed in (3.69) and (3.70), the conclusions follow if we can show $e_n \to_p 0$. Since $E\left(e_n^2\right) \to_p 0$ implies $e_n \to_p 0$ (see exercise in Chapter 1), we prove $E\left(e_n^2\right) \to_p 0$ below. To this end, we first prove the following lemma:

**Lemma**. For $1 \le l \le m$, let

$$h_l(y_1, \ldots, y_l) = E(h \mid \mathbf{y}_1, \ldots, \mathbf{y}_l), \quad \widetilde{h}_l = h_l - \theta$$
$$\sigma_l^2 = Var(h_l(\mathbf{y}_1, \ldots, \mathbf{y}_l)) = E\left[\widetilde{h}_l(\mathbf{y}_1, \ldots, \mathbf{y}_l)\right]^2$$

Then, we have

$$Var(U_n) = \sum_{l=0}^{n} \binom{n}{m}^{-1} \binom{m}{l} \binom{n-m}{m-l} \sigma_l^2 \tag{3.71}$$
$$= \frac{m^2}{n} Var(h_1(\mathbf{y}_1)) + O\left(n^{-2}\right)$$

where $\sigma_0^2 \equiv 0$.

Proof of Lemma. By expanding $Var(U_n)$, we obtain

$$Var(U_n) = \binom{n}{m}^{-2} \sum_{(i_1,\ldots,i_m),(j_1,\ldots,j_m) \in C_m^n} E\left[\widetilde{h}(\mathbf{y}_{i_1}, \ldots, \mathbf{y}_{i_m}) \widetilde{h}(\mathbf{y}_{j_1}, \ldots, \mathbf{y}_{j_m})\right]$$

Let $(k_1, \ldots, k_l)$ be the indices common to $(i_1, \ldots, i_m)$ and $(j_1, \ldots, j_m)$. It then follows that

$$I_{i_1,\ldots,i_m;j_1,\ldots,j_m} = E\left[\widetilde{h}(\mathbf{y}_{i_1}, \ldots, \mathbf{y}_{i_m}) \widetilde{h}(\mathbf{y}_{j_1}, \ldots, \mathbf{y}_{j_m})\right] \tag{3.72}$$
$$= E\left\{E\left[\widetilde{h}(\mathbf{y}_{i_1}, \ldots, \mathbf{y}_{i_m}) \widetilde{h}(\mathbf{y}_{j_1}, \ldots, \mathbf{y}_{j_m}) \mid \mathbf{y}_{k_1}, .., \mathbf{y}_{k_l}\right]\right\}$$
$$= E\left[E\left(\widetilde{h} \mid \mathbf{y}_{k_1}, .., \mathbf{y}_{k_l}\right) E\left(\widetilde{h} \mid \mathbf{y}_{k_1}, .., \mathbf{y}_{k_l}\right)\right]$$
$$= E\left[\widetilde{h}_l(\mathbf{y}_{k_1}, .., \mathbf{y}_{k_l})\right]^2$$
$$= \sigma_l^2$$

Since the number of distinct choices for two such sets having exactly $l$ elements in common is: $\binom{n}{m}\binom{m}{l}\binom{n-m}{m-l}$ (see exercise), it follows that

$$\begin{aligned} Var\left(U_n\right) &= \binom{n}{m}^{-2} \sum_{l=0}^{n} \binom{n}{m}\binom{m}{l}\binom{n-m}{m-l}\sigma_l^2 \qquad (3.73) \\ &= \sum_{l=0}^{n} \binom{n}{m}^{-1} \binom{m}{l}\binom{n-m}{m-l}\sigma_l^2. \end{aligned}$$

This proves the first equality in (3.71). Note that since $\sigma_0^2 = 0$ in (3.73), the summation may start from 0.

To show the second equality in (3.71), let $c\left(n,m,l\right) = \binom{n}{m}^{-1}\binom{m}{l}\binom{n-m}{m-l}$. Then, we have:

$$\begin{aligned} c\left(n,m,l\right) &= \binom{n}{m}^{-1} \binom{m}{l}\binom{n-m}{m-l} = \binom{m}{l}^2 \frac{\left[(n-m)!\right]^2}{n!\,(n-2m+l)!} \\ &= l!\binom{m}{l}^2 \frac{(n-m)^2\cdots(n-2m+l+1)^2}{\left[n\cdots(n-m+1)\right]} \\ &\quad \frac{\left[(n-2m+l)!\right]^2}{\left[(n-m)\cdots(n-2m+l+1)\right]\left[(n-2m+l)!\right]^2} \\ &= l!\binom{m}{l}^2 \frac{(n-m)\cdots(n-2m+l+1)}{n\cdots(n-m+1)} \\ &= l!\binom{m}{l}^2 \frac{1}{n^l} + o\left(\frac{1}{n^l}\right) \end{aligned}$$

where $o\left(\cdot\right)$ denotes the deterministic small $o$ (see Chapter 1 for definition and properties). The last equality above follows from the fact that as $n \to \infty$,

$$\begin{aligned} (n-m)\cdots(n-2m+l+1) &= n^m + o\left(n^m\right) \qquad (3.74) \\ n\cdots(n-m+1) &= n^{m-l} + o\left(n^m\right) \end{aligned}$$

Thus, it follows from (3.72) and (3.73) that

$$\begin{aligned} Var\left(U_n\right) &= \sum_{l=0}^{n} c\left(n,m,l\right)\sigma_l^2 = \frac{m^2}{n}\sigma_1^2 + \sum_{l=2}^{n} c\left(n,m,l\right)\sigma_l^2 \\ &= \frac{m^2}{n}\sigma_1^2 + O\left(n^{-2}\right) \end{aligned}$$

Now, consider proving $E\left(e_n^2\right) \to 0$ for the theorem. By writing out $E\left(e_n^2\right)$, we have

$$E\left(e_n^2\right) = nVar\left(\widehat{U}_n\right) - 2nCov\left(U_n, \widehat{U}_n\right) + nVar\left(U_n\right)$$

The proof involves straightforward calculations of each of three three terms above. First, let us find $nVar\left(\widehat{U}_n\right)$. Since $\widehat{U}_n$ is a sum of i.i.d. random terms as expressed in (3.57), it follows immediately that

$$nVar\left(\widehat{U}_n\right) = n\left(\frac{m}{n}\right)^2 \sum_{i=1}^{n} Var\left(h_1\left(\mathbf{y}_i\right)\right) = m^2 Var\left(h_1\left(\mathbf{y}_i\right)\right) \qquad (3.75)$$

Next, find $nCov\left(U_n, \widehat{U}_n\right)$. Let $I_{i_1,\ldots,i_m,i} = E\left[\widetilde{h}\left(\mathbf{y}_{i_1},\ldots,\mathbf{y}_{i_m}\right)\widetilde{h}_1\left(\mathbf{y}_i\right)\right]$. Then, we have

$$\begin{aligned} nCov\left(U_n, \widehat{U}_n\right) &= nE\left[\left(U_n - \theta\right)\left(\widehat{U}_n - \theta\right)\right] \\ &= m\binom{n}{m}^{-1} \sum_{i=1}^{n} \sum_{(i_1,\ldots,i_m)\in C_m^n} I_{i_1,\ldots,i_m,i} \end{aligned}$$

If $i \notin (i_1, \ldots, i_m)$, then

$$\begin{aligned} I_{i_1,\ldots,i_m,i} &= E\left[\widetilde{h}\left(\mathbf{y}_{i_1},\ldots,\mathbf{y}_{i_m}\right)\widetilde{h}_1\left(\mathbf{y}_i\right)\right] \\ &= E\left[\widetilde{h}\left(\mathbf{y}_{i_1},\ldots,\mathbf{y}_{i_m}\right)\right] E\left[\widetilde{h}_1\left(\mathbf{y}_i\right)\right] = 0 \end{aligned}$$

If $i \in (i_1, \ldots, i_m)$, we have

$$\begin{aligned} I_{i_1,\ldots,i_m,i} &= E\left[\widetilde{h}\left(\mathbf{y}_{i_1},\ldots,\mathbf{y}_{i_m}\right)\widetilde{h}_1\left(\mathbf{y}_i\right) \mid \mathbf{y}_i\right] \\ &= E\left[\widetilde{h}_1\left(\mathbf{y}_i\right) E\left(\widetilde{h}\left(\mathbf{y}_{i_1},\ldots,\mathbf{y}_{i_m}\right) \mid \mathbf{y}_i\right)\right] = E\left[\widetilde{h}_1\left(\mathbf{y}_i\right)\right]^2 \\ &= Var\left(h_1\left(\mathbf{y}_i\right)\right) \end{aligned}$$

Since for each $i$, the event $\{i \in (i_1, \ldots, i_m)\}$ occurs $\binom{n-1}{m-1}$ times (see exercise), it follows that

$$\begin{aligned} nCov\left(U_n, \widehat{U}_n\right) &= \sum_{i=1}^{n} m\binom{n}{m}^{-1}\binom{n-1}{m-1}\zeta_1 \qquad (3.76) \\ &= nm\binom{n}{m}^{-1}\binom{n-1}{m-1}\zeta_1 \\ &= m^2 Var\left(h_1\left(\mathbf{y}_1\right)\right) \end{aligned}$$

Finally, consider $nVar\left(U_n\right)$. By applying the lemma to $Var\left(U_n\right)$, we have

$$\begin{aligned} nVar\left(U_n\right) &= m^2 Var\left(h_1\left(\mathbf{y}_1\right)\right) + O\left(n^{-1}\right) \qquad (3.77) \\ &= m^2 Var\left(h_1\left(\mathbf{y}_1\right)\right) + o\left(1\right) \end{aligned}$$

Thus, it follows from (3.75), (3.76) and (3.77) that

$$E\left(e_n^2\right) = m^2 Var\left(h_1\left(\mathbf{y}_1\right)\right) - 2m^2 Var\left(h_1\left(\mathbf{y}_1\right)\right) + m^2 Var\left(h_1\left(\mathbf{y}_1\right)\right) + o\left(1\right) \to_p 0$$

With the asymptotic results in place, we can now make inference for statistical parameters of interest modeled by U-statistics. We demonstrated how to find the asymptotic distributions for the projections of the U-statistics for the sample mean and variance in Examples 3 and 4. For these two relative sample statistics, we are also able to derive their asymptotic distributions by direct applications of standard asymptotic methods. Below are some additional examples to show how to evaluate the asymptotic variance of the U-statistic. For the statistics in these examples, however, it is not possible to find their asymptotic distributions by directly applying the conventional asymptotic techniques.

**Example 5** (***One-sample Wilcoxon signed rank test***). In Section 3.1, we showed that one-sample Wilcoxon signed rank test is asymptotically equivalent to the following U-statistic:

$$U_n = \binom{n}{2}^{-1} \sum_{\{i,j\} \in C_2^n} h\left(y_i, y_j\right) = \binom{n}{2}^{-1} \sum_{\{i,j\} \in C_2^n} I_{\{y_i + y_j \leq 0\}}$$

By Theorem 1, $U_n$ is asymptotically normal. To find the asymptotic variance, first note that

$$h_1\left(y_1\right) = E\left[h\left(y_1, y_2\right) \mid y_1\right] = E\left(I_{\{y_1 + y_2 \leq 0\}} \mid y_1\right)$$

Thus, the asymptotic variance of $U_n$ is given by

$$\begin{aligned} \sigma_h^2 &= Var\left(h_1\left(y_1\right)\right) = E\left(h_1^2\left(y_1\right)\right) - \theta^2 \\ &= E\left[E\left(I_{\{y_1 + y_2 \leq 0\}} \mid y_1\right)\right]^2 - \theta^2 \end{aligned} \tag{3.78}$$

Now, consider testing the null that $y_i$ are symmetrically distribution around 0, $H_0 : \theta = \frac{1}{2}$. Let $F(y)$ denote the CDF of $y_i$. Since $y_i$ are symmetric about 0 under $H_0$, $y_i$ and $-y_i$ have the same CDF $F(y)$. To evaluate the first term in (3.78), note that given $y_1 = y$,

$$E\left(I_{\{y_1 + y_2 \leq 0\}} \mid y_1 = y\right) = E\left(I_{\{y_1 \leq -y_2\}} \mid y_1 = y\right) = 1 - F\left(y\right)$$

Let $u_i = F(y_i)$ be the probability integral transformation of $y_i$. As shown in Chapter 1, under $H_0$, $u_i$ follows $U[0,1]$, a uniform distribution on $(0,1)$.

Thus, under $H_0$, we have

$$\begin{aligned}\sigma_h^2 &= E\left[E\left(I_{\{y_1+y_2\leq 0\}} \mid y_1\right)\right]^2 - \theta^2 = E\left[1 - F\left(y_1\right)\right]^2 - \theta^2 \\ &= E\left(1-u_1\right)^2 - \theta^2 = 1 - 2E\left(u_1\right) + E\left(u_1^2\right) - \theta^2 \\ &= \frac{1}{3} - \theta^2 = \frac{1}{3} - \left(\frac{1}{2}\right)^2 = \frac{1}{12} \\ \sigma_U^2 &= 4\sigma_h^2 = \frac{1}{3}\end{aligned}$$

Under $H_0$, $U_n$ has the following asymptotic distribution:

$$\sqrt{n}\left(U_n - \frac{1}{2}\right) \to_d N\left(0, \sigma_U^2 = \frac{1}{3}\right)$$

**Example 6** (***U-statistic for covariance***). Let $\mathbf{z}_i = (x_i, y_i)^\top$ be i.i.d. pairs of continuous random variables ($1 \leq i \leq n$). Consider inference for $\theta = Cov(x_i, y_i)$, the covariance between $x_i$ and $y_i$. In Section 3.1, we introduced a U-statistic $U_n$ in (3.6) to estimate $\theta$ defined by the kernel function: $h(\mathbf{z}_i, \mathbf{z}_j) = \frac{1}{2}(x_i - x_j)(y_i - y_j)$. It follows from the Theorem 1 that $U_n$ is unbiased and asymptotically normal, with the asymptotic variance given by $\sigma^2 = 4Var\left(\widetilde{h}_1(\mathbf{z}_1)\right)$.

To find $\sigma^2$, $p_i(\mathbf{z}_i, \boldsymbol{\mu}) = (x_i - \mu_x)(y_i - \mu_y)$. Then, it is readily checked that (see exercise)

$$\begin{aligned}\widetilde{h}_1(\mathbf{z}_1) &= E\left[h(\mathbf{z}_1, \mathbf{z}_2) \mid \mathbf{z}_1\right] = E\left[\frac{1}{2}(x_1 - x_2)(y_1 - y_2) \mid \mathbf{z}_1\right] \\ &= \frac{1}{2}\left[(x_1 - \mu_x)(y_1 - \mu_y) + \theta\right]\end{aligned} \tag{3.79}$$

It follows that

$$\widetilde{h}_1(\mathbf{z}_1) = \widetilde{h}_1(\mathbf{z}_1) - \theta = \frac{1}{2}\left(p_i(\mathbf{z}_i, \boldsymbol{\mu}) - \theta\right) \tag{3.80}$$

Thus, the asymptotic variance of $U_n$ is $\sigma^2 = Var(p_1(\mathbf{z}_1, \boldsymbol{\mu}))$. Under the assumption that $\mathbf{z}_i$ is normally distributed, $\sigma^2$ can be expressed in closed form in terms of the elements of the variance matrix of $\mathbf{z}_i$ (see exercise). Otherwise, a consistent estimate of $\sigma^2$ is given by

$$\widehat{\sigma}^2 = \frac{1}{n}\sum_{i=1}^{n}\left(p_i(\mathbf{z}_i, \widehat{\boldsymbol{\mu}}) - \widehat{\theta}\right)^2, \quad \widehat{\boldsymbol{\mu}} = \left(\widehat{\mu}_x, \widehat{\mu}_y\right)^\top$$

where $\widehat{\mu}_x$ $\left(\widehat{\mu}_y\right)$ denotes some consistent estimate of $\mu_x$ $\left(\mu_y\right)$ such as the sample mean.

**Example 7** (***Kendall's*** $\tau$). Let $\mathbf{z}_i = (x_i, y_i)^\top$ be i.i.d. pairs of continuous bivariate random variables $(1 \leq i \leq n)$. In Section 3.1, we discussed Kendall's $\tau$ as a measure of association between $x_i$ and $y_i$ and estimating this correlation measure using U-statistics. Unlike the product–moment correlation, Kendall's $\tau$ defines association based on the notion of concordance and discordance and as such provides a more robust association measure. In this example, we consider inference for this important correlation index.

Kendall's $\tau$ is defined by $p_c = E\left[I_{\{(x_i-x_j)(y_i-y_j)>0\}}\right]$. It is often normalized so that it ranges between 0 and 1 to have the interpretation as a correlation coefficient. For convenience, we illustrate the consideration with $p_c$.

The U-statistic based estimate of $\theta = p_c$ is given by

$$\widehat{\theta} = \binom{n}{2}^{-1} \sum_{(i,j)\in C_2^n} h(\mathbf{z}_i, \mathbf{z}_j) = \binom{n}{2}^{-1} \sum_{(i,j)\in C_2^n} I_{\{(x_i-x_j)(y_i-y_j)>0\}}$$

It follows from Theorem 1 that $\widehat{\theta}$ is consistently and asymptotically normal. The asymptotic variance has the form:

$$\sigma_\theta^2 = 4Var\left(\widetilde{h}_1(\mathbf{z}_1)\right) = 4\left[E\left(h_1^2(\mathbf{z}_1)\right) - \theta^2\right]$$

Unlike the previous example, it is not possible to evaluate $\sigma_\theta^2$ in closed form. However, we can readily find a consistent estimate of $\sigma_\theta^2$.

Since

$$\begin{aligned} E\left(h_1^2(\mathbf{z}_1)\right) &= E\left[E\left(h(\mathbf{z}_1, \mathbf{z}_2) \mid \mathbf{z}_1\right)\right]^2 \\ &= E\left[E\left(h(\mathbf{z}_1, \mathbf{z}_2) \mid \mathbf{z}_1\right) E\left(h(\mathbf{z}_1, \mathbf{z}_2) \mid \mathbf{z}_1\right)\right] \\ &= E\left[E\left(h(\mathbf{z}_1, \mathbf{z}_2) h(\mathbf{z}_1, \mathbf{z}_3) \mid \mathbf{z}_1\right)\right] \\ &= E\left(h(\mathbf{z}_1, \mathbf{z}_2) h(\mathbf{z}_1, \mathbf{z}_3)\right) \end{aligned} \tag{3.81}$$

we can express $\sigma_\theta^2$ as:

$$\begin{aligned} \sigma_\theta^2 &= 4\left[E\left(h(\mathbf{z}_1, \mathbf{z}_2) h(\mathbf{z}_1, \mathbf{z}_3)\right) - \theta^2\right] \\ &= 4\left[E\left(I_{\{(x_1-x_2)(y_1-y_2)>0\}} I_{\{(x_1-x_3)(y_1-y_3)>0\}}\right) - \theta^2\right] \end{aligned}$$

Let $g(\mathbf{z}_1, \mathbf{z}_2, \mathbf{z}_3)$ be a symmetric version of $h(\mathbf{z}_1, \mathbf{z}_2) h(\mathbf{z}_1, \mathbf{z}_3)$. As noted in Section 3.1, such a symmetric function is readily constructed. Then, we can estimate $\zeta = E\left(h_1^2(\mathbf{z}_1)\right)$ using the following U-statistic:

$$\widehat{\zeta} = \binom{n}{3}^{-1} \sum_{(i,j,k)\in C_3^n} g(\mathbf{z}_i, \mathbf{z}_j, \mathbf{z}_k)$$

By Slutsky's theorem, we can estimate the asymptotic variance $\sigma_\theta^2$ by $\widehat{\sigma}_\theta^2 = 4\left(\widehat{\zeta} - \widehat{\theta}^2\right)$.

This example shows that we can estimate the asymptotic variance of $\widehat{\theta}$ using U-statistics. This is true for general one-sample U-statistics. If $U_n$ is defined by a symmetric kernel $h(\mathbf{y}_1, \ldots, \mathbf{y}_m)$, then $E\left(h_1^2(\mathbf{y}_1)\right)$ can be expressed as (see exercise):

$$\begin{aligned} E\left(h_1^2(\mathbf{y}_1)\right) &= E\left[h(\mathbf{y}_1, \mathbf{y}_2, \ldots, \mathbf{y}_m) h(\mathbf{y}_1, \mathbf{y}_{m+1}, \ldots, \mathbf{y}_{2m-1})\right] \\ &= E\left[g(\mathbf{y}_1, \mathbf{y}_2, \ldots, \mathbf{y}_{2m-1})\right] \end{aligned} \tag{3.82}$$

Thus, by defining a symmetric kernel based on $g(\mathbf{y}_1, \mathbf{y}_2, \ldots, \mathbf{y}_{2m-1})$, we can use U-statistics to estimate $E\left(h_1^2(\mathbf{y}_1)\right)$.

### 3.2.3 Asymptotic Distribution of K-Group U-Statistic

The projection and the asymptotic results in Theorem 1 are readily generalized to $K$-sample U-statistics. First, consider a two-sample U-statistic with $m_1$ and $m_2$ arguments for the first and second samples:

$$U_n = \left[\binom{n_1}{m_1}\binom{n_2}{m_2}\right]^{-1} \sum_{l=1}^{2} \sum_{\left(i_1,\ldots,i_{m_l}\right)\in C_{m_l}^{n_l}} h\left(\mathbf{y}_{1i_1}, \ldots, \mathbf{y}_{1i_{m_1}}; \mathbf{y}_{2i_1}, \ldots, \mathbf{y}_{2i_{m_2}}\right)$$

where $h$ is a symmetric kernel with respect to arguments within each sample and $\mathbf{y}_{ki}$ is an i.i.d. sample of $p \times 1$ column vectors $(1 \leq i \leq n_k,\ 1 \leq k \leq 2)$. The projection of $U_n$ is defined as follows:

$$\widehat{U}_n = \sum_{k=1}^{2}\sum_{i=1}^{n_k} E\left(U_n \mid \mathbf{y}_{ki}\right) - \left[(n_1 + n_2) - 1\right]\theta \tag{3.83}$$

where $\theta = E(h)$. By applying arguments similar to those used for one-sample the projection in Section 3.2.1, we can readily express the projection and its centered version in the following form (see exercise):

$$\widehat{U}_n = \sum_{k=1}^{2} \frac{m_k}{n_k} \sum_{i=1}^{n_k} \widetilde{h}_{1k}(\mathbf{y}_{ki}) + \theta, \quad \widehat{U}_n - \theta = \sum_{k=1}^{2} \frac{m_k}{n_k} \sum_{i=1}^{n_k} \widetilde{h}_{1k}(\mathbf{y}_{ki}) \tag{3.84}$$

where $\widetilde{h}_{1k}(\mathbf{y}_{ki}) = h_{1k}(\mathbf{y}_{ki}) - \theta$ and $h_{1k}(\mathbf{y}_{ki})$ is defined by

$$h_{1k}(\mathbf{y}_{k1}) = E\left[h(\mathbf{y}_{11}, \ldots, \mathbf{y}_{1m_1}; \mathbf{y}_{21}, \ldots, \mathbf{y}_{2m_2}) \mid \mathbf{y}_{k1}\right], \quad k = 1, 2$$

Using an argument similar to that for the one-sample U-statistic, we can also show that $U_n$ and $\widehat{U}_n$ have the same asymptotic distribution (see exercise).

To find the asymptotic distribution of $\widehat{U}_n$, let $n = n_1 + n_2$ and assume $\lim_{n\to\infty} \frac{n}{n_k} = \rho_k^2 < \infty$ $(k = 1, 2)$. Then,

$$\begin{aligned}\sqrt{n}\left(\widehat{U}_n - \theta\right) &= m_1 \frac{\sqrt{n}}{\sqrt{n_1}} \frac{\sqrt{n_1}}{n_1} \sum_{i=1}^{n_1} \widetilde{h}_{11}(\mathbf{y}_{1i}) + m_2 \frac{\sqrt{n}}{\sqrt{n_2}} \frac{\sqrt{n_2}}{n_2} \sum_{i=1}^{n_2} \widetilde{h}_{12}(\mathbf{y}_{2i}) \\ &= S_{n1} + S_{n2}\end{aligned}$$

Let $\sigma_{h_k}^2 = Var\left(\widetilde{h}_{k1}(\mathbf{y}_{ki})\right)$. By applying CLT to $S_{nk}$, we have

$$S_{nk} \to_d N\left(0, \rho_k^2 m_k^2 Var\left(\widetilde{h}_{k1}(\mathbf{y}_{k1})\right)\right), \quad k = 1, 2$$

Since $S_{n1}$ and $S_{n2}$ are independent, it follows from Slutsky theorem that

$$\widehat{U}_n \to_p \theta, \quad \sqrt{n}\left(\widehat{U}_n - \theta\right) = S_{n1} + S_{n2} \to_d N\left(0, \rho_1^2 m_1^2 \sigma_{h_1}^2 + \rho_2^2 m_2^2 \sigma_{h_2}^2\right)$$

We summarize the asymptotic results for the two-sample $U_n$ in a theorem below.

**Theorem 2** (Univariate two-sample). Let $n = n_1 + n_2$ and assume $\lim_{n\to\infty} \frac{n}{n_k} = \rho_k^2 < \infty$ $(k = 1, 2)$. Then, under mild regularity conditions, we have:

$$U_n \to_p \theta, \quad \sqrt{n}(U_n - \theta) \to_d N\left(0, \sigma_\theta^2 = \rho_1^2 m_1^2 \sigma_{h_1}^2 + \rho_2^2 m_2^2 \sigma_{h_2}^2\right) \tag{3.85}$$

Note that the assumption $\lim_{n\to\infty} \frac{n}{n_k} = \rho_k^2 < \infty$ in the theorem is to ensure that that $n_k$ increase at similar rates so that $\frac{n_1}{n_2} \to c \in (0, \infty)$ as $n \to \infty$. We may define $n$ differently without affecting the expression for the asymptotic distribution of $U_n$. For example, if $n = \min(n_1, n_2)$, then we require $\lim_{n\to\infty} \frac{n}{n_k} = \varsigma_k^2 < \infty$ to ensure $\frac{n_1}{n_2} \to c \in (0, \infty)$ $(k = 1, 2)$. The asymptotic distribution of $U_n$ remains the same form as given in (3.85) except for replacing $\rho_k^2$ by $\varsigma_k^2$ in the asymptotic variance $\sigma_\theta^2$.

**Example 8 (*Model for comparing variance between two groups*).** In Example 10 of Section 3.1, we discussed a U-statistic for comparing the variances of two i.i.d. samples $y_{ki}$ $(1 \le i \le n_k,\ 1 \le k \le 2)$ defined by the following kernel:

$$h(y_{1i}, y_{1j}; y_{2l}, y_{2m}) = \frac{1}{2}(y_{1i} - y_{1j})^2 - \frac{1}{2}(y_{2l} - y_{2m})^2$$

The mean of the kernel is $\theta = E(h) = \sigma_1^2 - \sigma_2^2$, the difference between the two samples' variances.

To find the asymptotic variance of this U-statistic, first note that

$$\begin{aligned} h_{11}(y_{1i}) &= E\left[h(y_{1i}, y_{1j}; y_{2l}, y_{2m}) \mid y_{1i}\right] = \frac{1}{2}E\left[(y_{1i} - y_{1j})^2 \mid y_{1i}\right] - \sigma_2^2 \\ &= \frac{1}{2}(y_{1i} - \mu_1)^2 + \frac{1}{2}\sigma_1^2 - \sigma_2^2 \end{aligned}$$

Thus, $\widetilde{h}_{11}(y_{1i}) = h_{11}(y_{1i}) - \theta = \frac{1}{2}\left[(y_{1i} - \mu_1)^2 - \sigma_1^2\right]$ and its variance is

$$\sigma_{h_1}^2 = E\left[\widetilde{h}_{11}(y_{1i})\right]^2 = \frac{1}{4}E\left[(y_{1i} - \mu_1)^2 - \sigma_1^2\right]^2 = \frac{1}{4}\left(\mu_{14} - \sigma_1^4\right)$$

where $\mu_{k4}$ denotes the centered fourth moment of $y_{ki}$ $(k = 1, 2)$. Similarly, we find

$$\sigma_{h_2}^2 = E\left[\widetilde{h}_{12}(y_{2j})\right]^2 = \frac{1}{4}E\left[(y_{2j} - \mu_2)^2 - \sigma_2^2\right]^2 = \frac{1}{4}\left(\mu_{24} - \sigma_2^4\right)$$

Thus, the asymptotic variance of the U-statistic is given by

$$\sigma_U^2 = \rho_1^2\left(\mu_{14} - \sigma_1^4\right) + \rho_2^2\left(\mu_{24} - \sigma_2^4\right) \tag{3.86}$$

Under the null of no difference in variance between the two groups, $H_0 : \theta = \sigma_1^2 - \sigma_2^2 = 0$, (3.86) reduces to

$$\sigma_U^2 = \rho_1^2\mu_{14} + \rho_2^2\mu_{24} - \left(\rho_1^2 + \rho_2^2\right)\sigma^4$$

where $\sigma^2$ denotes the pooled variance over the two samples. We can estimate $\sigma_U^2$ in (3.86) by substituting moment estimates for $\mu_{k4}$ and $\sigma_k^2$ and $\frac{n_1+n_2}{n_k}$ as estimates for $\rho_k^2$ $(k = 1, 2)$.

**Example 9** (***Two-sample Mann–Whitney–Wilcoxon rank sum test***). In Example 11 of Section 3.1, we discussed the U-statistic version of the two-sample Mann–Whitney–Wilcoxon rank sum test:

$$U_n = \left[\binom{n_1}{1}\binom{n_2}{1}\right]^{-1} \sum_{i=1}^{n_1}\sum_{j=1}^{n_2} I_{\{y_{1i} \le y_{2j}\}}$$

It follows from Theorem 2 that

$$\sqrt{n}\left(U_n - \theta\right) \to_d N\left(0, \rho_1^2\sigma_{h_1}^2 + \rho_2^2\sigma_{h_2}^2\right)$$

To find $\sigma^2_{h_1}$, first note that

$$h_{11}(y_{1i}; y_{2j}) = E[h(y_{1i}; y_{2j}) \mid y_{1i}] = E\left(I_{\{y_{1i} \leq y_{2j}\}} \mid y_{1i}\right)$$
$$h_{12}(y_{1i}; y_{2j}) = E[h(y_{1i}; y_{2j}) \mid y_{2j}] = E\left(I_{\{y_{1i} \leq y_{2j}\}} \mid y_{2j}\right)$$

Thus,

$$\sigma^2_{h_1} = E\left[E\left(\widetilde{h}(y_{1i}; y_{2j}) \mid y_{1i}\right)\right]^2 = E\left[E\left(h(y_{1i}; y_{2j}) \mid y_{1i}\right)\right]^2 - \theta^2$$

Let $F_k(y)$ denote the CDF of $y_{ki}$. Under the null $H_0 : \theta = \frac{1}{2}$ or $F_k(y) = F(y)$, the probability integral transformation, $F(y_{ki}) \sim U[0,1]$ $(k = 1, 2)$. It follows that

$$\begin{aligned}
\sigma^2_{h_1} &= E\left[E\left(I_{\{y_{1i} \leq y_{2j}\}} \mid y_{1i}\right)\right]^2 - \frac{1}{4} = E[1 - F(y_{1i})]^2 - \frac{1}{4} \\
&= E(1-U)^2 - \frac{1}{4} = 1 - 2E(U) + E\left(U^2\right) - \frac{1}{4} = \frac{1}{3} - \frac{1}{4} \\
&= \frac{1}{12}
\end{aligned}$$

Similarly, $\sigma^2_{h_2} = \frac{1}{12}$. Thus, we have

$$\sqrt{n}\left(U_n - \frac{1}{2}\right) \to_d N\left(0, \sigma^2_U = \frac{1}{12}(\rho_1 + \rho_2)\right)$$

Given the sample sizes $n_k$, the asymptotic variance of $U_n$ is estimated by $\widehat{\sigma}^2_U = \frac{1}{12}\left(\frac{n}{n_1} + \frac{n}{n_2}\right)$.

Now consider the general $K$-sample U-statistic $U_n$ defined in (3.26) of Section 3.1.2, we define the projection of $U_n$ as follows:

$$\widehat{U}_n = \sum_{k=1}^{K} \sum_{i=1}^{n_k} E\left(U_n \mid \mathbf{y}_{ki}\right) - \left(\sum_{k=1}^{K} n_k - 1\right)\theta \tag{3.87}$$

The centered projection can be expressed as:

$$\widehat{U}_n - \theta = \sum_{k=1}^{K} \frac{m_k}{n_k} \sum_{i=1}^{n_k} \left(h_{1k}(\mathbf{y}_{ki}) - \theta\right) = \sum_{k=1}^{K} \frac{m_k}{n_k} \sum_{i=1}^{n_k} \widetilde{h}_{1k}(\mathbf{y}_{ki}) \tag{3.88}$$

where $\theta = E(h)$ and

$$h_{1k}(\mathbf{y}_{k1}) = E\left[h\left(\mathbf{y}_{11}, .., \mathbf{y}_{1m_1}; \ldots; \mathbf{y}_{K1}, .., \mathbf{y}_{Km_K}\right) \mid \mathbf{y}_{k1}\right]$$
$$\widetilde{h}_{1k}(\mathbf{y}_{k1}) = h_{1k}(\mathbf{y}_{k1}) - \theta, \quad 1 \leq k \leq K$$

As before, it can be shown that $U_n$ and $\widehat{U}_n$ have the same asymptotic distribution (see exercise). By employing an argument similar to the one used for the two-sample case, we can also readily determine the asymptotic distribution of the projection $\widehat{U}_n$ and therefore the asymptotic distribution of $U_n$. We state these results in a theorem below.

**Theorem 3** (Univariate $K$-sample). Let $n = \sum_{k=1}^{K} n_k$ and $\lim_{n\to\infty} \frac{n}{n_k} = \rho_k^2 < \infty$ $(1 \le k \le K)$. Let

$$\sigma_{h_k}^2 = Var\left(\widetilde{h}_{1k}\left(\mathbf{y}_{k1}\right)\right) = E\left[\widetilde{h}_{1k}\left(\mathbf{y}_{k1}\right)\right]^2, \quad 1 \le k \le K$$

Then, under mild regularity conditions, we have

$$U_n \to_p \theta, \quad \sqrt{n}\left(U_n - \theta\right) \to_d N\left(0, \sigma_U^2 = \sum_{k=1}^{K} \rho_k^2 m_k^2 \sigma_{h_k}^2\right)$$

where $m_k$ is the number of arguments corresponding to the $k$th group.

As noted in the two-sample U-statistics case, $n$ may be defined differently so long as the requirement $\frac{n_k}{n_l} \to c_{kl} \in (0, \infty)$ is ensured for any $n_l$ and $n_k$ $(1 \le k, l \le K)$. The asymptotic distribution of $U_n$ retains the same form regardless of how $n$ is defined.

**Example 10 (*U-statistic for comparing AUCs with independent samples*).** Consider the problem of comparing the AUCs of ROC curves for two diagnostic tests with four independent samples discussed in Example 13 of Section 3.1.1. The four-sample U-statistic is defined by the kernel:

$$h\left(y_{1i}; y_{2j}; y_{3k}; y_{4l}\right) = I_{\{y_{2j} \le y_{1i}\}} - I_{\{y_{4l} \le y_{3k}\}} \tag{3.89}$$

It follows that

$$\delta = E\left(h\right) = E\left(I_{\{y_{2j} \le y_{1i}\}}\right) - E\left(I_{\{y_{4l} \le y_{3k}\}}\right) = \theta_1 - \theta_2$$

where $\theta_l$ is the AUC for the $l$th test $(l = 1, 2)$. Under the null of no difference between the two test kits, $\delta = \theta_1 - \theta_2 = 0$ and $\theta_1 = \theta_2 = \theta$.

In this case,

$$\widetilde{h}_{1k}\left(y_{k1}\right) = E\left[h\left(y_{11}; y_{21}; y_{31}; y_{41}\right) \mid y_{k1}\right] - \delta, \quad 1 \le k \le 4$$

It follows that for each $1 \le k \le 4$,

$$\sigma_{h_k}^2 = Var\left(\widetilde{h}_{1k}\left(y_{k1}\right)\right) = E\left[E\left(h\left(y_{11}; y_{21}; y_{31}; y_{41}\right) \mid y_{k1}\right)^2\right] \tag{3.90}$$

and the asymptotic variance of the U-statistic is $\sigma_U^2 = \sum_{k=1}^4 \rho_k^2 \sigma_{h_k}^2$.

To find $\sigma_{h_k}^2$, first consider $k = 1$. Then,

$$\begin{aligned} \widetilde{h}_{11}(y_{k1}) &= E\left(I_{\{y_{21} \leq y_{11}\}} - I_{\{y_{41} \leq y_{31}\}} \mid y_{11}\right) \\ &= E\left(I_{\{y_{21} \leq y_{11}\}} \mid y_{11}\right) - \theta_2 \end{aligned} \tag{3.91}$$

It follows from (3.90) and (3.91) that

$$\begin{aligned} \sigma_{h_1}^2 &= E\left[\left(E\left(I_{\{y_{21} \leq y_{11}\}} \mid y_{11}\right) - \theta_2\right)^2\right] \\ &= E\left[E\left(I_{\{y_{21} \leq y_{11}\}} \mid y_{11}\right)^2\right] - 2\theta E\left[E\left(I_{\{y_{21} \leq y_{11}\}} \mid y_{11}\right)\right] + \theta_2^2 \\ &= E\left[E\left(I_{\{y_{21} \leq y_{11}\}} \mid y_{11}\right) E\left(I_{\{y_{21} \leq y_{11}\}} \mid y_{11}\right)\right] - \\ &\quad - 2\theta E\left[E\left(I_{\{y_{21} \leq y_{11}\}} \mid y_{11}\right)\right] + \theta_2^2 \\ &= E\left[E\left(I_{\{y_{21} \leq y_{11}\}} I_{\{y_{21} \leq y_{12}\}} \mid y_{11}\right)\right] - 2\theta_2\theta_1 + \theta_2^2 \\ &= E\left(I_{\{y_{21} \leq y_{11}\}} I_{\{y_{21} \leq y_{12}\}}\right) - 2\theta_1\theta_2 + \theta_2^2 \end{aligned} \tag{3.92}$$

Similarly, we can find $\sigma_{h_k}^2$ for $k = 2, 3, 4$ (see exercise):

$$\begin{aligned} \sigma_{h_2}^2 &= E\left(I_{\{y_{21} \leq y_{11}\}} I_{\{y_{22} \leq y_{11}\}}\right) - 2\theta_1\theta_2 + \theta_2^2 \\ \sigma_{h_3}^2 &= E\left(I_{\{y_{41} \leq y_{31}\}} I_{\{y_{41} \leq y_{32}\}}\right) - 2\theta_1\theta_2 + \theta_1^2 \\ \sigma_{h_4}^2 &= E\left(I_{\{y_{41} \leq y_{31}\}} I_{\{y_{42} \leq y_{31}\}}\right) - 2\theta_1\theta_2 + \theta_1^2 \end{aligned} \tag{3.93}$$

Under the null, (3.92) and (3.93) simplify with $\theta_1 = \theta_2$.

Since $\sigma_{h_k}^2$ involve unknown parameters, we must estimate these parameters to compute the asymptotic variance $\sigma_U^2$ of the U-statistic. First, consider estimating $\theta_k$ ($k = 1, 2$). As noted in Example 13 of Section 3.1.1, the four-sample U-statistic is actually the difference between two U-statistics:

$$\begin{aligned} U_{1n} &= (n_1 n_2)^{-1} \sum_{i=1}^{n_1} \sum_{j=1}^{n_2} I_{\{y_{2j} \leq y_{1i}\}}, \\ U_{2n} &= (n_3 n_4)^{-1} \sum_{k=1}^{n_3} \sum_{l=1}^{n_4} I_{\{y_{4l} \leq y_{3k}\}} \end{aligned} \tag{3.94}$$

Since the mean of the kernel of $U_{kn}$ in (3.94) is $\theta_k$, we can estimate each $\theta_k$ using $U_{kn}$ ($k = 1, 2$). Alternatively, we can pool the two samples for each disease status to estimate the common $\theta$ under the null $H_0 : \theta_1 = \theta_2 = \theta$ using the following U-statistic:

$$U_n = \left[(n_1 + n_3)(n_2 + n_4)\right]^{-1} \sum_{i=1}^{n_1+n_3} \sum_{j=1}^{n_2+n_4} I_{\{y_{2j} \leq y_{1i}\}}$$

where $n_1 + n_3$ and $n_2 + n_4$ are the sample sizes for the combined diseased and disease-free samples, respectively.

Now, consider estimating the first term in the expression for $\sigma_{h_k}^2$ as shown in (3.92) or (3.93). For convenience, consider $\zeta_1 = E\left(I_{\{y_{21} \leq y_{11}\}} I_{\{y_{21} \leq y_{12}\}}\right)$. Let $h(y_{1i}, y_{1j}; y_{2k}) = I_{\{y_{2k} \leq y_{1i}\}} I_{\{y_{2k} \leq y_{1j}\}}$. Then, $h(y_{1i}, y_{1j}; y_{2l})$ is a symmetric kernel and $E(h(y_{1i}, y_{1j}; y_{2k})) = \zeta_1$. Thus, the $U$-statistic defined by:

$$\widehat{\zeta}_1 = \left[\binom{n_1}{2}\binom{n_2}{1}\right]^{-1} \sum_{k=1}^{n_2} \sum_{(i,j) \in C_2^{n_1}} I_{\{y_{2k} \leq y_{1i}\}} I_{\{y_{2k} \leq y_{1j}\}}$$

is a consistent estimate of $\zeta_1$. Similarly, we can find the U-statistics for estimating the first term for the other three $\sigma_{h_k}^2$ (see exercise).

Like Kendall's $\tau$, it is not possible to evaluate $\sigma_{h_k}^2$ in the asymptotic variance in closed form $(1 \leq k \leq 4)$. As in the case of one-sample U-statistics, we can U-statistics to construct a consistent estimate of $\sigma_{h_k}^2$ and consequently a consistent estimate of the asymptotic variance $\sigma_U^2$.

## 3.3 Exercises

### Section 3.1

1. Let $\mathbf{z}_i = (x_i, y_i)^\top$ be an i.i.d. sample with mean $\boldsymbol{\mu} = (\mu_x, \mu_y)^\top$, variance $\sigma_x^2$, $\sigma_y^2$ and covariance $\sigma_{xy}$ $(1 \leq i \leq n)$. Show that the sample covariance $\widehat{\sigma}_{xy}$ has two representations:

$$\widehat{\sigma}_{xy} = \frac{1}{n-1} \sum_{i=1}^{n} (x_i - \overline{x}_n)(y_i - \overline{y}_n) = \binom{n}{2}^{-1} \sum_{i,j \in C_2^n} \frac{1}{2}(x_i - x_j)(y_i - y_j)$$

where $\overline{x}_n = \frac{1}{n}\sum_{i=1}^n x_i$ and $C_2^n$ denotes the set of all distinct pairs $(i,j)$ from the integer set $\{1, 2, \ldots, n\}$.

2. Let $y_i$ be an i.i.d. sample from a continuous distribution $(1 \leq i \leq n)$. Let $R_i$ denote the rank of $y_i$. For the Wilcoxon signed rank test defined in (3.14), show

(a) $W_n^+$ ranges between 0 and $\frac{n(n+1)}{2}$.
(b) $E(W_n^+) = \frac{n(n+1)}{4}$ and $Var(W_n^+) = \frac{n(n+1)(2n+1)}{24}$.
(c) Under the null $H_0: \theta = 0$, the sampling distribution of $W_n^+$ is given by

$$\Pr\left(W_n^+ = k\right) = \begin{cases} \frac{Q_n^k}{2^n} & \text{if } 0 \leq k \leq \frac{n(n+1)}{2} \\ 0 & \text{if otherwise} \end{cases}$$

where $Q_n^k$ is the number of subjects from the integer set $\{1, 2, \ldots, n\}$ for which the sum is $k$.

3. Within the context of Problem 2, verify the steps in (3.16) and (3.17).

4. For the ordinal categorical outcomes in Example 8 of Section 3.1.1, show that both indices, $\tau_b$ and $\gamma$, range from $-1$ to 1.

5. Symmetrize the kernel $g(y_{1i}, y_{1j}, y_{2m}, y_{2l})$ in (3.22) and show that $h(y_{1i}, y_{1j}, y_{2m}, y_{2l})$ in (3.21) is a symmetric version of $g(y_{1i}, y_{1j}, y_{2m}, y_{2l})$.

5. Verify (3.24).

6. For the four-sample U-statistic in (3.27) for ROC analysis, show that it can be expressed as the difference between two two-sample U-statistics as shown in (3.28).

7. Consider the proof of Theorem 1.

(a) Verify (3.29).

(b) We can also prove Theorem 1/1 using a simpler, heuristic argument as follows. Consider a small interval around $t$, $(t - \Delta, t + \Delta)$, and let

$$A = \left\{y_{(l)} < t - \Delta; 1 \leq l \leq j - 1\right\}, \quad B = \left\{y_{(j)} \in (t - \Delta, t + \Delta)\right\}$$
$$C = \left\{y_{(m)} > t + \Delta; j + 1 \leq m \leq n\right\}$$

Then, since

$$\Pr(A) = \binom{n-1}{j-1} F^{j-1}(t - \Delta), \quad \Pr(B) = \binom{n}{1} f(t)\Delta$$
$$\Pr(C) = [1 - F(t + \Delta)]^{n-1-(j-1)}$$

it follows that

$$\begin{aligned} g_j(t)\Delta &= \Pr(A \cap B \cap C) + o(\Delta) = \Pr(A)\Pr(B)\Pr(C) + o(\Delta) \\ &= \frac{(n-1)!}{(j-1)!(n-j)!} n F^{j-1}(t - \Delta)[1 - F(t + \Delta)]^{n-j} f(t)\Delta + o(\Delta) \end{aligned}$$

where $o(\cdot)$ denotes the deterministic $o$. We obtain the density $g_j(u)$ in (3.29) by dividing both sides of the above equation by $\Delta$ and letting $\Delta \to 0$.

(c) Prove the Lemma.

8. Verify (3.30) and (3.31).

9. Consider the two forms of the two-sample Mann–Whitney–Wilcoxon rank sum test in (3.32).

(a) Use the expression in (3.33) to show (3.34).

(b) Show $E\left(I_{\{y_{(2j)} - y_{(1i)} \leq 0\}}\right) = \frac{1}{2}$. [*Hint*: Consider taking expectation of both sides in (3.34).]

10. For the Wilcoxon rank form of the test in (3.32), show

(c) $E(W_n) = \frac{n_1(n_1+n_2+1)}{2}$ and $Var(W_n) = \frac{n_1 n_2 (n_1+n_2+1)}{12}$.

(d) Under the null of identical CDFs between the two samples, the sampling distribution of $W_n$ is given by:

$$\Pr(W_n = k) = \begin{cases} \binom{n}{n_1}^{-1} \frac{Q^k_{n_1,n_2}}{2^n} & \text{if } \frac{n_1(n_1+1)}{2} \leq k \leq \frac{n_1(n_1+2n_2+1)}{2} \\ 0 & \text{if otherwise} \end{cases}$$

where $n = n_1 + n_2$ and $Q^k_{n_1,n_2}$ is the number of unordered subsets of $n_2$ integers taken without replacement from $\{1, 2, \ldots, n\}$ for which the sum is $k$.

11. Consider the proof of Theorem 2.

(a) Verify (3.48)

(b) Verify (3.51) for the case when $i > l$ $(1 \leq i, l \leq n)$.

(c) Extend the proof of Theorem 2 for the general case with $k \geq 3$, that is, for each $k$, when considered as a random sequence of $n$, $S_{kn}$ is a martingale with respect to the filtration, $\{\mathcal{F}_n = \sigma(y_1, \ldots, y_n); n \geq k\}$.

**Section 3.2**

1. Consider the integer set $\{1, 2, \ldots, n\}$. For $1 \leq m \leq n$, let $(i_1, \ldots, i_m)$ be a permutation of $(1, 2, \ldots, n)$. Show

(a) The number of choices for two such permutations $(i_1, \ldots, i_m)$ and $(j_1, \ldots, j_m)$ to have exactly $l$ elements in common is given by: $\binom{n}{m}\binom{m}{l}\binom{n-m}{m-l}$.

(b) For a given integer $i$, the event $\{i \in (i_1, \ldots, i_m)\}$, that is, $i$ is one of $i_l$, occurs $\binom{n-1}{m-1}$ times $(1 \leq l \leq m)$.

2. Verify the two equalities in (3.74).

3. Let $\mathbf{z}_i = (x_i, y_i)^\top$ be i.i.d. pairs of continuous random variables $(1 \leq i \leq n)$. Consider the U-statistic $U_n$ defined by the kernel $h(\mathbf{z}_i, \mathbf{z}_j) = \frac{1}{2}(x_i - x_j)(y_i - y_j)$.

(a) Verify (3.70).

(b) If $\mathbf{z}_i \sim$i.i.d.$N(\boldsymbol{\mu}, \boldsymbol{\Sigma})$, show that $\sigma^2 = Var(p_1(\mathbf{z}_1, \boldsymbol{\mu}))$ has a closed form expression defined by the elements of $\boldsymbol{\Sigma}$, where $p_1(\mathbf{z}_1, \boldsymbol{\mu})$ is given by (3.75).

4. Consider a one-sample U-statistic $U_n$ with $m$ arguments defined by a symmetric kernel $h(\mathbf{y}_1, \ldots, \mathbf{y}_m)$. Show that then $E\left(h_1^2(\mathbf{y}_1)\right)$ can be expressed as the mean of a U-statistic in (3.82).

5. Let $y_{ki}$ be two independent samples $(1 \leq i \leq n_k; 1 \leq k \leq 2)$. For the projection of the two-sample U-statistic defined in (3.83) of Section 3.2.3, show that the centered projection is given in (3.84).

6. Prove Theorem 2 for two-sample U-statistics by extending the argument used in the proof of Theorem 1.

7. For a general $K$-sample $U$-statistic defined in (3.87) of Section 3.2.3, show

(a) The centered projection is given in (3.88).

(b) Under the assumptions of Theorem 3, the projection $\widehat{U}_n$ has the asymptotic distribution:

$$\sqrt{n}\left(\widehat{U}_n - \theta\right) \to_d N\left(0, \sum_{k=1}^{K} \rho_k^2 m_k^2 \sigma_k^2\right)$$

(c) Show that $\widehat{U}_n$ and $U_n$ have the same asymptotic distribution.

8. Consider the four-sample U-statistic defined by (3.89).

(a) Verify (3.93).

(b) Find the U-statistic for estimating the first term of each of the $\sigma_{h_k}^2$ in (3.93).

# Chapter 4

# Models for Clustered Data

This chapter introduces models for clustered data. Longitudinal studies are the most popular examples of such data, as outcomes from repeated assessments of an individual over time are not stochastically independent. Clustered data also arise from cross-sectional study designs in survey and epidemiological research, where subjects sampled from within a common habitat (cluster), such as families, schools and communities, are more similar than those sampled across different habitats, leading to correlated responses within a cluster. We touched upon the notion of data clustering in Chapter 2 when discussing the issue of overdispersion that arises from modeling count responses using standard Poisson-based log-linear models. Recall that a major problem with the Poisson-based log-linear model is that under the model assumption, the variance of the response equals the mean response. In practical studies, the variance is often larger than the mean, making this model inappropriate for such overdispersed data. A common cause of overdispersion is data clustering. In this chapter, we provide a systematic treatment of models for clustered data, especially those arising from longitudinal study designs, and discuss both parametric and distribution-free (semiparametric) approaches for modelling such data.

In Section 4.1, we contrast longitudinal study data with clustered data from cross-sectional studies and highlight the often subtle, yet fundamental differences between the two types of data clustering. In Section 4.2, we describe parametric models for longitudinal data. Within the parametric modeling paradigm, the mixed-effects (or latent variable) modeling approach is the most popular for modeling such data. However, our discussion of parametric models, particularly mixed-effects models, will not be in-depth given the large number of excellent textbooks on this topic and the focus of this book on distribution-free and U-statistics based models. In Section 4.3, we discuss the mean response based distribution-free model or GEE I for longitudinal data analysis. One major drawback of the parametric models for longitudinal study data is the multiple layers of distribution assumptions that are often at odds with real study data, particularly those from research in the behavioral and social sciences, making inference prone to error. This popular distribution-free alternative GEE I addresses this fundamental problem with parametric modeling. In Section 4.4, we address missing data. Again, we will concentrate on distribution-free models given the focus of the book and the fact that inferences for mixed-effects models are robust to missing data mechanisms in most applications. In Section 4.5, we will discuss an extension of the mean based distribution-free model, known as GEE II for modeling variance parameters. As noted in Chapters 2 and 3, many important indices and models of interest in the behavioral and social sciences are defined by second order moments and multiple responses, which are beyond the scope of the mean based distribution-free models. GEE II can address some of the problems in a limited and ad hoc fashion. We will provide a systematic treatment of second and higher order moments modeling in Chapters 5 and 6 by extending the univariate U-statistics theory developed in Chapter 3 to a multivariate and longitudinal data setting. In Section 4.6, we discuss structural equations models (SEM), which are widely used in the behavioral and social sciences for modeling causal mechanisms. Although similar in appearance, linear SEM are fundamentally different from multiple regression models. Since there are good textbooks out there on the general theory and applications of linear SEM, we will not provide an in-depth treatment of the models. Rather, we will highlight the major distinctions and unique features of SEM and discuss the use of GEE II as an alternative, distribution-free inference procedure for SEM. Since linear SEM is essentially a class of models involving second order moments, it can be treated as a special case of theU-statistics based functional response models for high order moments in Chapter 6. Thus, we will take up SEM again in Chapter 6.

## 4.1 Longitudinal versus Cross-Sectional Designs

In statistics, cluster is a generic term used to describe groups of subjects and/or data that are more similar within each group than across groups. Longitudinal study data is the most common example of data clustering, although many cross-sectional study designs in epidemiologic and survey research and psychosocial interventions also produce such data. In a longitudinal study, each subject receives multiple assessments over time and data from such repeated assessments are correlated and clustered around the subject. Studies in the fields of epidemiology, behavioral and social sciences often sample subjects from a common habitat, yielding clustered responses. For example, in a cross-section study on academic performance with students sampled from multiple schools, those within the same school are likely to have their grades on some standardized test clustered together because of similar curriculums received. In psychosocial research, many psychotherapy and behavioral intervention studies also result in clustered data because of the unique mode of treatment delivery. In psychotherapy studies, multiple subjects receive treatment from a common therapist and the heterogeneity among therapists due to different training programs and level of experience makes subjects' responses cluster around each therapist. In group based treatment studies, interventions are delivered in group formats with each group led by a trained instructor(s). The between-instructor heterogeneity again creates clustered responses. Although both types of study designs give rise to clustered responses, there are important differences between them as we highlight below.

First, clustered responses arising in cross-sectional studies that sample multiple subjects from a common habitat such as school usually have a relatively simple correlation structure. For example, in most cases, any pair of responses within a cluster will have the same correlation. However, clustered responses from most longitudinal studies do not have such a symmetric correlation structure; correlations between two responses from a subject at different times are likely to depend on the length of time between the two assessments.

Second, although varying cluster sizes are common to both study designs, they arise for quite different reasons and therefore must be differentially addressed. Varying cluster size in cross-sectional studies as in survey research and epidemiologic research is often the result of design consideration and generally has no impact on model estimation and inference. For example, in an epidemiologic study on estimating disease prevalence, varying cluster size is employed to ensure sufficiently large sample sizes to reliably estimate the disease prevalence in regions where the prevalence is relatively low. For longitudinal study data, however, different cluster sizes are the product of missing data, which must be carefully addressed in order to provide valid inference.

Shown in Figure 4.1 are two possible scenarios with observed responses that may arise for a hypothetical longitudinal study with six subjects and

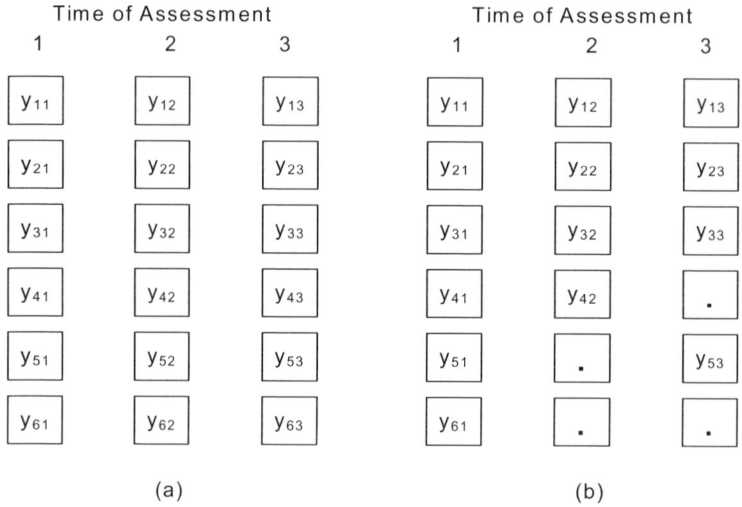


Figure 4.1. Observed responses for a longitudinal study with six subjects and three assessment points with $y_{it}$ denoting $i$th subject's response at time $t$ and dots missing data.

three assessments. In scenario (a), we observed all three responses ($y_{it}$) for each of the subjects, while in (b), each subject (cluster) had a varying number of observed responses due to missing data (dots). Missing data are commonplace for almost all longitudinal studies, no matter how well the study is planned, designed and executed. Thus, scenario (a) is atypical while pattern (b) is commonly encountered in data analysis. One approach is to perform analysis based only on the complete responses from the first three subjects. Alternatively, we may want to use all available data from the six subjects. It is obvious that an approach using only those subjects with complete responses is likely to be less powerful; what is not so obvious is the serious flaw in such an approach. We will take up these issues in Section 4.4. For now, it suffices to say that the presence of missing data in longitudinal studies makes the analysis of such study data more complex than the analysis of clustered data arising from cross-sectional study designs. Thus, from the model development perspective, we will focus on methods for longitudinal data analysis since they may be adapted to and/or modified for cross-sectional clustered data analysis.

## 4.2 Parametric Models

As in the analysis of cross-sectional study data, parametric models have been and still are a primary analysis tool for modeling longitudinal data. For continuous outcomes, multivariate normal distribution-based models remain the most popular in applications, especially in the behavioral and social sciences. In particular, the multivariate analysis of variance (MANOVA) and multivariate linear models have played a significant role in shaping the landscape of quantitative analysis and methodological research in the behavioral and social sciences that we see today. Although the proliferation in recent years of new and cutting-edge methods for longitudinal and other multi-level clustered data designs, such as the mixed-effects and structural equations models, has addressed some fundamental limitations of such classic models and enabled researchers to study complex individual, social and cultural processes that unfold over time, most of these methods have roots in the multivariate normal distribution based linear models. Thus, whether from a historic perspective or for teaching purposes, it is important to understand the theory, applications and limitations of the classic multivariate linear models.

### 4.2.1 Multivariate Normal Distribution Based Models

The multivariate normal distribution forms the basis for many parametric models that have been developed for longitudinal and other clustered data analysis. Although the classic MANOVA and multivariate linear models have been phased out and replaced by state-of-the-art approaches, such as mixed-effects and structural equations models, the theory of multivariate normal distribution still plays a significant role in developing new models in many research studies especially in the behavioral and social sciences. In this section, we introduce some of the normal-based popular models. These models will be revisited in later sections of this chapter, as well as in Chapters 5 and 6 as we relax the normal distribution assumption and address other limitations such as missing data.

**Example 1 (*MANOVA/Repeated measures ANOVA*).** Consider a longitudinal study for comparing treatment differences with $m$ assessment times. Suppose that there are $g$ $(> 1)$ different treatment conditions and we are interested in modeling differences in the mean response over time across the different treatment groups. At each assessment time $t$, the analysis of variance (ANOVA) model discussed in Chapter 2 can be used to compare the mean response across treatment conditions. However, in most longitudinal studies, it is also of primary interest to see how differences in group means change over time. Inferences about such time trend by group interactions require joint modeling the group means across the different assessment times, which introduces data clustering among the repeated responses collected over time.

Let $y_{kit}$ denote the response from the $i$th subject within the $k$th group at the $t$th assessment time for $1 \le i \le n_k$, $1 \le k \le g$ and $1 \le t \le m$. For each subject $i$ within a treatment group $k$, $y_{kit}$ are correlated over different assessment times $t$ since they are obtained from the same subject. Note that throughout the book, $t = 1$ will be referred to as *baseline*. For randomized clinical study trials, baseline typically designates the time when subjects are assessed before treatment initiation. Thus, interest usually lies in treatment differences at follow-up times such as $t = 2$, 3, and so on.

Let

$$\mathbf{y}_{ki} = \begin{pmatrix} y_{ki1} & \cdots & y_{kim} \end{pmatrix}^\top, \quad \boldsymbol{\mu}_k = E(\mathbf{y}_{ki}), \quad \Sigma_k = Var(\mathbf{y}_{ki}), \qquad (4.1)$$
$$1 \le i \le n_k, 1 \le k \le g$$

The classic multivariate analysis of variance (MANOVA) model is to assume that $\mathbf{y}_{ki}$ follow a multivariate normal distribution with mean $\boldsymbol{\mu}_k$ and variance $\Sigma_k$, that is, $\mathbf{y}_{ki} \sim N(\boldsymbol{\mu}_k, \Sigma_k)$ $(1 \le k \le g)$. As in the case of ANOVA, treatment differences under MANOVA can be assessed by testing appropriate linear hypotheses of the form:

$$H_0 : C\boldsymbol{\mu} = \mathbf{a} \quad \text{versus} \quad H_a : C\boldsymbol{\mu} \neq \mathbf{a} \qquad (4.2)$$

where $\boldsymbol{\mu} = \begin{pmatrix} \boldsymbol{\mu}_1^\top & \cdots & \boldsymbol{\mu}_g^\top \end{pmatrix}^\top$ and $C$ is some full rank $p \times (gm)$ matrix with known elements. In most applications, $\mathbf{a} = \mathbf{0}$ and $H_0$ in (4.2) is known as a linear contrast.

For example, suppose that the study has three treatment conditions and three assessment times. Then, $g = m = 3$. If we are interested in testing the hypothesis of no mean difference across the groups at the two follow-up times $t = 2$, 3, we can express the null hypothesis $H_0$ as a linear contrast as follows:

$$H_0 : \mu_{12} - \mu_{22} = 0, \quad \mu_{12} - \mu_{32} = 0, \quad \mu_{13} - \mu_{23} = 0, \quad \mu_{13} - \mu_{33} = 0$$

with $C$ given by

$$C = \begin{pmatrix} 0 & 1 & 0 & 0 & -1 & 0 & 0 & 0 & 0 \\ 0 & 1 & 0 & 0 & 0 & 0 & 0 & -1 & 0 \\ 0 & 0 & 1 & 0 & 0 & -1 & 0 & 0 & 0 \\ 0 & 0 & 1 & 0 & 0 & 0 & 0 & 0 & -1 \end{pmatrix}$$

A fundamental factor that sets research in behavioral and social sciences apart from their biomedical counterpart is outcome assessment. Many mental health disorders, such as depression and anxiety, are latent constructs and thereby not directly observable. Outcomes for measuring such disorders in psychosocial research are generally derived based on instruments (questionnaires) that delve into multifaceted personal feelings and private life rather than unidimensional, quantitative changes, such as blood pressure, heart rate and CD4 counts in drug treatment studies. For example, the popular Hamilton rating scale for depression severity consists of a total of 24 items to assess a wide range of emotional and physical problems, such as guilty feelings, suicide ideation and insomnia. Most items are rated either by a binary response to indicate the presence or absence of some psychological and physical phenomena and symptoms or by a *Likert scale* with integer values such as 0,1,2, etc. to quantify the extent of suffering, such as from "Not at all" to "A little", "A moderate amount", "Very much", and "An extreme amount". Item scores are then totaled to provide an assessment of the underlying latent construct.

Note that items in some instruments may be grouped to form measures of different domains of the construct, such as suicide ideation and insomnia, as in the case of Hamilton rating scale for depression severity. Item scores may be totaled within each domain to provide domain-specific assessment. Also, when an instrument (domain) contains both *positive* and *negative* items, one of the two types of the items must be reverse scored before the total scores within the instrument (domain) are computed. For example, in the 26-item version of the instrument for quality of life assessment, WHOQOL-BRIEF, developed by the World Health Organization (WHO), two of the questions in the domain of physical health are as follows:

*1. How much do you need any medical treatment to function in your daily life ?*

*2. Do you have enough energy for everyday life ?*

Both are scored by a five-item Likert scale: 1 for "Not at all", 2 for "A little", 3 for "A moderate amount", 4 for "Very much" and 5 for "An extreme amount". The first is a negative, while the second is a positive item. If the negative item is reverse scored, the response Q1 for the first question will be rescored as 6−Q1 when totaling the domain score for each respondent.

For the total score of an instrument (domain) to be a meaningful dimensional measure, the item scores within the instrument (domain) must be positively correlated. A common index for assessing such *internal consistency* is the Cronbach coefficient alpha.

**Example 2 (*Cronbach coefficient alpha*).** Consider an instrument or a subscale of an instrument with $m$ number of individual items, which is administered to a group of $n$ subjects. Let $y_{ik}$ denote the response from the $i$th subject to the $k$th item of the instrument.

Let

$$\mathbf{y}_i = \left(y_{i1}, \ldots, y_{im}\right)^{\top}, \quad \boldsymbol{\mu} = E\left(\mathbf{y}_i\right), \quad \Sigma = Var\left(\mathbf{y}_i\right) = \left[\sigma_{kl}\right] \tag{4.3}$$

where $[\sigma_{kl}]$ denotes the $m \times m$ variance matrix of $\mathbf{y}_i$ with $\sigma_{kl}$ denoting the $kl$th element. Let $\rho_{kl}$ denote the correlation between the $k$th and $l$th item scores. Cronbach's coefficient $\alpha$ is defined as:

$$\begin{aligned} \alpha &= \frac{m}{m-1}\left(1 - \frac{\sum_{k=1}^{m}\sigma_k^2}{\sum_{k,l}\sigma_{kl}}\right) = \frac{m}{m-1}\left(\frac{\sum_{k\neq l}\sigma_{kl}}{\sum_{k=1}^{m}\sigma_k^2 + \sum_{k\neq l}\sigma_{kl}}\right) \\ &= \frac{m}{m-1}\left(\frac{\sum_{k\neq l}\sigma_k\sigma_l\rho_{kl}}{\sum_{k=1}^{m}\sigma_k^2 + \sum_{k\neq l}\sigma_k\sigma_l\rho_{kl}}\right) \end{aligned} \tag{4.4}$$

where $\rho_{kl}$ denote the between-item score correlations. Thus, Cronbach's alpha compares the sum of the covariances to that of the total elements of the variance matrix $\Sigma$.

In the theory of measurement for item-based instrumentation, the true score for measuring some latent construct of interest is assumed to be the limit of the average of item scores with items randomly selected from an infinite pool of possible items (Nunnally and Bernstein, 1994). Thus, measurement error tends to zero as the number of items grows unbounded. The between-item correlations $\rho_{kl}$ indicate the extend to which a common core exists among the different items. If all such between-item score correlations are zero, then $\alpha = 0$. In this extreme case, the items are totally unrelated and share nothing in common, and thus should not be used together to measure the latent construct. At the other end of the spectrum, if all the item scores are perfectly correlated with a common variance, then $\alpha = 1$. In this special case, all the items are identical and the construct is not really latent as it can be measured by a single item without error. Of course, these extreme cases are hypothetical scenarios. In practice, $0 < \alpha < 1$ and $\alpha \to 1$ as $m \to \infty$ provided that $\rho_{kl} > 0$ for all $(1 \leq k, l \leq m)$. Thus, Cronbach alpha is an index of quality of measurement for an instrument with $k$ addable items. In fact, it can be shown that it is an estimate of the squared product–moment correlation between the true score and the averaged item scores of the instrument (Nunnally and Bernstein, 1994).

Although the index $\alpha$ above is well defined regardless of the data distribution, $\mathbf{y}_i$ is typically assumed to follow a multivariate normal with mean $\boldsymbol{\mu}$ and variance $\Sigma$ in most applications because of the difficulty in making inference about $\alpha$ without such a parametric distribution assumption. However, item scores in most instruments used in the behavioral and social sciences are defined by a binary or Likert scale and as such the normal distribution assumption is fundamentally at odds with the data distribution of item scores for most instruments. We will discuss nonparametric inference about $\alpha$ by removing the normal assumption in Chapter 5.

In many longitudinal studies, regression analysis is often required to evaluate treatment effects even though such studies only involve comparisons of treatment conditions. For example, it is often necessary to control for baseline covariates such as age and gender. When including covariates that contain continuous variables, we must use regression models for comparing treatment conditions. Regression models are also necessary for modeling relationship between a continuous response and a set of independent variables.

**Example 3 (*Multivariate linear regression*).** Consider a longitudinal study with $n$ subjects assessed at $m$ different time points. Let $y_{it}$ denote the response and $\mathbf{x}_{it}$ the vector of independent variables from the $i$th subject at assessment time $t$. Let

$$\mathbf{y}_i = \left(y_{i1}, \ldots, y_{im}\right)^\top, \quad \mathbf{x}_i = \left(\mathbf{x}_{i1}^\top, \ldots, \mathbf{x}_{im}^\top\right)^\top, \quad X_i = \left(\mathbf{x}_{i1}, \ldots, \mathbf{x}_{im}\right)^\top$$

Then, we model the linear relation between $y_{it}$ and $\mathbf{x}_{it}$ as follows:

$$\begin{aligned} y_{it} &= \mathbf{x}_{it}^\top \boldsymbol{\beta} + \epsilon_{it}, \quad \text{or} \quad \mathbf{y}_i = X_i \boldsymbol{\beta} + \boldsymbol{\epsilon}_i \\ \boldsymbol{\epsilon}_i &= \left(\epsilon_{i1}, \ldots, \epsilon_{im}\right)^\top \sim \text{i.i.d.} N\left(0, \Sigma\right) \end{aligned} \tag{4.5}$$

where $\epsilon_{it}$ ($\boldsymbol{\epsilon}_i$) denotes the model error. Note that if different $\boldsymbol{\beta}_t$ are used for different times, the model is known as the *seemingly unrelated regression*.

In many controlled, randomized longitudinal trials, we are only interested in treatment difference at posttreatment. In this case, we can apply

ANOVA discussed in Chapter 2 for comparing different treatment conditions or ANCOVA if we want to control for covariates at baseline. These methods yield valid inference in the absence of missing data. However, when there is missing data, both generally gives rise to biased estimates when implemented within the distribution-free inference framework discussed in Chapter 2. We address this missing data issue in Section 4.4.

Although multivariate normality based approaches are widely used in biomedical and psychosocial research, many studies have indicated problems with using estimates derived from such models because of the strong distribution assumption imposed on the response. As noted earlier, most instruments used in psychosocial research are based on item scores, and so are intrinsically discrete. Thus, the treatment of such variables as though they were continuous is purely for analytic simplicity. For variables with a relatively large range, such as the total Hamilton rating scale for depression severity, such a treatment is convenient and sensible. However, because these variables are inherently discrete, normal-based parametric models are fundamentally flawed when used for such outcomes. Thus, whenever possible, distribution-free alternatives should be considered.

### 4.2.2 Linear Mixed-Effects Model

The two major limitations of the classic multivariate normal distribution based approach are (1) its limited ability to deal with missing data and (2) its requirement of a set of common assessment times for all subjects. The mixed-effects (or latent variable) modeling approach and distribution-free models provide an effective solution to both issues. We start with the linear mixed-effects model for continuous response, which is a direct extension of the classic multivariate linear models.

The linear mixed-effects model (LMM) is a general class of models widely used in biomedical and psychosocial research to model the linear relationship between a response and a set of independent variables within a longitudinal and clustered data setting. As data clustering arises in research studies employing longitudinal study designs and multilevel sampling strategies (e.g., sampling subjects in classes nested within schools) across a wide range of disciplines, various applications of this general class of models are found under different guises, such as random coefficient models, random regression, hierarchical linear models (HLM), latent variable models, mixed models and multilevel linear models (Goldstein, 1987; Raudenbush and Bryk, 2002; Laird and Ware, 1982). The linear mixed-effects model employs random effects or latent variables to account for within-cluster correlations, rather than directly correlating the clustered responses as in the classic multivariate linear model. As a result, this approach enables one to address the difficulty in modeling correlated responses arising from varying assessment times as in some longitudinal cohort studies.

**Example 4** (***Linear growth curve model with constant assessment times***). Consider a longitudinal study with $n$ subjects. Assume first a set of fixed assessment times for all subjects, denoted by $t = 1, 2, \ldots, m$. If $y_{it}$ is a linear function of time $t$, then the classic linear model has the form:

$$y_{it} = \beta_0 + \beta_1 t + \epsilon_{it}, \quad \boldsymbol{\epsilon}_{it} = (\epsilon_{i1}, \cdots, \epsilon_{im})^\top \sim \text{i.i.d.} N(\mathbf{0}, \Sigma), \qquad (4.6)$$
$$1 \leq i \leq n, \quad 1 \leq t \leq m$$

In the above, $\Sigma$ is the variance of $\mathbf{y}_i = (y_{i1}, \ldots, y_{im})^\top$, which contains both the between- and within-subject variation. The idea behind the linear mixed-effects model is to break up the two sources of variation by modeling each separately.

For each subject $i$, let $b_{i0}$ and $b_{i1}$ denote the intercept and slope of the response $y_{it}$ and let $\mathbf{b}_i = (b_{i0}, b_{i1})^\top$. Then, for each subject $i$, we model the within-subject variation as follows:

$$y_{it} \mid \gamma_i = \beta_0 + \beta_1 t + b_{i0} + b_{i1} t + \epsilon_{it} \qquad (4.7)$$
$$\epsilon_{it} \sim \text{i.i.d. } N\left(0, \sigma^2\right), \quad 1 \leq t \leq m$$

In other words, for the $i$th individual, the response $y_{it}$ is modeled as a linear function of time with intercept, $\beta_0 + b_{i0}$, and slope, $\beta_1 + b_{i1}$. Thus, by modifying the population mean $\boldsymbol{\beta} = (\beta_0, \beta_1)^\top$ using the individual specific $\mathbf{b}_i$ to account for between-subject differences in the linear predictor, the error terms $\epsilon_{i1}, \epsilon_{i2}, \ldots, \epsilon_{im}$, in (4.7) can be assumed to be i.i.d. This is in stark contrast to the assumption of the error term for the classic multivariate linear regression in (4.6) in which the model error $\epsilon_{it}$ are correlated over $t$ to account for individual differences. By letting $\mathbf{b}_i$ vary across the subjects, we obtain a linear mixed-effects model that accounts for both between- and within-subject variations. In many applications, $\mathbf{b}_i$ are assumed to follow a multivariate normal $N(\mathbf{0}, D)$.

By combining the two-level specifications, we can express this linear mixed-effects model in a hierarchical form:

$$y_{it} = \beta_0 + t\beta_1 + b_{i0} + b_{i1} t + \epsilon_{it} = \mathbf{x}_{it}^\top \boldsymbol{\beta} + \mathbf{z}_{it}^\top \mathbf{b}_i + \epsilon_{it} \qquad (4.8)$$
$$\mathbf{y}_i = X_i \boldsymbol{\beta} + Z_i \mathbf{b}_i + \boldsymbol{\epsilon}_i$$
$$\epsilon_{it} \sim \text{i.i.d.} N\left(0, \sigma^2\right), \quad \mathbf{b}_i = N(b_{i0}, b_{i1})^\top \sim \text{i.i.d. } N(\mathbf{0}, D)$$

where $\mathbf{x}_{it} = \mathbf{z}_{it} = (1, t)^\top$ and $X_i = Z_i = (\mathbf{x}_{i1}, \mathbf{x}_{i2}, \ldots, \mathbf{x}_{im})^\top$. The linear predictor of the LMM in (4.8) has two parts; the first part $\mathbf{x}_{it}^\top \boldsymbol{\beta}$ describes the variation of the population mean over time, while the second $\mathbf{z}_{it}^\top \mathbf{b}_i$ models the deviation of each individual subject from the population mean. Since $\mathbf{b}_i$ is random, $\mathbf{b}_i$ (or $\mathbf{z}_{it}^\top \mathbf{b}_i$) is called the *random effect*. The population vector of parameters $\boldsymbol{\beta}$ (or $\mathbf{x}_{it}^\top \boldsymbol{\beta}$) is called the *fixed effect*, and hence the name of the linear mixed-effects model. An alternative perspective is to view $\mathbf{b}_i$ as a latent quantity since it is unobservable, which gives rise to the term of *latent variable model*.

The covariance between the $s$th and $t$th assessment points for the classic (4.6) and linear mixed-effects (4.7) models are given by

$$Cov(y_{is}, y_{it}) = \sigma_{st}, \quad Cov(y_{is}, y_{it}) = \mathbf{z}_{is}^\top D \mathbf{z}_{it} + \sigma^2, \quad 1 \le s, t \le m$$

If assessment times vary across individuals, $\sigma_{ist}$ will depend on $i$ and become inestimable except in special cases with some particular covariance structures. For example, if the covariance between $y_{is}$ and $y_{it}$ follows the uniform compound symmetry assumption, $\sigma_{st} = \sigma$ and are estimable. However, this issue does not arise for the linear mixed-effects model since $D$ and $\sigma^2$ are well-defined regardless of the structure of assessment.

Unlike the classic multivariate linear model, the variance parameters for the random effect in LMM has an interesting and useful interpretation. The diagonals of $D$ measure the variability in individual intercepts and slopes among different subjects in the study population. Thus, in addition to $\boldsymbol{\beta}$, inference about $D$ is also often of interest to assess the variability of individual intercepts or slopes or both.

**Example 5** (***Linear growth curve model with varying assessment times***). In Example 4, suppose that each subject has a varying number as well as times of assessments. For each subject $i$, let $m_i$ denote the number of assessments and $t_{ij}$ the assessment times ($1 \le j \le m_i$). Then, the LMM in this case can be expressed as:

$$\begin{aligned} y_{it_i} &= \beta_0 + t_i\beta_1 + b_{i0} + b_{i1}t_i + \epsilon_{it_i} = \mathbf{x}_{it_i}^\top \boldsymbol{\beta} + \mathbf{z}_{it_i}^\top \mathbf{b}_i + \epsilon_{it_i} \qquad (4.9) \\ \mathbf{y}_i &= X_i\boldsymbol{\beta} + Z_i\mathbf{b}_i + \boldsymbol{\epsilon}_i, \ X_i = Z_i = \left(\mathbf{x}_{it_{i1}}, \ldots, \mathbf{x}_{it_{im_i}}\right)^\top \\ \epsilon_{it} &\sim \text{i.i.d. } N\left(0, \sigma^2\right), \quad \mathbf{b}_i = (b_{i0}, b_{i1})^\top \sim \text{i.i.d. } N(\mathbf{0}, D) \end{aligned}$$

In comparison to (4.8), $\mathbf{y}_i$ ($X_i$) above is a $m_i \times 1$ ($m_i \times 2$) vector (matrix) of varying dimension, rather than a fixed $m \times 1$ ($m \times 2$) vector (matrix) as in the case with fixed assessment times.

A general LMM has the following form:

$$y_{it_{ij}} = \mathbf{x}_{it_{ij}}^{\top}\boldsymbol{\beta} + \mathbf{z}_{it_{ij}}^{\top}\mathbf{b}_i + \epsilon_{it_{ij}}, \quad \mathbf{y}_i = X_i\boldsymbol{\beta} + Z_i\mathbf{b}_i + \boldsymbol{\epsilon}_i, \quad \mathbf{b}_i \perp \boldsymbol{\epsilon}_i \qquad (4.10)$$
$$\mathbf{b}_i \sim \text{i.i.d.} N(0, D), \quad \boldsymbol{\epsilon}_i \sim \text{i.i.d.} N\left(0, \sigma^2 I_{m_i}\right), \quad 1 \leq j \leq m_i$$

where $\mathbf{x}_{it}^{\top}\boldsymbol{\beta}$ is the fixed and $\mathbf{z}_{it}^{\top}\mathbf{b}_i$ the random effect, $\perp$ denotes stochastic independence and $W_i = \left(\mathbf{w}_{it_{i1}}, \ldots, \mathbf{w}_{it_{im_i}}\right)^{\top}$ ($W = X$ or $Z$). For growth-curve analysis (modeling change of $y_{it}$ over time as in longitudinal study), $\mathbf{z}_{it}$ is often equal to $\mathbf{x}_{it}$. It follows from the assumptions of the LMM that:

$$E(\mathbf{y}_i \mid X_i, Z_i) = X_i\boldsymbol{\beta}, \quad Var(\mathbf{y}_i \mid X_i, Z_i) = Z_i D Z_i^{\top} + \sigma^2 I_{m_i} \qquad (4.11)$$

Clustered data also often arise from cross-sectional studies. For example, psychotherapy or behavioral intervention programs in the behavioral and social sciences are often delivered in a group format. Because therapists may differ in their skill set or ability to form a therapeutic bond and alliance with their patients, which are important considerations for intervention effect, there are often real differences between therapists. Such variability leads to different treatment outcomes in patients who are treated by different therapists and should be accounted for in assessing treatment effect in such studies.

**Example 6 (*LMM for controlling for therapists' effect*).** Let $n$ denote the number of therapists and $m_i$ the number of patients seen by the $i$th therapist in a cross-sectional study comparing two treatment conditions $(1 \leq i \leq n)$. By treating $n$ as the number of clusters and $m_i$ as the size of each cluster, patient's response, $y_{ij}$, can be modeled using the following LMM:

$$y_{ij} = \beta_0 + x_i\beta_1 + b_i + \epsilon_{ij}, \quad b_i \sim \text{i.i.d.} N\left(0, \sigma_b^2\right) \qquad (4.12)$$
$$\epsilon_{ij} \sim \text{i.i.d.} N\left(0, \sigma^2\right), \quad 1 \leq i \leq n, \quad 1 \leq j \leq m_i$$

where the binary $x_i$ indicates the treatment conditions. In the above, $b_i$ accounts for the individual effect of the $i$th therapist with $\sigma_b^2$ measuring the variability across therapists. Thus, concerns about therapists' variability can be examined by testing the null hypothesis: $H_0 : \sigma_b^2 = 0$. Note that in this example, $\mathbf{x}_{it} = (1, x_i)^{\top} \neq \mathbf{z}_{it} = 1$.

**Example 7.** For the LMM in (4.12) of Example 6, we have:

$$Var(\mathbf{y}_i \mid x_i) = \left(\sigma_b^2 + \sigma^2\right) C_{m_i}\left(\frac{\sigma_b^2}{\sigma_b^2 + \sigma^2}\right) \qquad (4.13)$$

where $C_m(\rho)$ denotes the uniform compound symmetry correlation matrix of dimension $m$ with a constant correlation $\rho = Corr(y_{ij}, y_{ik})$. The *intraclass* correlation (ICC), $\rho = \frac{\sigma_b^2}{\sigma_b^2 + \sigma^2}$, measures the proportion of therapists' variability in the total variability of patients' outcomes. If all therapists are the same, then $\sigma_b^2 = 0$ and $\rho = 0$. In this case, $y_{ij}$ become uncorrelated across $j$ within each $i$ and there is no cluster in the data. At the other extreme, if $\sigma_b^2 = \infty$, that is, all therapists are really different, then $\rho = 1$. In this case, $y_{it}$ only reflects the large variation in the therapists. In this example, $\rho$ has a more interesting interpretation than $\sigma_b^2$.

**Example 8 (*LMM for site difference in multisite trial*).** Consider a randomized, multicenter study trial on comparing two treatment conditions. For simplicity, we only consider modeling treatment difference at one time point only, such as the posttreatment assessment. Let $y_{ij}$ denote some response of interest from the $j$th subject within the $i$th site. Then, an appropriate LMM is given by

$$y_{ij} = \beta_0 + x_{ij}\beta_1 + b_i + \epsilon_{ij}, \quad b_i \sim \text{i.i.d.} N\left(0, \sigma_b^2\right), \quad \epsilon_{it} \sim \text{i.i.d.} N\left(0, \sigma^2\right)$$

where $x_{ij}$ indicates the treatment received by subject $j$ at site $i$ and $b_i$ denotes the (random) effect of site. We can assess whether there is significant site effect by testing the null: $H_0 : \sigma_b^2 = 0$. If this null is not rejected, we can simplify the model by dropping $b_i$.

When the number of sites is small such as two or three, we may want to model potential site difference using a fixed effect. In this case, site difference is interpreted with respect to the particular sites in the study. If there is a large number of randomly selected sites, it is sensible to model site difference using random effects. A significant random effect implies differences not only among the participating sites, but also across similar sites not included in the study. This difference in interpretation is usually not important in applications.

In Section 4.2.1, we discussed a popular index, Cronbach coefficient alpha, for measuring *interitem* consistency. Another major concern with respect to quality of outcome in psychosocial research is the *interrater* or between-observer agreement when an instrument is rated by different evaluators to assess mental disorders, disease conditions and behavior patterns in a person. Such agreement is also known as test–retest reliability when such an instrument is administered repeatedly over time.

Different approaches are employed for assessing interrater consistency, depending on whether the outcome is continuous or discrete (ordinal or categorical) or treated as such. The product–moment and intraclass correlations (ICC) are the most widely used measures of interrater agreement and test–retest reliability for continuous outcomes. For two raters, the product–moment correlation is used to indicate agreement, while for more than two judges, the ICC measures overall agreement by averaging all such pair-wise correlations among the observers.

Consider a sample of $n$ subjects, each of which is assessed by $M$ independent observers. Let $y_{im}$ denote the rating from the $m$th judge on the $i$th subject and $\mathbf{y}_i = (y_{i1}, \ldots, y_{iM})^\top$. Let $\boldsymbol{\mu} = E(\mathbf{y}_i)$ and $\Sigma = Var(\mathbf{y}_i)$ and assume that $\mathbf{y}_i \sim N(\boldsymbol{\mu}, \Sigma = [\sigma_k \sigma_l \rho_{kl}])$. Then, $\rho_{kl} = Corr(y_{ik}, y_{il})$, is the product–moment correlation between judges $k$ and $l$ $(1 \leq k < l \leq M)$. The product–moment based ICC as an index for overall agreement is defined as:

$$\rho_{PMICC} = \binom{m}{2}^{-1} \sum_{(k,l) \in C_2^m} Corr(y_{ik}, y_{il}) = \binom{m}{2}^{-1} \sum_{(k,l) \in C_2^m} \rho_{kl} \tag{4.14}$$

As we discuss shortly, this product–moment correlation based ICC has a serious drawback and may provide misleading information in some applications.

**Example 9 (*LMM for ICC with no judges' bias*).** A popular alternative to the product–moment correlation based ICC above is the model-based ICC. This approach is more appealing since it frames the problem under the more familiar regression framework. Consider the following LMM:

$$\begin{aligned} y_{im} &= \mu + \lambda_i + \epsilon_{im}, \quad \lambda_i \sim \text{i.i.d.} N\left(0, \sigma_\lambda^2\right), \quad \lambda_i \perp \epsilon_{im} \\ \epsilon_{im} &\sim \text{i.i.d.} N\left(0, \sigma^2\right), \quad 1 \leq i \leq n, \quad 1 \leq m \leq M \end{aligned} \tag{4.15}$$

where the fixed effect $\mu$ denotes the mean rating over the raters, the random effect $\lambda_i$ explains individual differences across the subjects, and the model error $\epsilon_{im}$ accounts for variability among the judges.

Under (4.15), $\mathbf{y}_i = (y_{i1}, \ldots, y_{iM})^\top$ has a multivariate normal with mean and variance:

$$\boldsymbol{\mu} = E(\mathbf{y}_i) = \mu \mathbf{1}_m, \quad \Sigma = Var(\mathbf{y}_i) = \sigma_\lambda^2 \mathbf{J}_m + \sigma^2 \mathbf{I}_m \tag{4.16}$$

where $\mathbf{1}_m$ is an $m \times 1$ column vector of 1's and $\mathbf{J}_m$ an $m \times m$ matrix of 1's. It then follows that the product–moment correlation between any two raters is, $\rho_{LMMICC} = Corr(y_{im}, y_{il}) = \frac{\sigma_\lambda^2}{\sigma_\lambda^2 + \sigma^2}$, a constant across the raters. This model-based between-rater correlation $\rho_{LMMICC}$ is often used as the definition of intraclass correlation as it elegantly explicates the effect of the interrater variability on the variability of the subject's ratings. If $\sigma^2$ is small (larger) relative to $\sigma_\lambda^2$, $\rho_{LMMICC}$ will be close to 1 (0), indicating good (poor) interrater agreement.

It follows from (4.16) that the LMM in (4.15) implies a common mean rating $E(y_{im}) = \mu$ across the $M$ observers. When there is consistent difference or bias between judges' ratings, this LMM no longer applies. By allowing judges' mean ratings to vary, we can readily revise the LMM to address such judges' bias.

**Example 10 (*LMM for ICC with judges' bias*).** In Example 9, consider replacing the constant mean $\mu$ by the individual judge's mean rating $\mu_m$ in (4.15). The resulting LMM is defined by

$$\begin{aligned} &y_{im} = \mu_m + \lambda_i + \epsilon_{im}, \quad \lambda_i \sim \text{i.i.d.} N\left(0, \sigma_\lambda^2\right), \quad \lambda_i \perp \epsilon_{im} \\ &\epsilon_{im} \sim \text{i.i.d.} N\left(0, \sigma^2\right), \quad 1 \leq i \leq n, \quad 1 \leq m \leq M \end{aligned} \tag{4.17}$$

Under this revised $M$-fixed factor LMM, the mean rating, $\mu_m = E(y_{im})$, is allowed to differ across the judges. However, the correlation between any two raters is still a constant, $\rho_{LMMICC} = Corr(y_{im}, y_{il}) = \frac{\sigma_\lambda^2}{\sigma_\lambda^2+\sigma^2}$. Thus, although the above model extends the single-factor LMM in (4.15), it still assumes constant variance and covariance between the judges' ratings, a structure not imposed by the product–moment correlation based model in (4.14).

When there is judges' bias, both versions of ICC in (4.14) and (4.17) may yield incorrect information about interrater agreement. Consider a hypothetical study with six subjects and suppose that ratings from two judges are 3, 4, 5, 6, 7, 8 and 5, 6, 7, 8, 9, 10, respectively. Both the product–moment and multifactor LMM based ICC is 1, indicating perfect agreement between the judges. However, there clearly exists judges' difference in rating the subjects and neither version of ICC accounts for such a difference. We will discuss another agreement index in Chapter 5 to address the limitation of ICC.

Note that an application of the single-factor LMM in Example 6 to the data yields an ICC close to 0.56. Although this low ICC provides a more reasonable conclusion about the discrepancy between the two judges' ratings in this example, it does not provide the right interpretation of the poor agreement between the two raters. Since this single-factor model assumes a common mean rating $\mu$, it too ignores judges' bias. The low ICC does not indicate large *random* variability between the raters as it is designed to measure, but rather the *fixed* bias in the judges' ratings.

### 4.2.3 Generalized Linear Mixed-Effects Models

A major limitation of the linear mixed-effects model is that it only applies to continuous response. By extending the generalized linear models (GLM) for cross-sectional study data considered in Chapter 2 to a longitudinal and clustered-data setting using random effects, we can develop similar mixed-effects models for other types of response such as binary and categorical outcomes.

Consider a longitudinal data study with $n$ subjects and $m$ assessments. For notational brevity, we assume a set of fixed assessment times, $1 \leq t \leq m$. Let $y_{it}$ denote the response and $\mathbf{x}_{it}$ a vector of covariates of interest from the $i$th subject at time $t$ ($1 \leq i \leq n$, $1 \leq t \leq m$). The principle of extending GLM to a longitudinal data setting is the same as in generalizing the classic univariate linear regression to LMM. For each subject, we first model the within-subject variability using a generalized linear model:

$$y_{it} \mid \mathbf{x}_{it}, \mathbf{z}_{it_i}, \mathbf{b}_i \sim \text{i.d. } f(\mu_{it}), \quad g(\mu_{it}) = \mathbf{x}_{it}^{\top}\boldsymbol{\beta} + \mathbf{z}_{it_i}^{\top}\mathbf{b}_i \tag{4.18}$$

where $g(\cdot)$ is a link function, $\mathbf{z}_{it}$ is a subvector of $\mathbf{x}_{it}$, $\mathbf{b}_i$ denotes the random effects, and $f(\mu)$ some probability distribution with mean $\mu$. Note that the model specification in (4.18) is quite similar to the cross-sectional GLM except for the additional individual effect $\mathbf{b}_i$. Then, by adding a distribution for $\mathbf{b}_i$ to explain the between-subject variation, we obtain from (4.18) the class of *generalized linear mixed-effects model* (GLMM). As in the case of LMM, $\mathbf{b}_i$ is often assumed to follow a multivariate normal $\mathbf{b}_i \sim N(\mathbf{0}, D)$ (a common assumption in most studies), although other types of more complex distributions such mixtures of normals may also be specified.

**Example 11 (*GLMM for binary response*).** In (4.18), if $y_{it}$ is a binary response, then $f(\mu)$ is a binomial distribution with mean $\mu$ and sample size 1, $\text{BI}(\mu; 1)$ ($1 \leq i \leq n$, $1 \leq t \leq m$). The most popular link for modeling such a response is the logit function. If we model the trajectory of $y_{it}$ over time as a linear function of $t$ with a bivariate normal random effect for the mean and slope, the GLMM becomes:

$$y_{it} \sim \text{i.d. BI}(\mu_{it}; 1), \quad \log\left(\frac{\mu_{it}}{1-\mu_{it}}\right) = \beta_0 + \beta t + b_{i0} + b_{i1}t$$
$$\mathbf{b}_i = (b_{i0}, b_{i1})^{\top} \sim \text{i.i.d.} N(\mathbf{0}, D), \quad 1 \leq i \leq n, 1 \leq t \leq m$$

As in the LMM case, $\beta_0 + \beta t$ describes the change over time for the population average, while the random effect $b_{i0} + b_{i1}t$ accounts for individual deviations.

**Example 12 (*Poisson log-linear model*).** In Example 11, let $y_{it}$ be a count response. By assuming a Poisson distribution $f(\mu)$, we obtain a random-effects based log-linear model for the trajectory of a count response $y_{it}$ over time:

$$\begin{aligned} y_{it} &\sim \text{i.d. } Poisson\left(\mu_{it}\right), \quad \log\left(\mu_{it}\right) = \beta_0 + \beta t + b_{i0} + b_{i1} t \\ \mathbf{b}_i &= \left(b_{i0}, b_{i1}\right)^\top \sim \text{i.i.d.} N\left(\mathbf{0}, D\right), \quad 1 \leq i \leq n, 1 \leq t \leq m \end{aligned} \tag{4.19}$$

The fixed and random effects have the same interpretation as in Example 11.

In Chapter 2, we discussed some major limitations of the Poisson based log-linear model when data exhibit overdispersion and discussed alternative models for addressing the underlying problems causing such overdispersion, such as the negative binomial and zero-inflated Poisson regression models. Overdispersion can also occur in longitudinal data analysis and thus it is of interest to generalize such models to a longitudinal data setting.

**Example 13 (*Negative binomial based log-linear model*).** Under the Poisson log-linear model in Example 9, the mean and variance of $y_{it}$ conditional on the random effect $\mathbf{b}_i$ are identical. In many applications, the variance is often larger than the mean, a phenomenon known as overdispersion. By substituting a negative binomial distribution, $NB(\mu, \alpha)$ in place of the Poisson model in (4.19), we obtain a GLMM that addresses over dispersion. Note that the negative binomial distribution has an extra parameter $\alpha$ that describes the degree of overdispersion in the conditional variance of $y_{it}$ given $\mathbf{b}_i$: $Var(y_{it} \mid \mathbf{b}_i) = \mu_{it}(1 + \alpha\mu_{it})$, which is larger than the mean $\mu_{it}$ by an amount of $\alpha\mu_i^2$. This extra term accounts for overdispersion in modeling $y_{it}$.

Overdispersion is not the only problem with count response. As discussed in Chapter 2, count response arising in biomedical and psychosocial research often contains structural zeros. Such structural zeros not only cause overdispersion in the variance, but also affect the mean of the response. For example, under the Poisson or negative binomial model, the presence of positive structural zeros (excess amount of zeros) will pull the mean response down towards zero, biasing model estimates. Thus, even estimates for descriptive analyses, such as the mean response, are incorrect. The zero-inflated Poisson (ZIP) model considered in Chapter 2 can be extended to address structural zeros within a longitudinal data setting.

**Example 14 (*Zero-inflated Poisson model*).** Within the context of Example 13, consider the following mixed-effects ZIP:

$$\begin{aligned} y_{it} \mid \mathbf{x}_{1it}, \mathbf{z}_{1it}, \mathbf{x}_{2it}, \mathbf{z}_{2it}, \mathbf{b}_{1i}, \mathbf{b}_{2i} &\sim \text{i.d. ZIP}\left(\mu_{it}, \rho_{it}\right) \qquad (4.20) \\ \log\left(\frac{\rho_{it}}{1-\rho_{it}}\right) &= \mathbf{x}_{1it}^{\top}\boldsymbol{\beta}_1 + \mathbf{z}_{1it}^{\top}\mathbf{b}_{1i}, \quad \log\left(\mu_{it}\right) = \mathbf{x}_{2it}^{\top}\boldsymbol{\beta}_2 + \mathbf{z}_{2it}^{\top}\mathbf{b}_{2i} \\ \mathbf{b}_{ki} &\sim \text{i.i.d.} N\left(\mathbf{0}, D_k\right), \quad 1 \leq k \leq 2, \ 1 \leq i \leq n, \ 1 \leq t \leq m \end{aligned}$$

where ZIP$(\mu, \rho)$ denotes the mixture distribution for ZIP with parameters $\mu$ and $\rho$, $\mathbf{x}_{kit}$ $(\mathbf{z}_{kit})$ denotes the covariates for the fixed (random) effect for modeling the occurrence of structural zero and the Poisson mean, respectively.

In addition to binary and count response, we can similarly generalize the cross-sectional generalized logit and proportional odds models for multilevel nominal and ordinal responses discussed in Chapter 2 to a longitudinal data setting (see exercise).

### 4.2.4 Maximum Likelihood Inference

Maximum likelihood is the most popular inference procedure for the classic multivariate linear model as well as the modern GLMM. Given our focus on distribution-free models and the large number of references on multivariate linear models and GLMM, we only provide a brief review and highlight the main steps in performing such inference. We begin with the classic multivariate linear model.

The multivariate linear models considered in Section 4.2.1 have the following general form:

$$\begin{aligned} \mathbf{y}_i &= \boldsymbol{\mu}_i + \mathbf{e}_i, \quad \mathbf{y}_i = \left(y_{i1}, \ldots, y_{im}\right)^{\top} \qquad (4.21) \\ \boldsymbol{\epsilon}_i &= \left(\epsilon_{i1}, \ldots, \epsilon_{im}\right)^{\top} \sim \text{i.i.d.} N\left(0, \Sigma\right), \quad 1 \leq i \leq n \end{aligned}$$

where $\boldsymbol{\mu}_i = \boldsymbol{\mu}$, $\boldsymbol{\mu}_k$ or $X_i\boldsymbol{\beta}$ depending on whether we have a homogeneous one-sample, MANOVA or multivariate linear regression model. Note that in the MANOVA case, $\mathbf{y}_i$ are expressed in the form of $\mathbf{y}_{ki}$ with $\boldsymbol{\mu}_k = E\left(\mathbf{y}_{ki}\right)$, where $\mathbf{y}_{ki}$ denotes the response vector from the $i$th subject within the $k$th group $(1 \leq i \leq n_k, 1 \leq k \leq g)$. We discuss maximum likelihood inference for the general linear model defined by (4.21) with $\boldsymbol{\mu}_i = X_i\boldsymbol{\beta}$. The general theory is readily adapted and applied to other specific models such as MANOVA.

The likelihood and log-likelihood functions for the multivariate linear regression model with $\boldsymbol{\mu}_i = X_i\boldsymbol{\beta}$ are given by

$$L_n(\boldsymbol{\beta},\Sigma) = \prod_{i=1}^{n} f(\mathbf{y}_i \mid \boldsymbol{\theta}) = (2\pi)^{-\frac{nm}{2}} |\Sigma|^{-\frac{n}{2}} \exp\left(\frac{-1}{2}\sum_{i=1}^{n} \mathbf{e}_i^{\top}\Sigma^{-1}\mathbf{e}_i\right) \quad (4.22)$$

$$l_n(\boldsymbol{\beta},\Sigma) = \log(L_n(\boldsymbol{\beta},\Sigma)) = -\frac{1}{2}n\log|\Sigma| - \frac{1}{2}\sum_{i=1}^{n}\mathbf{e}_i^{\top}\Sigma^{-1}\mathbf{e}_i$$

where the constant $c = -\frac{nm}{2}\log(2\pi)$ is dropped from the log-likelihood since it is irrelevant for estimation of $\boldsymbol{\beta}$ and $\Sigma$. By taking the first and second order derivatives of the log-likelihood $l_n$, we obtain the score vector equation for the parameters of interest $\boldsymbol{\beta}$ and $\Sigma$. The classic multivariate linear model is among the few classes of multivariate models for which the maximum likelihood estimate (MLE) can be expressed in closed form.

By taking derivatives of the log-likelihood $l$ with respect to $\boldsymbol{\beta}$ and $\Sigma$, we obtain the score vector (see exercise):

$$\frac{\partial}{\partial\boldsymbol{\beta}}l_n = -\sum_{i=1}^{n}\left(\frac{\partial\mathbf{e}_i}{\partial\boldsymbol{\beta}}\right)^{\top}(\Sigma^{-1}\mathbf{e}_i) = -\sum_{i=1}^{n}\left(-\frac{\partial}{\partial\boldsymbol{\beta}}(X_i\boldsymbol{\beta})\right)(\Sigma^{-1}\mathbf{e}_i) \quad (4.23)$$

$$= \sum_{i=1}^{n} X_i^{\top}\Sigma^{-1}(\mathbf{y}_i - X_i\boldsymbol{\beta})$$

$$\frac{\partial}{\partial\Sigma}l_n = -\frac{1}{2}n\Sigma^{-1} + \frac{1}{2}\sum_{i=1}^{n}\Sigma^{-1}\mathbf{e}_i\mathbf{e}_i^{\top}\Sigma^{-1}$$

where $\frac{\partial}{\partial\Sigma}l_n = \left[\frac{\partial}{\partial\sigma_{st}}l_n\right]$ is an $m \times m$ matrix. By setting the score vector above to 0 and solving for $\boldsymbol{\beta}$ and $\Sigma$, we obtain

$$\widehat{\boldsymbol{\beta}} = \left(\sum_{i=1}^{n} X_i^{\top}\widehat{\Sigma}^{-1}X_i\right)^{-1}\left(\sum_{i=1}^{n} X_i^{\top}\widehat{\Sigma}^{-1}y_i\right) \quad (4.24)$$

$$\widehat{\Sigma} = \frac{1}{n}\sum_{i=1}^{n}\left(\mathbf{y}_i - X_i\widehat{\boldsymbol{\beta}}\right)\left(\mathbf{y}_i - X_i\widehat{\boldsymbol{\beta}}\right)^{\top}$$

The above solution can be viewed as maximizing the log-likelihood by conditioning on either $\boldsymbol{\beta}$ or $\Sigma$, that is, given $\Sigma$, $\widehat{\boldsymbol{\beta}}$ is the MLE of $\boldsymbol{\beta}$, while given $\boldsymbol{\beta}$, $\widehat{\Sigma}$ is the MLE of $\Sigma$. Such a conditional maximization also converges, albeit at a slower rate than the Newton–Raphson algorithm. However, the benefit of using this alternative algorithm is that it allows us to derive closed-form expressions of the MLE of $\boldsymbol{\beta}$ and $\Sigma$. In most regression analyses primary interest lies in inference about the parameters $\boldsymbol{\beta}$ in the mean response. In this case, we are only interested in the MLE of $\boldsymbol{\beta}$. However, many quantities of interest in the behavioral and social sciences are also defined by second order moments in which case it is also of interest to make inference about $\Sigma$.

To find the MLE of $\boldsymbol{\beta}$ and $\Sigma$ in (4.7), we can also apply the Newton-Raphson algorithm directly to find the MLE and its asymptotic variance. Let $\boldsymbol{\theta} = \left(\boldsymbol{\beta}^\top, vec^\top(\Sigma)\right)^\top$ with $vec(\Sigma)$ denoting the $\frac{m(m-1)}{2} \times 1$ column vector containing the unique elements of $\Sigma$, that is,

$$vec(\Sigma) = \left(\sigma_{i11}, \sigma_{i12}, \ldots, \sigma_{i1m}, \sigma_{i22}, \sigma_{i23}, \ldots, \sigma_{i2m}, \ldots, \sigma_{imm}\right)^\top$$

Starting with some initial value $\boldsymbol{\theta}^{(0)}$, we iterate the following until convergence to obtain the MLE of $\boldsymbol{\theta}$:

$$\boldsymbol{\theta}^{(k+1)} = \boldsymbol{\theta}^{(k)} + \left[-\frac{\partial^2}{\partial\boldsymbol{\theta}\partial\boldsymbol{\theta}^\top} l(\boldsymbol{\theta})\right]^{-1} \frac{\partial}{\partial\boldsymbol{\theta}} l\left(\boldsymbol{\theta}^{(k)}\right)$$

The asymptotic variance is calculated and estimated the same way as before.

Now, consider inference for the LMM defined in (4.10). Let $\boldsymbol{\theta} = \left(\boldsymbol{\beta}^\top, vec^\top(D), \boldsymbol{\sigma}^2\right)^\top$. The log-likelihood is given by

$$l_n(\boldsymbol{\theta}) = \sum_{i=1}^{n} \log\left[\int f_{y|x,z,b}(\mathbf{y}_i \mid X_i, Z_i, \mathbf{b}_i) f_b(\mathbf{b}_i) d\mathbf{b}_i\right] \tag{4.25}$$

where $f_{y|x,z,b}(\mathbf{y}_i \mid X_i, Z_i, \mathbf{b}_i)$ denotes the density function of a multivariate normal $N(\boldsymbol{\mu}_i, \Sigma_i)$ with mean $X_i\boldsymbol{\beta} + Z_i\mathbf{b}_i$ and variance $\Sigma_i = Z_i D Z_i^\top + \sigma^2 I_{m_i}$ and $f_b(\mathbf{b}_i)$ the density of the normal random effect $\mathbf{b}_i$. The integral in (4.25) is the result of integrating out the latent $\mathbf{b}_i$. One major technical problem with inference for mixed-effect models is how to deal with such an integral. Fortunately for LMM, under the normal distribution assumptions, this integral can be completed in closed form. In fact, the marginal of $\mathbf{y}_i$ is again normal: $N\left(X_i\boldsymbol{\beta}, V_i = \left(Z_i D Z_i^\top + \sigma_b^2 \mathbf{I}_{m_i}\right)\right)$. By this property, the log-likelihood function is given by

$$l_n(\boldsymbol{\theta}) = -\frac{1}{2}\left[N \ln\sigma^2 + \sum_{i=1}^{n} \log|V_i| + \boldsymbol{\sigma}^2 \sum_{i=1}^{n} \mathbf{e}_i^\top V_i \mathbf{e}_i\right] \tag{4.26}$$

where $N = \sum_{i=1}^{n} m_i$. Note that although the dimension of $D$ is fixed, $V_i$ and $\mathbf{e}_i = \mathbf{y}_i - X_i\boldsymbol{\beta}$ both have a varying dimension from subject to subject. The MLE of $\boldsymbol{\theta}$ is obtained by maximizing the log-likelihood $l_n(\boldsymbol{\theta})$.

Although straightforward in principle, it is actually quite difficult to maximize $l_n(\boldsymbol{\theta})$ in (4.26), since obtaining the derivatives of $l(\boldsymbol{\theta})$ in closed form is quite a daunting task. Because of the analytic complexity in calculating the derivatives, several algorithms have been proposed and used. Among them, a popular approach is to use the expectation/maximization (EM) algorithm and its various enhanced versions, such as the expectation/conditional maximization either (ECME) algorithm. These approaches aim to obtain the maximum likelihood estimate (MLE) by maximizing the alternative, expected rather than the observed-data log-likelihood. Because expected log-likelihood functions can often be evaluated in closed form, applications of such algorithms bypass the difficulty of evaluating high dimensional integrals when maximizing the observed-data log-likelihood. However, it is well known that the convergence of EM is extremely slow in many applications, while the enhanced versions such as ECME are often difficult to implement, with unknown convergence rates. Further, these algorithms only produce the estimate of $\boldsymbol{\theta}$ and require additional methods to compute its asymptotic variance, the latter of which is often quite complex and a problem in its own right. Thus, while using EM-type algorithms for a particular application at hand may be convenient and effective, such an approach is not practical for developing models for routine use and wide applications. Most software packages use the Newton–Raphson algorithm (Lindstrom and Bates, 1988; Wolfinger, 1993; Demidenko, 2004). By deriving the first and second order derivatives of the log-likelihood, the MLE $\widehat{\boldsymbol{\theta}}$ of $\boldsymbol{\theta}$ is obtained by applying the Newton–Raphson method.

Inference for GLMM faces similar difficulties. In particular, it is generally not possible to obtain analytic derivatives. Even with the most popular normal random effects, integration cannot be completed in closed form for most models. The approach implemented in SAS by the NLMIXED procedure utilizes advanced Gaussian quadrature methods for high dimensional integration to maximize the observed data log-likelihood and can produce estimates as close to MLE as required by improving precision in the numerical integration (SAS Institute Inc. 2006). Thus, estimates so obtained can be treated as MLE for any practical purposes with well known asymptotic properties. The SAS NLMIXED procedure not only provides default routines for fitting some popular GLMM models, such as linear mixed-effects models, but also offers a flexible and powerful programming environment for developing inference procedures for very general random-effects models, such as those with nonlinear random effects.

**Example 15.** Consider the GLMM for binary response in Example 8.

Let $\boldsymbol{\alpha} = vec^{\top}(D)$ and $\boldsymbol{\theta} = \left(\boldsymbol{\beta}^{\top}, \boldsymbol{\alpha}^{\top}\right)^{\top}$. Since $\mathbf{b}_i$ is latent, the log-likelihood is given by

$$l_n(\boldsymbol{\theta}) = \sum_{i=1}^{n} \log\left[\int f_{y|x,z,b}(\mathbf{y}_i \mid \mathbf{b}_i, \boldsymbol{\beta}) f_b(\mathbf{b}_i \mid \boldsymbol{\alpha})\, d\mathbf{b}_i\right]$$
$$= \sum_{i=1}^{n} \log\left[\int \prod_{i=1}^{m} \mu_{it}^{y_{it}} (1-\mu_{it})^{1-y_{it}} f_b(\mathbf{b}_i \mid \boldsymbol{\alpha})\, d\mathbf{b}_i\right]$$

where $f_{y|x,z,b}(\mathbf{y}_i \mid \mathbf{b}_i, \boldsymbol{\beta}) = \prod_{i=1}^{m} \mu_{it}^{y_{it}} (1-\mu_{it})^{1-y_{it}}$ is the conditional joint density of $m$ independent binary components of $\mathbf{y}_i = (y_{i1}, \ldots, y_{im})^{\top}$ given the random effect $\mathbf{b}_i$ with mean $\mu_{it} = \exp(\beta_0 + \beta t + b_{i0} + b_{i1} t)$ and $f_b(\mathbf{b}_i \mid \boldsymbol{\alpha})$ is the joint density of the binary normal $\mathbf{b}_i = (b_{i0}, b_{i1})^{\top}$. The MLE of $\boldsymbol{\theta}$ is defined by the solution to the score vector equation $\frac{\partial}{\partial \boldsymbol{\theta}} l_n(\boldsymbol{\theta}) = \mathbf{0}$. We can use the NLMIXED procedure in SAS to numerically obtain the MLE of $\boldsymbol{\theta}$ and the observed information matrix.

As in the univariate case, hypotheses concerning $\boldsymbol{\theta}$ for most applications can be expressed in the following form:

$$H_0 : C\boldsymbol{\theta} = \mathbf{b} \quad \text{versus} \quad H_a : C\boldsymbol{\theta} \neq \mathbf{b} \tag{4.27}$$

where $C$ is some known full rank $k \times p$ matrix with $p$ $(\geq k)$ denoting the dimension of $\boldsymbol{\theta}$. If $\mathbf{b} = \mathbf{0}$, $H_0$ becomes a linear contrast. As discussed in Chapter 2, both the Wald and likelihood ratio tests can be used to examine the general linear hypothesis in (4.25). If $\boldsymbol{\theta}$ only consists of parameters for the fixed-effect, that is, $\boldsymbol{\theta} = \boldsymbol{\beta}$, we can reexpress the linear hypothesis in terms of a linear contrast by performing the transformation: $\boldsymbol{\gamma} = \boldsymbol{\beta} - C^{\top}\left(C^{\top}C\right)^{-1}\mathbf{b}$. When expressed in the new parameter vector $\boldsymbol{\gamma}$, the linear predictor will contain an offset term. For example, the linear predictor for the GLMM in (4.18) under this transformation becomes

$$\eta_{it} = \mathbf{x}_{it}^{\top}\boldsymbol{\beta} + \mathbf{z}_{it_i}^{\top}\mathbf{b}_i = c_{it} + \mathbf{x}_{it}^{\top}\boldsymbol{\gamma} + \mathbf{z}_{it_i}^{\top}\mathbf{b}_i$$

where $c_{it} = \mathbf{x}_{it}^{\top}\left(C^{\top}\left(C^{\top}C\right)^{-1}\mathbf{b}\right)$ is the offset. Except for LMM, where $c_{it}$ can be absorbed into the response by redefining the dependent variable as $z_{it} = y_{it} - c_{it}$, the offset must be specified when fitting GLMM using software packages.

## 4.3 Distribution-Free Models

The GLMM is developed by introducing random effects to model the within-subject variation. A fundamental problem with using random effects to account for correlated responses is the difficulty in empirically validating the distribution assumption for the random effects since they are latent variables. To further exacerbate this problem, GLMM also relies on parametric assumptions for the distribution of the response for inference. If either of these assumptions is violated, estimates will be biased and inconsistent. Unfortunately, such assumptions are often taken for granted and unacknowledged, laying the basis for inconsistent and spurious findings. A remarkable breakthrough underlying cutting-edge distribution-free modeling is the elimination of both sets of assumptions, leading to estimates that are robust to data distributions and within-cluster correlation structures.

### 4.3.1 Distribution-Free Models for Longitudinal Data

Since the seminal work of Liang and Zeger (1986), distribution-free regression models have been widely used as a robust alternative to mixed-effects models. In comparison to LMM and GLMM, such models offer more robust estimates and inference. By modeling the marginal mean of the response at each assessment time, this approach eliminates both layers of assumptions and thereby provides consistent estimates regardless of the complexity of the correlation structure and the distribution of the response. Unfortunately, many applications of GLMM in studies in the behavioral and social sciences fail to acknowledge the impact of distribution assumptions and the issue of robustness. Outcome data from most instruments in these studies are especially prone to violations of distribution assumption, because they are not even continuous let alone do they follow the normal distribution. In Chapter 2, we discussed distribution-free models for nonclustered data and introduced estimating equations for inference. Such cross-sectional models are readily extended to a longitudinal data setting.

Consider a longitudinal study with $n$ subjects and $m$ assessment times. Again, for notational brevity, we assume a set of fixed assessment times $1 \leq t \leq m$. Let $y_{it}$ denote a response and $\mathbf{x}_{it}$ a set of independent variables of interest from the $i$th subject and at time $t$. By applying the distribution-free GLM discussed in Chapter 2 to each time $t$, we obtain a class of distribution-free regression models for the current context of longitudinal data:

$$E\left(y_{it} \mid \mathbf{x}_{it}\right) = \mu_{it}, \quad g\left(\mu_{it}\right) = \mathbf{x}_{it}^{\top}\boldsymbol{\beta}, \quad 1 \leq t \leq m, \quad 1 \leq i \leq n \tag{4.28}$$

where $g(\cdot)$ is some link function and $\boldsymbol{\beta}$ is a $p \times 1$ vector of parameters of interest. Note that at each time $t$, the distribution-free GLM above is exactly the same as the distribution-free GLM for cross-sectional data.

**Example 1 (*MANOVA/repeated measures ANOVA*).** Consider a longitudinal study for comparing $g$ treatment groups, with each group involving $n_k$ number of subjects $(1 \leq k \leq g)$. Let $y_{kit}$ denote the continuous response of interest from the $i$th subject within the $k$th group at the $t$th assessment $(1 \leq i \leq n_k,\ 1 \leq k \leq g,\ 1 \leq t \leq m)$. The distribution-free MANOVA/repeated measures ANOVA models the mean of each group as follows:

$$E(y_{kit}) = \mu_{kt}, \quad 1 \leq i \leq n_k, \quad 1 \leq k \leq g, \quad 1 \leq t \leq m \tag{4.29}$$

In comparison to the normal based MANOVA/repeated measures ANOVA in Section (4.2.1), the distribution-free version only specifies the mean response of each group at each time $t$ with no assumption regarding the joint distribution of the within-subject response vector $\mathbf{y}_{ki} = (y_{ki1}, \ldots, y_{kim})^{\top}$. As a result, inference is valid regardless of how $\mathbf{y}_{ki}$ are distributed.

**Example 2 (*Group comparison for binary response over time*).** In Example 1, let $y_{kit}$ be a binary response. In this case, the mean response $\mu_{kt} = E(y_{kit})$ represents the probability of the occurrence of the event $y_{kit} = 1$ for each group over time $t$.

**Example 3.** In Example 1, if $y_{kit}$ is a count response, then the model in (4.28) can be used to compare the mean frequency of the response across the different groups. In parametric analysis, $y_{kit}$ is assumed to follow an analytic model, such as the Poisson. In comparison, the distribution-free model only specifies the mean of the response $y_{kit}$. Thus, the distribution-free model provides valid inference even if $y_{kit}$ does not follow a Poisson distribution. For example, in Section 4.2.3 we discussed the negative binomial model to account for overdispersion. The negative binomial distribution has the same mean as the Poisson, but differs from Poisson in that it has a larger variance. Since the distribution-free model only involves the mean response, it yields valid inference regardless of whether $y_{kit}$ follows the Poisson, negative binomial or any other distribution.

**Example 4 (*Linear model for growth-curve analysis*).** Consider a growth curve analysis for a longitudinal study with $n$ subjects and $m$ assessments. Assume a linear growth with one baseline covariate $x$. Let $\mathbf{x}_{it} = (1, x_i, t, x_i t)^{\top}$. Then, (4.28) yields

$$E(y_{it} \mid x_i) = \beta_0 + x_i\beta_1 + t\beta_2 + x_i t\beta_3, \quad 1 \leq i \leq n, 1 \leq t \leq m \tag{4.30}$$

If $x$ is a binary indicator for two treatment conditions, the above models the mean response over time with $\beta_2$ indicating the growth rate for one treatment group and $\beta_2+\beta_3$ the growth rate for the other group. Treatment difference is indicated by a significant group by time interaction, $\beta_3 \neq 0$.

If $x$ represents some demographic variable, such as gender and age, the above models the growth curve of a single treatment condition with the group by time interaction denoting the moderating effect of this covariate; $x$ moderates the growth rate if $\beta_3 \neq 0$ and has no effect if otherwise.

Note that $y_{it}$ in (4.28) can be a continuous, binary or count response. However, as in the case of cross-sectional data analysis, this model does not apply to multi-level nominal or ordinal response. We can readily generalize the distribution-free models for such types of response in Chapter 2 to the longitudinal data setting.

**Example 5 (*Distribution-free models for multilevel categorical response*).** Within the longitudinal study context in Example 4, consider a categorical response with $J$ distinct outcomes $r_1, \ldots, r_J$. Let $\mathbf{x}_{it}$ be a vector of independent variables from the $i$th subject at time $t$. For each time $t$, let $\mathbf{y}_{it} = (y_{i1t}, \ldots, y_{iJt})^\top$ be a vector of binary responses $y_{ijt}$ from the $i$th subject at time $t$ with $y_{ijt} = 1$ if the response is in the $j$th category and $y_{ijt} = 0$ if otherwise $(1 \leq j \leq J)$.

For nominal response, we discussed in Chapter 2 the generalized logit or multinomial response model for such a response type. Within the current longitudinal data setting, let

$$\mu_{ijt} = \mu_{jt}(\mathbf{x}_{it}) = E(y_{ijt} \mid \mathbf{x}_{it}) = \Pr(y_{ijt} = 1 \mid \mathbf{x}_{it}), \quad j = 1, \ldots, J$$

the conditional probability of response in the $j$th category given $\mathbf{x}_{it}$. The generalized logit model is defined as:

$$\log\left(\frac{\mu_{ijt}}{\mu_{iJt}}\right) = \eta_{ijt} = \alpha_j + \boldsymbol{\beta}_j^\top \mathbf{x}_{it}, \quad j = 1, \ldots, J-1 \tag{4.31}$$

where $\boldsymbol{\alpha} = (\alpha_1, \ldots, \alpha_{J-1})^\top$ is the vector of parameters of the *baseline* generalized logits in the absence of $\mathbf{x}$ and $\boldsymbol{\beta} = \left(\boldsymbol{\beta}_j^\top, \ldots, \boldsymbol{\beta}_{J-1}^\top\right)^\top$ the parameter vector of interest.

Similarly, we generalize the proportion odds model for ordinal response in Chapter 2 to the longitudinal data context. Let $\gamma_{ijt}(\mathbf{x}_{it}) = \sum_{k=1}^{j} \mu_{ikt}(\mathbf{x}_{it})$ denote the cumulative probability of response up to and including category $j$ conditional on the vector of independent variables $\mathbf{x}_{it}$ at time $t$. The proportional odds model is defined by

$$\log\left(\frac{\gamma_{ijt}}{1-\gamma_{ijt}}\right) = \alpha_j + \boldsymbol{\beta}^\top \mathbf{x}_{it}, \quad \alpha_1 \leq \alpha_2 \leq \ldots \leq \alpha_{J-1} \tag{4.32}$$

where $\boldsymbol{\alpha} = (\alpha_1, \ldots, \alpha_{J-1})^\top$ and $\boldsymbol{\beta}$ have the same interpretation as the generalized logit model.

For ordinal response, it is preferable to model the cumulative response because of the *invariance* property when response categories are combined. For example, suppose that there are five ordered responses, $r_1 < r_2 < \cdots < r_5$. If the first two categories are combined, we will have four categories, $\widetilde{r}_1 < r_2 < \cdots < r_4$. This change in the response categories has a negligible effect on the ordinal model as it only reduces the number of parameters without affecting the interpretation of the remaining parameters, particularly $\boldsymbol{\beta}$.

### 4.3.2 Inference for Distribution-Free Models

Consider the class of distribution-free models in (4.28). At each time $t$, (4.28) reduces to the distribution-free model for cross-sectional data discussed in Chapter 2. For such a model, the most popular inference approach for inference is the estimating equation:

$$\mathbf{w}_{nt}(\boldsymbol{\beta}) = \sum_{i=1}^{n} G_{it}(\mathbf{x}_{it})(y_{it} - \mu_{it}) = \mathbf{0} \tag{4.33}$$

where $G_{it}(\mathbf{x}_{it})$ is some known matrix function of $\mathbf{x}_{it}$. As noted in Chapter 2, estimates obtained as solutions to the equation in (4.33) are consistent regardless of the choice of $G_{it}(\mathbf{x}_{it})$ so long as $E(y_{it} - \mu_{it}) = 0$. In most applications, $G_{it}(\mathbf{x}_{it})$ is selected based on the following to achieve efficiency:

$$G_{it}(\mathbf{x}_{it}) = D_{it} v^{-1}(\mu_{it}), \quad D_{it} = \frac{\partial}{\partial \boldsymbol{\beta}} \mu_{it} \tag{4.34}$$

If $y_{it}$ given $\mathbf{x}_{it}$ follows the exponential family of distributions, $v(\cdot)$ is readily determined and the estimate equation in (4.33) yields the MLE of $\boldsymbol{\beta}$.

The *generalized estimating equation* (GEE) is developed by extending the estimating equation (4.33) at a single time $t$ to multiple times across all assessments. This is achieved by capitalizing on the fact that the use of a wrong correlation matrix has no impact on the consistency of the GEE estimate of $\boldsymbol{\beta}$, just as in the univariate case the misspecification of $v(\cdot)$ does not affect the consistency of the estimating equation estimate.

Let $S_i = \mathbf{y}_i - \boldsymbol{\mu}_i$ with

$$\mathbf{y}_i = (y_{i1}, \ldots, y_{im})^\top, \quad \boldsymbol{\mu}_i = (\mu_{i1}, \ldots, \mu_{im})^\top, \quad \mathbf{x}_i = \left(\mathbf{x}_{i1}^\top, \ldots, \mathbf{x}_{im}^\top\right)^\top$$

In analogy to (4.33), the generalized estimating equation (GEE) is defined by

$$\mathbf{w}_n = \sum_{i=1}^{n} G_i(\mathbf{x}_i) S_i(\boldsymbol{\beta}) = \sum_{i=1}^{n} G_i(\mathbf{x}_i)(\mathbf{y}_i - \boldsymbol{\mu}_i) = \mathbf{0} \tag{4.35}$$

where $G_i(\mathbf{x}_i)$ is some matrix function of $\mathbf{x}_i$. The GEE estimate $\widehat{\boldsymbol{\beta}}$ is obtained by solving the above vector equation. As in the univariate case, $\widehat{\boldsymbol{\beta}}$ is consistent independent of the choice of $G_i(\mathbf{x}_i)$ provided that $E(\mathbf{y}_i - \boldsymbol{\mu}_i) = \mathbf{0}$ (see Theorem 1 below). It is readily checked that under the model in (4.28), this condition holds true (see exercise). In addition, $E(\mathbf{w}_n) = \mathbf{0}$, that is, the GEE is *unbiased*.

In most applications, $G_i(\mathbf{x}_i)$ has the form:

$$G_i(\mathbf{x}_i) = D_i V_i^{-1}, \quad D_i = \frac{\partial}{\partial \boldsymbol{\beta}} \boldsymbol{\mu}_i, \quad A_i = \operatorname{diag}(v(\mu_{it})) \tag{4.36}$$
$$V_i = A_i^{\frac{1}{2}} R(\boldsymbol{\alpha}) A_i^{\frac{1}{2}}$$

where $R(\boldsymbol{\alpha})$ denotes a *working* correlation matrix parameterized by $\boldsymbol{\alpha}$ and diag$(v(\mu_{it}))$ a diagonal matrix with $v(\mu_{it})$ on the $t$th diagonal. The phrase "working correlation" is used to emphasize the fact that $R(\boldsymbol{\alpha})$ is not necessarily the true correlation matrix. For example, the simplest choice is $R = I_m$. In this case, the correlated components of $\mathbf{y}_i$ are treated as if they are independent. In addition, there is no parameter associated with this particular *working independence model.* Another popular choice is the uniform compound symmetry correlation matrix, $R(\boldsymbol{\alpha}) = C_m(\rho)$, which assumes a common correlation $\rho$ for any pair of the component responses of $\mathbf{y}_i$. The working correlation matrix involves a single parameter, $\boldsymbol{\alpha} = \rho$.

Under the specification in (4.36), (4.35) can be expressed as

$$\mathbf{w}_n(\boldsymbol{\beta}) = \sum_{i=1}^{n} D_i V_i^{-1} S_i = \sum_{i=1}^{n} D_i V_i^{-1}(\mathbf{y}_i - \boldsymbol{\mu}_i) = \mathbf{0} \tag{4.37}$$

The above is identical to the estimating equation in (4.33) for cross-sectional data analysis except that $D_i$ in (4.36) is a $p \times m$ matrix rather than a $p \times 1$ vector. Although the GEE in (4.37) is a function of both $\boldsymbol{\beta}$ and $\boldsymbol{\alpha}$, we express it explicitly as a function of $\boldsymbol{\beta}$ to emphasize the fact that (4.37) is used to obtain the estimate of the parameter vector of interest $\boldsymbol{\beta}$. If $\boldsymbol{\alpha}$ is known (e.g., for the working independence model), we can obtain the solution of $\boldsymbol{\beta}$ to (4.37) by either of the recursive algorithms:

$$\boldsymbol{\beta}^{(k+1)} = \boldsymbol{\beta}^{(k)} - \left[\frac{\partial}{\partial \boldsymbol{\beta}} \mathbf{w}_n(\boldsymbol{\beta})\right]^{-\top} \mathbf{w}_n\left(\boldsymbol{\beta}^{(k)}\right) \tag{4.38}$$

$$\boldsymbol{\beta}^{(k+1)} = \boldsymbol{\beta}^{(k)} + \left(\sum_{i=1}^{n} D_i V_i^{-1} D_i\right)^{-1} \mathbf{w}_n\left(\boldsymbol{\beta}^{(k)}\right)$$

where $\boldsymbol{\beta}^{(0)}$ denotes some initial value (see exercise). As in the univariate case, the second equation above is easier to compute. The GEE estimate $\widehat{\boldsymbol{\beta}}$ is obtained by iterating the selected algorithm until convergence. When $\boldsymbol{\alpha}$ is unknown, we must estimate it so that (4.37) can be used to find estimates of $\boldsymbol{\beta}$. We illustrate these considerations below.

**Example 6 (*MANOVA/repeated measures ANOVA*).** Consider the distribution-free model for group comparison in Example 1. In this case,

$$\mathbf{y}_{ki} = (y_{ki1}, \ldots, y_{kim})^\top, \quad \boldsymbol{\mu}_k = \begin{pmatrix} \mu_{k1} & \cdots & \mu_{km} \end{pmatrix}^\top \tag{4.39}$$

$$D_k = \frac{\partial}{\partial \boldsymbol{\mu}_k} \boldsymbol{\mu}_k = I_m, \quad S_{ki} = \mathbf{y}_{ki} - \boldsymbol{\mu}_k, \quad 1 \le k \le g$$

To find $v(\mu_{it})$, assume $y_{kit} \sim N(\mu_{k1}, \sigma^2)$. Then, $v(\mu_{it}) = 1$ and $A_{ki} = I_m$. Thus, $V_{ki} = A_{ki}^{\frac{1}{2}} R_k(\boldsymbol{\alpha}_k) A_{ki}^{\frac{1}{2}} = R_k(\boldsymbol{\alpha}_k)$.

$$\mathbf{w}_n(\boldsymbol{\mu}) = \sum_{k=1}^{g} \sum_{i=1}^{n_k} R_k^{-1}(\boldsymbol{\alpha}_k) S_{ki} = \sum_{k=1}^{g} \sum_{i=1}^{n_k} R_k^{-1}(\boldsymbol{\alpha}_k)(\mathbf{y}_{ki} - \boldsymbol{\mu}_k) = \mathbf{0} \tag{4.40}$$

Regardless of the choice of $R_k(\boldsymbol{\alpha}_k)$, the GEE estimate of $\boldsymbol{\mu}_k$ is given by $\widehat{\boldsymbol{\mu}}_k = \frac{1}{n_k} \sum_{i=1}^{n_k} \mathbf{y}_{ki}$ $(1 \le k \le g)$. Further, if $\mathbf{y}_{ki} \sim N(\mathbf{0}, \sigma^2 R_k(\boldsymbol{\alpha}_k))$, $\widehat{\boldsymbol{\mu}}_k$ is actually the MLE of $\boldsymbol{\mu}_k$ (see exercise).

**Example 7 (*Distribution-free linear regression model*).** Let $y_{it}$ and $\mathbf{x}_{it}$ be the response and vector of independent variables of interest from the $i$th subject assessed at time $t$ in a longitudinal study with $n$ subjects and $m$ assessments. Consider the distribution-free linear regression model:

$$E(y_{it} \mid \mathbf{x}_{it}) = \mathbf{x}_{it}^\top \boldsymbol{\beta}, \quad 1 \le i \le n, 1 \le t \le m \tag{4.41}$$

As in Example 6, if we assume $y_{it} \mid \mathbf{x}_{it} \sim N(\mathbf{x}_{it}^\top \boldsymbol{\beta}, \sigma^2)$, then $v(\mu_{it}) = 1$. Thus,

$$A_i = I_m, \quad V_i = R(\boldsymbol{\alpha}), \quad \boldsymbol{\mu}_i = X_i \boldsymbol{\beta}, \quad D_i = \frac{\partial}{\partial \boldsymbol{\beta}} \boldsymbol{\mu}_i = X_i^\top$$

where $X_i = (\mathbf{x}_{i1}, \ldots, \mathbf{x}_{im})^\top$. It follows that

$$\mathbf{w}_n(\boldsymbol{\beta}) = \sum_{i=1}^{n} X_i^\top R^{-1}(\boldsymbol{\alpha}) S_i = \sum_{i=1}^{n} X_i^\top R^{-1}(\boldsymbol{\alpha})(\mathbf{y}_i - X_i\boldsymbol{\beta}) = 0 \qquad (4.42)$$

The above is readily solved to obtain

$$\widehat{\boldsymbol{\beta}} = \left(\sum_{i=1}^{n} X_i^\top R^{-1}(\boldsymbol{\alpha}) X_i\right)^{-1} \left(\sum_{i=1}^{n} X_i^\top R^{-1}(\boldsymbol{\alpha}) \mathbf{y}_i\right) \qquad (4.43)$$

If $\boldsymbol{\alpha}$ is known, $\widehat{\boldsymbol{\beta}}$ is the GEE estimate of $\boldsymbol{\beta}$. Otherwise, $\boldsymbol{\alpha}$ must be estimated in order to obtain estimates of $\boldsymbol{\beta}$. As an example, consider the completely unstructured working correlation matrix, $R(\boldsymbol{\alpha}) = [\rho_{st}]$. We can estimate $\rho_{st}$ by the Pearson correlation:

$$\widehat{\rho}_{st} = \frac{\sum_{i=1}^{n}(r_{is} - \overline{r}_{\cdot s})(r_{it} - \overline{r}_{\cdot t})}{\sqrt{\sum_{i=1}^{n}(r_{is} - \overline{r}_{\cdot s})^2 \sum_{i=1}^{n}(r_{it} - \overline{r}_{\cdot t})^2}}, \qquad r_{it} = y_{it} - \mathbf{x}_{it}^\top \widehat{\boldsymbol{\beta}} \qquad (4.44)$$

where $\overline{r}_{\cdot t} = \frac{1}{n}\sum_{i=1}^{n} r_{it}$. If $\mathbf{y}_i \mid X_i \sim N\left(X_i\boldsymbol{\beta}, \sigma^2 R(\boldsymbol{\alpha})\right)$ and $\sigma^2$ is known or estimated by (4.44), $\widehat{\boldsymbol{\beta}}$ is the MLE of $\boldsymbol{\beta}$ (see exercise).

**Example 8** (***Distribution-free logistic regression model***). In Example 7, assume a binary response $y_{it}$ and consider the logistic model:

$$E(y_{it} \mid \mathbf{x}_{it}) = \mu_{it}, \quad \log\left(\frac{\mu_{it}}{1 - \mu_{it}}\right) = \mathbf{x}_{it}^\top \boldsymbol{\beta}, \quad 1 \le i \le n, 1 \le t \le m \qquad (4.45)$$

As discussed for the univariate distribution-free GLM in Chapter 2, for each $t$, the estimating equation in (4.33) with $G_{it}$ and $D_{it}$ given in (4.34) and $v(\mu_{it}) = \mu_{it}(1 - \mu_{it})$ yields the MLE of $\boldsymbol{\beta}$ if $y_{it} \mid \mathbf{x}_{it} \sim$BI$(\mu_{it}, 1)$. Thus, we may choose $A_i =$diag$(\mu_{it}(1 - \mu_{it}))$ for the GEE to estimate $\boldsymbol{\beta}$ for the multivariate model in (4.45).

As in Example 7, if $\boldsymbol{\alpha}$ is unknown, it must be estimated. However, since the correlation $\rho_{i,st} = Corr(y_{is}, y_{it} \mid \mathbf{x}_{is}, \mathbf{x}_{it})$ for binary response generally varies across subjects, a constant $R(\boldsymbol{\alpha})$ is hardly ever the true correlation matrix except for some special cases. In addition, $\rho_{i,st}$ must also satisfy an additional set of *Frechet bounds* (see exercise):

$$\max\left\{\left[\frac{\mu_{is}\mu_{it}}{(1-\mu_{is})(1-\mu_{it})}\right]^{1/2}, -\left[\frac{(1-\mu_{is})(1-\mu_{it})}{\mu_{is}\mu_{it}}\right]^{1/2}\right\} \qquad (4.46)$$

$$\le \rho_{i,st} \le \min\left\{\left[\frac{\mu_{is}(1-\mu_{it})}{(1-\mu_{is})\mu_{it}}\right]^{1/2}, \left[\frac{(1-\mu_{is})\mu_{it}}{\mu_{is}(1-\mu_{it})}\right]^{1/2}\right\}$$

Because of these constraints, $\boldsymbol{\alpha}$ is typically selected by some ad hoc rules rather than estimated. For example, under the uniform compound symmetry assumption, $R(\rho) = C_m(\rho)$, we may select $\rho$ that satisfies the *Frechet bounds* in (4.46).

Although quite similar in form, GEE differs from the univariate estimating equation considered for the distribution-free GLM discussed in Chapter 2 in that $G_i$ or $V_i$ may involve an unknown vector of parameters $\boldsymbol{\alpha}$. While primary interest lies in $\boldsymbol{\beta}$, $\boldsymbol{\alpha}$ must be estimated to proceed with the computation of the GEE estimate of $\boldsymbol{\beta}$. Although consistency of the GEE estimate $\widehat{\boldsymbol{\beta}}$ does not depend on how $\boldsymbol{\alpha}$ is estimated, judicious choices of the type of estimates of $\boldsymbol{\alpha}$ used in (4.35) not only ensure the asymptotic normality but also simplify the asymptotic variance of $\widehat{\boldsymbol{\beta}}$.

**Theorem 1** (GEE estimates). Let $\widehat{\boldsymbol{\beta}}$ denote the estimate obtained by solving the generalized estimating equation in (4.35). Let $B = E\left(G_i D_i^\top\right)$. Then, under mild regularity conditions, we have

1. $\widehat{\boldsymbol{\beta}}$ is consistent.
2. If $\sqrt{n}\left(\widehat{\boldsymbol{\alpha}} - \boldsymbol{\alpha}\right) = \mathbf{O}_p(1)$, $\widehat{\boldsymbol{\beta}}$ is asymptotically normal with its asymptotic variance $\boldsymbol{\Sigma}_\beta$ given by

$$\boldsymbol{\Sigma}_\beta = B^{-1} E\left(G_i S_i S_i^\top G_i^\top\right) B^{-\top} \tag{4.47}$$

where $B^{-\top} = \left(B^{-1}\right)^\top$. A consistent estimate of $\boldsymbol{\Sigma}_\beta$ is given by

$$\widehat{\boldsymbol{\Sigma}}_\beta = \frac{1}{n} \widehat{B}^{-1} \sum_{i=1}^{n} \left(\widehat{G}_i \widehat{S}_i \widehat{S}_i^\top \widehat{G}_i^\top\right) \widehat{B}^{-\top} \tag{4.48}$$

where $\widehat{A}$ denotes the estimated $A$ obtained by replacing $\boldsymbol{\beta}$ and $\boldsymbol{\alpha}$ with their respective estimates. If $G_i = D_i V_i^{-1}$, then $B = E\left(D_i V_i^{-1} D_i^\top\right)$. In this case, (4.47) and (4.48) reduce to

$$\boldsymbol{\Sigma}_\beta = B^{-1} E\left(D_i V_i^{-1} S_i S_i^\top V_i^{-1} D_i^\top\right) B^{-1} \tag{4.49}$$

$$\widehat{\boldsymbol{\Sigma}}_\beta = \frac{1}{n} \widehat{B}^{-1} \sum_{i=1}^{n} \left(\widehat{D}_i \widehat{V}_i^{-1} \widehat{S}_i \widehat{S}_i^\top \widehat{V}_i^{-1} \widehat{D}_i^\top\right) \widehat{B}^{-1}$$

Proof. The proof follows from an argument similar to the one used for the estimating equation estimates for the univariate distribution-free GLM studied in Chapter 2.

Let $\mathbf{w}_n(\boldsymbol{\beta}) = \frac{1}{n}\sum_{i=1}^{n} G_i S_i$. Since $G_i S_i = G_i(\mathbf{x}_i, \boldsymbol{\beta}, \boldsymbol{\alpha}) S_i(\mathbf{y}_i, \mathbf{x}_i, \boldsymbol{\beta})$, we have

$$\begin{aligned} E(G_i S_i) &= E(G_i E(S_i \mid \mathbf{x}_i)) = E[G_i E(S_i \mid \mathbf{x}_i)] = \mathbf{0} \qquad (4.50) \\ Var(G_i S_i) &= E\left[G_i E\left(S_i S_i^\top \mid \mathbf{x}_i\right) G_i^\top\right] = E\left(G_i S_i S_i^\top G_i^\top\right) \end{aligned}$$

Thus, it follows from CLT that

$$\mathbf{w}_n(\boldsymbol{\beta}) = \frac{1}{n}\sum_{i=1}^{n} G_i S_i \rightarrow_d N\left(\mathbf{0}, E\left(G_i S_i S_i^\top G_i^\top\right)\right) \qquad (4.51)$$

By an argument similar to the proof of consistency for the univariate estimating equation estimate, we can show that the GEE estimate is consistent regardless of the choice of $G_i$ (see exercise).

By a Taylor series expansion, we have

$$\begin{aligned} \mathbf{w}_n\left(\widehat{\boldsymbol{\beta}}, \widehat{\boldsymbol{\alpha}}\right) &= \mathbf{w}_n(\boldsymbol{\beta}, \boldsymbol{\alpha}) + \left(\frac{\partial}{\partial \boldsymbol{\beta}} \mathbf{w}_n(\boldsymbol{\beta}, \boldsymbol{\alpha})\right)^\top \left(\widehat{\boldsymbol{\beta}} - \boldsymbol{\beta}\right) + \\ &+ \left(\frac{\partial}{\partial \boldsymbol{\alpha}} \mathbf{w}_n(\boldsymbol{\beta}, \boldsymbol{\alpha})\right)^\top (\widehat{\boldsymbol{\alpha}} - \boldsymbol{\alpha}) + \mathbf{o}_p\left(n^{-\frac{1}{2}}\right) \end{aligned}$$

Thus,

$$\begin{aligned} \sqrt{n}\mathbf{w}_n(\boldsymbol{\beta}, \boldsymbol{\alpha}) &= -\left(\frac{\partial}{\partial \boldsymbol{\beta}} \mathbf{w}_n(\boldsymbol{\beta}, \boldsymbol{\alpha})\right)^\top \sqrt{n}\left(\widehat{\boldsymbol{\beta}} - \boldsymbol{\beta}\right) - \qquad (4.52) \\ &- \left(\frac{\partial}{\partial \boldsymbol{\alpha}} \mathbf{w}_n(\boldsymbol{\beta}, \boldsymbol{\alpha})\right)^\top \sqrt{n}(\widehat{\boldsymbol{\alpha}} - \boldsymbol{\alpha}) + \mathbf{o}_p(1) \end{aligned}$$

Now, let

$$\begin{aligned} &\boldsymbol{\beta} = (\beta_1, \ldots, \beta_p)^\top, \quad \boldsymbol{\alpha} = (\alpha_1, \ldots, \alpha_q)^\top, \quad \frac{\partial}{\partial \boldsymbol{\alpha}}(G_i S_i) = [e_{ilk}]_{q \times p} \\ &G_i = [g_{ikt}]_{p \times m} \end{aligned}$$

It is readily checked that (see exercise):

$$\begin{aligned} e_{ilk}(\mathbf{x}_i, \boldsymbol{\beta}, \boldsymbol{\alpha}) &= \sum_{t=1}^{m} (y_{it} - \mu_{it}) \frac{\partial}{\partial \alpha_l} g_{ikt}(\mathbf{x}_i, \boldsymbol{\beta}, \boldsymbol{\alpha}) \qquad (4.53) \\ E[e_{ilk}(\mathbf{x}_i, \boldsymbol{\beta}, \boldsymbol{\alpha})] &= \sum_{t=1}^{m} E\left[E[(y_{it} - \mu_{it}) \mid \mathbf{x}_i] \frac{\partial}{\partial \alpha_l} g_{ikt}(\mathbf{x}_i, \boldsymbol{\beta}, \boldsymbol{\alpha})\right] = 0 \end{aligned}$$

It follows that $\frac{\partial}{\partial \boldsymbol{\alpha}} \mathbf{w}_n(\boldsymbol{\beta}, \boldsymbol{\alpha}) = \mathbf{o}_p(1)$ and thereby

$$\left(\frac{\partial}{\partial \boldsymbol{\alpha}} \mathbf{w}_n(\boldsymbol{\beta}, \boldsymbol{\alpha})\right)^\top \sqrt{n}(\widehat{\boldsymbol{\alpha}} - \boldsymbol{\alpha}) = \mathbf{o}_p(1) \mathbf{O}_p(1) = \mathbf{o}_p(1) \tag{4.54}$$

Thus, (4.52) reduces to:

$$\sqrt{n}\left(\widehat{\boldsymbol{\beta}} - \boldsymbol{\beta}\right) = \left(-\frac{\partial}{\partial \boldsymbol{\beta}} \mathbf{w}_n(\boldsymbol{\beta}, \boldsymbol{\alpha})\right)^{-\top} \sqrt{n} \mathbf{w}_n(\boldsymbol{\beta}, \boldsymbol{\alpha}) + \mathbf{o}_p(1). \tag{4.55}$$

It is readily checked that (see exercise):

$$\frac{\partial}{\partial \boldsymbol{\beta}} \mathbf{w}_n(\boldsymbol{\beta}, \boldsymbol{\alpha}) = \frac{1}{n} \sum_{i=1}^{n} \left(\frac{\partial}{\partial \boldsymbol{\beta}} S_i\right) G_i^\top + \mathbf{o}_p(1) \to_p -E\left(D_i G_i^\top\right) \tag{4.56}$$

The asymptotic normality of $\widehat{\boldsymbol{\beta}}$ follows from (4.51), (4.55), (4.56) and Slutsky's theorem with the asymptotic variance $\boldsymbol{\Sigma}_\beta$ given in (4.47).

Note that if $\sqrt{n}(\widehat{\boldsymbol{\alpha}} - \boldsymbol{\alpha}) = \mathbf{O}_p(1)$, that is, $\sqrt{n}(\widehat{\boldsymbol{\alpha}} - \boldsymbol{\alpha})$ is stochastically bounded, $\widehat{\boldsymbol{\alpha}}$ is called $\sqrt{n}$*-consistent*. As seen from the proof of the theorem, this assumption allows us to ignore the term $\left(\frac{\partial}{\partial \boldsymbol{\alpha}} \mathbf{w}_n(\boldsymbol{\beta}, \boldsymbol{\alpha})\right)^\top \sqrt{n}(\widehat{\boldsymbol{\alpha}} - \boldsymbol{\alpha})$ in the asymptotic expansion of $\sqrt{n}\mathbf{w}_n(\boldsymbol{\beta}, \boldsymbol{\alpha})$ in (4.52) to obtain the asymptotic normality and compute the asymptotic variance of $\widehat{\boldsymbol{\beta}}$. As discussed in Chapter 1, $\sqrt{n}(\widehat{\boldsymbol{\alpha}} - \boldsymbol{\alpha}) = \mathbf{O}_p(1)$ holds true if $\sqrt{n}(\widehat{\boldsymbol{\alpha}} - \boldsymbol{\alpha})$ converges in distribution. In most applications, $\widehat{\boldsymbol{\alpha}}$ is asymptotically normal and thus is $\sqrt{n}$-consistent.

**Example 9 (*MANOVA/repeated measures ANOVA*).** For the distribution-free model in Example 6, it follows from Theorem 1 that the GEE estimate $\widehat{\boldsymbol{\mu}}$ is consistent and asymptotically normal. In addition, it follows from (4.49) that the asymptotic variance for the GEE estimate $\widehat{\boldsymbol{\mu}}_k$ is given by

$$\begin{aligned}
B_k &= E\left(D_{ki} V_{ki}^{-1} D_{ki}^\top\right) = R_k^{-1}(\boldsymbol{\alpha}) \\
\Sigma_k &= R_k(\boldsymbol{\alpha})\left(R_k^{-1}(\boldsymbol{\alpha}) E\left(S_i S_i^\top\right) R_k^{-1}(\boldsymbol{\alpha})\right) R_k(\boldsymbol{\alpha}) = E\left(S_i S_i^\top\right) \\
&= E\left[(\mathbf{y}_{ki} - \boldsymbol{\mu}_k)(\mathbf{y}_{ki} - \boldsymbol{\mu}_k)^\top\right]
\end{aligned}$$

In this particular case, $\Sigma_k$ is the variance of $\mathbf{y}_{ki}$ and independent of the choice of working correlation matrix $R_k(\boldsymbol{\alpha})$. Further, if $\mathbf{y}_{ki} \sim N(\mathbf{0}, \Sigma_k)$, $\Sigma_k$ is the asymptotic variance of the MLE of $\widehat{\boldsymbol{\mu}}_k$ $(1 \le k \le g)$.

**Example 10 (*Linear regression*).** In Example 7, the GEE estimate $\widehat{\boldsymbol{\beta}}$ given in (4.43) is consistent. If $\boldsymbol{\alpha}$ is known, $\widehat{\boldsymbol{\beta}}$ is also asymptotically normal. The asymptotic variance is given by

$$\Sigma = B^{-1} E\left(X_i^\top R^{-1}(\boldsymbol{\alpha}) S_i S_i^\top R^{-1}(\boldsymbol{\alpha}) X_i^\top\right) B^{-1}$$

where $B = E\left(X_i^\top R^{-1}(\boldsymbol{\alpha}) X_i^\top\right)$. If $\mathbf{y}_i \mid X_i \sim N\left(X_i \boldsymbol{\beta}, \sigma^2 R(\boldsymbol{\alpha})\right)$, then

$$\begin{aligned} \Sigma &= B^{-1} E\left(X_i^\top R^{-1}(\boldsymbol{\alpha}) S_i S_i^\top R^{-1}(\boldsymbol{\alpha}) X_i^\top\right) B^{-1} \\ &= \sigma^2 E\left(X_i^\top R^{-1}(\boldsymbol{\alpha}) X_i^\top\right) \end{aligned} \tag{4.57}$$

In this case, $\widehat{\boldsymbol{\beta}}$ is the MLE of $\boldsymbol{\beta}$ and $\Sigma$ is the asymptotic variance of $\widehat{\boldsymbol{\beta}}$ (see exercise).

If $\boldsymbol{\alpha}$ is known and estimated by (4.44), $\boldsymbol{\alpha}$ is $\sqrt{n}$-consistent and therefore by Theorem 1, $\Sigma$ in (4.57) is the asymptotic variance of $\widehat{\boldsymbol{\beta}}$. If $\mathbf{y}_i \mid X_i \sim N\left(X_i \boldsymbol{\beta}, \sigma^2 R(\boldsymbol{\alpha})\right)$, $\widehat{\boldsymbol{\beta}}$ becomes of the MLE and $\Sigma$ is the asymptotic variance of the MLE $\widehat{\boldsymbol{\beta}}$.

**Example 11.** In Example 10, let $\Sigma = Var(\mathbf{y}_i \mid X_i) = [\sigma_s \sigma_t \rho_{st}]$ with $Var(y_{it} \mid \mathbf{x}_{it}) = \sigma_t^2$ and $Corr(y_{it} \mid \mathbf{x}_{it}) = \rho_{st}$. If $\sigma_t^2 \neq \sigma^2$, that is, a non-constant across time, then $\widehat{\boldsymbol{\beta}}$ given in (4.43) will not be the MLE of $\boldsymbol{\beta}$ when $\mathbf{y}_i \mid X_i \sim N(X_i \boldsymbol{\beta}, \Sigma)$. For the GEE in (4.37) to yield the MLE of $\boldsymbol{\beta}$ in this case, we can set $R(\boldsymbol{\alpha}) = \Sigma = [\sigma_s \sigma_t \rho_{st}]$ and $A_i = \mathbf{I}_m$. Of course, $R(\boldsymbol{\alpha})$ no longer has the interpretation of being the correlation matrix. We can estimate the entries of $R(\boldsymbol{\alpha})$ by the residuals $r_{it} = y_{it} - \mathbf{x}_{it}^\top \widehat{\boldsymbol{\beta}}$.

The GEE in (4.35) only applies to the distribution-free model defined in (4.28). We must develop a different approach for inference about the model for multilevel categorical response discussed in Example 5. We consider the generalized logit model for nominal response below. Similar considerations apply to the proportion odds for ordinal response.

**Example 12 (*Generalized logit model*).** For the nominal response model in Example 5, we can equivalently express (4.31) as follows:

$$\begin{aligned} E(y_{ijt} \mid \mathbf{x}_{it}) &= \mu_{ijt} = \frac{\exp\left(\alpha_j + \boldsymbol{\beta}_j^\top \mathbf{x}_{it}\right)}{1 + \sum_{j=1}^{J-1} \exp\left(\alpha_j + \boldsymbol{\beta}_j^\top \mathbf{x}_{it}\right)}, \\ j &= 1, \ldots, J-1 \end{aligned} \tag{4.58}$$

Let

$$\mathbf{y}_{it} = \left(y_{i1t}, \ldots, y_{i(J-1)t}\right)^{\top}, \quad \boldsymbol{\mu}_{it} = \left(\mu_{i1t}, \ldots, \mu_{i(J-1)t}\right)^{\top} \tag{4.59}$$
$$\mathbf{y}_{i} = \left(\mathbf{y}_{i1}^{\top}, \ldots, \mathbf{y}_{im}^{\top}\right)^{\top}, \quad \boldsymbol{\mu}_{i} = \left(\boldsymbol{\mu}_{i1}^{\top}, \ldots, \boldsymbol{\mu}_{im}^{\top}\right)^{\top}$$
$$\mathbf{x}_{i} = \left(\mathbf{x}_{i1}^{\top}, \ldots, \mathbf{x}_{im}^{\top}\right)^{\top}, \quad D_i = \frac{\partial}{\partial \boldsymbol{\beta}} \boldsymbol{\mu}_i, \quad S_i = \mathbf{y}_i - \boldsymbol{\mu}_i$$

For each $t$, $\mathbf{y}_{it}$ has a multinomial distribution with mean $\boldsymbol{\mu}_{it}$ and variance $A_{itt}$ given by

$$A_{itt} = \begin{pmatrix} \mu_{i1t}\left(1 - \mu_{i1t}\right) & \cdots & -\mu_{i1t}\mu_{i(J-1)t} \\ \vdots & \ddots & \vdots \\ -\mu_{i1t}\mu_{i(J-1)t} & \cdots & \mu_{i(J-1)t}\left(1 - \mu_{i(J-1)t}\right) \end{pmatrix}$$

Thus, let $V_i = A_i^{\frac{1}{2}} R(\boldsymbol{\alpha}) A_i^{\frac{1}{2}}$ with $R(\boldsymbol{\alpha})$ and $A_i$ defined by

$$R(\boldsymbol{\alpha}) = \begin{pmatrix} \mathbf{I}_{J-1} & R_{12} & \cdots & R_{1m} \\ R_{12} & \mathbf{I}_{J-1} & \cdots & R_{2m} \\ \vdots & \cdots & \ddots & \vdots \\ R_{1m}^{\top} & R_{2m}^{\top} & \cdots & \mathbf{I}_{J-1} \end{pmatrix}, \quad A_i = \operatorname{diag}_t\left(A_{itt}\right) \tag{4.60}$$

where $R_{jk}$ are some $(J-1) \times (J-1)$ matrices parameterized by $\boldsymbol{\alpha}$. Define the GEE in the same form as in (4.37) except for using this newly defined $V_i$ above, and $D_i$ and $S_i$ in (4.59). As in the case of binary response, a constant working correlation matrix $R(\boldsymbol{\alpha})$ is generally not the true correlation matrix and similar Frechet bounds exist for the elements of $R(\boldsymbol{\alpha})$ (see exercise). Thus, unless we use the working independence model, $R(\boldsymbol{\alpha}) = \mathbf{I}_{m(J-1)\times m(J-1)}$, we must be mindful about the bounds when selecting $R(\boldsymbol{\alpha})$.

## 4.4 Missing Data

Two key issues arising in the analysis of longitudinal data are the within-subject correlation among repeated assessments and missing data. Up to this point, we have discussed how to address the former issue under both parametric and distribution-free modeling frameworks. In this section, we address the latter missing data issue.

In most longitudinal studies, missing data is inevitable, even in well designed and executed clinical trials. In longitudinal studies, subjects may simply quit the study or they may not show up at follow-up visits because of problems with transportation, weather conditions, health status, relocation, and so on. In clinical trials, missing data may also be the results of patients' deteriorated or improved health conditions due to treatment, treatment-related complications, treatment response, and so on. Some of the reasons for missing data are clearly treatment related while others are not.

In statistics, we characterize the impact of missing data on model estimates through assumptions or missing data mechanisms. Such assumptions allow statisticians to ignore the multitude of reasons for missing data and focus on addressing their impact on estimation of model parameters. The *missing completely at random* (MCAR) assumption is used to define a class of missing data that does not affect model estimates when completely ignored. For example, in a treatment study, missing data resulting from patient's relocation and conflict of schedules fall into this category. MCAR corresponds to a lay person's notion of random missing, that is, missing data are completely random with absolutely nothing to do with treatment effect.

The next category is the *missing at random* (MAR) assumption, which generalizes MCAR to deal with a popular class of treatment-related missing data. In many clinical trials, missing data is often associated with the treatment interventions under study. For example, a patient may quit the study if he/she feels that the study treatment has deteriorated his/her health conditions and any further treatment will only worsen the medical or psychological problems. Or, a patient may feel that he/she has completely responded to the treatment and therefore does not see any additional benefit in continuing the treatment. In such cases, missing data does not follow the MCAR model since they are predicted by treatment related response. By positing that the occurrence of a missing response at an assessment time depends on the response history or observed pattern prior to the assessment point, MAR constitutes a plausible and applicable statistical approach to model this class of treatment related missing data.

Missing data that satisfies either the MCAR or MAR model is also known as *ignorable missing data*. *Nonignorable missing data* is defined as a type of missing data the occurrence of which can depend on unobserved or future response (in a longitudinal setting). Since the mechanism of non-ignorable missing data involves unobserved response, it is generally quite difficult to model such missing data in real studies without additional data or information regarding the relationship between the missing and observed data.

Note that the term "ignorable missing data" may be a misnomer. For parametric models, we can indeed ignore such missing data since maximum likelihood estimates are consistent when obtained based on the observed-data. For distribution-free models, however, GEE estimates will generally not be consistent when missing data follows the MAR model. Thus, alternative estimating equations must be constructed to provide consistent model estimates.

In this book, we focus on MCAR and MAR, which apply to most studies in biomedical and psychosocial research. In addition, we only consider missing data that occurs in the response variable. Although missing independent variables are also common in real study applications, it is more complex to model them, especially in the presence of missing response.

### 4.4.1 Inference for Parametric Models

Consider a longitudinal study design with $n$ subjects and $m$ assessments. Let $y_{it}$ be the response and $\mathbf{x}_{it}$ a vector of independent variables of interest. For each subject, we define a missing (or rather, observed) data indicator as:

$$r_{it} = \begin{cases} 1 \text{ if } y_{it} \text{ is observed} \\ 0 \text{ if } y_{it} \text{ is missing} \end{cases}, \quad \mathbf{r}_i = (r_{i1}, ..., r_{im})^\top \tag{4.61}$$

We assume no missing data at baseline $t = 1$ such that $r_{i1} = 0$ for all $1 \leq i \leq n$.

Let $\mathbf{y}_i = (y_{i1}, ..., y_{im})^\top$ and $\mathbf{x}_i = (\mathbf{x}_{i1}^\top, ..., \mathbf{x}_{im}^\top)^\top$. Let $\mathbf{y}_i^o$ and $\mathbf{y}_i^m$ denote the observed and unobserved responses, respectively. Thus, $\mathbf{y}_i^o$ and $\mathbf{y}_i^m$ form a partition of $\mathbf{y}_i$. Under likelihood based parametric inference, we jointly model the response $\mathbf{y}_i$ and missing data indicator $\mathbf{r}_i$. The joint density or probability distribution, $f(\mathbf{y}_i, \mathbf{r}_i \mid \mathbf{x}_i)$, can be factored into the product of marginal and conditional distributions using two different approaches, giving rise to two distinct classes of models known as the *selection* and *mixture models*. We outline the two approaches below.

One way to factor the joint distribution is as follows:

$$f(\mathbf{y}_i, \mathbf{r}_i \mid \mathbf{x}_i) = f(\mathbf{y}_i \mid \mathbf{x}_i) f(\mathbf{r}_i \mid \mathbf{y}_i, \mathbf{x}_i) \tag{4.62}$$

Selection models are developed based on the above factorization with the term *selection* reflecting the probability of observing a response or a selection process. Under MAR, the distribution of $\mathbf{r}_i$ only depends on the observed responses, $\mathbf{y}_i^o$, and we thus have

$$f(\mathbf{r}_i \mid \mathbf{y}_i, \mathbf{x}_i) = f(\mathbf{r}_i \mid \mathbf{y}_i^o, \mathbf{y}_i^m, \mathbf{x}_i) = f(\mathbf{r}_i \mid \mathbf{y}_i^o, \mathbf{x}_i) \tag{4.63}$$

It follows from (4.62) and (4.63) that:

$$\begin{aligned} f(\mathbf{y}_i^o, \mathbf{r}_i \mid \mathbf{x}_i) &= \int f(\mathbf{y}_i^o, \mathbf{y}_i^m \mid \mathbf{x}_i) f(\mathbf{r}_i \mid \mathbf{y}_i^o, \mathbf{x}_i)\, d\mathbf{y}_i^m \\ &= f\left(\mathbf{r}_i \mid \mathbf{y}_i^o, \mathbf{x}_i, \boldsymbol{\theta}_{y|r}\right) \int f(\mathbf{y}_i^o, \mathbf{y}_i^m \mid \mathbf{x}_i)\, d\mathbf{y}_i^m \\ &= f(\mathbf{y}_i^o \mid \mathbf{x}_i, \boldsymbol{\theta}_y) f\left(\mathbf{r}_i \mid \mathbf{y}_i^o, \mathbf{x}_i, \boldsymbol{\theta}_{y|r}\right) \end{aligned} \tag{4.64}$$

If $\boldsymbol{\theta}_y$ and $\boldsymbol{\theta}_{y|r}$ are disjoint, then following (4.64) the log-likelihood based on the joint observations $(\mathbf{y}_i^o, \mathbf{r}_i)$ is given by

$$\begin{aligned} l(\boldsymbol{\theta}) &= l_1(\boldsymbol{\theta}_y) + l_2\left(\boldsymbol{\theta}_{y|r}\right) \\ &= \sum_{i=1}^{n} \log\left(f(\mathbf{y}_i^o \mid \mathbf{x}_i, \boldsymbol{\theta}_y)\right) + \sum_{i=1}^{n} \log\left(f\left(\mathbf{r}_i \mid \mathbf{y}_i^o, \mathbf{x}_i, \boldsymbol{\theta}_{y|r}\right)\right) \end{aligned} \tag{4.65}$$

Thus, inference about the regression model can simply be based on $l_1(\boldsymbol{\theta}_y)$. In other words, missing data can be "ignored" if interest is centered on modeling the relationship between $\mathbf{y}_i$ and $\mathbf{x}_i$. It should be emphasized, however, that if $\boldsymbol{\theta}_y$ and $\boldsymbol{\theta}_{y|r}$ are not disjoint, inference based on $f(\mathbf{y}_i^o \mid \mathbf{x}_i, \boldsymbol{\theta}_y)$ may be incorrect. In practice, it is difficult to validate this disjoint assumption between $\boldsymbol{\theta}_y$ and $\boldsymbol{\theta}_{y|r}$, which makes it a potential weakness for applications of selection models.

Under MCAR, it follows from (4.63) and (4.64) that

$$f\left(\mathbf{r}_i \mid \mathbf{y}_i^o, \mathbf{x}_i, \boldsymbol{\theta}_{y|r}\right) = f\left(\mathbf{r}_i \mid \mathbf{x}_i, \boldsymbol{\theta}_{y|r}\right)$$

Unlike the MAR case, $f\left(\mathbf{r}_i \mid \mathbf{x}_i, \boldsymbol{\theta}_{y|r}\right)$ is independent of $\mathbf{y}_i^o$, lending support to the disjoint assumption between $\boldsymbol{\theta}_y$ and $\boldsymbol{\theta}_{y|r}$. Thus, only under MCAR can missing data be completely ignored when making inference concerning the relationship between between $\mathbf{y}_i$ and $\mathbf{x}_i$.

Alternatively, we can factor $f(\mathbf{y}_i, \mathbf{r}_i \mid \mathbf{x}_i)$ as

$$f(\mathbf{y}_i, \mathbf{r}_i \mid \mathbf{x}_i) = f(\mathbf{y}_i \mid \mathbf{r}_i, \mathbf{x}_i) f(\mathbf{r}_i \mid \mathbf{x}_i) \tag{4.66}$$

Since the marginal distribution of $\mathbf{y}_i$ obtained by integrating out $\mathbf{r}_i$ in the factorization above is in the form of a mixture with mixing probability

$f(\mathbf{r}_i \mid \mathbf{x}_i)$, models developed based on (4.66) are known as *mixture models*. This approach is frequently used when the missing pattern itself is also of interest. For example, in a *saturated pattern-mixture model*, the relationship between $\mathbf{y}_i$ and $\mathbf{x}_i$ are modeled separately for each pattern. The overall relationship is then a mixture of the different missing data patterns.

Let $p$ denote the number of distinct missing data patterns. Then, it follows that (see exercise):

$$\begin{aligned} f(\mathbf{y}_i^o, \mathbf{r}_i \mid \mathbf{x}_i) &= \int f(\mathbf{y}_i^m \mid \mathbf{y}_i^o, \mathbf{r}_i, \mathbf{x}_i) f(\mathbf{y}_i^o \mid \mathbf{r}_i, \mathbf{x}_i) f(\mathbf{r}_i \mid \mathbf{x}_i) \, d\mathbf{y}_i^m \qquad (4.67) \\ &= f(\mathbf{r}_i \mid \mathbf{x}_i) f(\mathbf{y}_i^o \mid \mathbf{r}_i, \mathbf{x}_i) \int f(\mathbf{y}_i^m \mid \mathbf{y}_i^o, \mathbf{r}_i, \mathbf{x}_i) \, d\mathbf{y}_i^m \\ &= f(\mathbf{r}_i \mid \mathbf{x}_i, \boldsymbol{\phi}_r) f\left(\mathbf{y}_i^o \mid \mathbf{r}_i, \mathbf{x}_i, \boldsymbol{\phi}_y^k\right) \end{aligned}$$

where $\boldsymbol{\phi}_y^k$ denotes the parameter vector for modeling the relationship between $\mathbf{y}_i$ and $\mathbf{x}_i$ for the $k$th pattern $(1 \leq k \leq p)$. If $\boldsymbol{\phi}_y^k = \boldsymbol{\phi}_y$, that is, a common relationship across all patterns, then missing data follows MCAR and may be totally ignored if interest lies in $\boldsymbol{\phi}_y$.

Under MAR, $f\left(\mathbf{y}_i^o \mid \mathbf{r}_i, \mathbf{x}_i, \boldsymbol{\phi}_y^k\right)$ depends on missing data patterns. If missing response can occur at any time $t$, there are potentially $2^{m-1}$ different patterns, making it difficult to model this relationship. Fortunately, as noted in the beginning of Section 4.4, missing data in most longitudinal studies typically occurs as the result of subject dropout, reflecting the subject's deteriorated/improved health conditions and other related conditions. In this case, missing data follows the *monotone missing data pattern* (MMDP) and has $p = m$ distinct patterns. Under MMDP, $y_{it}$ is observed only if $y_{is}$ are all observed prior to time $t$ $(1 \leq s \leq t \leq m)$. Such structured missing data patterns greatly simplify modeling $f\left(\mathbf{y}_i^o \mid \mathbf{r}_i, \mathbf{x}_i, \boldsymbol{\phi}_y^k\right)$ and other relationships involving $r_{it}$.

Suppose that missing data satisfies the MMDP assumption. Then, there are $m$ distinct patterns defined by the last observed response at $t = 1, 2, \ldots, m$. Thus, we can index the $m$ patterns by $p = t$ $(1 \leq p \leq m)$. Let $\widetilde{\mathbf{y}}_{it} = \left(y_{i1}, ..., y_{i(t-1)}\right)^\top$. It is readily shown that MAR is equivalent to the condition:

$$f(y_{it} \mid \widetilde{\mathbf{y}}_{it}, \mathbf{x}_i, p = j) = f(y_{it} \mid \widetilde{\mathbf{y}}_{it}, \mathbf{x}_i, p \geq t) \qquad (4.68)$$

for all $t \geq 2$ and $j < t$. The above condition is known as the *available case missing value* restriction.

**Example 1 (*Multivariate linear regression*).** Consider a longitudinal study with two assessments. Let $y_{it}$ be a response and $\mathbf{x}_{it}$ a vector of independent variables of interest and assume a multivariate linear regression:

$$\mathbf{y}_i = X_i\boldsymbol{\beta} + \epsilon_i, \quad \boldsymbol{\epsilon}_i = (\epsilon_{i1}, \epsilon_{i2})^\top \sim \text{i.i.d.}N(0, \Sigma) \tag{4.69}$$

where $\mathbf{y}_i = (y_{i1}, y_{i2})^\top$ and $X_i = (\mathbf{x}_{i1}, \mathbf{x}_{i2})^\top$. There are two missing data patterns. Under the saturated pattern-mixture model assumption, we have

$$\mathbf{y}_i \mid p = k, X_i \sim \text{i.d.}N(X_i\boldsymbol{\beta}_k, \Sigma_k), \quad 1 \le k \le 2$$

It follows from the properties of multivariate normal distribution that

$$y_{i2} \mid y_{i2}, p = k, X_i \sim \text{i.d.}N\left(\mathbf{x}_{i2}^\top\boldsymbol{\beta}_k + \frac{\sigma_{12k}}{\sigma_{11k}}\left(y_{i1} - \mathbf{x}_{i1}^\top\boldsymbol{\beta}_k\right), \sigma_{22k} - \frac{\sigma_{12k}^2}{\sigma_{11k}}\right)$$

where $\sigma_{jlk}$ denotes the $jl$th element of $\Sigma_k$. Under MAR, it follows from (4.68) that $\boldsymbol{\beta}_k$ and $\Sigma_k$ must satisfy:

$$\mathbf{x}_{i2}^\top\boldsymbol{\beta}_1 + \frac{\sigma_{121}}{\sigma_{111}}\left(y_{i1} - \mathbf{x}_{i1}^\top\boldsymbol{\beta}_1\right) = \mathbf{x}_{i2}^\top\boldsymbol{\beta}_2 + \frac{\sigma_{122}}{\sigma_{112}}\left(y_{i1} - \mathbf{x}_{i1}^\top\boldsymbol{\beta}_2\right)$$

$$\sigma_{222} - \frac{\sigma_{121}^2}{\sigma_{111}} = \sigma_{221} - \frac{\sigma_{122}^2}{\sigma_{112}}$$

Under MCAR, $\boldsymbol{\beta}_1 = \boldsymbol{\beta}_2$ and $\Sigma_1 = \Sigma_2$, and the above holds true.

### 4.4.2 Inference for Distribution-Free Models

For parametric models, missing data can be ignored even when it follows the MAR model. Unfortunately, this is not true for distribution-free models. More specifically, the generalized estimating equation discussed in Section 4.3.2 only guarantees consistent model estimates when missing data follows MCAR. In this Section, we discuss inference using a revised estimating equation known as the weighted generalized estimating equation (WGEE). We motivate the development of WGEE by a relatively simple longitudinal study design with only two assessment points.

**Example 2 (*One-sample repeated measures ANOVA under MCAR*).** Consider a longitudinal study with two assessments and a homogeneous study group with $n$ subjects. Let $y_{it}$ denote a continuous response from the $i$th subject at time $t$ $(t = 1, 2)$. We are interested in estimating the mean of $\mathbf{y}_i = (y_{1t}, y_{2t})^\top$. By treating this one-sample problem as a special case of the repeated measures ANOVA, it follows from the discussion in Section 4.3 that the GEE estimate of the mean $\boldsymbol{\mu} = E(\mathbf{y}_i)$ of $\mathbf{y}_i$ is the sample mean.

Under our assumption, $y_{i1}$ is always observed, but $y_{i2}$ may be missing. Let $r_{i2}$ be a missing data indicator for $y_{i2}$, with $r_{i2} = 1$ if $y_{i2}$ is observed and 0 if otherwise. Consider the following estimate of $\boldsymbol{\mu}$:

$$\widehat{\mu}_1 = \frac{1}{n}\sum_{i=1}^{n} y_{i1}, \quad \widehat{\mu}_2 = \frac{1}{n_2}\sum_{i=1}^{n} r_{i2}y_{i2}, \quad n_2 = \sum_{i=1}^{n} r_{i2} \tag{4.70}$$

It is readily seen that the above is simply a fancy way to express the sample mean of $y_{i2}$ based on observed data. If $y_{i2}$ is not subject to missingness, $\widehat{\boldsymbol{\mu}}$ above yields the GEE estimate of $\boldsymbol{\mu}$ in the complete data case. Otherwise, if missing $y_{i2}$ follows MCAR, $r_{i2}$ is independent of $y_{i1}$ as well as $y_{i2}$. Thus, we have

$$\pi_{i2} = \Pr\left(r_i = 1 \mid y_{i1}, y_{i2}\right) = \Pr\left(r_i = 1\right) \tag{4.71}$$

Intuitively, $\widehat{\mu}_2$ should be consistent. This is indeed the case; it follows from LLN, Slutsky's theorem and (4.71) that

$$\widehat{\mu}_2 = \frac{\frac{1}{n}\sum_{i=1}^{n} r_{i2}y_{i2}}{\frac{1}{n}\sum_{i=1}^{n} r_{i2}} \to_p \frac{1}{E\left(r_{i2}\right)} E\left(r_{i2}y_{i2}\right) = \mu_2 \tag{4.72}$$

**Example 3** (***One-sample repeated measures ANOVA under MAR***). In Example 2, if the missingness depends on $y_{i1}$, (4.71) no longer holds and $\widehat{\mu}_2$ in general will not be consistent. For example, if $y_{i1}$ and $y_{i2}$ are positively correlated with higher values of $y_{i1}$ more likely leading to missing $y_{i2}$, $\widehat{\mu}_2$ will be downwardly biased. In treatment studies, this type of response-dependent missingness may happen if a patient feels that he/she has responded to the treatment and decides not to undergo additional treatment. To construct a consistent estimate, we may somehow "recover" the missing response $y_{i2}$ for these study dropouts and include them in the estimation of $\mu_2$. Of course, such a recovery process is not possible in most studies. However, with a probability model for the missingness data indicator $r_i$, we can "statistically recover" such missing $y_{i2}$.

Since the missingness of $y_{i1}$ only depends on $y_{i1}$ as guaranteed by the MAR model, we have:

$$\pi_{i2} = \Pr\left(r_{i2} = 1 \mid y_{i1}, y_{i2}\right) = \Pr\left(r_{i2} = 1 \mid y_{i1}\right) \tag{4.73}$$

For each subject $i$, we observe $y_{i2}$ with probability $\pi_{i2}$. Thus, this subject represents a subgroup of $\frac{1}{\pi_{i2}}$ subjects with response $y_{i1}$ at time 1 who are unobserved at time 2. By augmenting each observed subject with the weight function $\frac{1}{\pi_{i2}}$ in the following estimate, we are in effect statistically recovering and including such "ghosts" in the estimation of $\mu_2$:

$$\widehat{\mu}_2 = \frac{1}{\sum_{i=1}^n \pi_{i2}^{-1} r_{i2}} \sum_{i=1}^n \frac{r_{i2}}{\pi_{i2}} y_{i2} \tag{4.74}$$

We can think of the denominator as representing the total number of subjects (observed plus ghosts) and the numerator the sum of response at time 2 over all such subjects. The weighted estimate $\widehat{\mu}_2$ is indeed consistent. To see this, observe first that

$$\begin{aligned} E\left(\frac{r_{i2}}{\pi_{i2}} y_{i2}\right) &= E\left[E\left(\frac{r_{i2}}{\pi_{i2}} y_{i2}\right) \mid \mathbf{y}_i\right] = E\left[y_{i2}\frac{1}{\pi_{i2}} E\left(r_{i2} \mid y_{i1}, y_{i2}\right)\right] \\ &= E\left[y_{i2}\frac{1}{\pi_{i2}} E\left(r_{i2} \mid y_{i1}\right)\right] = E\left(y_{i2}\right) = \mu_2 \end{aligned} \tag{4.75}$$

Then, by applying an argument similar to (4.72), it follows from (4.75) that $\widehat{\mu}_2$ is consistent (see exercise).

**Example 4 (*One-sample repeated measures ANOVA under non-ignorable missingness*).** In Example 3, suppose that the missingness also depends on $y_{i2}$, that is,

$$\pi_{i2} = \Pr\left(r_{i2} = 1 \mid y_{i1}, y_{i2}\right) \tag{4.76}$$

In this case, $\widehat{\mu}_2$ defined in (4.74) is still consistent (see exercise). However, unlike the model in (4.73), it is generally not possible to estimate $\pi_{i2}$ defined in (4.76) based on observed data when $\pi_{i2}$ is unknown as in most applications. For example, we can model and estimate $\pi_{i2}$ as a function of $y_{i1}$ in (4.73) using a logistic model. But, this approach does not work for $\pi_{i2}$ in (4.76) as $y_{i2}$ is not observed $r_{i2} = 1$. Additional information and/or data source (other than the data at hand) are needed to help identify the parameters in such a logistic model.

Examples 2-4 cover the three major missing data mechanisms employed in statistical modeling of missing data and illustrate how to construct consistent model estimates under each situation. The weighted generalized estimating equation (WGEE) is developed based on the same principles by using weight functions to account for the contribution of missing response from the ghost.

**Example 5 (*WGEE for one-sample repeated measures ANOVA*

***under MAR***). Within the context of Example 3, let

$$\Delta_i = \begin{pmatrix} 1 & 0 \\ 0 & \frac{r_{i2}}{\pi_{i2}} \end{pmatrix}, \quad A_i = \mathbf{I}_2, \quad R(\rho) = \begin{pmatrix} 1 & \rho \\ \rho & 1 \end{pmatrix}$$

$$D_i = \frac{\partial}{\partial \boldsymbol{\mu}} \boldsymbol{\mu} = \mathbf{I}_2, \quad S_i = \mathbf{y}_i - \boldsymbol{\mu}, \quad V = A_i^{\frac{1}{2}} R(\rho) A_i^{\frac{1}{2}} = R(\rho)$$

Define a weighted generalized estimating equation (WGEE) as follows:

$$\begin{aligned} \mathbf{w}_n(\boldsymbol{\mu}) &= \sum_{i=1}^{n} D_i V_i^{-1} \Delta_i S_i \\ &= \sum_{i=1}^{n} R^{-1}(\rho) \begin{pmatrix} 1 & 0 \\ 0 & \frac{r_{i2}}{\pi_{i2}} \end{pmatrix} \begin{pmatrix} y_{i1} - \mu_1 \\ y_{i2} - \mu_2 \end{pmatrix} = \mathbf{0} \end{aligned} \tag{4.77}$$

The WGEE above is quite similar to the GEE in (4.37) except for the extra term weight function $\Delta_i$. The presence of $\Delta_i$ is essential to ensure the unbiasedness of the above equation, that is, $E(\mathbf{w}_n(\boldsymbol{\mu})) = \mathbf{0}$, under MAR, which in turn provides consistent estimates of $\boldsymbol{\mu}$. This readily follows from (see exercise):

$$E(\Delta_i S_i) = E[E(\Delta_i S_i \mid \mathbf{r}_i, \mathbf{y}_i)] = E[S_i E(\Delta_i \mid y_{i1})] = \mathbf{0} \tag{4.78}$$

The equation in (4.77) is readily solved to yield

$$\widehat{\mu}_1 = \frac{1}{n} \sum_{i=1}^{n} y_{i1}, \quad \widehat{\mu}_2 = \frac{1}{\sum_{i=1}^{n} \pi_{i2}^{-1} r_{i2}} \sum_{i=1}^{n} \frac{r_{i2}}{\pi_{i2}} y_{i2} \tag{4.79}$$

which is identical to the weighted estimate of $\boldsymbol{\mu}$ in (4.74) constructed in Example 3. Thus, the estimate of $\boldsymbol{\mu}$ obtained from the WGEE in (4.77) is consistent. Note that $R^{-1}(\rho)$ in this special case does not play any role in inference for $\boldsymbol{\mu}$.

The WGEE for this relatively simple one-sample ANOVA with a pre- and posttreatment study design is readily generalized to the general setting with more complex models and multiple assessments. Consider a longitudinal study with $n$ subjects and $m$ assessments. Let $y_{it}$ be a response and $\mathbf{x}_{it}$ a vector of independent variables of interest. Consider the distribution-free regression model in (4.28), which for convenience is copied below:

$$E(y_{it} \mid \mathbf{x}_{it}) = \mu_{it}, \quad g(\mu_{it}) = \mathbf{x}_{it}^{\top} \boldsymbol{\beta}, \quad 1 \leq t \leq m, \quad 1 \leq i \leq n \tag{4.80}$$

where $g(\cdot)$ is some known link function and $\boldsymbol{\beta}$ a $p \times 1$ vector. Let $r_{it}$ denote a binary indicator with the value 1 if $y_{it}$ is observed and 0 if otherwise. Let

$$\pi_{it} = \Pr(r_{it} = 1 \mid \mathbf{x}_i, \mathbf{y}_i), \quad \Delta_{it} = \frac{r_{it}}{\pi_{it}}, \quad \Delta_i = \text{diag}_t(\Delta_{it}) \tag{4.81}$$

The WGEE for the class of models in (4.80) has the form:

$$\mathbf{w}_n(\boldsymbol{\beta}) = \sum_{i=1}^{n} G_i(\mathbf{x}_i)\Delta_i S_i = \sum_{i=1}^{n} G_i(\mathbf{x}_i)\Delta_i(\mathbf{y}_i - \mathbf{h}_i) = \mathbf{0} \tag{4.82}$$

where $G_i(\mathbf{x}_i)$, $S_i$, $\mathbf{y}_i$ and $\mathbf{h}_i$ are defined the same way as for the unweighted GEE in (4.35) in the absence of missing data. The WGEE in (4.82) is unbiased, that is, $E(\mathbf{w}_n(\boldsymbol{\beta})) = \mathbf{0}$, if $E(\Delta_i S_i) = \mathbf{0}$. By using an argument similar to proving (4.78) in Example 5, it is readily checked that $E(\Delta_i S_i) = \mathbf{0}$ is guaranteed if the model in (4.81) (and of course, the model for regression in (4.80) as well) is specified correctly (see exercise).

To be able to observe representatives to statistically recover all ghosts, $\pi_{it} > 0$ must hold true for all $1 \le t \le m$ and $1 \le i \le n$. From an inference point of view, this requirement ensures that representatives of all subgroups are observed. Otherwise, some subgroups will be left out, yielding biased estimates of $\boldsymbol{\beta}$. If $\pi_{it}$ is very small, the WGEE estimate obtained from (4.82) may be unstable. Thus, we assume that $\pi_{it} \ge c > 0$. Also, $G_i$ may depend on some parameter vector $\boldsymbol{\alpha}$, and for most applications, $G_i(\mathbf{x}_i) = \left(\frac{\partial}{\partial \beta}\Delta_i \mathbf{h}_i\right) V_i^{-1} = D_i \Delta_i V_i^{-1}$, where $D_i$ and $V_i$ are defined the same way as for the unweighted GEE in (4.36). As in the case of unweighted GEE, the consistency of the estimate from (4.82) is independent of the type of estimates of $\boldsymbol{\alpha}$ used, but the asymptotic normality of the estimate is ensured when $\boldsymbol{\alpha}$ is substituted by some $\sqrt{n}$-consistent estimate. We summarize these asymptotic properties of the WGEE estimate in a theorem below. The proof of the theorem follows by applying an argument similar to the proof of Theorem 1 for unweighted GEE in Section 4.3.2 (see exercise).

**Theorem 1** (WGEE estimates). Let $\widehat{\boldsymbol{\beta}}$ denote the estimate obtained by solving the WGEE in (4.82). Assume that $\pi_{it}$ are known and $\pi_{it} \ge c > 0$ $(1 \le i \le n,\ 1 \le t \le m)$. Let $B = E\left(G_i \Delta_i D_i^\top\right)$. Then, under mild regularity conditions, we have

1. $\widehat{\boldsymbol{\beta}}$ is consistent.
2. If $\widehat{\boldsymbol{\alpha}}$ is $\sqrt{n}$-consistent, $\widehat{\boldsymbol{\beta}}$ is asymptotically normal with the asymptotic variance $\boldsymbol{\Sigma}_\beta$ given by

$$\boldsymbol{\Sigma}_\beta = B^{-1} E\left(G_i \Delta_i S_i S_i^\top \Delta_i G_i^\top\right) B^{-\top} \tag{4.83}$$

A consistent estimate of $\mathbf{\Sigma}_\beta$ is given by

$$\widehat{\mathbf{\Sigma}}_\beta = \frac{1}{n}\widehat{B}^{-1}\sum_{i=1}^{n}\left(\widehat{G}_i\Delta_i\widehat{S}_i\widehat{S}_i^\top\Delta_i\widehat{G}_i^\top\right)\widehat{B}^{-\top} \tag{4.84}$$

where $\widehat{A}$ denotes the estimate of $A$ obtained by substituting the estimates of $\boldsymbol{\beta}$ and $\boldsymbol{\alpha}$ in place of the respective parameter vectors. If $G_i(\mathbf{x}_i) = D_i\Delta_iV_i^{-1}$, then $B = E\left(D_i\Delta_iV_i^{-1}\Delta_iD_i^\top\right)$, and (4.83) and (4.84) reduce to

$$\mathbf{\Sigma}_\beta = B^{-1}E\left(D_i\Delta_iV_i^{-1}S_iS_i^\top V_i^{-1}\Delta_iD_i^\top\right)B^{-\top} \tag{4.85}$$

$$\widehat{\mathbf{\Sigma}}_\beta = \frac{1}{n}\widehat{B}^{-1}\sum_{i=1}^{n}\left(\widehat{D}_i\Delta_i\widehat{V}_i^{-1}\widehat{S}_i\widehat{S}_i^\top\widehat{V}_i^{-1}\Delta_i\widehat{D}_i^\top\right)\widehat{B}^{-\top}$$

In most applications, $\pi_{it}$ is unknown and must be estimated. Under MCAR, $\mathbf{r}_i$ is independent of $\mathbf{x}_i$ and $\mathbf{y}_i$ and thus $\pi_{it} = \Pr(r_{it} = 1) = \pi_t$. In this case, $\pi_t$ is a constant independent of $\mathbf{x}_i$ and $\mathbf{y}_i$ and is readily estimated by the sample moment: $\widehat{\pi}_t = \frac{1}{n}\sum_{i=1}^{n} r_{it}$ $(2 \leq t \leq m)$. Under MAR, $\pi_{it}$ becomes dependent on the observed $\mathbf{x}_i$ and $\mathbf{y}_i$. As discussed in Section 4.4.1, it is generally difficult to model and estimate $\pi_{it}$ without imposing the MMDP assumption because of the large number of missing data patterns. The structured patterns under MMDP greatly reduce not only the number of missing data patterns, but also the complexity in modeling $\pi_{it}$.

Let $\widetilde{\mathbf{x}}_{it} = \left(\mathbf{x}_{i1}^\top, \ldots, \mathbf{x}_{i(t-1)}^\top\right)^\top$ and $\widetilde{\mathbf{y}}_{it} = \left(y_{i1}, \ldots, y_{i(t-1)}\right)^\top$ $(2 \leq t \leq m)$, denoting the vectors containing the independent and response variables prior to time $t$, respectively. Let $H_{it} = \{\widetilde{\mathbf{x}}_{it}, \widetilde{\mathbf{y}}_{it}; 2 \leq t \leq m\}$ denote the observed data prior to time $t$ $(2 \leq t \leq m,\ 1 \leq i \leq n)$. Then, under MAR we have

$$\pi_{it} = \Pr(r_{it} = 1 \mid \mathbf{x}_i, \mathbf{y}_i) = \Pr(r_{it} = 1 \mid H_{it}) \tag{4.86}$$

Under MMDP $y_{it}$ is observed only if all $y_{is}$ prior to time $t$ are observed. We can first model the one-step transition probability $p_{it}$ of the occurrence of missing data and then compute $\pi_{it}$ as a function of $p_{it}$ by invoking the MMDP assumption.

Let $p_{it} = E\left(r_{it} = 1 \mid r_{i(t-1)} = 1, H_{it}\right)$ denote the one-step transition probability from observing the response at $t-1$ to $t$. We can readily model $p_{it}$ using a logistic regression model:

$$\begin{aligned} \operatorname{logit}(p_{it}) &= \operatorname{logit}\left[E\left(r_{it} = 1 \mid r_{i(t-1)} = 1, H_{it}\right)\right] \\ &= \xi_t + g_t\left(\boldsymbol{\eta}_t, \widetilde{\mathbf{x}}_{it}, \widetilde{\mathbf{y}}_{it}\right), \quad 2 \leq t \leq m, \quad 1 \leq i \leq n \end{aligned} \tag{4.87}$$

where $\boldsymbol{\gamma}_t = \left(\xi_t, \boldsymbol{\eta}_t^\top\right)^\top$ denotes the model parameters and $g_t\left(\boldsymbol{\eta}_t, \widetilde{\mathbf{x}}_{it}, \widetilde{\mathbf{y}}_{it}\right)$ some functions of $\left(\boldsymbol{\eta}_t, \widetilde{\mathbf{x}}_{it}, \widetilde{\mathbf{y}}_{it}\right)$. For example, if $y_{it}$ and $\mathbf{x}_{it}$ are all continuous, we can set $g_t\left(\boldsymbol{\eta}_t, \widetilde{\mathbf{x}}_{it}, \widetilde{\mathbf{y}}_{it}\right) = \boldsymbol{\eta}_{xt}^\top \widetilde{\mathbf{x}}_{it} + \boldsymbol{\eta}_{yt}^\top \widetilde{\mathbf{y}}_{it}$ with $\boldsymbol{\eta}_t = \left(\boldsymbol{\eta}_{xt}^\top, \boldsymbol{\eta}_{yt}^\top\right)^\top$. More complex forms of $g_t\left(\boldsymbol{\eta}_t, \widetilde{\mathbf{x}}_{it}, \widetilde{\mathbf{y}}_{it}\right)$ as well as cases with other types of $y_{it}$ and $\mathbf{x}_{it}$, such as binary, are similarly considered. Under MMDP, we have (see exercise):

$$\pi_{it}(\boldsymbol{\gamma}) = p_{it} \Pr\left(r_{i(t-1)} = 1 \mid H_{i(t-1)}\right) = \prod_{s=2}^{t} p_{is}\left(\boldsymbol{\gamma}_s\right), \qquad (4.88)$$
$$2 \leq t \leq m, \quad 1 \leq i \leq n$$

where $\boldsymbol{\gamma} = \left(\boldsymbol{\gamma}_1^\top, \ldots, \boldsymbol{\gamma}_m^\top\right)^\top$. Thus, we can estimate $\pi_{it}$ from $p_{it}$ in (4.87) using the above relationship.

Note that when modeled by (4.87) and (4.88), $\pi_{it}$ in theory ranges between 0 and 1. However, $\mathbf{x}_{it}$ can be assumed to be bounded and the requirement $\pi_{it} \geq c > 0$ is satisfied for all practical purposes. Note also that when $\pi_{it}$ is estimated by (4.88), the asymptotic variance in (4.83) or (4.85) is likely to underestimate the variability of $\widehat{\boldsymbol{\beta}}$. Unlike the parameter vector $\boldsymbol{\alpha}$ in $G_i$, the variability of $\widehat{\boldsymbol{\gamma}}$ cannot be ignored even for $\sqrt{n}$-consistent estimate. If $\widehat{\boldsymbol{\gamma}}$ is obtained from GEE or maximum likelihood method, we can readily find the asymptotic variance of $\widehat{\boldsymbol{\beta}}$ obtained from (4.82) with $\boldsymbol{\gamma}$ substituted by $\widehat{\boldsymbol{\gamma}}$.

Let $\widehat{\boldsymbol{\gamma}}$ be the solution to the equation: $\mathbf{u}_n(\boldsymbol{\gamma}) = \sum_{i=1}^n \mathbf{u}_{ni}(\boldsymbol{\gamma}) = \mathbf{0}$, where $\mathbf{u}_{ni}$ is the score (for maximum likelihood estimation) or score-like vector (for GEE estimation) for the $i$th subject ($1 \leq i \leq n$). As in the proof of Theorem 1 of Section 4.3.2, let $\mathbf{u}_n$ and $\mathbf{w}_n$ denote the normalized versions of $\mathbf{u}_n$ and $\mathbf{w}_n$ in (4.82), that is,

$$\mathbf{u}_n = \frac{1}{n} \sum_{i=1}^n \mathbf{u}_{ni}, \quad \mathbf{w}_n = \frac{1}{n} \sum_{i=1}^n G_i\left(\mathbf{x}_i\right) \Delta_i S_i$$

Then, by a Taylor series expansion, we have

$$\sqrt{n}\left(\widehat{\boldsymbol{\gamma}} - \boldsymbol{\gamma}\right) - \left(-\frac{\partial}{\partial \boldsymbol{\gamma}} \mathbf{u}_n\right)^{-\top} \sqrt{n} \mathbf{u}_n + \mathbf{o}_p(1)$$
$$= -H^{-1} \frac{\sqrt{n}}{n} \sum_{i=1}^n \mathbf{u}_{ni} + \mathbf{o}_p(1)$$

By applying an argument similar to the proof of Theorem 1 of Section 4.3.2, it follows from the above that (see exercise):

$$\sqrt{n}\left(\widehat{\boldsymbol{\beta}}-\boldsymbol{\beta}\right) = \left(-\frac{\partial}{\partial \boldsymbol{\beta}}\mathbf{w}_n\right)^{-\top} \sqrt{n}\left[\mathbf{w}_n + \left(\frac{\partial}{\partial \boldsymbol{\gamma}}\mathbf{w}_n\right)^{\top}(\widehat{\gamma}-\gamma)\right] + \qquad (4.89)$$
$$+\,\mathbf{o}_p(1)$$
$$= B^{-1}\frac{\sqrt{n}}{n}\sum_{i=1}^{n}\left(\mathbf{w}_{ni} - CH^{-1}\mathbf{u}_{ni}\right) + \mathbf{o}_p(1)$$

where

$$B = E\left(G_i \Delta_i D_i^{\top}\right), \quad C = E\left[\frac{\partial}{\partial \gamma}\left(G_i\left(\mathbf{x}_i\right)\Delta_i S_i\right)\right]^{\top}, \quad H = E\left(\frac{\partial}{\partial \gamma}\mathbf{u}_{ni}\right)^{\top}$$

It immediately follows from (4.89) that $\widehat{\boldsymbol{\beta}}$ is asymptotically normal with the asymptotic variance given by (see exercise):

$$\Sigma_{\beta} = B^{-1}\left[E\left(G_i \Delta_i S_i S_i^{\top} \Delta_i G_i^{\top}\right) + \Phi\right] B^{-\top} \qquad (4.90)$$
$$\Phi = CH^{-1}Var\left(\mathbf{u}_{ni}\right)H^{-\top}C^{\top} - 2E\left(\mathbf{w}_{ni}\mathbf{u}_{ni}^{\top}H^{-\top}C^{\top}\right)$$

A consistent estimate of $\Sigma_{\beta}$ is readily obtained by substituting consistent estimates of the respective quantities in (4.90). In comparison to (4.83), the asymptotic variance has an additional correction factor.

So far, we have assumed that we can model $\pi_{it}$ correctly when it is unknown. If the model for $\pi_{it}$ is wrong, that is, $\pi_{it} \neq E\left(r_{it} = 1 \mid H_{it}\right)$, the condition $E\left(\Delta_i S_i\right) = \mathbf{0}$ may not be true and the WGEE may yield biased estimates of $\boldsymbol{\beta}$. In some cases, however, we can still consistently estimate $\boldsymbol{\beta}$ if we have a correct model for the missing $y_{it}$. We illustrate this *double robustness* idea again with a relatively simple pre-post study design involving two assessment times.

As noted in Section 4.2.1, ANOVA and ANCOVA may be applied to compare treatment difference at posttreatment for controlled randomized longitudinal trials. Under MAR, these models generally yield biased estimates when applied directly to the observed data at posttreatment. By generalizing Example 5 to multisample repeated measures ANOVA, we can immediately address MAR using WGEE.

**Example 6** (***Multisample repeated measures ANOVA under MAR***). Within the pre-post study design in Example 5, consider comparing $g$ treatment conditions with $n_k$ subjects in the $k$th treatment condition

$(1 \leq k \leq g)$. Let $y_{kit}$ denote a continuous response from the $i$th subject within the $k$th treatment at time $t$ $(t = 1, 2)$. Let

$$\mathbf{y}_{ki} = (y_{ki1}, y_{ki2})^\top, \quad \boldsymbol{\mu}_k = E(\mathbf{y}_{ki}) = (\mu_{k1}, \mu_{k2})^\top \tag{4.91}$$
$$\pi_{ki2} = E(r_{ki2} \mid y_{ki1}), \quad \Delta_{ki1} = 1, \quad \Delta_{ki2} = \frac{r_{ki2}}{\pi_{ki2}}, \quad \Delta_{ki} = \text{diag}_t(\Delta_{kit})$$

By substituting $\mathbf{I}_2$, $\Delta_{ki}$ and $\mathbf{y}_{ki} - \boldsymbol{\mu}_k$ in place of $G_i$, $\Delta_i$ and $S_i$ in (4.82) and solving for $\boldsymbol{\mu}_k$, we immediately obtain

$$\widehat{\mu}_{k1} = \frac{1}{n_k}\sum_{i=1}^{n_k} y_{ki1}, \quad \widehat{\mu}_{k2} = \frac{1}{\sum_{i=1}^{n_k} \pi_{ki2}^{-1} r_{ki2}} \sum_{i=1}^{n_k} \frac{r_{ki2}}{\pi_{ki2}} y_{ki2}, \quad 1 \leq k \leq g \tag{4.92}$$

The WGEE estimate of $\boldsymbol{\mu}_k$ has the same form as in (4.79) except for the obvious generalization to the $g$ groups.

The estimate $\widehat{\mu}_{k2}$ in (4.92) may be biased if $\pi_{ki2}$ is not modeled correctly. As an alternative, we may use an ANCOVA-like model to address MAR without positing any model for $\pi_{ki2}$.

**Example 7 (*ANCOVA for pre-post study design under MAR*).** Within the context of Example 6, suppose that we have a linear model for $y_{ki2}$ as a function of $y_{ki1}$ as follows:

$$y_{ki2} = \mu_{ki2} + \epsilon_{ki}, \quad \mu_{ki2} = \tau_k + \beta_k y_{ki1} = \mathbf{x}_{ki}^\top \boldsymbol{\theta}_k \tag{4.93}$$
$$\epsilon_{ki} \sim \text{i.i.d.}\left(0, \sigma^2\right), \quad 1 \leq i \leq n_k, \quad 1 \leq k \leq g$$

where $\mathbf{x}_{ki} = (1, y_{ki1})^\top$ and $\boldsymbol{\theta}_k = (\tau_k, \beta_k)^\top$. If we apply (4.93) to the observed response $y_{ki2}$, we can estimate $\boldsymbol{\theta}_k$ by the following unweighted GEE:

$$\sum_{i=1}^{n_k} G_i(y_{ki1}) r_{ki2} S_i = \sum_{i=1}^{n_k} G_i(y_{ki1}) r_{ki2} (y_{ki2} - \mu_{ki2}) = 0 \tag{4.94}$$

It is readily checked that $E(r_{ki2} S_i) = 0$ so that the above GEE is unbiased (see exercise). In addition, by setting $G_i(y_{ki1}) = D_{ki} = \frac{\partial \mu_{ki2}}{\partial \boldsymbol{\theta}_k}$, we obtain from (4.94) the GEE estimate of $\boldsymbol{\theta}_k$:

$$\widehat{\boldsymbol{\theta}}_k = \left(\sum_{i=1}^{n_k} r_{ki2} \mathbf{x}_{ki} \mathbf{x}_{ki}^\top\right)^{-1} \left(\sum_{i=1}^{n_k} r_{ki2} \mathbf{x}_{ki} y_{ki2}\right), \quad 1 \leq k \leq g \tag{4.95}$$

The consistency of $\widehat{\boldsymbol{\theta}}_k$ above follows immediately from the unbiasedness of the GEE in (4.94), but can also be checked directly by using its expression in

(4.95) (see exercise). Note that $\widehat{\boldsymbol{\theta}}_k$ in (4.95) is also the maximum likelihood estimate of $\boldsymbol{\theta}_k$ if $\epsilon_{ki} \sim$i.i.d.$N\left(0, \sigma^2\right)$. Thus, we can estimate $\mu_{k2}$ by $\widehat{\mu}_{k2} = \frac{1}{n_k}\sum_{i=1}^{n_k} \mathbf{x}_{ki}^\top \widehat{\boldsymbol{\theta}}_k$.

Example 7 shows that we can estimate $\mu_{k2}$ in Example 6 without relying on modeling $r_{ki2}$ if we have a correct model for relating $y_{ki2}$ to $y_{k1i}$. In practice, we may not know which model is correct. By combining the approaches in the two examples, we can develop a modified WGEE that will deliver consistent estimates of $\boldsymbol{\beta}$ if one of these models is correctly specified.

**Example 8 (*Double-robust estimate for pre-post study design under MAR*).** Consider estimating $\mu_{k2}$ in Example 6. Let

$$S_{ki} = \begin{pmatrix} y_{ki1} - \mu_{k1} \\ \frac{r_{ki2}}{\pi_{ki2}}\left(y_{ki2} - \mu_{k2}\right) + \left(1 - \frac{r_{ki2}}{\pi_{ki2}}\right)\left(\widetilde{y}_{ki2} - \mu_{k2}\right) \end{pmatrix} \tag{4.96}$$

where $\widetilde{y}_{ki2} = \mathbf{x}_{ki}^\top \boldsymbol{\theta}_k$ with $\mathbf{x}_{ki}$ and $\boldsymbol{\theta}_k$ defined in (4.93) of Example 7. Assume that $\boldsymbol{\theta}_k$ is known or substituted by a consistent estimate. It is readily checked that $E\left(S_{ki}\right) = 0$ if $E\left(r_{ki2} \mid y_{ki1}\right) = \pi_{ki2}$ or the model in (4.93) is correct. Thus, by substituting $S_{ki}$ in (4.96) together with $\mathbf{I}_2$ and $\Delta_{ki}$ in (4.92) in place of $S_i$, $G_i$ and $\Delta_i$ in (4.82), we obtain an unbiased WGEE for estimating $\boldsymbol{\mu}_k$ if one of the models for $r_{ki2}$ and $y_{ki2}$ is correct.

Note that if $\beta_k = \beta$ and $E\left(y_{ki1}\right) = \mu_1$ in Example 7, (4.93) becomes ANCOVA. In this case, we can use this model directly to examine post-treatment difference by comparing $\tau_k$ since $\tau_k - \tau_l = \mu_{k2} - \mu_{l2}$ $(1 \leq k, l \leq g)$.

As in the univariate distribution-free GLM case, the Wald statistic can be used to test the general linear hypothesis of the form in (4.27). Similarly, general linear hypotheses concerning the parameter vector $\boldsymbol{\beta}$ in the linear predictor can be tested through linear contrasts by including an offset term in the linear predictor or by performing a linear transformation on the response $y_{it}$ for linear models.

## 4.5 GEE II for Modeling Mean and Variance

A major drawback of the distribution-free model discussed in Section 4.3 and Section 4.4 is that it only models the conditional mean of the response $y$ given the vector of independent variables $\mathbf{x}$. However, as noted in Section 4.2.1, it is also of great interest to model second and even higher order moments. For example, Cronbach coefficient alpha and intraclass correlations are all defined by the variance parameters of the response variable. As a result, the mean based distribution-free model is unable to provide inference for such indices of interest.

To overcome this limitation, we can generalize the mean response based distribution-free model in (4.28) to model simultaneously the mean and variance of a response variable. By augmenting the model for the mean response with a variance component, we can jointly model the conditional mean and variance of $y$ given $\mathbf{x}$.

**Example 1 (*Linear mixed-effects model*).** Consider the linear mixed-effects model:

$$y_{it} = \mathbf{x}_{it}^\top \boldsymbol{\beta} + \mathbf{z}_{it}^\top \mathbf{b}_i + \epsilon_{it}, \quad \mathbf{y}_i = X_i \boldsymbol{\beta} + Z_i \mathbf{b}_i + \boldsymbol{\epsilon}_i \tag{4.97}$$
$$\mathbf{b}_i \sim \text{i.i.d.} N(0, D), \quad \boldsymbol{\epsilon}_i \sim \text{i.i.d.} N\left(0, \sigma^2 I_m\right), \quad \mathbf{b}_i \perp \boldsymbol{\epsilon}_i, \quad 1 \leq i \leq n$$

For notational brevity, we have assumed fixed assessment times. If we only model the mean response, we then obtain from (4.97) the distribution-free counterpart as follows:

$$E(y_{it} \mid \mathbf{x}_{it}) = \mathbf{x}_{it}^\top \boldsymbol{\beta}, \quad 1 \leq t \leq m, \quad 1 \leq i \leq n \tag{4.98}$$

The above model only involves the parameter vector $\boldsymbol{\beta}$ for the mean response or fixed effects. However, in many applications, we are also interested in the parameters $D$ for the random effects and $\sigma^2$. This is not a problem for the parametric LMM in (4.97) since $\boldsymbol{\beta}$, $D$ and $\sigma^2$ are simultaneously estimated from the likelihood function. But the distribution-free version in (4.98) only involves $\boldsymbol{\beta}$ and as such does not provide inference about $D$ and $\sigma^2$.

To address this limitation, we can add to (4.98) a component for modeling the variance of $y_{it}$ to obtain

$$E(y_{it} \mid \mathbf{x}_{it}) = \mu_{it} = \mathbf{x}_{it}^\top \boldsymbol{\beta} \tag{4.99}$$
$$E(s_{ist} \mid \mathbf{x}_{is}, \mathbf{x}_{it}) = \mathbf{z}_{is}^\top D \mathbf{z}_{it} + \sigma^2 \delta_{st}, \quad 1 \leq i \leq n, \ 1 \leq s < t \leq m$$

where $s_{ist} = (y_{is} - \mu_{is})(y_{it} - \mu_{it})$ and $\delta_{st} = 1$ if $s = t$ and 0 if otherwise. Thus, by complementing the mean response based distribution-free model with a component for the conditional variance, we are able to conduct distrubution-free inference for all the parameters in the parametric LMM.

**Example 2 (*Models for count response*).** In Example 1, assume a count response $y_{it}$. Then, the distribution-free log-linear model has the following form:

$$E(y_{it} \mid \mathbf{x}_{it}) = \mu_{it}, \ \log\left(\frac{\mu_{it}}{1 - \mu_{it}}\right) = \mathbf{x}_{it}^\top \boldsymbol{\beta}, \quad 1 \leq i \leq n, \ 1 \leq t \leq m \tag{4.100}$$

The above model provides robust inference about $\boldsymbol{\beta}$ regardless of whether there is overdispersion in the data. For example, this model applies if conditional on $\mathbf{x}_{it}$ $y_{it}$ follows either a Poisson or negative binomial (NB) distribution.

Suppose now that we want to test over- or underdispersion in the data distribution, which occurs when $Var(y_{it} \mid \mathbf{x}_{it}) \neq \mu_{it}$ for some $t$ $(1 \leq t \leq m)$. We can add to (4.100) a component for modeling this conditional variance to obtain the following model:

$$\begin{aligned} E(y_{it} \mid \mathbf{x}_{it}) &= \mu_{it} \quad \log\left(\frac{\mu_{it}}{1-\mu_{it}}\right) = \mathbf{x}_{it}^{\top}\boldsymbol{\beta} \\ Var(y_{it} \mid \mathbf{x}_{it}) &= \sigma(\mu_{it}), \quad 1 \leq i \leq n, \quad 1 \leq t \leq m \end{aligned} \tag{4.101}$$

where $\sigma(\mu_{it})$ is some function of $\mu_{it}$. For example, we may assume $\sigma(\mu_{it}) = \phi_t^2 \mu_{it}$ and test for over- or underdispersion by considering the null $H_0: \phi_t^2 = 1$ for all $1 \leq t \leq m$. If we believe that $y_{it}$ given $\mathbf{x}_{it}$ follows an NB like distribution, we may posit a different model, $\sigma(\mu_{it}) = \mu_{it}(1 + \lambda_t \mu_{it})$, and assess for over- or underdispersion by testing $H_0: \lambda_t = 0$ for all $1 \leq t \leq m$. Note that unlike the parametric formulation, we have the flexibility to allow $\lambda_t$ to be either positive or negative so that we can use the variance form of the NB model to test either over- or underdispersion. Of course, only a positive $\lambda_t$ corresponds to a NB model for the conditional distribution of $y_{it}$ given $\mathbf{x}_{it}$.

In general, consider the distribution-free model for both the mean and variance given by

$$\begin{aligned} E(y_{it} \mid \mathbf{x}_{it}) &= \mu_{it}, \quad g(\mu_{it}) = \mathbf{x}_{it}^{\top}\boldsymbol{\beta} \\ E(s_{ist} \mid \mathbf{x}_{is}, \mathbf{x}_{it}) &= \sigma_{ist}(\mu_s, \mu_t), \quad 1 \leq i \leq n, \ 1 \leq s, t \leq m \end{aligned} \tag{4.102}$$

As illustrated in Examples 1 and 2, for some applications we may only specify the conditional variance $Var(y_{it} \mid \mathbf{x}_{it})$ in which case we can set $s = t$ in (4.102). The generalized estimating equation discussed for mean response based distribution-free models in Sections 4.3 is readily extended to provide inference for the model defined in (4.102).

Let $\boldsymbol{\beta}$ be the parameter vector for the regression component, $\boldsymbol{\zeta}$ the vector containing the remaining parameters in the conditional variance in (4.102), and $\boldsymbol{\theta} = \left(\boldsymbol{\beta}^\top, \boldsymbol{\zeta}^\top\right)^\top$ the $q \times 1$ column vector of all model parameters. Let

$$\begin{aligned}
\boldsymbol{\mu}_i &= \left(\mu_{i1}, \ldots, \mu_{im}\right)^\top, \quad \mathbf{x}_i = \left(\mathbf{x}_{i1}^\top, \ldots, \mathbf{x}_{im}^\top\right)^\top \\
\mathbf{s}_i &= \left(s_{i11}, s_{i12}, \ldots, s_{i1m}, s_{i22}, s_{i23}, \ldots, s_{i2m}, \ldots, s_{imm}\right)^\top \\
\boldsymbol{\sigma}_i &= \left(\sigma_{i11}, \sigma_{i12}, \ldots, \sigma_{i1m}, \sigma_{i22}, \sigma_{i23}, \ldots, \sigma_{i2m}, \ldots, \sigma_{imm}\right)^\top \\
S_i &= \left(\left(\mathbf{y}_i - \boldsymbol{\mu}_i\right)^\top, \left(\mathbf{s}_i - \boldsymbol{\sigma}_i\right)^\top\right)^\top
\end{aligned} \tag{4.103}$$

where $\mathbf{s}_i$ and $\boldsymbol{\sigma}_i$ are both $\frac{(m-1)m}{2} \times 1$ column vectors with $s_{ist}$ and $\sigma_{ist}$ defined in (4.102) and $S_i$ is a $\frac{(m+1)m}{2} \times 1$ column vector. In analogy to the GEE for the mean response based model, we define the generalized estimating equation for (4.102) as follows

$$\mathbf{w}_n(\boldsymbol{\theta}) = \sum_{i=1}^n G_i S_i = \sum_{i=1}^n G_i \begin{pmatrix} \mathbf{y}_i - \boldsymbol{\mu}_i \\ \mathbf{s}_i - \boldsymbol{\sigma}_i \end{pmatrix} = \mathbf{0} \tag{4.104}$$

where $G_i$ is some $q \times \frac{(m+1)m}{2}$ matrix function of $\mathbf{x}_i$. Again, the Newton–Raphson algorithm can be applied to compute the estimate of $\boldsymbol{\theta}$. The generalized estimating equation in (4.104) is known as GEE II to differentiate it from GEE or GEE I used for the mean response based model.

As in GEE I, the estimating equation estimate $\widehat{\boldsymbol{\theta}}$ obtained from (4.104) is consistent independent of the choice of $G_i$. Further, $\widehat{\boldsymbol{\theta}}$ has an asymptotic normal distribution (see exercise):

$$\widehat{\boldsymbol{\theta}} \sim AN\left(\boldsymbol{\theta}, \frac{1}{n}\boldsymbol{\Sigma}_\theta\right), \quad \boldsymbol{\Sigma}_\theta = \frac{1}{n} B^{-1} E\left(G_i S_i S_i^\top G_i^\top\right) B^{-\top} \tag{4.105}$$

where $B = E\left(G_i D_i^\top\right)$. The above still holds true if $G_i$ is defined by some parameter vector $\boldsymbol{\alpha}$ and $\widehat{\boldsymbol{\theta}}$ is obtained from (4.104) with $\boldsymbol{\alpha}$ substituted by some $\sqrt{n}$-consistent estimate. A consistent estimate of $\boldsymbol{\Sigma}_\theta$ is similarly obtained as in the case of GEE I by substituting consistent estimates of $\boldsymbol{\theta}$, $B$ and $E\left(G_i S_i S_i^\top G_i^\top\right)$ in place of the respective parameters.

Although similar, there is still some important difference between GEE I in (4.35) and GEE II in (4.104). When $G_i = D_i V_i^{-1}$ in GEE I with $D_i = \frac{\partial}{\partial \beta}\boldsymbol{\mu}_i$, $B = E\left(D_i V_i^{-1} D_i^\top\right)$ and the asymptotic variance of the GEE estimate is

$$\boldsymbol{\Sigma}_\beta = \frac{1}{n} B^{-1} E\left(D_i V_i^{-1} S_i S_i^\top V_i^{-1} D_i^\top\right) B^{-\top} \tag{4.106}$$

For GEE II, if we set $G_i = D_i V_i^{-1}$ with $D_i = \frac{\partial}{\partial \boldsymbol{\theta}} \left(\boldsymbol{\mu}_i^\top, \boldsymbol{\sigma}_i^\top\right)^\top$, the asymptotic variance $\boldsymbol{\Sigma}_\theta$ of $\widehat{\boldsymbol{\theta}}$ still has the above form. However, $B = E\left(F_i V_i^{-1} D_i^\top\right)$, where $F_i = -\frac{\partial}{\partial \boldsymbol{\theta}} S_i$ (see exercise). Since $\frac{\partial}{\partial \boldsymbol{\theta}} S_i \neq -\frac{\partial}{\partial \boldsymbol{\theta}} \left(\boldsymbol{\mu}_i^\top, \boldsymbol{\sigma}_i^\top\right)^\top$, $B \neq E\left(D_i V_i^{-1} D_i^\top\right)$. More importantly, the use of $D_i = -\frac{\partial}{\partial \boldsymbol{\theta}} S_i$ in GEE II may yield inconsistent estimates of $\boldsymbol{\theta}$. Unlike GEE I, $S_i$ is not linear in $y_{it}$ and thus $\frac{\partial}{\partial \boldsymbol{\theta}} S_i$ generally depends on $y_{it}$. As a result, $G_i = D_i V_i^{-1}$ may not be a function of $\mathbf{x}_i$ only as required to ensure the unbiasedness of the GEE II in (4.104).

**Example 3.** Consider the LMM in Example 1. It is readily checked that (see exercise)

$$D_i = \frac{\partial}{\partial \boldsymbol{\theta}} \left(\boldsymbol{\mu}_i^\top, \boldsymbol{\sigma}_i^\top\right)^\top = \begin{pmatrix} \frac{\partial}{\partial \beta} \boldsymbol{\mu}_i & \frac{\partial}{\partial \beta} \boldsymbol{\sigma}_i \\ \frac{\partial}{\partial \zeta} \boldsymbol{\mu}_i & \frac{\partial}{\partial \zeta} \boldsymbol{\sigma}_i \end{pmatrix} = \begin{pmatrix} D_{i11} & \mathbf{0} \\ \mathbf{0} & D_{i22} \end{pmatrix} \tag{4.107}$$

where $\boldsymbol{\zeta} = \left(vec^\top (D), \sigma^2\right)^\top$. Let

$$V_i = \begin{pmatrix} Z_i D Z_i^\top + \sigma^2 I_m & \mathbf{0} \\ & V_{i22} \end{pmatrix} = \begin{pmatrix} V_{i11} & \mathbf{0} \\ & V_{i22} \end{pmatrix} \tag{4.108}$$

where $Z_i = \left(\mathbf{z}_{i1}, \ldots, \mathbf{z}_{im}\right)^\top$ and $V_{i22}(\boldsymbol{\alpha})$ is some $\frac{(m-1)m}{2} \times \frac{(m-1)m}{2}$ working variance matrix parameterized by $\boldsymbol{\alpha}$. By setting $G_i = D_i V_i^{-1}$, it follows from (4.104) that

$$\mathbf{w}_n(\boldsymbol{\theta}) = \sum_{i=1}^n \begin{pmatrix} D_{i11} V_{i11}^{-1} \left(\mathbf{y}_i - \boldsymbol{\mu}_i\right) \\ D_{i22} V_{i22}^{-1} \left(\mathbf{s}_i - \boldsymbol{\sigma}_i\right) \end{pmatrix} = \mathbf{0} \tag{4.109}$$

If $\boldsymbol{\alpha}$ is known, such as by setting $V_{i22} = \mathbf{I}_{\frac{(m-1)m}{2} \times \frac{(m-1)m}{2}}$, we can readily solve the above equation to obtain the estimate of $\boldsymbol{\theta}$. Otherwise, a $\sqrt{n}$-consistent estimate of $\boldsymbol{\alpha}$ is required to use (4.106) for inference about $\boldsymbol{\theta}$. As noted earlier, the asymptotic variance of this GEE II estimate of $\boldsymbol{\theta}$ is still given by (4.106) except for a revised $B = E\left(F_i V_i^{-1} D_i^\top\right)$ with $F_i = -\frac{\partial}{\partial \boldsymbol{\theta}} S_i$.

In this example, $-\frac{\partial}{\partial \boldsymbol{\theta}} S_i \neq \frac{\partial}{\partial \boldsymbol{\theta}} \left(\boldsymbol{\mu}_i^\top, \boldsymbol{\sigma}_i^\top\right)^\top$ and GEE IIs defined by setting $D_i = -\frac{\partial}{\partial \boldsymbol{\theta}} S_i$ is no longer guaranteed to be unbiased to yield consistent estimates of $\boldsymbol{\theta}$. Note also that the selection of a block diagonal $V_i$ in (4.108) has an advantage to ensure the consistency of estimate of $\boldsymbol{\beta}$ from (4.109) if only the mean response in (4.99) is modeled correctly, that is, if

$E(y_{it} \mid \mathbf{x}_{it}) = \mathbf{x}_{it}^\top \boldsymbol{\beta}$ holds true. If a nonblock diagonal $V_i$ is used, the GEE II has the following form:

$$\mathbf{w}_n(\boldsymbol{\theta}) = \sum_{i=1}^{n} \begin{pmatrix} G_{i11}(\mathbf{y}_i - \boldsymbol{\mu}_i) + G_{i12}(\mathbf{s}_i - \boldsymbol{\sigma}_i) \\ G_{i21}(\mathbf{y}_i - \boldsymbol{\mu}_i) + G_{i22}(\mathbf{s}_i - \boldsymbol{\sigma}_i) \end{pmatrix} = \mathbf{0}$$

The above GEE II may yield biased estimates of $\boldsymbol{\beta}$ if the variance component in (4.99) is not specified correctly.

**Example 4.** Consider the model in (4.100) from Example 2 with $\sigma(\mu_{it}) = \phi_t^2 \mu_{it}$. In this case,

$$s_{it} = (y_{it} - \mu_{it})^2, \quad \mathbf{s}_i = (s_{i1}, s_{i2}, \ldots, s_{im})^\top, \quad \sigma_{it} = \phi_t^2 \mu_{it}$$
$$\boldsymbol{\sigma}_i = (\sigma_{i1}, \sigma_{i2}, \ldots, \sigma_{im})^\top, \quad \boldsymbol{\zeta} = (\phi_1^2, \ldots, \phi_m^2)^\top$$

It follows that (see exercise)

$$D_i = \frac{\partial}{\partial \boldsymbol{\theta}} \left( \boldsymbol{\mu}_i^\top, \boldsymbol{\sigma}_i^\top \right)^\top = \begin{pmatrix} \frac{\partial}{\partial \boldsymbol{\beta}} \boldsymbol{\mu}_i & \frac{\partial}{\partial \boldsymbol{\beta}} \boldsymbol{\mu}_i \text{diag}_t(\phi_t^2) \\ \mathbf{0} & \text{diag}_t(\mu_{it}) \end{pmatrix} \tag{4.110}$$

Let $A_i = \text{diag}_t(\phi_t^2 \mu_{it})$ and $R(\boldsymbol{\alpha}_1)$ $(V_{i22}(\boldsymbol{\alpha}_2))$ be some $m \times m$ working correlation (variance) matrix parameterized by $\boldsymbol{\alpha}_1$ $(\boldsymbol{\alpha}_2)$. Set

$$V_i = \begin{pmatrix} A_i^{\frac{1}{2}} R(\boldsymbol{\alpha}_1) A_i^{\frac{1}{2}} & \mathbf{0} \\ & V_{i22}(\boldsymbol{\alpha}_2) \end{pmatrix} = \begin{pmatrix} V_{i11} & \mathbf{0} \\ & V_{i22} \end{pmatrix} \tag{4.111}$$

By substituting (4.110) and (4.111) into (4.109), we obtain a GEE II for simultaneous inference about $\boldsymbol{\beta}$ and $\boldsymbol{\zeta}$. In particular, we can use it to formally test for overdispersion, that is, whether $\phi_t^2 = 1$ for all $1 \le t \le m$. As in Example 3, the asymptotic normality of the estimate follows from (4.105) provided that $\boldsymbol{\alpha} = \left(\boldsymbol{\alpha}_1^\top, \boldsymbol{\alpha}_2^\top\right)^\top$ is known or substituted by a $\sqrt{n}$-consistent estimate.

Note that for this example,

$$D_i = -\frac{\partial}{\partial \boldsymbol{\theta}} S_i = \begin{pmatrix} \frac{\partial}{\partial \boldsymbol{\beta}} \boldsymbol{\mu}_i & \frac{\partial}{\partial \boldsymbol{\beta}} \boldsymbol{\mu}_i \text{diag}_t(\phi_t^2) - \frac{\partial}{\partial \boldsymbol{\beta}} \mathbf{s}_i \\ \mathbf{0} & \text{diag}_t(\mu_{it}) \end{pmatrix} \tag{4.112}$$

Thus, as in Example 3, $D_i$ is a not function of $\mathbf{x}_i$ only and GEE IIs defined by (4.112) may yield inconsistent estimates of $\boldsymbol{\theta}$.

In both Examples 3 and 4 above, the GEE IIs yields

When there is missing data, the approach in Section 4.4.2 for GEE I is readily generalized to the current setting. For each subject $i$, define a missing data indicator $r_{it}$ as in (4.61) and let

$$\pi_{ist} = \Pr\left[r_{is} = 1, r_{it} = 1 \mid \mathbf{x}_i, \mathbf{y}_i\right], \quad \Delta_{ist} = \frac{r_{is} r_{it}}{\pi_{ist}} \tag{4.113}$$
$$1 \leq s \leq t \leq m, \quad 1 \leq i \leq n$$

As a special case, if $s = t$, we obtain from the above $\pi_{it} = \Pr\left(r_{it} = 1 \mid \mathbf{x}_i, \mathbf{y}_i\right)$ and $\Delta_{it}$, which are defined in (4.81) of Section 4.4.2 in the discussion of WGEE I in the presence of missing data. For the mean response $y_{it} - \mu_{it}$, it follows from an argument similar to (4.78) that $E\left[\Delta_{it}\left(y_{it} - \mu_{it}\right)\right] = 0$. For the variance component $s_{ist}$, we have for $s \leq t$,

$$\begin{aligned} E\left[\Delta_{ist}\left(s_{ist}\right)\right] &= E\left[E\left(\frac{r_{is}}{\pi_{ist}}\left(y_{is} - \mu_{is}\right)\left(y_{it} - \mu_{it}\right) \mid \mathbf{r}_i, \mathbf{x}_i, \mathbf{y}_i\right)\right] \\ &= E\left[\left(y_{is} - \mu_{is}\right)\left(y_{it} - \mu_{it}\right)\frac{1}{\pi_{ist}} E\left(r_{is} r_{it} \mid \mathbf{r}_i, \mathbf{x}_i, \mathbf{y}_i\right)\right] \\ &= E\left(s_{ist}\right) = \sigma_{ist} \end{aligned} \tag{4.114}$$

Let

$$\begin{aligned} \boldsymbol{\delta}_i^y &= \left(\Delta_{i1}, \Delta_{i2}, \ldots, \Delta_{im}\right)^\top \\ \boldsymbol{\delta}_i^s &= \left(\Delta_{i1}, \Delta_{i12}, \ldots, \Delta_{i1m}, \Delta_{i2}, \Delta_{i23}, \ldots, \Delta_{i2m}, \ldots, \Delta_{im}\right)^\top \\ \Delta_i &= \begin{pmatrix} \operatorname{diag}\left(\boldsymbol{\delta}_i^y\right) & 0 \\ 0 & \operatorname{diag}\left(\boldsymbol{\delta}_i^s\right) \end{pmatrix}, \quad S_i = \begin{pmatrix} \mathbf{y}_i - \boldsymbol{\mu}_i \\ \mathbf{s}_i - \boldsymbol{\sigma}_i \end{pmatrix} \end{aligned} \tag{4.115}$$

where diag($\boldsymbol{\delta}$) denotes the diagonal matrix with the vector $\boldsymbol{\delta}$ forming the diagonal. As in the case of GEE I, we revise the estimating equation in (4.109) in the presence of missing data as follows:

$$\mathbf{w}_n\left(\boldsymbol{\theta}\right) = \sum_{i=1}^{n} G_i \Delta_i S_i = 0 \tag{4.116}$$

The weighted GEE (WGEE) II above is unbiased, that is, $E\left(\mathbf{w}_n\left(\boldsymbol{\theta}\right)\right) = \mathbf{0}$, if $E\left(\Delta_i S_i\right) = \mathbf{0}$. Thus, as in WGEE I, the estimating equation in (4.116) yields consistent estimates of $\boldsymbol{\theta}$ if the regression model in (4.102) and missing data indicator model in (4.109) are both specified correctly. Further, the asymptotic results in Theorem 1 of Section 4.4.2 for WGEE I are similarly generalized to WGEE II estimates obtained from (4.116) (see exercise).

As in WGEE I, we must estimate $\pi_{it}$ in most applications. Under MCAR, $\mathbf{r}_i$ is independent of $\mathbf{x}_i$ and $\mathbf{y}_i$ and thereby

$$\pi_{ist} = \Pr\left(r_{is} = 1, r_{it} = 1\right) = \pi_{st}$$

The constant weight function $\pi_{st}$ is readily estimated by the sample moment: $\widehat{\pi}_{st} = \frac{1}{n}\sum_{i=1}^{n} r_{ist}$ $(2 \leq s \leq t \leq m)$. Under MAR, $\pi_{ist}$ becomes dependent on $\mathbf{x}_i$ and $\mathbf{y}_i$. As in WGEE I, under MMDP, $\pi_{ist} = \pi_{it}$ and $\Delta_{ist} = \frac{r_{it}}{\pi_{it}} = \Delta_{it}$ for $1 \leq s \leq t \leq m$. Thus, we can obtain estimates of $\pi_{ist}$ by modeling $\pi_{it}$ as discussed in detail in Section 4.4.2.

## 4.6 Structural Equations Models

In treatment studies, it is often of great interest to identify and study mechanisms by which an intervention achieves its effect. By understanding the mediational process through which treatment affects study outcomes, not only can we further our understanding of the pathology of the disease and treatment, but also provide alternative treatments for the disease with efficient use of resources. However, regression analysis is ill-suited for modeling such a causal relationship. The structural equations model (SEM) provides a formal modeling and inference framework for this and other types of causal problems. In this book, we discuss only the classic linear SEM for continuous response. For convenience, we refer to linear SEM as SEM throughout the rest of the book. Note that our intention for this chapter is to provide the necessary background for SEM so that we can discuss distribution-free inference for such models in Section 4.6.3 as well as in Chapter 6. Thus, detailed considerations about many of the important issues, such as latent variables and model identifiability, are not provided. Readers may want to consult other texts such as Bollen (1989) for an in-depth treatment of the theory and application of linear SEM.

### 4.6.1 Path Diagrams and Models

SEM is a system of linear models linked together to capture the complex and dynamic relationships among a web of variables. Although similar in appearance, SEM is fundamentally different from linear regression. In a regression model, there exists a clear distinction between dependent and independent variables. In SEM, such concepts only apply in relative terms since a dependent variable in one model equation can become an independent variable in other components of the SEM system. It is precisely this type of reciprocal role a variable plays that enables SEM to infer causal relationships.

SEMs are best represented by *path diagrams*. A path diagram consists of nodes representing the variables and arrows showing relations among the variables. SEM, or path analysis, frequently employs latent variables to represent abstract constructs that are not directly observable, but are measured by multiple variables, or *indicators*, in the SEM literature. For example, depression is a latent construct in mental health research and popular measures include the structured clinical interview for depression, the Hamilton rating scale and the Beck depression inventory. In a path diagram, latent variables are distinguished from their observed counterparts in convention by using a circle or ellipse rather than the rectangular or square box used for the observed variables. Error terms are not observed, but since they are not part of the causal chain, they are not treated as latent variables. When we want to indicate that a variable is measured with error, we add an error term with an arrow from it to the variable. Arrows are generally used to represent relationships among the variables (observed or latent). A single straight arrow indicates a causal relation from the base of the arrow to the head of the arrow. Two straight single-headed arrows in opposing directions connecting two variables indicate a reciprocal causal relationship. A curved two-headed arrow indicates there may be some association between the two variables. For example, in most path diagrams error terms are not connected, indicating stochastic independence across the error terms. But if we suspect association between them, they should be connected by curved two-headed arrows.

The path diagrams in Figure 4.2 depict a mediational process involving a predictor $x$, a mediator $z$ and a response depending on whether the response is directly observable or not. The path diagram in (a) shows a mediational process with an observable response $y$, while the one in (b) represents a more complex situation involving a latent response $\eta$ measured by the observable indicator $y$. The two arrows emanating from $x$ indicate the causal effect of $x$ on both the mediator and response and the single arrow stemming from $z$ defines the effect of the moderator on the response variable. The strengths of the relationships are specified by the Greek letters, $\gamma_{11}$, $\gamma_{21}$, $\beta$, next to the arrow with the error in relating the variables denoted by $\epsilon_g$ $(g = z, y, \eta)$.

**Example 1** (***SEM for mediational process without latent variable***). Consider the diagram in (a) of Figure 4.2. The SEM for this

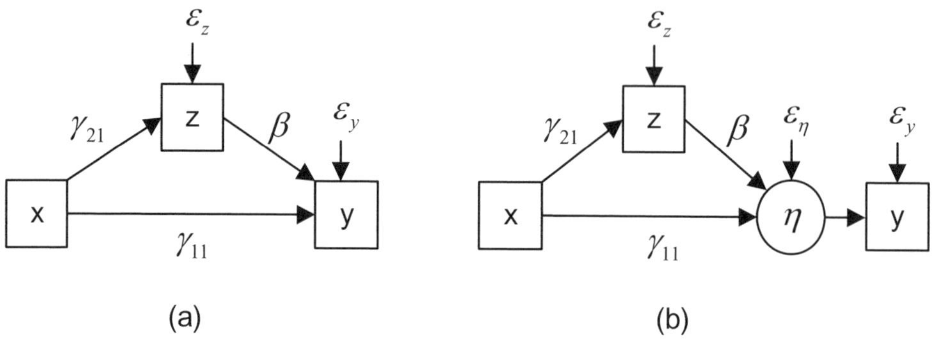


Figure 4.2. Path diagrams for a mediational process with a single predictor, mediator and observed response in (a) and latent response in (b).

mediation model is given by

$$y = \gamma_{10} + \gamma_{11}x + \beta z + \epsilon_y, \quad z = \gamma_{20} + \gamma_{21}x + \epsilon_z \tag{4.117}$$
$$E(\epsilon_y) = 0, \quad Cov(x, \epsilon_y) = 0, \quad Cov(z, \epsilon_y) = 0$$
$$E(\epsilon_z) = 0, \quad Cov(x, \epsilon_z) = 0$$

In the above, $x$ is called an *exogenous* variable, while $z$ and $y$ are called *endogenous* variables, as the latter two are determined by other variables within this SEM. As a special case, if $\beta_1 = 0$, (4.117) yields a *full* mediational process with no residual direct effect of $x$ on $y$.

Note that in the nomenclature of regression, $y$ is a response variable for the first and $z$ a response variable for the second model equations in (4.117). However, such terms do not apply to SEM since a variable can serve as both a dependent and independent variable, such as $z$ in (4.117).

Since a primary goal of SEM is to make causal inference, stochastic independence is not taken for granted. The usual independence assumption is replaced by zero correlation (or covariance), which unlike independence can be empirically checked and validated. For example, to assess the causal effect of $x$ and $z$ in (4.117), it is critical that neither be correlated with $\epsilon_y$ in the first equation and that $x$ not be correlated with $\epsilon_z$ in the second equation. Such a zero-correlation assumption is known as *pseudo-isolation* in SEM.

**Example 2 (*SEM for mediational process with latent response*).** Similarly, by using the path diagram in Figure 4.2 (b) as a guide, we can express the SEM for the mediational process involving a latent *endogenous* response $\eta$ as

$$\begin{aligned} \eta &= \gamma_{10} + \gamma_{11}x + \beta z + \epsilon_\eta, \\ z &= \gamma_{20} + \gamma_{21}x + \epsilon_z, \quad y = \eta + \delta_y \\ E(\epsilon_\eta) &= 0, \quad E(\delta_y) = 0, \quad Cov(x, \epsilon_\eta) = 0 \\ E(\epsilon_z) &= 0, \quad Cov(z, \epsilon_\eta) = 0, \quad Cov(x, \epsilon_z) = 0, \quad Cov(\eta, \delta_y) = 0 \end{aligned} \tag{4.118}$$

In (4.118), $\delta_y$ denotes the error in measuring $\eta$ with $y$, which is assumed to be uncorrelated with $\eta$ with mean 0.

As illustrated by the next example, the pseudo-isolation assumption plays a critical role to ensure consistent estimation of model parameters.

**Example 3.** In Example 1, assume $\beta \neq 0$. Suppose that $\beta z$ is mistakenly dropped from the first equation in (4.117), that is,

$$\begin{aligned} y_i &= \alpha_0 + \alpha_1 x_i + \epsilon^*_{y_i}, \quad z_i = \gamma_{20} + \gamma_{21}x_i + \epsilon_{z_i} \\ E\left(\epsilon^*_{y_i}\right) &= 0, \quad E(\epsilon_{z_i}) = 0, \quad Cov(x_i, \epsilon_{z_i}) = 0 \end{aligned} \tag{4.119}$$

By comparing (4.117) and (4.119), it is seen that

$$\alpha_0 = \gamma_{10} + \beta E(z_i), \quad \alpha_1 = \gamma_{11}, \quad \epsilon^*_{y_i} = [\beta z_i - \beta E(z_i) + \epsilon_{y_i}]$$

If $\gamma_{21} \neq 0$, we have (see exercise):

$$Cov\left(x_i, \epsilon^*_{y_i}\right) = Cov(x_i, \beta z_i - \beta E(z_i) + \epsilon_{y_i}) = \beta\gamma_{21} Var(x_i) \neq 0 \tag{4.120}$$

In other words, $x_i$ and $\epsilon^*_{y_i}$ in (4.119) are correlated. Let $\widehat{\alpha}_1$ denote the least squares estimate of $\alpha_1$. It then follows that (see exercise)

$$\begin{aligned} \widehat{\alpha}_1 &= \left[\sum_{i=1}^{n} (x_i - \overline{x}_n)^2\right]^{-1} \left[\sum_{i=1}^{n} (x_i - \overline{x}_n)(y_i - \overline{y}_n)\right] \to_p \frac{Cov(y_i, x_i)}{Var(x_i)} \\ &= \gamma_{11} + \beta\gamma_{21} \end{aligned} \tag{4.121}$$

Thus, $\alpha_1$ in the misspecified model in (4.119) does not estimate the direct effect, but rather the (biased) total effects of $x$ on $y$.

Path diagrams not only indicate causal paths, but between-variable correlations as well. For example, the connectors between $x_1$ and $x_2$ in both path diagrams in Figure 4.3 indicate that these exogenous predictors are correlated. Similarly, unlike the error terms in part (a), $\epsilon_{y_1}$ and $\epsilon_{y_2}$ in Diagram (b) are correlated.

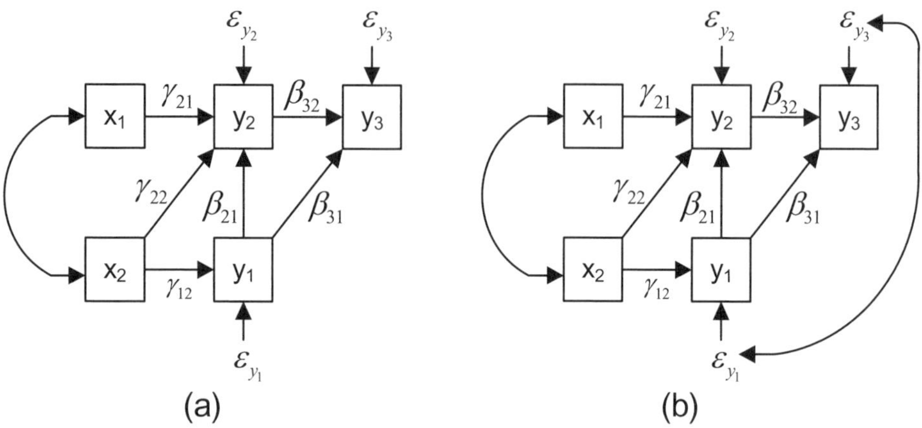


Figure 4.3. Path diagrams with correlated and uncorrelated error terms.

**Example 4.** The SEMs for the two path diagrams in Figure 4.3 share the following core components:

$$\begin{aligned} y_1 &= \gamma_{12}x_2 + \epsilon_{y_1}, \quad y_2 = \beta_{21}y_1 + \gamma_{21}x_1 + \gamma_{22}x_2 + \epsilon_{y_2} \\ y_3 &= \beta_{31}y_1 + \beta_{32}y_2 + \epsilon_{y_3} \\ E(\epsilon_{y_k}) &= 0, \quad Cov(x_1, \epsilon_{y_k}) = 0, \quad Cov(x_2, \epsilon_{y_k}) = 0, \quad k = 1, 2, 3 \end{aligned} \tag{4.122}$$

The difference between the two lies in the assumptions about the error terms:

$$\begin{aligned} &\text{For (a)}: Cov\left(\epsilon_{y_j}, \epsilon_{y_k}\right) = 0, \quad 1 \le j, k \le 3 \\ &\text{For (b)}: Cov(\epsilon_{y_1}, \epsilon_{y_2}) = 0, \quad Cov(\epsilon_{y_2}, \epsilon_{y_3}) = 0, \quad Cov(\epsilon_{y_1}, \epsilon_{y_3}) \neq 0 \end{aligned} \tag{4.123}$$

In SEM analysis, path diagrams are also useful for calculating *direct*, *indirect* and *total* effects of a causal variable. For example, for both diagrams in Figure 4.3, $x_1$ has a direct effect $\gamma_{21}$ on $y_2$, while $x_2$ has a direct effect $\gamma_{12}$ on $y_1$, and a direct effect $\gamma_{22}$ and an indirect effect $\beta_{21}\gamma_{12}$ on $y_2$. Thus, the total effect of $x_2$ on $y_2$ is $\gamma_{22} + \beta_{21}\gamma_{12}$. We can also derive these effects through the model form. For example, by substituting the first equation in place of $y_1$ in the second in (4.122), we obtain

$$y_2 = \beta_{21}(\gamma_{12}x_2 + \epsilon_{y_1}) + \gamma_{22}x_2 + \epsilon_{y_2} = (\gamma_{22} + \beta_{21}\gamma_{12})x_2 + (\beta_{21}\epsilon_{y_1} + \epsilon_{y_2})$$

which gives the direct, indirect and total effects of $x_2$ on $y_2$.

**Example 5 (*SEM for linear regression with measurement error*).** Consider a simple linear regression model with measurement errors in both the response and predictor:

$$\begin{aligned} \eta_i = \gamma\xi_i + \epsilon_{\eta_i}, \quad x_i = \xi_i + \delta_{x_i}, \quad y_i = \eta_i + \delta_{y_i} \\ E\left(\epsilon_{\eta_i}\right) = 0, \quad Cov\left(\xi_i, \epsilon_{\eta_i}\right) = 0, \quad E\left(\delta_{x_i}\right) = 0 \\ E\left(\delta_{y_i}\right) = 0, \quad Cov\left(\xi_i, \delta_{x_i}\right) = 0, \quad Cov\left(\eta_i, \delta_{y_i}\right) = 0 \end{aligned} \tag{4.124}$$

where the latent response $\eta_i$ (predictor $\xi_i$) is observed by $y_i$ ($x_i$). Note that for convenience, we have set the intercept to 0 in (4.124). In addition, we assume that $\epsilon_{\eta_i}$, $\delta_{y_i}$ and $\delta_{x_i}$ are uncorrelated with each other as well as with $\xi_i$ and $\eta_i$. Primary interest lies in estimating $\gamma$.

If we ignore measurement error and fit a linear regression based on the observed variables, we have

$$y_i = \gamma^* x_i + \epsilon_{y_i}, \quad E\left(\epsilon_{y_i}\right) = 0, \quad Cov\left(x_i, \epsilon_{y_i}\right) = 0 \tag{4.125}$$

It is readily shown that (see exercise)

$$\gamma^* = \frac{Cov\left(x_i, y_i\right)}{Var\left(x_i\right)} = \frac{\sigma_\xi^2}{\sigma_\xi^2 + \sigma_{\delta_x}^2}\gamma = \tau\gamma \tag{4.126}$$

where $\sigma_\xi^2 = Var\left(\xi_i\right)$ and $\sigma_{\delta_x}^2 = Var\left(\delta_{x_i}\right)$. The ICC, $\tau = \frac{\sigma_\xi^2}{\sigma_\xi^2 + \sigma_{\delta_x}^2}$, is also known as the *reliability ratio* in measurement error models. Unless $\sigma_{\delta_x}^2 = 0$, that is, $\xi$ is observed without error, the regression model in (4.125) based on the observed $x$ and $y$ will yield downwardly biased estimates of $\gamma$. It is interesting to note that measurement error in the response $\eta$ does not create any bias in the estimation of $\gamma$.

Consider the mediation model with no latent variable in (4.117) of Example 1. Let

$$\begin{aligned} \mathbf{y} = \left(y_1, y_2\right)^\top = \left(y, z\right)^\top, \quad \mathbf{x} = \left(x_1, x_2\right)^\top = \left(1, x\right)^\top \\ B = \begin{pmatrix} 0 & \beta \\ 0 & 0 \end{pmatrix}, \quad \Gamma = \begin{pmatrix} \gamma_{10} & \gamma_{11} \\ \gamma_{20} & \gamma_{21} \end{pmatrix}, \quad \boldsymbol{\epsilon}_y = \begin{pmatrix} \epsilon_{y1} \\ \epsilon_{y2} \end{pmatrix} \end{aligned} \tag{4.127}$$

We can then express this SEM in a matrix form as follows:

$$\mathbf{y} = B\mathbf{y} + \Gamma\mathbf{x} + \boldsymbol{\epsilon}_y, \quad E\left(\boldsymbol{\epsilon}_y\right) = \mathbf{0}, \quad Cov\left(\mathbf{x}, \boldsymbol{\epsilon}_y\right) = \mathbf{0} \tag{4.128}$$

In the above, $\mathbf{y}$ is a vector consisting of the endogenous variables, while the vector $\mathbf{x}$ contains the intercept and the exogenous variable $x$. The matrices

$B$ and $\Gamma$ contain the parameters of interest. The matrix form also applies to a general SEM with more endogenous and exogenous variables. More generally, when either or both endogenous and exogenous variables contain latent variables, SEM can be expressed as

$$\begin{aligned} \boldsymbol{\eta} &= B\boldsymbol{\eta} + \Gamma\boldsymbol{\xi} + \boldsymbol{\epsilon}_\eta \\ E(\boldsymbol{\epsilon}_\eta) &= \mathbf{0}, \quad Cov(\boldsymbol{\xi}, \boldsymbol{\epsilon}_\eta) = \mathbf{0}, \quad E(\boldsymbol{\xi}) = \mathbf{0}, \quad E(\boldsymbol{\eta}) = \mathbf{0} \\ \mathbf{y} &= \Lambda_y\boldsymbol{\eta} + \boldsymbol{\delta}_y, \quad \mathbf{x} = \Lambda_x\boldsymbol{\xi} + \boldsymbol{\delta}_x \\ E(\boldsymbol{\epsilon}_y) &= \mathbf{0}, \quad Cov(\boldsymbol{\eta}, \boldsymbol{\delta}_y) = \mathbf{0}, \quad E(\boldsymbol{\epsilon}_x) = \mathbf{0}, \quad Cov(\boldsymbol{\xi}, \boldsymbol{\delta}_x) = \mathbf{0} \\ Cov(\boldsymbol{\epsilon}_\eta, \boldsymbol{\delta}_x) &= \mathbf{0}, \quad Cov(\boldsymbol{\epsilon}_\eta, \boldsymbol{\delta}_y) = \mathbf{0}, \quad Cov(\boldsymbol{\delta}_y, \boldsymbol{\delta}_x) = \mathbf{0} \end{aligned} \tag{4.129}$$

where $\boldsymbol{\eta}$ ($\boldsymbol{\xi}$) denotes the column vectors containing all endogenous (exogenous) variables and $\mathbf{y}$ ($\mathbf{x}$) the observed indicators for the latent endogenous (exogenous) variables. The parameters of primary interest are the matrices $B$ and $\Gamma$.

It is important to emphasize that SEM is not a multivariate linear model. For example, consider the SEM in (4.129) with all variables observed, that is, $\boldsymbol{\eta} = \mathbf{y}$ and $\boldsymbol{\xi} = \mathbf{x}$. If $\mathbf{y}$ is an $m \times 1$ vector, (4.129) can be equivalently expressed as:

$$\begin{aligned} \mathbf{y} &= (\mathbf{I}_m - B)^{-1}(\Gamma\mathbf{x} + \boldsymbol{\epsilon}_y) = (\mathbf{I}_m - B)^{-1}\Gamma\mathbf{x} + \boldsymbol{\epsilon}_y^*, \\ E(\boldsymbol{\epsilon}_y^*) &= (\mathbf{I}_m - B)^{-1}E(\boldsymbol{\epsilon}_y) = \mathbf{0}, \quad Cov(\mathbf{x}, \boldsymbol{\epsilon}_y^*) = Cov(\mathbf{x}, \boldsymbol{\epsilon}_y) = \mathbf{0}. \end{aligned} \tag{4.130}$$

Since $(\mathbf{I}_m - B)^{-1}\boldsymbol{\Gamma}$ is generally a nonlinear function of parameters, the above is not a multivariate linear model, although the equation is linear in $\mathbf{x}$.

**Example 6.** We can express the SEM in (4.117) from Example 1 in the matrix form of (4.130) as below:

$$\begin{aligned} \begin{pmatrix} y \\ z \end{pmatrix} &= \begin{pmatrix} 1 & -\beta \\ 0 & 1 \end{pmatrix}^{-1} \left[ \begin{pmatrix} \gamma_{10} & \gamma_{11} \\ \gamma_{20} & \gamma_{21} \end{pmatrix} \begin{pmatrix} 1 \\ x \end{pmatrix} + \begin{pmatrix} \epsilon_y \\ \epsilon_z \end{pmatrix} \right] \\ &= \begin{pmatrix} \gamma_{10} + \beta\gamma_{20} + (\gamma_{11} + \beta\gamma_{21})x + \epsilon_{y^*} \\ \gamma_{20} + \gamma_{21}x + \epsilon_{z^*} \end{pmatrix} \end{aligned}$$

It is clear that the above SEM is not linear in the parameters.

### 4.6.2 Maximum Likelihood Inference

First, consider the SEM in (4.128) with all observed variables. The classic approach utilizes the second order moments (or variance–covariance matrix) for estimation.

Let $\mathbf{u}_i = \left(\mathbf{y}_i^\top, \mathbf{x}_i^\top\right)^\top$, $\Phi = Var(\mathbf{x}_i)$ and $\Psi = Var(\boldsymbol{\epsilon}_{y_i})$. Then, it follows that (see exercise)

$$\Sigma(\boldsymbol{\theta}) = Var(\mathbf{u}_i) = \begin{pmatrix} \Sigma_{yy}(\boldsymbol{\theta}) & \Sigma_{yx}(\boldsymbol{\theta}) \\ \Sigma_{yx}^\top(\boldsymbol{\theta}) & \Sigma_{xx}(\boldsymbol{\theta}) \end{pmatrix} \tag{4.131}$$

$$= \begin{pmatrix} (\mathbf{I}_m - B)^{-1}\left(\Gamma\Phi\Gamma^\top + \Psi\right)(\mathbf{I}_m - B)^{-\top} & (\mathbf{I}_m - B)^{-1}\boldsymbol{\Gamma}\Phi \\ & \Phi \end{pmatrix}$$

Note that although primary interest lies in $B$ and $\Gamma$, we must also simultaneously estimate the nuisance parameters $\Phi$ and $\Psi$ in (4.131). Let $\boldsymbol{\theta}$ denote the vector containing all these parameters.

If we assume $\mathbf{u}_i \sim N(\mathbf{0}, \Sigma(\boldsymbol{\theta}))$, then $\mathbf{u}_i$ has the following density function:

$$f(\mathbf{u}_i, \Sigma(\boldsymbol{\theta})) = (2\pi)^{\frac{-p}{2}} |\Sigma(\boldsymbol{\theta})|^{\frac{-1}{2}} \exp\left[\frac{-1}{2}\mathbf{u}_i^\top \Sigma(\boldsymbol{\theta})\mathbf{u}_i\right] \tag{4.132}$$

where $p$ denotes the length of the column vector $\mathbf{u}_i$. Given a sample of $n$ i.i.d. observations, the log-likelihood function is given by (see exercise):

$$l_n(\boldsymbol{\theta}) = -\frac{n}{2}\left[\log|\Sigma(\boldsymbol{\theta})| + tr\left(S_n\Sigma^{-1}(\boldsymbol{\theta})\right)\right] \tag{4.133}$$

where $S_n$ denotes the sample variance of $\mathbf{u}_i$ and $tr(\cdot)$ the trace of a matrix. Aside from technical details, inference about $\boldsymbol{\theta}$ is straightforward based on the asymptotic distribution of the MLE of $\boldsymbol{\theta}$.

In addition to maximum likelihood, other alternative inference procedures have also been proposed. For example, the unweighted (ULS) and weighted (WLS) least-squares methods minimize the following objective functions:

$$F_{ULS} = tr(S_n - \Sigma(\boldsymbol{\theta}))^2, \quad F_{WLS} = tr\left((S_n - \Sigma(\boldsymbol{\theta}))W^{-1}\right)^2 \tag{4.134}$$

where $W$ is some symmetric matrix with known elements. Thus, ULS minimizes the sum of squares of elements in the residual matrix $S_n - \Sigma(\boldsymbol{\theta})$, while WLS minimizes a similar sum with such squared terms weighted differently according to $W$. For example, by setting $W = S_n$, elements of $S_n - \Sigma(\boldsymbol{\theta})$ are weighted by the corresponding entries of $S_n$ in the weighted sum of $F_{WLS}$.

**Example 7 (*SEM for mediational process without latent variable*).** For the mediation model in Example 1, it follows from (4.127) that (see exercise):

$$\Phi = \begin{pmatrix} 0 & 0 \\ 0 & \phi_{22} \end{pmatrix}, \quad \Sigma_{yx} = \begin{pmatrix} 0 & \phi_{22}(\gamma_{11} + \beta\gamma_{21}) \\ 0 & \gamma_{21}\phi_{22} \end{pmatrix} \tag{4.135}$$

$$\Psi = \begin{pmatrix} \psi_{11} & 0 \\ 0 & \psi_{22} \end{pmatrix}, \quad \Gamma\Phi\Gamma^\top + \Psi = \begin{pmatrix} \psi_{11} + \gamma_{11}^2\phi_{22} & \gamma_{11}\gamma_{21}\phi_{22} \\ \gamma_{11}\gamma_{21}\phi_{22} & \psi_{22} + \gamma_{21}^2\phi_{22} \end{pmatrix}$$

Thus, we have

$$\begin{aligned} \Sigma_{yy} &= \begin{pmatrix} 1 & \beta \\ 0 & 1 \end{pmatrix} \left(\Gamma\Phi\Gamma^\top + \Psi\right) \begin{pmatrix} 1 & 0 \\ \beta & 1 \end{pmatrix} \\ &= \begin{pmatrix} G_0 + 2\beta G_1 + \beta^2 G_2 & G_1 + \beta G_2 \\ G_1 + \beta G_2 & G_2 \end{pmatrix} \end{aligned} \tag{4.136}$$

where $G_0 = \psi_{11} + \gamma_{11}^2\phi_{22}$, $G_1 = \gamma_{11}\gamma_{21}\phi_{22}$ and $G_2 = \psi_{22} + \gamma_{21}^2\phi_{22}$. So, the variance matrix $\Sigma(\boldsymbol{\theta})$ is given by

$$\Sigma(\boldsymbol{\theta}) = \begin{pmatrix} G_0 + 2\beta G_1 + \beta^2 G_2 & G_1 + \beta G_2 & 0 & \phi_{22}(\gamma_{11} + \beta\gamma_{21}) \\ & G_2 & 0 & \gamma_{21}\phi_{22} \\ & & 0 & 0 \\ & & 0 & \phi_{22} \end{pmatrix} \tag{4.137}$$

By removing the row and column containing only zeros, we have:

$$\Sigma(\boldsymbol{\theta}) = \begin{pmatrix} G_0 + 2\beta G_1 + \beta^2 G_2 & G_1 + \beta G_2 & \phi_{22}(\gamma_{11} + \beta\gamma_{21}) \\ & G_2 & \gamma_{21}\phi_{22} \\ & & \phi_{22} \end{pmatrix}. \tag{4.138}$$

In (4.138), $\Sigma(\boldsymbol{\theta})$ has six unique elements and $\boldsymbol{\theta} = (\beta, \gamma_{11}, \gamma_{21}, \phi_{22}, \psi_{11}, \psi_{22})^\top$ has six parameters. Thus, the MLE, ULS, and WLS estimates of $\boldsymbol{\theta}$ are all identical and given by the unique solution to the equation $\Sigma(\boldsymbol{\theta}) = S_n$.

Note that the row and column with only zeros in (4.137) are the result of including the intercept terms, $\gamma_{10}$ and $\gamma_{20}$, in the model in (4.127), which are not estimated. To avoid such rows and columns in $\Sigma(\boldsymbol{\theta})$ at the outset, we can center the SEM in (4.128) on $E(\mathbf{x}_i)$ to obtain

$$\mathbf{y}_i - E(\mathbf{y}_i) = B(\mathbf{y}_i - E(\mathbf{y}_i)) + \Gamma(\mathbf{x}_i - E(\mathbf{x}_i)) + \boldsymbol{\epsilon}_{y_i}$$

For example, when applied to this mediation model, (4.127) can be expressed as:

$$\begin{pmatrix} y - E(y) \\ z - E(z) \end{pmatrix} = \begin{pmatrix} (\gamma_{11} + \beta\gamma_{21})(x - E(x)) + \epsilon_{y^*} \\ \gamma_{21}(x - E(x)) + \epsilon_{z^*} \end{pmatrix} \tag{4.139}$$

When calculated based on this centered model, $\Sigma(\boldsymbol{\theta})$ has the form in (4.138) (see exercise).

Now consider inference for the SEM in (4.129) with latent variables. Let

$$\Phi = Var(\boldsymbol{\xi}), \quad \Delta_x = Var(\boldsymbol{\delta}_x), \quad \Delta_y = Var(\boldsymbol{\delta}_y), \quad \Psi = Var(\boldsymbol{\epsilon}_\eta)$$

Let $\boldsymbol{\theta}$ denote the vector consisting of all the parameters. If we assume $\mathbf{u} = (\mathbf{y}^\top, \mathbf{x}^\top)^\top \sim N(\mathbf{0}, \Sigma(\boldsymbol{\theta}))$, then $\Sigma(\boldsymbol{\theta})$ has the same partition as in (4.131), but with the individual components given by

$$\begin{aligned} &\Sigma_{xx}(\boldsymbol{\theta}) = \Lambda_x \Phi \Lambda_x^\top + \Delta_x, \quad \Sigma_{yx}(\boldsymbol{\theta}) = \Lambda_y (\mathbf{I}_m - B)^{-1} \boldsymbol{\Gamma} \Phi \Lambda_x^\top \\ &\Sigma_{yy}(\boldsymbol{\theta}) = \Lambda_y (\mathbf{I}_m - B)^{-1} \left(\boldsymbol{\Gamma} \Phi \boldsymbol{\Gamma}^\top + \Psi\right) (\mathbf{I}_m - B)^{-\top} \Lambda_y^\top + \Delta_y \end{aligned} \tag{4.140}$$

where $m$ is the length of $\mathbf{y}$. The density and log-likelihood functions have the same form as those in (4.132) and (4.133). Also, inference proceeds in the same way as before using either the maximum likelihood or least-squares approach.

**Example 8 (*SEM for linear regression with measurement error*).** Consider the SEM in (4.124) for linear regression with measurement error in Example 5. Assume that only the predictor $\xi$ is measured with error by $r$ replicates $x_{ij}$, that is,

$$\begin{gathered} y_i = \gamma \xi_i + \epsilon_{y_i}, \quad x_{ij} = \xi_i + \delta_{x_{ij}} \\ E(\epsilon_{y_i}) = 0, \quad Cov(\xi_i, \epsilon_{y_i}) = 0, \quad E(\delta_{x_{ij}}) = 0 \\ Cov(\xi_i, \delta_{x_{ij}}) = 0, \quad Cov(\delta_{x_{ij}}, \delta_{x_{ik}}) = 0, \quad 1 \le j < k \le r \end{gathered}$$

Let $\mathbf{x} = (x_1, \ldots, x_r)^\top$ and $\boldsymbol{\delta}_{x_i} = (\delta_{x_{i1}}, \ldots, \delta_{x_{ir}})^\top$. Then, $\mathbf{x}_i = \xi_i \mathbf{1}_r + \boldsymbol{\delta}_{x_i}$, where $\mathbf{1}_r$ denotes an $r \times 1$ vector of 1's. It is readily checked that (see exercise)

$$Var(y_i) = \gamma^2\sigma_\xi^2 + \psi^2, \quad Var(\mathbf{x}_i) = \sigma_\xi^2\mathbf{J}_r + \sigma_{\delta_x}^2\mathbf{I}_r \qquad (4.141)$$
$$Cov(y_i, \mathbf{x}_i) = \gamma\sigma_\xi^2\mathbf{1}_r^\top$$

where $\mathbf{J}_m$ is an $m \times m$ matrix of 1's, $\sigma_\xi^2 = Var(\xi_i)$, $\sigma_{\delta_x}^2 = Var(\delta_{x_{ij}})$ and $\psi^2 = Var(\epsilon_{y_i})$. Thus,

$$\Sigma(\boldsymbol{\theta}) = \begin{pmatrix} \gamma^2\sigma_\xi^2 + \psi^2 & \gamma\sigma_\xi^2\mathbf{1}_r^\top \\ & \sigma_\xi^2\mathbf{J}_r + \sigma_{\delta_x}^2\mathbf{I}_r \end{pmatrix}$$

The matrix $\Sigma(\boldsymbol{\theta})$ has a total of $\frac{1}{2}(r+1)(r+2)$ unique elements, which is $> 4$ only if $r \geq 2$. Thus, we must have at least two replicates to identify the parameter vector $\boldsymbol{\theta} = \left(\gamma, \sigma_\xi^2, \psi^2, \sigma_{\delta_x}^2\right)^\top$ for this model.

**Example 9 (*SEM for mediational process with latent response*).** Consider the SEM in (4.118) from Example 2. Let $y_1 = y$, $y_2 = z$, $x_1 = 1$ and $x_2 = x$. By expressing the model in matrix form, we have

$$\begin{pmatrix} \eta_1 \\ \eta_2 \end{pmatrix} = \begin{pmatrix} 1 & -\beta \\ 0 & 1 \end{pmatrix}^{-1} \left[ \begin{pmatrix} \gamma_{10} & \gamma_{11} \\ \gamma_{20} & \gamma_{21} \end{pmatrix} \begin{pmatrix} x_1 \\ x_2 \end{pmatrix} + \begin{pmatrix} \epsilon_{\eta_1} \\ \epsilon_{\eta_2} \end{pmatrix} \right]$$
$$\mathbf{y} = \begin{pmatrix} y_1 \\ y_2 \end{pmatrix} = \begin{pmatrix} \eta_1 \\ \eta_2 \end{pmatrix} + \begin{pmatrix} \delta_{y_1} \\ 0 \end{pmatrix}, \quad \mathbf{x} = \boldsymbol{\xi} = \begin{pmatrix} \xi_1 \\ \xi_2 \end{pmatrix}$$

Thus,

$$\Lambda_x = \Lambda_y = \mathbf{I}_2, \quad \Delta_x = \mathbf{0}, \quad \Delta_y = Var\left(\begin{pmatrix} \delta_{y_1} \\ 0 \end{pmatrix}\right) = \begin{pmatrix} \sigma_{\delta_{y_1}}^2 & 0 \\ 0 & 0 \end{pmatrix}$$
$$\Phi = Var(\mathbf{x}) = \begin{pmatrix} 0 & 0 \\ 0 & \phi_{22} \end{pmatrix}, \quad \Psi = Var(\boldsymbol{\epsilon}_\eta) = \begin{pmatrix} \psi_{22} & 0 \\ 0 & \psi_{22} \end{pmatrix}$$

where $\sigma^2_{\delta_{y1}} = Var(\delta_{y1})$. It follows that

$$\Sigma_{yx} = \begin{pmatrix} 0 & \phi_{22}(\gamma_{11} + \beta\gamma_{21}) \\ 0 & \gamma_{21}\phi_{22} \end{pmatrix}, \tag{4.142}$$

$$\Sigma_{yy} = \begin{pmatrix} 1 & \beta \\ 0 & 1 \end{pmatrix} \left( \mathbf{\Gamma\Phi\Gamma}^\top + \Psi \right) \begin{pmatrix} 1 & 0 \\ \beta & 1 \end{pmatrix} + \begin{pmatrix} \sigma^2_{\delta_{y1}} & 0 \\ 0 & 0 \end{pmatrix}$$
$$= \begin{pmatrix} G_0 + 2\beta G_1 + \beta^2 G_2 + \sigma^2_{\delta_{y1}} & G_1 + \beta G_2 \\ & G_2 \end{pmatrix}$$

where $G_0$, $G_1$ and $G_2$ have the same interpretation as those in (4.136). By removing the row and column containing only zeros, we can express $\Sigma(\boldsymbol{\theta})$ as:

$$\Sigma(\boldsymbol{\theta}) = \begin{pmatrix} G_0 + 2\beta G_1 + \beta^2 G_2 + \sigma^2_{\delta_{y1}} & G_1 + \beta G_2 & \phi_{22}(\gamma_{11} + \beta\gamma_{21}) \\ & G_2 & \gamma_{21}\phi_{22} \\ & & \phi_{22} \end{pmatrix}$$

The above matrix has 6 unique elements, but the vector of parameters, $\boldsymbol{\theta} = \left( \beta, \gamma_{11}, \gamma_{21}, \phi_{22}, \psi_{11}, \psi_{22}, \sigma^2_{\delta_{y1}} \right)^\top$ has seven parameters. Thus, the model is unidentifiable. As in Example 8, one way to identify the model is to have replicate observations $y$.

### 4.6.3 GEE-Based Inference

The problem with the maximum likelihood estimation is the lack of robustness. Although least-squares based alternatives such as ULS attempt to address this issue, they do not consider missing data. By framing the inference problems under GEE, we can address both issues.

Consider first the SEM in (4.128) with all observed variables. We can express the model in terms of the mean and variance as follows:

$$E(\mathbf{y}_i | \mathbf{x}_i) = \boldsymbol{\mu}_i = (\mathbf{I}_m - B)^{-1} \mathbf{\Gamma x}_i, \tag{4.143}$$
$$Var(\mathbf{y}_i | \mathbf{x}_i) = \Sigma_i = (\mathbf{I}_m - B)^{-1} \Psi (\mathbf{I}_m - B)^{-\top}$$

Let $\boldsymbol{\theta}$ denote the vector containing all the parameters in $B$, $\mathbf{\Gamma}$, and $\Psi$. Note that since we condition on $\mathbf{x}_i$, we no longer need to consider estimating the variance matrix $\Phi$ of $\mathbf{x}_i$ as in the classic approaches discussed in Section 4.6.2. Since $E(\mathbf{y}_i|\,\mathbf{x}_i)$ and $Var(\mathbf{y}_i|\,\mathbf{x}_i)$ both contain the parameters of interest, we can use GEE II for inference about $\boldsymbol{\theta}$.

As in Section 4.5, let

$$\begin{aligned}\mathbf{s}_i &= (s_{i11}, s_{i12}, \ldots, s_{i1m}, s_{i22}, s_{i23}, \ldots, s_{i2m}, \ldots, s_{imm})^\top \\ \boldsymbol{\mu}_i &= (\mu_{i1}, \ldots, \mu_{im})^\top, \quad \boldsymbol{\sigma}_i = vec(\Sigma_i) \\ S_i &= \left((\mathbf{y}_i - \boldsymbol{\mu}_i)^\top, (\mathbf{s}_i - \boldsymbol{\sigma}_i)^\top\right)^\top\end{aligned} \tag{4.144}$$

where $\boldsymbol{\sigma}_i$ is an $\frac{(m-1)m}{2} \times 1$ column vector containing the unique elements of $\Sigma_i$ and $S_i$ an $\frac{(m+1)m}{2} \times 1$ column vector. The GEE II for estimating $\boldsymbol{\theta}$ is given by

$$\mathbf{w}_n(\boldsymbol{\theta}) = \sum_{i=1}^n G_i S_i = \sum_{i=1}^n G_i \begin{pmatrix} \mathbf{y}_i - \boldsymbol{\mu}_i \\ \mathbf{z}_i - \boldsymbol{\sigma}_i \end{pmatrix} = \mathbf{0} \tag{4.145}$$

where $G_i$ is some $q \times \frac{(m+1)m}{2}$ matrix function of $\mathbf{x}_i$ with $q$ denoting the length of $\boldsymbol{\theta}$. Inference for $\boldsymbol{\theta}$ follows from the discussion in Section 4.5. In particular, if $G_i$ is parameterized by $\boldsymbol{\alpha}$, the estimate of $\boldsymbol{\theta}$ that solves the equation in (4.145) is asymptotically normal with the asymptotic variance given in (4.105) if $\boldsymbol{\alpha}$ is known or substituted by a $\sqrt{n}$-consistent estimate.

**Example 10 (*SEM for mediational process without latent variable*).** For the mediation model in Example 1, we have

$$\begin{aligned}\boldsymbol{\mu}_i &= \begin{pmatrix} \gamma_{10} + \beta\gamma_{20} + (\gamma_{11} + \beta\gamma_{21}) x_i \\ \gamma_{20} + \gamma_{21} x_i \end{pmatrix} = \begin{pmatrix} \mu_{i1} \\ \mu_{i2} \end{pmatrix} \\ \Sigma_i &= \begin{pmatrix} \beta^2\psi_{22} + \psi_{11} & \beta\psi_{22} \\ & \psi_{22} \end{pmatrix} = \begin{pmatrix} v_{i11} & v_{i12} \\ & v_{i22} \end{pmatrix} \\ \boldsymbol{\theta} &= (\beta, \gamma_{10}, \gamma_{11}, \gamma_{20}, \gamma_{21}, \phi_{22}, \psi_{11}, \psi_{22})^\top\end{aligned} \tag{4.146}$$

Let

$$G_i = D_i V_i^{-1}, \quad D_i = \frac{\partial}{\partial \boldsymbol{\theta}} \boldsymbol{\mu}_i, \quad V_i = \begin{pmatrix} \Sigma_i & \mathbf{0} \\ & V_{i22}(\boldsymbol{\alpha}) \end{pmatrix} \tag{4.147}$$

where $V_{i22}(\boldsymbol{\alpha})$ is parameterized by $\boldsymbol{\alpha}$ and $\frac{\partial}{\partial \boldsymbol{\theta}} \boldsymbol{\mu}_i$ is readily calculated (see exercise). By substituting $G_i$ given in (4.147) into (4.145), we can estimate $\boldsymbol{\theta}$.

For example, if setting $V_{i22}(\boldsymbol{\alpha}) = \mathbf{I}_{\frac{(m-1)m}{2} \times \frac{(m-1)m}{2}}$, the estimate $\widehat{\boldsymbol{\theta}}$ obtained also follows an asymptotic normal distribution given in (4.105).

Now, consider the SEM involving latent variables defined in (4.129). Let

$$\mathbf{u}_i = \left(\mathbf{y}_i^\top, \mathbf{x}_i^\top\right)^\top, \quad \Sigma = Var(\mathbf{u}_i) = [\sigma_{st}], \quad \boldsymbol{\sigma} = vec(\Sigma)$$
$$s_{ist} = (u_{is} - \sigma_s)(u_{it} - \sigma_t)$$
$$\mathbf{s}_i = (s_{i11}, s_{i12}, \ldots, s_{i1p}, s_{i22}, s_{i23}, \ldots, s_{i2p}, \ldots, s_{ipp})^\top$$

where $p$ denotes the length of the column vector $\mathbf{u}_i$ and $\Sigma$ has the form in (4.131) with the component matrices $\Sigma_{xx}$, $\Sigma_{xy}$ and $\Sigma_{yy}$ defined by (4.140). The GEE II for estimating $\boldsymbol{\theta}$ is given by:

$$\mathbf{w}_n(\boldsymbol{\theta}) = \sum_{i=1}^{n} G_i S_i = \sum_{i=1}^{n} G_i(\mathbf{s}_i - \boldsymbol{\sigma}) = \mathbf{0} \tag{4.148}$$

As before, $G_i$ may contain a vector of unknown parameters $\boldsymbol{\alpha}$. The estimate of $\boldsymbol{\theta}$ obtained by solving the above equation has an asymptotic normal distribution given in (4.105) if $\boldsymbol{\alpha}$ is known or substituted by a $\sqrt{n}$-consistent estimate.

**Example 11 (*Least squares estimate*).** Consider the weighted least squares estimate (WLS) of $\boldsymbol{\theta}$ for the SEM model in (4.129) obtained by minimizing the objective function $F_{WLS}$ in (4.134) with some weight matrix $W$. Let $\boldsymbol{\omega} = vec\left(W^{-1}\right)$ denote $\frac{p(p-1)}{2} \times 1$ column vector containing the unique elements of $W$ and $V^{-1} = \text{diag}(\boldsymbol{\omega})$, the $\frac{p(p-1)}{2} \times \frac{p(p-1)}{2}$ diagonal matrix with $\boldsymbol{\omega}$ forming the diagonal. Then, by setting $G_i = \left(\frac{\partial}{\partial \boldsymbol{\theta}} \boldsymbol{\sigma}\right) V_i^{-1}$, we obtain from (4.148) the same WLS estimate of $\boldsymbol{\theta}$ as defined by $F_{WLS}$ in (4.134) (see exercise).

Inference under missing data follows from the discussions in Sections 4.4 and 4.5 for parametric and distribution-free models. The procedures are again straightforward for parametric models under both MCAR and MAR. For distribution-free inference, we can apply WGEE II. Under MAR, if the weight function is unknown, the feasibility of applying such an approach depends on whether the weight function can be estimated.

## 4.7 Exercises

### Section 4.2

1. Consider Cronbach's alpha defined in (4.4). Show that for a standardized instrument, that is, all items have the same standard deviation, or

$\Sigma = I_m$, $\alpha = 1$ if all items are perfectly correlated and $\alpha = 0$ if all items are stochastically independent.

2. Verify (4.11) for the LMM defined in (4.10).
3. Verify (4.16) for the LMM defined in (4.15).
4. Let $y$ follow the Poisson distribution:

$$f_P(y \mid \mu) = \frac{\mu^y}{y!} \exp(-\mu), \quad \mu > 0, \quad y = 0, 1, \ldots$$

Let $b$ be a random variable such that $\exp(b)$ follows a gamma distribution with mean 1 and variance $\alpha$. Consider a mixed-effects model:

$$y \mid b \sim Poisson(\mu), \quad \log(\mu) = \eta + b, \quad y = 0, 1, \ldots$$

Show that $y$ has a marginal negative binomial distribution $NB(\mu, \alpha)$ with the following density function:

$$f_{NB}(y \mid \mu, \alpha) = \frac{\Gamma\left(y + \frac{1}{\alpha}\right)}{y!\Gamma\left(\frac{1}{\alpha}\right)} \left(\frac{1}{1 + \alpha\mu}\right)^{\frac{1}{\alpha}} \left(\frac{\mu}{\frac{1}{\alpha} + \mu}\right)^{y_i}$$
$$\mu > 0, \quad \alpha > 0, \quad y = 0, 1, \ldots$$

5. Let $M$ be some $p \times p$ full rank constant matrix and $\mathbf{a}$ some $p \times 1$ column vector. Show
   (a) $\frac{\partial}{\partial M} \log|M| = \left(M^\top\right)^{-1}$.
   (b) $\frac{\partial}{\partial M}\left(\mathbf{a}^\top M^{-1}\mathbf{a}\right) = -M^{-1}\mathbf{a}\mathbf{a}^\top M^{-1}$.
6. Consider the multivariate normal linear regression model in (4.21) with $\boldsymbol{\mu}_i = X_i\boldsymbol{\beta}$.
   (a) Verify the likelihood and log-likelihood expressions in (4.22).
   (b) Use the identities in Problem 6 and those regarding differentiation of vector-valued function discussed in Chapter 2 to verify the score equation in (4.23).
   (c) Show that the estimates of $\widehat{\boldsymbol{\beta}}$ and $\widehat{\Phi}$ in (4.24) are the maximum likelihood estimates of $\boldsymbol{\beta}$ and $\Phi$.
7. Consider the LMM defined in (4.10). Show that

$$\mathbf{y}_i \mid X_i, Z_i \sim N\left(X_i\boldsymbol{\beta}, V_i = \left(Z_i D Z_i^\top + \sigma_b^2 \mathbf{I}_{m_i}\right)\right)$$

8. By using the normal random effects, extend the generalized logit and proportional odds models for nominal and ordinal response to a longitudinal study setting.
9. Consider the Cronbach's alpha defined in (4.4).

(a) Find the MLE $\widehat{\alpha}$ of $\alpha$.

(b) Show that the asymptotic variance $\sigma_{\alpha}^2$ of $\widehat{\alpha}$ is given by

$$\sigma_{\alpha}^2 = \frac{2m^2\left[\left(\mathbf{1}_m^\top \Phi \mathbf{1}_m\right)\left(tr\left(\Phi^2\right) + tr^2\left(\Phi\right)\right) - 2tr\left(\Phi\right)\left(\mathbf{1}_m^\top \Phi^2 \mathbf{1}_m\right)\right]}{(m-1)\left(\mathbf{1}_m^\top \Phi \mathbf{1}_m\right)^3}$$

where $tr(\cdot)$ denotes the trace of a matrix. [*Hint*: See van Zyl et al., 2000.]

**Section 4.3**

1. Assume that the distribution-free model in (4.28) is specified correctly. Show

(a) $E(\mathbf{y}_i - \boldsymbol{\mu}_i) = \mathbf{0}$.

(b) $E(\mathbf{w}_n) = \mathbf{0}$.

2. Show that at convergence, both recursive equations in (4.38) yield the GEE estimate of $\boldsymbol{\beta}$.

3. Consider the repeated measures ANOVA in Example 6. Show

(a) $\widehat{\boldsymbol{\mu}}_k$ satisfies the GEE defined in (4.40) under the working independence model.

(b) $\widehat{\boldsymbol{\mu}}_k$ is the MLE of $\boldsymbol{\mu}_k$ if $\mathbf{y}_{ki} \sim N(\mathbf{0}, \Sigma_k)$ for $1 \leq k \leq g$.

4. Consider the GEE estimate $\widehat{\boldsymbol{\beta}}$ in (4.43) for the distribution-free linear model defined in (4.41) and assume $\mathbf{y}_i \mid X_i \sim N\left(X_i\boldsymbol{\beta}, \sigma^2 R(\boldsymbol{\alpha})\right)$. Show

(a) $\widehat{\boldsymbol{\beta}}$ is the MLE of $\boldsymbol{\beta}$ if $\boldsymbol{\alpha}$ is known.

(b) $\widehat{\boldsymbol{\beta}}$ is the MLE of $\boldsymbol{\beta}$ if $\boldsymbol{\alpha}$ is unknown and estimated by (4.44).

5. Verify the Frechet bounds in (4.46) for the distribution-free model (4.45) with a binary response $y_{it}$. [Hint: see Prentice, 1988.]

6. Verify (4.50), (4.53) and (4.56) in the proof of Theorem 1.

7. Use the asymptotic normality of the score-like vector $\mathbf{w}_n(\boldsymbol{\beta})$ in (4.51) in the proof of Theorem 1 to show that the GEE estimate $\widehat{\boldsymbol{\beta}}$ is consistent regardless of the choice of $G_i$.

8. Derive the Frechet bounds for the correlation matrices $R_{jk}$ in (4.60) for the generalized logit model defined in (4.58).

**Section 4.4**

1. Consider the general LMM defined in (4.10). Under MAR, find

(a) the observed-data log-likelihood $l_1(\boldsymbol{\theta}_y)$ in (4.65).

(b) the MLE of $\boldsymbol{\beta}$.

2. Show that under MMDP MAR is equivalent to the available case missing value restriction condition in (4.68). [*Hint*: See Verbeke and Molenberghs, 2000.]

3. Extend the available case missing value restriction condition for the linear regression model in (4.69) to a longitudinal study with three assessment points.

4. Consider the estimate $\widehat{\mu}_2$ of $\mu_2$ defined in (4.74). Show

(a) $\widehat{\mu}_2$ is consistent under MAR, that is, if (4.73) holds true.

(b) $\widehat{\mu}_2$ is consistent if missing data is non-ignorable, that is, if $\pi_{i2}$ in (4.76) is a function of both $y_{i1}$ and $y_{i2}$.

(c) Verify (4.78).

5. Consider the WGEE I defined in (4.82). Show

(a) $E(\mathbf{w}_n(\boldsymbol{\beta})) = \mathbf{0}$, if $E(\Delta_i S_i) = \mathbf{0}$.

(b) $E(\Delta_i S_i) = \mathbf{0}$ if the models in (4.80) and (4.81) for regression and $\mathbf{r}_i$ are both specified correctly.

6. Prove Theorem 1 by employing an argument similar to the proof of Theorem 1 in Section 4.3.2.

7. Verify (4.88), (4.89) and (4.90). [*Hint*: For verifying (4.89), note that $\sqrt{n}(\widehat{\gamma} - \gamma) = \left(-\frac{\partial}{\partial \gamma^\top}\mathbf{u}_n\right)^{-1} \sqrt{n}\mathbf{u}_n + \mathbf{o}_p(1)$.]

8. Consider the model in (4.93).

(a) Show that $E[r_{ki2}(y_{ki2} - \mu_{ki2})] = 0$ $(1 \leq k \leq g)$. Thus, the estimating equation in (4.94) is unbiased.

(b) Let $\pi_{ki2} = E(r_{ki2} \mid y_{ki1})$ and $\widehat{\boldsymbol{\theta}}_k$ be defined in (4.95). Show

$$\widehat{\boldsymbol{\theta}}_k \to_p \begin{pmatrix} E(\pi_{ki2}) & E(\pi_{ki2}y_{ki1}) \\ E(\pi_{ki2}y_{ki1}) & E(\pi_{ki2}y_{ki1}^2) \end{pmatrix}^{-1} \begin{pmatrix} E(\pi_{ki2}y_{ki2}) \\ E(\pi_{ki2}y_{ki1}y_{ki2}) \end{pmatrix}$$

and

$$E(\pi_{ki2}y_{ki2}) = \beta_k E(\pi_{ki2}y_{ki1}) + \tau_k E(\pi_{ki2})$$
$$E(\pi_{ki2}y_{ki1}y_{ki2}) = \beta_k E(\pi_{ki2}y_{ki1}^2) + \tau_k E(\pi_{ki2}y_{ki1})$$

(c) Use the results in (b) to show $\widehat{\boldsymbol{\theta}}_k \to_p \boldsymbol{\theta}_k$.

(d) Assume $\beta_k = \beta$ $(1 \leq k \leq g)$. Show the GEE below defines a consistent estimate of $\boldsymbol{\theta} = (\tau_1, \ldots, \tau_g, \beta_k)^\top$:

$$\sum_{k=1}^{g}\sum_{i=1}^{n_k} \frac{\partial \mu_{ki2}}{\partial \boldsymbol{\theta}} r_{ki2} S_i = \sum_{k=1}^{g}\sum_{i=1}^{n_k} \frac{\partial \mu_{ki2}}{\partial \boldsymbol{\theta}} r_{ki2}(y_{ki2} - \mu_{ki2}) = 0$$

(e) Find the estimate $\widehat{\boldsymbol{\theta}}$ by solving the equation in (d).

**Section 4.5**

1. Let $\widehat{\boldsymbol{\theta}}$ denote the estimate of $\boldsymbol{\theta}$ obtained from the GEE II defined in (4.104). By employing an argument similar to the proof of Theorem 1 in Section 4.3.2, show

(a) $\widehat{\boldsymbol{\theta}}$ is consistent regardless of the choice of $G_i$.

(b) $\widehat{\boldsymbol{\theta}}$ has the asymptotic distribution in (4.105) if $\boldsymbol{\alpha}$ is either known or substituted by a $\sqrt{n}$-consistent estimate.

(c) $B = E\left(F_i V_i^{-1} D_i^\top\right)$ if $G_i = D_i V_i^{-1}$ with $D_i = \frac{\partial}{\partial \boldsymbol{\theta}}\left(\boldsymbol{\mu}_i^\top, \boldsymbol{\sigma}_i^\top\right)^\top$.

2. Consider the distribution-free model defined in (4.101) with $\sigma\left(\mu_{it}\right) = \mu_{it}\left(1 + \lambda_t \mu_{it}\right)$. Discuss inference for $\boldsymbol{\theta} = \left(\boldsymbol{\beta}^\top, \boldsymbol{\lambda}^\top\right)^\top$ with $\boldsymbol{\lambda} = \left(\lambda_1, \ldots, \lambda_m\right)^\top$ using GEE II.

3. Verify (4.107), (4.109), (4.110) and (4.112).

4. Consider the LMM for therapists' variability in (4.12) of Section 4.2.2. It follows from (4.12) and (4.13) that

$$E\left(y_{ij} \mid x_i\right) = \mu_i = \beta_0 + x_i \beta_1, \quad E\left(s_{ijk} \mid x_i\right) = \sigma_T^2 \delta_{jk} + \sigma_T^2 \rho \left(1 - \delta_{jk}\right)$$

where $s_{ijk} = \left(y_{ij} - \mu_i\right)\left(y_{ik} - \mu_i\right)$, $\sigma_T^2 = \sigma_b^2 + \sigma^2$, $\rho = \frac{\sigma_b^2}{\sigma_b^2 + \sigma^2}$ and $\delta_{jk} = 1$ if $j = k$ and $\delta_{jk} = 0$ if otherwise. Discuss inference about $\boldsymbol{\theta} = \left(\sigma_T^2, \rho\right)^\top$ using GEE II.

5. Consider the WGEE II defined in (4.116). Show

(a) $E\left(\mathbf{w}_n\left(\boldsymbol{\beta}\right)\right) = \mathbf{0}$, if $E\left(\Delta_i S_i\right) = \mathbf{0}$.

(b) $E\left(\Delta_i S_i\right) = \mathbf{0}$ if the models in (4.102) and (4.113) for regression and $\mathbf{r}_i$ are both specified correctly.

6. Within the context of Problem 5, generalize the asymptotic results in Theorem 1 of Section 4.4.2 for WGEE I to WGEE II estimates obtained from (4.116).

**Section 4.6**

1. Verify (4.120) and (4.121).

2. The SEM for a multiple linear regression with two predictors is by

$$y = \gamma_0 + \gamma_1 x_1 + \gamma_2 x_2 + \epsilon_y$$
$$E\left(\epsilon_y\right) = 0, \quad Cov\left(x_1, \epsilon_y\right) = 0, \quad Cov\left(x_1, \epsilon_y\right) = 0$$

For convenience, we have set the intercept to 0 in the above model. Suppose that $x_2$ is mistakenly dropped, that is,

$$y = \alpha_0 + \alpha_1 x_1 + \epsilon_y^*, \quad E\left(\epsilon_y^*\right) = 0, \quad Cov\left(x_1, \epsilon_y^*\right) \neq 0$$

(a) Show

$$Cov\left(x_1, y\right) = \left(\gamma_1 + \gamma_2 \frac{Cov\left(x_1, x_2\right)}{Var\left(x_1\right)}\right) Var\left(x_1\right)$$

(b) Let $\widehat{\alpha}_1$ be the least-squares estimate of $\alpha_1$. Show

$$\widehat{\alpha}_1 \rightarrow_p \frac{Cov\left(y, x_1\right)}{Var\left(x_1\right)} = \gamma_1 + \gamma_2 \frac{Cov\left(x_1, x_2\right)}{Var\left(x_1\right)}$$

Thus, unless $x_1$ and $x_2$ are uncorrelated, that is, $Cov(x_1, x_2) = 0$, $\widehat{\alpha}_1$ is not a consistent estimate of $\gamma_1$.

3. Consider the simple linear regression model with measurement errors in (4.124) and the model based on the observed variables in (4.125).

(a) Show

$$Cov(\xi, \eta) = \gamma\phi^2, \quad Cov(x, y) = \gamma\phi^2 = \gamma^* \sigma_x^2$$

(b) Use (a) to confirm (4.126).

4. Verify (4.131), (4.132) and (4.133).

5. Verify (4.135) and (4.136).

6. Find the variance matrix $\Sigma(\boldsymbol{\theta}) = Var\left((y, z, x)^\top\right)$ for the centered model in (4.139) and show that it is given by (4.138).

7. Verify (4.141) and (4.142).

8. Find $D_i = \frac{\partial}{\partial \boldsymbol{\theta}} \boldsymbol{\mu}_i$ in (4.147) by using the expressions for $\mu_{it}$ and $v_{ist}$ in (4.146) $(1 \leq s < t \leq 2)$.

9. Verify the claim in Example 11 that with $G_i = \left(\frac{\partial}{\partial \boldsymbol{\theta}} \boldsymbol{\sigma}\right) V_i^{-1}$ the GEE II in (4.148) yields the same weighted least squares estimate $\boldsymbol{\theta}$ as defined by $F_{WLS}$ in (4.134).

# Chapter 5

# Multivariate U-Statistics

In Chapter 3, we introduced univariate U-statistics and studied the properties of such a class of statistics, especially with respect to their large sample behaviors, such as consistency and asymptotic distributions. We also highlighted their applications to an array of important problems including the classic Kendall's tau and the Mann–Whitney–Wilcoxon rank based statistics as well as modern applications in comparing variances between two independent samples and modeling areas under ROC curves between two test kits. The focus of this chapter is twofold. First, we generalize univariate U-statistics and the associated inference theory to a multivariate setting so that the classical U-statistic-based statistics can be applied to a wider class of problems requiring such multivariate U-statistics. Second and more importantly, we want to extend the U-statistics theory to longitudinal data analysis.

As highlighted in Chapter 4, longitudinal study designs are employed in virtually all scientific investigations in the fields of biomedical, behavioral and social sciences. Such study designs not only afford an opportunity to study disease progression and model changes over time, but also provide a systematic approach for inference of causal mechanisms. In Chapter 4, we discussed the primary issues in the analysis of data arising from such study designs and how to address them for regression analysis, particularly within the distribution-free inference paradigm. As in Chapter 2, these models are primarily developed for modeling the mean, or first order moment, of a response variable. Many quantities of interest in applications also involve higher order moments and even functions of responses from multiple subjects, such as the product–moment correlation and Mann–Whitney–Wilcoxon rank based tests. Although GEE II may be applied to address second order moments, models and statistics defined by multiple subjects, such as Kendall's tau and rank based tests, are amenable to such treatment, as we have seen in Chapter 3. By extending the univariate U-statistics theory to a longitudinal data setting, we will be able to apply such classic statistics to modern study designs.

In addition to generalizing the univariate U-statistics models in Chapter 3 to longitudinal data analysis, we also introduce some additional applications, particularly those pertaining to reliability of measurement. As discussed in Chapter 4, measurement in psychosocial research hinges critically on the reliability of instruments designed to capture the often latent constructs that define a disorder or a condition. Internal instrument consistency and external rater validity are of critical importance for deriving reliable outcomes and valid study findings. Most available methods for such reliability assessments are developed based on the parametric modeling paradigm. As departures from parametric distributions are quite common, applications of such methods are liable for spurious results and misleading conclusions.

In Section 5.1, we consider multivariate U-statistics arising from cross-sectional study designs and develop the asymptotic theory for such statistics by generalizing the univariate theory of Chapter 3. In Section 5.2, we apply multivariate U-statistics and the inference theory to longitudinal study data. Within longitudinal data applications, we first consider the complete data case in Section 5.2.1, and then address missing data in Sections 5.2.2 and 5.2.3.

## 5.1 Models for Cross-Sectional Study Designs

### 5.1.1 One Sample Multivariate U-Statistics

As discussed in Chapter 3, many problems of interest in biomedical and psychosocial research involve second and even higher order moments, which are often defined by multiple correlated outcomes of different constructs and/or dimensions. Thus, even with a single assessment point, such models involve multiple correlated outcomes that are generally different from the focus of regression analysis, since the latter is primarily concerned with the mean of a single-subject response. Although regression may sometimes be used to model the quantity of interest, it is at best cumbersome. We illustrate this key difference and thereby motivate the development of the theory of multivariate U-statistics and its applications by considering the product–moment correlation.

**Example 1 (*Pearson correlation*).** Consider a study with $n$ subjects and let $\mathbf{z}_i = (x_i, y_i)^\top$ be a pair of variables from the $i$th subject $(1 \le i \le n)$. Let

$$\sigma_{xy} = Cov(x_i, y_i), \quad \sigma_x^2 = Var(x_i), \quad \sigma_y^2 = Var(y_i)$$

For the product–moment correlation between $x_i$ and $y_i$, $\rho = Corr(x_i, y_i) = \frac{\sigma_{xy}}{\sigma_x \sigma_x}$, the Pearson estimate of $\rho$ is defined by

$$\widehat{\rho} = \frac{\sum_{i=1}^{n} (x_i - \overline{x}_n)(y_i - \overline{y}_n)}{\sqrt{\sum_{i=1}^{n} (x_i - \overline{x}_n)^2 \sum_{i=1}^{n} (y_i - \overline{y}_n)^2}} \tag{5.1}$$

Although the above estimate can be obtained from a linear regression model relating $y_i$ to $x_i$ or vice versa, inference still requires separate treatment since the theory of linear regression does not provide inference for $\rho$ (see exercise).

To develop distribution-free inference for $\widehat{\rho}$ using multivariate U-statistics, let

$$h_1(\mathbf{z}_i, \mathbf{z}_j) = \frac{1}{2}(x_i - x_j)(y_i - y_j), \quad h_2(\mathbf{z}_i, \mathbf{z}_j) = \frac{1}{2}(x_i - x_j)^2 \tag{5.2}$$
$$h_3(\mathbf{z}_i, \mathbf{z}_j) = \frac{1}{2}(y_i - y_j)^2, \quad \mathbf{h}(\mathbf{z}_i, \mathbf{z}_j) = (h_1, h_2, h_3)^\top$$

As we discussed in Chapter 3, the symmetric kernel $h_1(\mathbf{z}_i, \mathbf{z}_j)$ gives rise to a univariate U-statistic that provides an unbiased estimate of the covariance $Cov(x_i, y_i)$ between $x_i$ and $y_i$. Although useful for testing whether $x_i$ and $y_i$ are correlated, it does not provide a measure for the strength of the association.

Now consider a vector statistic defined by

$$\widehat{\boldsymbol{\theta}} = \binom{n}{2}^{-1} \sum_{(i,j)\in C_2^n} \mathbf{h}(\mathbf{z}_i, \mathbf{z}_j) = \begin{pmatrix} \binom{n}{2}^{-1} \sum_{(i,j)\in C_2^n} h_1(\mathbf{z}_i, \mathbf{z}_j) \\ \binom{n}{2}^{-1} \sum_{(i,j)\in C_2^n} h_2(\mathbf{z}_i, \mathbf{z}_j) \\ \binom{n}{2}^{-1} \sum_{(i,j)\in C_2^n} h_3(\mathbf{z}_i, \mathbf{z}_j) \end{pmatrix} = \begin{pmatrix} \widehat{\theta}_1 \\ \widehat{\theta}_2 \\ \widehat{\theta}_3 \end{pmatrix}$$

The above has the same form as a one-sample, univariate U-statistic except that $\mathbf{h}(\mathbf{z}_i, \mathbf{z}_j)$ is a vector rather than a variable. In addition, each component $\widehat{\theta}_k$ of $\widehat{\boldsymbol{\theta}}$ is a univariate U-statistic, estimating the corresponding component of the parameter vector $\boldsymbol{\theta}$; $\widehat{\theta}_1$ estimates $\sigma_{xy}$ while $\widehat{\theta}_2$ and $\widehat{\theta}_3$ are unbiased estimates of $\sigma_x^2$ and $\sigma_y^2$.

It is readily checked that $\widehat{\rho} = \frac{\widehat{\theta}_1}{\sqrt{\widehat{\theta}_2 \widehat{\theta}_3}}$. Thus, if we know the asymptotic distribution of $\widehat{\boldsymbol{\theta}}$, we can use the Delta method to find the asymptotic distribution of $\widehat{\rho}$. However, as $\widehat{\boldsymbol{\theta}}$ is a vector statistic, the theory for univariate

U-statistics discussed in Chapter 3 does not apply and must be generalized to derive the joint distribution of $\widehat{\boldsymbol{\theta}}$.

Note that in this particular application, we can also apply GEE II for inference about $\widehat{\rho}$ (see exercise). In this case, we first model both the mean and variance:

$$E(x_i) = \mu_x, \quad E(y_i) = \mu_y, \quad E(s_{xy}) = \sigma_{xy} \qquad (5.3)$$
$$s_{xy} = (x_i - \mu_x)(y_i - \mu_y), \quad 1 \le i \le n$$

where $s_x^2 = (x_i - \mu_x)^2$ and $\sigma_x^2 = E(s_{xy})$ if $x = y$. The model parameter, $\boldsymbol{\zeta} = \left(\mu_x, \mu_y, \sigma_x^2, \sigma_y^2, \sigma_{xy}\right)^\top$, can be estimated by GEE II. We can then apply the Delta method to find the asymptotic distribution of $\widetilde{\rho} = \frac{\widehat{\sigma}_{xy}}{\sqrt{\widehat{\sigma}_x^2 \widehat{\sigma}_y^2}}$ based on the WGEE II estimate $\widehat{\boldsymbol{\zeta}}$ (see exercise). Further, we can define a GEE II so that it yields the Pearson estimate in (5.1).

Before proceeding with more examples, a formal definition of a multivairate U-statistic vector is in order. Consider an i.i.d. sample of $p \times 1$ column vector of response $\mathbf{y}_i$ $(1 \le i \le n)$. Let $\mathbf{h}(\mathbf{y}_1, \ldots, \mathbf{y}_m)$ be a symmetric vector-valued function with $m$ arguments. We define a one-sample, $m$-argument multivariate U-statistic or U-statistic vector as follows:

$$\mathbf{U}_n = \binom{n}{m}^{-1} \sum_{(i_1,\ldots,i_m) \in C_m^n} \mathbf{h}(\mathbf{y}_{i_1}, \ldots, \mathbf{y}_{i_m}) \qquad (5.4)$$

where $C_m^n = \{(i_1, \ldots, i_m); 1 \le i_1 < \ldots < i_m \le n\}$ denotes the set of all distinct combinations of $m$ indices $(i_1, \ldots, i_m)$ from the integer set $\{1, 2, \ldots, n\}$. As in the univariate case, $\mathbf{U}_n$ is an unbiased estimate of $\boldsymbol{\theta} = E(\mathbf{U}_n) = E(\mathbf{h}_n)$.

In Chapter 3, we discussed two measures of association, the Goodman and Kruskal's $\gamma$ and Kendall's $\tau_b$, for two ordinal outcomes. Both measures are defined based on the concept of comparing probabilities between concordant and discordant pairs, but are normalized to range from $-1$ to 1 so that they can be interpreted as correlations like the product–moment correlation for association between two continuous outcomes. We developed a U-statistic as an unbiased estimate of the difference between the concordant and discordant pair probabilities and used it to test the null of no association between two ordinal categorical outcomes. In the next examples, we apply multivariate U-statistics to develop estimates of Goodman and Kruskal's $\gamma$ and Kendall's $\tau_b$.

**Example 2 (*Goodman and Kruskal's* $\gamma$).** Consider an i.i.d. sample with bivariate ordinal outcomes $\mathbf{z}_i = (u_i, v_i)^\top$ $(1 \le i \le n)$. Suppose $u_i$ has $K$ and $v_i$ has $M$ levels indexed by $k$ and $m$, respectively. For a pair of subjects $\mathbf{z}_i = (k, m)^\top$ and $\mathbf{z}_j = (k', m')^\top$, concordance and discordance are defined as follows:

$$(\mathbf{z}_i, \mathbf{z}_j) \equiv \begin{cases} \text{concordant} & \text{if } u_i < (>) u_j, \ v_i < (>) v_j \\ \text{discordant} & \text{if } u_i > (<)\, u_j, \ v_i < (>)\, v_j \\ \text{neither} & \text{if otherwise} \end{cases}$$

Let $p_{cp}$ and $p_{dp}$ denote the probability of concordant and discordant pairs. The Goodman and Kruskal's $\gamma$ is defined by $\gamma = \frac{p_{cp} - p_{dp}}{p_{cp} + p_{dp}}$. Let

$$h_1(\mathbf{z}_i, \mathbf{z}_j) = I_{\{(u_i - u_j)(v_i - v_j) > 0\}}, \quad h_2(\mathbf{z}_i, \mathbf{z}_j) = I_{\{(u_i - u_j)(v_i - v_j) < 0\}} \tag{5.5}$$
$$\mathbf{h}(\mathbf{z}_i, \mathbf{z}_j) = (h_1, h_2)^\top$$

where $I_{\{\cdot\}}$ denotes a set indicator. Then, since

$$p_{cp} = E\left(I_{\{(u_i - u_j)(v_i - v_j) > 0\}}\right), \quad p_{dp} = E\left(I_{\{(u_i - u_j)(v_i - v_j) < 0\}}\right)$$

it follows that the bivariate U-statistic,

$$\begin{aligned} \widehat{\boldsymbol{\theta}} &= \binom{n}{2}^{-1} \sum_{(i,j) \in C_2^n} \mathbf{h}(\mathbf{z}_i, \mathbf{z}_j) \\ &= \binom{n}{2}^{-1} \sum_{(i,j) \in C_2^n} \begin{pmatrix} I_{\{(u_i - u_j)(v_i - v_j) > 0\}} \\ I_{\{(u_i - u_j)(v_i - v_j) < 0\}} \end{pmatrix} \end{aligned} \tag{5.6}$$

is an unbiased estimate of $\boldsymbol{\theta} = (\theta_1, \theta_2)^\top = (p_{cp}, p_{dp})^\top$. Thus, as in the case of the product–moment correlation, once we determine the asymptotic distribution of $\widehat{\boldsymbol{\theta}}$, we can use the Delta method to find the asymptotic distribution $\widehat{\gamma} = \frac{\widehat{\theta}_1 - \widehat{\theta}_2}{\widehat{\theta}_1 + \widehat{\theta}_2}$ for inference about $\gamma$.

**Example 3 (*Kendall's* $\tau_b$ *for ordinal outcome*).** Within the setting of Example 2, consider Kendall's $\tau_b$. Unlike Goodman and Kruskal's $\gamma$, Kendall's $\tau_b$ normalizes $p_{cp} - p_{dp}$ by using a different scaling factor, $\sqrt{(1 - p_u)(1 - p_v)}$, that is, $\tau_b = \frac{p_{cp} - p_{dp}}{\sqrt{(1 - p_u)(1 - p_v)}}$, where

$$p_u = \Pr(u_i = u_j), \quad p_v = \Pr(v_i = v_j) \tag{5.7}$$

Let $h_1(\mathbf{z}_i, \mathbf{z}_j)$ and $h_2(\mathbf{z}_i, \mathbf{z}_j)$ be defined as in (5.5) and let

$$h_3(\mathbf{z}_i, \mathbf{z}_j) = I_{\{u_i = u_j\}}, \quad h_4(\mathbf{z}_i, \mathbf{z}_j) = I_{\{v_i = v_j\}}, \quad \mathbf{h}(\mathbf{z}_i, \mathbf{z}_j) = (h_1, h_2, h_3, h_4)^\top$$

Then, the U-statistic vector:

$$\widehat{\boldsymbol{\theta}} = \binom{n}{2}^{-1} \sum_{(i,j) \in C_2^n} \mathbf{h}(\mathbf{z}_i, \mathbf{z}_j) = \binom{n}{2}^{-1} \sum_{(i,j) \in C_2^n} \begin{pmatrix} I_{\{(u_i - u_j)(v_i - v_j) > 0\}} \\ I_{\{(u_i - u_j)(v_i - v_j) < 0\}} \\ I_{\{u_i = u_j\}} \\ I_{\{v_i = v_j\}} \end{pmatrix}$$

is an unbiased estimate of $\boldsymbol{\theta} = (\xi, p_u, p_v)^\top$. As in the earlier examples, we estimate $\tau_b$ by $\widehat{\tau}_b = \frac{\widehat{\theta}_1 - \widehat{\theta}_2}{\sqrt{(1-\widehat{\theta}_3)(1-\widehat{\theta}_4)}}$ and apply the Delta method to determine its asymptotic distribution from that of $\widehat{\boldsymbol{\theta}}$.

Multivariate U-statistics are usually defined by vector-valued outcomes $\mathbf{y}_i$ rather than scalar variables. As order statistics are not well defined for random vectors, most of the results and properties of univariate U-statistics derived based on order statistics in Chapter 3 are not generalizable to a multivariate setting. However, the concept of projection can be extended to our current setting along with the asymptotic relationship between a U-statistic and its projection. Thus, as in the univariate case, we can characterize asymptotic distributions of multivariate U-statistics by their projection counterparts.

Consider the general one-sample U-statistic vector defined in (5.4). Define the projection of $\mathbf{U}_n$ by

$$\widehat{\mathbf{U}}_n = \sum_{i=1}^{n} E(\mathbf{U}_n \mid \mathbf{y}_i) - \boldsymbol{\theta}(n-1) \tag{5.8}$$

Compared to projections defined for univariate U-statistics, the multivariate projection $\widehat{\mathbf{U}}_n$ above has exactly the same form except for the obvious difference in dimensionality. Analogously, the centered projection $\widehat{\mathbf{U}}_n - \boldsymbol{\theta}$ has a similar form:

$$\widehat{\mathbf{U}}_n - \boldsymbol{\theta} = \frac{m}{n} \sum_{i=1}^{n} (\mathbf{h}_1(\mathbf{y}_i) - \boldsymbol{\theta}) = \frac{m}{n} \sum_{i=1}^{n} \widetilde{\mathbf{h}}_1(\mathbf{y}_i) \tag{5.9}$$

where $\mathbf{h}_1(\mathbf{y}_1) = E(\mathbf{h}_1(\mathbf{y}_1, \ldots, \mathbf{y}_m) \mid \mathbf{y}_1)$ and $\widetilde{\mathbf{h}}_1(\mathbf{y}_1) = \mathbf{h}_1(\mathbf{y}_1) - \boldsymbol{\theta}$. As $\widehat{\mathbf{U}}_n$ is a sum of i.i.d. random vectors, an application of LLN and CLT immediately yields

$$\widehat{\mathbf{U}}_n \rightarrow_p \boldsymbol{\theta}, \quad \sqrt{n}\left(\widehat{\mathbf{U}}_n - \boldsymbol{\theta}\right) \rightarrow_d N\left(\mathbf{0}, m^2 Var\left[\widetilde{\mathbf{h}}_1\left(\mathbf{y}_1\right)\right]\right)$$

As in the univariate case, the next step is to show that $\widehat{\mathbf{U}}_n$ and $\mathbf{U}_n$ have the same asymptotic distribution so that the limiting distribution of $\widehat{\mathbf{U}}_n$ above can be used for inference about $\boldsymbol{\theta} = E\left(\mathbf{U}_n\right)$. We prove this and state the fact formally in a theorem below.

**Theorem 1** (Multivariate one-sample). Let $\Sigma_h = Var\left[\widetilde{\mathbf{h}}_1\left(\mathbf{y}_1\right)\right] = E\left[\widetilde{\mathbf{h}}_1\left(\mathbf{y}_1\right)\widetilde{\mathbf{h}}_1^\top\left(\mathbf{y}_1\right)\right]$. Then, under mild regularity conditions,

$$\mathbf{U}_n \rightarrow_p \boldsymbol{\theta}, \quad \sqrt{n}\left(\mathbf{U}_n - \boldsymbol{\theta}\right) \rightarrow_d N\left(\mathbf{0}, \Sigma_U = m^2 \Sigma_h\right)$$

Proof. The proof follows the same steps as in the univariate case. First, write

$$\sqrt{n}\left(\mathbf{U}_n - \boldsymbol{\theta}\right) = \sqrt{n}\left(\widehat{\mathbf{U}}_n - \boldsymbol{\theta}\right) + \sqrt{n}\left(\mathbf{U}_n - \widehat{\mathbf{U}}_n\right) = \sqrt{n}\left(\widehat{\mathbf{U}}_n - \boldsymbol{\theta}\right) + \mathbf{e}_n$$

Then, as in the univariate case, we show $\mathbf{e}_n \rightarrow_p \mathbf{0}$ so that $\sqrt{n}\left(\mathbf{U}_n - \boldsymbol{\theta}\right)$ and $\sqrt{n}\left(\widehat{\mathbf{U}}_n - \boldsymbol{\theta}\right)$ have the same asymptotic distribution. Again, as in the univariate case, we prove $\mathbf{e}_n \rightarrow_p \mathbf{0}$ by showing that $E\left(\mathbf{e}_n^\top \mathbf{e}_n\right) \rightarrow \mathbf{0}$ (see exercise).

**Lemma**. For $1 \leq l \leq m$, let

$$\begin{aligned}
&\widetilde{\mathbf{h}}_l = \mathbf{h}_l - \boldsymbol{\theta}, \quad \mathbf{h}_l\left(\mathbf{y}_1, \ldots, \mathbf{y}_l\right) = E\left(\mathbf{h} \mid \mathbf{y}_1, \ldots, \mathbf{y}_l\right) \\
&\Sigma_l = Var\left(\mathbf{h}_l\left(\mathbf{y}_1, \ldots, \mathbf{y}_l\right)\right) = E\left[\widetilde{\mathbf{h}}_l\left(\mathbf{y}_1, \ldots, \mathbf{y}_l\right)\widetilde{\mathbf{h}}_l^\top\left(\mathbf{y}_1, \ldots, \mathbf{y}_l\right)\right]
\end{aligned}$$

Then, we have

$$\begin{aligned}
Var\left(\mathbf{U}_n\right) &= \sum_{l=0}^{n} \binom{n}{m}^{-1} \binom{m}{l} \binom{n-m}{m-l} \Sigma_l \qquad (5.10) \\
&= \frac{m^2}{n} Var\left(\mathbf{h}_1\left(\mathbf{y}_1\right)\right) + O\left(n^{-2}\right)
\end{aligned}$$

Proof of Lemma. Let $\left(k_1, \ldots, k_l\right)$ be the indices that are common to

$(i_1, \ldots, i_m)$ and $(j_1, \ldots, j_m)$. It then follows that

$$\begin{aligned} \mathbf{I}_{i_1,\ldots,i_m;j_1,\ldots,j_m} &= E\left[\widetilde{\mathbf{h}}\left(\mathbf{y}_{i_1}, \ldots, \mathbf{y}_{i_m}\right)\widetilde{\mathbf{h}}^\top\left(\mathbf{y}_{j_1}, \ldots, \mathbf{y}_{j_m}\right)\right] \\ &= E\left[E\left(\widetilde{\mathbf{h}}\left(\mathbf{y}_{i_1}, \ldots, \mathbf{y}_{i_m}\right)\widetilde{\mathbf{h}}_l^\top\left(\mathbf{y}_{j_1}, \ldots, \mathbf{y}_{j_m}\right) \mid \mathbf{y}_{k_1}, .., \mathbf{y}_{k_l}\right)\right] \\ &= E\left[E\left(\widetilde{\mathbf{h}}\left(\mathbf{y}_{i_1}, \ldots, \mathbf{y}_{i_m}\right) \mid \mathbf{y}_{k_1}, .., \mathbf{y}_{k_l}\right) E\left(\widetilde{\mathbf{h}}^\top \mid \mathbf{y}_{k_1}, .., \mathbf{y}_{k_l}\right)\right] \\ &= E\left[\widetilde{\mathbf{h}}_l\left(\mathbf{y}_{k_1}, \ldots, \mathbf{y}_{k_l}\right)\widetilde{\mathbf{h}}_l^\top\left(\mathbf{y}_{k_1}, \ldots, \mathbf{y}_{k_l}\right)\right] \\ &= \Sigma_l \end{aligned} \tag{5.11}$$

As in the univariate case, the number of distinct choices for two such sets having exactly $l$ elements in common is $\binom{n}{m}\binom{m}{l}\binom{n-m}{m-l}$. Thus, it follows that

$$\begin{aligned} Var\left(\mathbf{U}_n\right) &= \binom{n}{m}^{-2} \sum_{l=0}^{n} \binom{n}{m}\binom{m}{l}\binom{n-m}{m-l}\Sigma_l \\ &= \sum_{l=0}^{n} \binom{n}{m}^{-1}\binom{m}{l}\binom{n-m}{m-l}\Sigma_l \end{aligned} \tag{5.12}$$

Also, as shown in the proof of the univariate case,

$$\binom{n}{m}^{-1}\binom{m}{l}\binom{n-m}{m-l} = l!\binom{m}{l}^2 \frac{1}{n^l} + o\left(\frac{1}{n^l}\right) \tag{5.13}$$

It follows from (5.11) and (5.12) that

$$Var\left(\mathbf{U}_n\right) = n\sum_{l=0}^{n} l!\binom{m}{l}^2 \frac{1}{n^l}\Sigma_l = \frac{m^2}{n}\Sigma_h + \mathbf{O}\left(n^{-2}\right) \tag{5.14}$$

where $\mathbf{O}(\cdot)$ is the vector version of $O(\cdot)$ (see Chapter 1 for definition and properties).

Now, consider proving $E\left(\mathbf{e}_n^\top \mathbf{e}_n\right) \to \mathbf{0}$ for the theorem. First, we have

$$E\left(\mathbf{e}_n^\top \mathbf{e}_n\right) = nVar\left(\widehat{\mathbf{U}}_n\right) - 2nCov\left(\mathbf{U}_n, \widehat{\mathbf{U}}_n\right) + nVar\left(\mathbf{U}_n\right).$$

As in the univariate case, we calculate each term above. Since $\widehat{\mathbf{U}}_n$ is a sum of i.i.d. random terms, it follows immediately that

$$nVar\left(\widehat{\mathbf{U}}_n\right) = n\left(\frac{m}{n}\right)^2 \sum_{i=1}^{n} Var\left(\mathbf{h}_1\left(\mathbf{y}_i\right)\right) = m^2 Var\left(\mathbf{h}_1\left(\mathbf{y}_i\right)\right) \tag{5.15}$$

To find $nCov\left(\mathbf{U}_n, \widehat{\mathbf{U}}_n\right)$, let $\mathbf{I}_{i_1,\ldots,i_m,i} = E\left[\widetilde{\mathbf{h}}\left(\mathbf{y}_{i_1}, \ldots, \mathbf{y}_{i_m}\right)\widetilde{\mathbf{h}}_1^\top\left(\mathbf{y}_i\right)\right]$. Then, we have:

$$\begin{aligned} nCov\left(\mathbf{U}_n, \widehat{\mathbf{U}}_n\right) &= nE\left[\left(\mathbf{U}_n - \boldsymbol{\theta}\right)\left(\widehat{\mathbf{U}}_n - \boldsymbol{\theta}\right)\right] \\ &= m\binom{n}{m}^{-1}\sum_{i=1}^{n}\sum_{(i_1,\ldots,i_m)\in C_m^n}\mathbf{I}_{i_1,\ldots,i_m,i} \end{aligned}$$

If $i \notin (i_1, \ldots, i_m)$, then

$$\begin{aligned} \mathbf{I}_{i_1,\ldots,i_m,i} &= E\left[\widetilde{\mathbf{h}}\left(\mathbf{y}_{i_1}, \ldots, \mathbf{y}_{i_m}\right)\widetilde{\mathbf{h}}_1^\top\left(\mathbf{y}_i\right)\right] = E\left[\widetilde{\mathbf{h}}\left(\mathbf{y}_{i_1}, \ldots, \mathbf{y}_{i_m}\right)\right]E\left[\widetilde{\mathbf{h}}_1^\top\left(\mathbf{y}_i\right)\right] \\ &= 0 \end{aligned}$$

If $i \in (i_1, \ldots, i_m)$, this term equals:

$$\begin{aligned} \mathbf{I}_{i_1,\ldots,i_m,i} &= E\left[\widetilde{\mathbf{h}}\left(\mathbf{y}_{i_1}, \ldots, \mathbf{y}_{i_m}\right)\widetilde{\mathbf{h}}_1^\top\left(\mathbf{y}_i\right) \mid \mathbf{y}_i\right] \\ &= E\left[\widetilde{\mathbf{h}}_1\left(\mathbf{y}_i\right)E\left(\widetilde{\mathbf{h}}^\top\left(\mathbf{y}_{i_1}, \ldots, \mathbf{y}_{i_m}\right) \mid \mathbf{y}_i\right)\right] = E\left(\widetilde{\mathbf{h}}_1\left(\mathbf{y}_i\right)\widetilde{\mathbf{h}}_1^\top\left(\mathbf{y}_i\right)\right) \\ &= Var\left(\mathbf{h}_1\left(\mathbf{y}_1\right)\right). \end{aligned}$$

Since for each $i$ the event $\{i \in (i_1, \ldots, i_m)\}$ occurs $\binom{n-1}{m-1}$ times, it follows that

$$\begin{aligned} nCov\left(\mathbf{U}_n, \widehat{\mathbf{U}}_n\right) &= \sum_{i=1}^{n} m\binom{n}{m}^{-1}\binom{n-1}{m-1}Var\left(\mathbf{h}_1\left(\mathbf{y}_1\right)\right) \qquad (5.16) \\ &= nm\binom{n}{m}^{-1}\binom{n-1}{m-1}Var\left(\mathbf{h}_1\left(\mathbf{y}_1\right)\right) \\ &= m^2 Var\left(\mathbf{h}_1\left(\mathbf{y}_1\right)\right) = m^2\Sigma_h \end{aligned}$$

Thus, it follows from (5.10), (5.15), and (5.16) that

$$E\left(\mathbf{e}_n^\top \mathbf{e}_n\right) = m^2\Sigma_h - 2m^2\Sigma_h + m^2\Sigma_h + \mathbf{O}\left(n^{-2}\right) \to_p \mathbf{0}$$

**Example 4.** Consider again the U-statistic vector for the prodoct-moment correlation in Example 1. In this case, $m = 2$ and it follows from Theorem 1 that

$$\sqrt{n}\left(\widehat{\boldsymbol{\theta}}_n - \boldsymbol{\theta}\right) \to_d N\left(0, \Sigma_\theta = 4Var\left(\widetilde{\mathbf{h}}_1\left(\mathbf{z}_1\right)\right)\right), \quad \text{as } n \to \infty$$

To find $\Sigma_\theta$, let

$$\mathbf{p}_i\left(\mathbf{z}_i, \boldsymbol{\mu}\right) = \left(\left(x_i - \mu_x\right)\left(y_i - \mu_y\right), \left(x_i - \mu_x\right)^2, \left(y_i - \mu_y\right)^2\right)^\top$$

where $\boldsymbol{\mu} = \left(\mu_x, \mu_y\right)^\top$. Then, it is readily checked that (see exercise)

$$\mathbf{h}_1(\mathbf{z}_1) = E\left[\mathbf{h}(\mathbf{z}_1, \mathbf{z}_2) \mid \mathbf{z}_1\right] = E\left[\frac{1}{2}\begin{pmatrix} (x_1 - x_2)(y_1 - y_2) \\ (x_1 - x_2)^2 \\ (y_1 - y_2)^2 \end{pmatrix} \mid \mathbf{z}_1\right] \tag{5.17}$$

$$= \frac{1}{2}\left[\begin{pmatrix} (x_1 - \mu_x)(y_1 - \mu_y) \\ (x_1 - \mu_x)^2 \\ \left(y_1 - \mu_y\right)^2 \end{pmatrix} + \boldsymbol{\theta}\right]$$

Thus,

$$\widetilde{\mathbf{h}}_1(\mathbf{z}_1) = \mathbf{h}_1(\mathbf{z}_1) - \boldsymbol{\theta} = \frac{1}{2}\left[\mathbf{p}_1(\mathbf{z}_1, \boldsymbol{\mu}) - \boldsymbol{\theta}\right] \tag{5.18}$$

It follows that

$$\Sigma_\theta = 4Var\left(\widetilde{\mathbf{h}}_1(\mathbf{z}_1)\right) = Var\left(\mathbf{p}_1(\mathbf{z}_1)\right)$$

Under the assumption that $\mathbf{z}_i$ is normally distributed, $Var\left(\mathbf{p}_1(\mathbf{z}_1)\right)$ can be expressed in closed form in terms of the elements of the variance matrix of $\mathbf{z}_i$ (see exercise). Otherwise, a consistent estimate of $\Sigma_\theta$ is given by:

$$\widehat{\Sigma}_\theta = \frac{1}{n}\sum_{i=1}^{n}\left(\mathbf{p}_i(\mathbf{z}_i, \widehat{\boldsymbol{\mu}}) - \widehat{\boldsymbol{\theta}}\right)\left(\mathbf{p}_i(\mathbf{z}_i, \widehat{\boldsymbol{\mu}}) - \widehat{\boldsymbol{\theta}}\right)^\top$$

where $\widehat{\boldsymbol{\mu}} = \left(\widehat{\mu}_x, \widehat{\mu}_y\right)^\top$ with $\widehat{\mu}_x$ $\left(\widehat{\mu}_y\right)$ denoting some consistent estimate of $\mu_x$ $\left(\mu_y\right)$, such as the sample mean.

By applying the Delta method, we obtain the asymptotic distribution of $\widehat{\rho}$:

$$\sqrt{n}\left(\widehat{\rho} - \rho\right) \to_p N\left(0, \sigma_\rho^2 = D(\boldsymbol{\theta})\,\Sigma_\theta D^\top(\boldsymbol{\theta})\right), \quad n \to \infty \tag{5.19}$$

$$D(\boldsymbol{\theta}) = \left(\frac{1}{\sigma_x\sigma_y}, \frac{-\sigma_{xy}}{2\sigma_x^3\sigma_y}, \frac{-\sigma_{xy}}{2\sigma_x\sigma_y^3}\right)$$

For multivariate normal $\mathbf{z}_i$, $\sigma_\rho^2 = \left(1 - \rho^2\right)^2$ (see exercise). Otherwise, we can estimate $\sigma_\rho^2$ by: $\widehat{\sigma}_\rho^2 = D\left(\widehat{\boldsymbol{\theta}}\right)\widehat{\Sigma}_\theta D^\top\left(\widehat{\boldsymbol{\theta}}\right)$.

In the above example, we estimated the asymptotic variance of the U-statistic vector by first evaluating the asymptotic variance in closed form and then estimating it by substituting consistent estimates in place of the respective parameters. Alternatively, we can estimate the asymptotic variance without evaluating the analytic form by constructing another U-statistic as discussed in Chapter 3 for the univariate case. We illustrate this alternative approach for the Goodman and Kruskal's $\gamma$ index discussed in Example 2.

**Example 5.** Consider again the U-statistic vector defined in (5.6) for estimating the numerator and denominator of the Goodman and Kruskal's $\gamma$ in Example 3. The kernel vector is

$$\mathbf{h}\left(\mathbf{z}_i, \mathbf{z}_j\right) = \left(h_1, h_2\right)^\top = \left(I_{\{(u_i-u_j)(v_i-v_j)>0\}}, I_{\{(u_i-u_j)(v_i-v_j)>0\}}\right)^\top$$

It follows that

$$\mathbf{h}_1\left(\mathbf{z}_1\right) = E\left[\mathbf{h}\left(\mathbf{z}_1, \mathbf{z}_2\right) \mid \mathbf{z}_1\right] = E\left[\begin{pmatrix} I_{\{(u_1-u_2)(v_1-v_2)>0\}} \\ I_{\{(u_1-u_2)(v_1-v_2)<0\}} \end{pmatrix} \mid \mathbf{z}_1\right]$$

By applying the law of iterated conditional expectation, we have:

$$\begin{aligned}
\Sigma_h &= E\left[\left(\mathbf{h}_1\left(\mathbf{z}_1\right) - \boldsymbol{\theta}\right)\left(\mathbf{h}_1\left(\mathbf{z}_1\right) - \boldsymbol{\theta}\right)^\top\right] \\
&= E\left(\mathbf{h}_1\left(\mathbf{z}_1\right)\mathbf{h}_1^\top\left(\mathbf{z}_1\right)\right) - 2\boldsymbol{\theta} E\left(\mathbf{h}_1^\top\left(\mathbf{z}_1\right)\right) + \boldsymbol{\theta}\boldsymbol{\theta}^\top \\
&= E\left[E\left(\mathbf{h}\left(\mathbf{z}_1, \mathbf{z}_2\right) \mid \mathbf{z}_1\right) E\left(\mathbf{h}^\top\left(\mathbf{z}_1, \mathbf{z}_2\right) \mid \mathbf{z}_1\right)\right] - 2\boldsymbol{\theta}\boldsymbol{\theta}^\top + \boldsymbol{\theta}\boldsymbol{\theta}^\top \\
&= E\left[E\left(\mathbf{h}\left(\mathbf{z}_1, \mathbf{z}_2\right)\mathbf{h}^\top\left(\mathbf{z}_1, \mathbf{z}_3\right) \mid \mathbf{z}_1\right)\right] - \boldsymbol{\theta}\boldsymbol{\theta}^\top \\
&= E\left[\mathbf{h}\left(\mathbf{z}_1, \mathbf{z}_2\right)\mathbf{h}^\top\left(\mathbf{z}_1, \mathbf{z}_3\right)\right] - \boldsymbol{\theta}\boldsymbol{\theta}^\top
\end{aligned}$$

To estimate $\Sigma_h$, we must estimate $\boldsymbol{\Phi}_h = E\left[\mathbf{h}\left(\mathbf{z}_1, \mathbf{z}_2\right)\mathbf{h}^\top\left(\mathbf{z}_1, \mathbf{z}_3\right)\right]$ and $\boldsymbol{\theta}$. For $\boldsymbol{\theta}$, we can use the U-statistic in (5.6). To estimate $\boldsymbol{\Phi}_h$, we can construct another multivariate U-statistic. Let $\mathbf{g}\left(\mathbf{z}_1, \mathbf{z}_2, \mathbf{z}_3\right) = \mathbf{h}\left(\mathbf{z}_1, \mathbf{z}_2\right)\mathbf{h}^\top\left(\mathbf{z}_1, \mathbf{z}_3\right)$ and $\widetilde{\mathbf{g}}\left(\mathbf{z}_1, \mathbf{z}_2, \mathbf{z}_3\right)$ be a symmetric version of $\mathbf{g}\left(\mathbf{z}_1, \mathbf{z}_2, \mathbf{z}_3\right)$. For example, given the special form of $\mathbf{h}\left(\mathbf{z}_1, \mathbf{z}_2\right)\mathbf{h}^\top\left(\mathbf{z}_1, \mathbf{z}_3\right)$, the following is such a symmetric function:

$$\widetilde{\mathbf{g}}\left(\mathbf{z}_1, \mathbf{z}_2, \mathbf{z}_3\right) = \frac{1}{3}\left(\mathbf{g}\left(\mathbf{z}_1, \mathbf{z}_2, \mathbf{z}_3\right) + \mathbf{g}\left(\mathbf{z}_2, \mathbf{z}_1, \mathbf{z}_3\right) + \mathbf{g}\left(\mathbf{z}_3, \mathbf{z}_2, \mathbf{z}_1\right)\right)$$

Then, the U-statistic vector,

$$\widehat{\boldsymbol{\Phi}}_h = \binom{n}{3}^{-1} \sum_{(i,j,k)\in C_3^n} \widetilde{\mathbf{g}}\left(\mathbf{z}_i, \mathbf{z}_j, \mathbf{z}_k\right)$$

is a consistent estimate of $\boldsymbol{\Phi}_h$.

| $u$ | $v$ | | | |
|---|---|---|---|---|
| | $q_1$ | $\cdots$ | $q_M$ | Marginal Total |
| $r_1$ | $\widehat{p}_{11} = \frac{n_{11}}{n}$ | $\cdots$ | $\widehat{p}_{1M} = \frac{n_{1M}}{n}$ | $\widehat{p}_{1\cdot} = \frac{n_{1+}}{n}$ |
| $\vdots$ | $\vdots$ | $\ddots$ | $\vdots$ | $\vdots$ |
| $r_K$ | $\widehat{p}_{K1} = \frac{n_{K1}}{n}$ | $\cdots$ | $\widehat{p}_{KM} = \frac{n_{KM}}{n}$ | $\widehat{p}_{K\cdot} = \frac{n_{K+}}{n}$ |
| Marginal Total | $\widehat{p}_{\cdot 1} = \frac{n_{+1}}{n}$ | $\cdots$ | $\widehat{p}_{\cdot M} = \frac{n_{+M}}{n}$ | 1 |

Table 5.1. Classification of two categorical variables with cell counts $n_{km}$ and cell proportions $p_{km}$

In Chapter 3, we discussed a U-statistic based approach to test independence between two binary outcomes. In the next example, we generalize this approach to consider testing independence between two categorical outcomes.

**Example 6** (***Test for independence between two categorical outcomes***). Consider two categorical variables, $u$ and $v$. Assume $u$ has $K$ and $v$ has $M$ levels indexed by $r_k$ and $q_m$, respectively. Denote the joint cell counts (probabilities) by $n_{km}$ ($p_{km}$) and the marginal cell counts (probabilities) by $n_{k+}$ and $n_{+j}$ ($p_{k\cdot}$ and $p_{\cdot j}$). The different combinations of the two categorical outcomes are typically displayed in a contingency table as in Table 5.1. A primary interest is in testing the stochastic independence between the two variables. Chi-squares or Fisher's exact tests are often used to test such a null hypothesis. In this example, we describe a similar asymptotic test based on multivariate $U$-statistics.

Let

$$x_{ik} = I_{\{u_i = r_k\}}, \quad y_{im} = I_{\{v_i = q_m\}}, \quad p_{k\cdot} = E(x_{ik}), \quad p_{\cdot m} = E(y_{im})$$
$$\mathbf{y}_i = (y_{i1}, \ldots, y_{iM})^\top, \quad \mathbf{x}_i = (x_{i1}, \ldots, x_{iK})^\top, \quad \mathbf{z}_i = \left(\mathbf{x}_i^\top, \mathbf{y}_i^\top\right)^\top$$

Then, we have:

$$\delta_{km} = E(x_{ik} y_{im}) - E(x_{ik}) E(y_{im}) = p_{km} - p_{k\cdot} p_{\cdot m}$$

Thus, $u$ and $v$ are independent iff the null hypothesis, $H_0 : \delta_{km} = 0$ ($1 \leq k \leq K$, $1 \leq m \leq M$), holds true. Note that since

$$\sum_k \delta_{km} \equiv \sum_m \delta_{km} \equiv \sum_{k,m} \delta_{km} \equiv 0$$

there are only $(K-1)(M-1)$ independent $\delta_{km}$ value.

Let

$$h_{km}(x_{ik}, x_{jk}, y_{im}, y_{jm}) = \frac{1}{2}(x_{ik} - x_{jk})(y_{im} - y_{jm})$$
$$\mathbf{h}(\mathbf{z}_i, \mathbf{z}_j) = \left(h_{11}, \ldots, h_{1(M-1)}, h_{21}, \ldots, h_{(K-1)(M-1)}\right)^\top$$
$$\boldsymbol{\delta} = \left(\delta_{11}, \ldots, \delta_{1(M-1)}, \delta_{21}, \ldots, \delta_{(K-1)(M-1)}\right)^\top$$

where $\boldsymbol{\delta}$ contains the $(K-1)(M-1)$ independent $\delta_{km}$ and $\mathbf{h}(\mathbf{z}_i, \mathbf{z}_j)$ the corresponding $h_{km}$. The null hypothesis of independence between $u$ and $v$ corresponds to the condition $E(\mathbf{h}(\mathbf{z}_i, \mathbf{z}_j)) = \boldsymbol{\delta} = \mathbf{0}$. Thus, we can use the following U-statistic defined by the kernel $\mathbf{h}(\mathbf{z}_i, \mathbf{z}_j)$ to test the null hypothesis:

$$\widehat{\boldsymbol{\delta}} = \binom{n}{2}^{-1} \sum_{(i,j)\in C_2^n} \mathbf{h}(\mathbf{z}_i, \mathbf{z}_j)$$

By Theorem 1, $\widehat{\boldsymbol{\delta}}$ has the following asymptotic distribution:

$$\sqrt{n}\left(\widehat{\boldsymbol{\delta}} - \boldsymbol{\delta}\right) \to_d N\left(\mathbf{0}, \Sigma_\delta = 4Var\left(\widetilde{\mathbf{h}}_1(\mathbf{z}_1)\right)\right)$$

To find $\Sigma_\delta$, note that $\mathbf{h}_1(\mathbf{z}_1) = E[\mathbf{h}(\mathbf{z}_1, \mathbf{z}_2) \mid \mathbf{z}_1]$. For a component $h_{km}$ of $\mathbf{h}(\mathbf{z}_i, \mathbf{z}_j)$, we have

$$E[h_{km}(\mathbf{z}_1, \mathbf{z}_2) \mid \mathbf{z}_1] = \frac{1}{2}E[(x_{1k} - x_{2k})(y_{1m} - y_{2m}) \mid \mathbf{z}_1]$$
$$= \frac{1}{2}[(x_{1k} - p_{k\cdot})(y_{1m} - p_{\cdot m}) + \delta_{km}]$$

It follows that the $km$th component $\widetilde{h}_{1,km}(\mathbf{z}_1)$ of $\widetilde{\mathbf{h}}_1(\mathbf{z}_1)$ is given by

$$\widetilde{h}_{1,km}(\mathbf{z}_1) = E[h_{km}(\mathbf{z}_1, \mathbf{z}_2) \mid \mathbf{z}_1] - \delta_{km}$$
$$= \frac{1}{2}[(x_{1k} - p_{k\cdot})(y_{1m} - p_{\cdot m}) - \delta_{km}]$$

Let

$$q_{i,km} = (x_{ik} - p_{k\cdot})(y_{im} - p_{\cdot m}),$$
$$\mathbf{q}_i = \left(q_{i,11}, \ldots, q_{i,1(M-1)}, q_{i,21}, \ldots, q_{i,(K-1)(M-1)}\right)^\top$$

Then, we can express the asymptotic variance as $\Sigma_\delta = Var(\mathbf{q}_1)$. A consistent estimate of $\Sigma_\delta$ is given by:

$$\widehat{\Sigma}_\delta = \frac{1}{n}\sum_{i=1}^{n}\left(\mathbf{q}_i - \widehat{\boldsymbol{\delta}}\right)\left(\mathbf{q}_i - \widehat{\boldsymbol{\delta}}\right)^\top$$

An advantage of the U-statistics derived test is that when the null is rejected, we can perform post hoc analyses to see which cells contribute to the rejection of $H_0$. Since

$$\delta_{km} = E(x_{ik} y_{im}) - E(x_{ik}) E(y_{im}) = Cov(x_{ik}, y_{im})$$

$\delta_{km}$ is interpreted as the covariance between $x_{ik}$ and $y_{im}$. Thus, relatively larger estimates of $\delta_{km}$ are indications of likely dependence for the corresponding cells.

Instrument reliability is a broad issue affecting scientific and research integrity in many areas of psychosocial and mental health research, such as in assessing accuracy of diagnoses of mental disorders, treatment adherence and therapist competence in psychotherapy research, and individuals' competence for providing informed consent for research study participation in ethics research. In Chapter 4, we discussed intraclass correlations for assessing interrater agreement or test–retest reliability for continuous outcomes. For discrete outcomes, different indices are used, with Kappa being the most popular for comparing agreement between two raters.

**Example 7** (***Interrater agreement between two categorical outcomes***). Consider a sample of $n$ subjects to be classified based on a categorical rating scale with $G$ categories: $r_1, r_2, \ldots, r_G$. Two judges independently rate each of the subjects based on the rating scale and their ratings form a total of $G^2$ combinations, resulting in a $G \times G$ contingency table. For example, if the ratings from raters 1 and 2 are identified as the variables $u$ and $v$ in Example 6, the joint distribution of their ratings can be displayed in a table similar to Table 5.1 with $n_{km}$ ($\widehat{p}_{km}$) denoting the number of subjects (proportions) classified in the $k$th and $m$th categories by the first and second raters. The marginal cell counts $n_{k+}$ ($n_{+m}$) and proportions $\widehat{p}_{1k}$ ($\widehat{p}_{2m}$) describe the number (proportions) of subjects classified by the first (second) rater according the rating categories.

The widely used Cohen's weighted kappa statistic $\kappa$ for two raters is defined by

$$\widehat{\kappa} = \frac{\sum_{k=1}^{G} \sum_{m=1}^{G} w_{km} \left( \widehat{p}_{km} - \widehat{p}_{1k} \widehat{p}_{2m} \right)}{1 - \sum_{k=1}^{G} \sum_{m=1}^{G} w_{km} \widehat{p}_{1k} \widehat{p}_{2m}} \tag{5.20}$$

where the weight $w_{km}$ is a known function with $0 \leq w_{km} \leq 1$. Note that $\widehat{p}_{1k}\widehat{p}_{2m}$ is the proportion under the assumption of independence between the two judges. As a special case, if $w_{km} = 1$ for $k = q$ and 0 otherwise, the above yields the unweighted $\kappa$:

$$\widehat{\kappa} = \frac{\sum_{k=1}^{G} (\widehat{p}_{kk} - \widehat{p}_{1k}\widehat{p}_{2k})}{1 - \sum_{k=1}^{G} \widehat{p}_{1k}\widehat{p}_{2k}} \tag{5.21}$$

Thus, the unweighted $\kappa$ requires absolute between-rater agreement, while the weighted version allows for between-rater difference weighted by $w_{km}$.

To define the population $\kappa$, let

$$y_{i1k} = I_{\{u_i = r_k\}}, \quad y_{i2m} = I_{\{v_i = r_m\}}, \quad \mathbf{y}_{i1} = (y_{i11}, \ldots, y_{i1G})^\top \tag{5.22}$$
$$\mathbf{y}_{i2} = (y_{i21}, \ldots, y_{i2G})^\top, \quad \mathbf{y}_i = \left(\mathbf{y}_{i1}^\top, \mathbf{y}_{i2}^\top\right)^\top, \quad E(y_{i1k}) = p_{1k}$$
$$E(y_{i2m}) = p_{2m}, \quad p_{km} = E(y_{i1k} y_{i2m})$$

Then, we can represent the observed proportions in the contingency table using the random variables $y_{i1k}$ and $y_{i2m}$ as follows:

$$\widehat{p}_{km} = \frac{1}{n}\sum_{i=1}^{n} y_{i1k}y_{i2m}, \quad \widehat{p}_{1k} = \frac{1}{n}\sum_{i=1}^{n} y_{i1k}, \quad \widehat{p}_{2m} = \frac{1}{n}\sum_{i=1}^{n} y_{i2m} \tag{5.23}$$

Thus, in light of (5.21), the population $\kappa$ is defined by

$$\kappa = \frac{\sum_{k=1}^{G}\sum_{m=1}^{G} w_{km}(p_{km} - p_{1k}p_{2m})}{1 - \sum_{k=1}^{G}\sum_{m=1}^{G} w_{km}p_{1k}p_{2m}} \tag{5.24}$$

To express $\widehat{\kappa}$ as a function of U-statistics, let

$$\boldsymbol{\zeta} = (\zeta_1, \zeta_2)^\top, \quad \zeta_1 = \sum_{k=1}^{G}\sum_{m=1}^{G} w_{km}(p_{km} - p_{1k}p_{2m}), \tag{5.25}$$
$$\zeta_2 = 1 - \sum_{k=1}^{G}\sum_{m=1}^{G} w_{km}p_{1k}p_{2m}$$

Then, the population kappa is a function of $\boldsymbol{\zeta}$: $\kappa = \frac{\zeta_1}{\zeta_2} = f(\boldsymbol{\zeta})$. Let

$$\widehat{\boldsymbol{\zeta}} = \binom{n}{2}^{-1} \sum_{(i,j)\in C_2^n} \mathbf{u}_{ij}, \quad \mathbf{u}_{ij} = (u_{ij1}, u_{ij2})^\top \tag{5.26}$$
$$u_{ij1} = \frac{1}{2}\sum_{k=1}^{G}\sum_{m=1}^{G} w_{km}(y_{i1k} - y_{j1k})(y_{i2m} - y_{j2m})$$
$$u_{ij2} = 1 - \frac{1}{2}\sum_{k=1}^{G}\sum_{m=1}^{G} w_{km}(y_{i1k}y_{j2m} + y_{i2m}y_{j1k}), \quad (i,j) \in C_2^n$$

Then, $\widehat{\boldsymbol{\zeta}}$ is a bivariate one-sample U-statistic with kernel $\mathbf{h}(\mathbf{y}_i, \mathbf{y}_j) = \mathbf{u}_{ij}$. Further, it is readily checked that $E\left(\widehat{\boldsymbol{\zeta}}\right) = \boldsymbol{\zeta}$ (see exercise). Thus, $\widehat{\kappa} = f\left(\widehat{\boldsymbol{\zeta}}\right) = \frac{\widehat{\zeta}_1}{\widehat{\zeta}_2}$ is a consistent estimate of $\kappa$.

It follows from Theorem 1 that

$$\widehat{\boldsymbol{\zeta}} = \binom{n}{2}^{-1} \sum_{(i,j)\in C_2^n} \mathbf{u}_{ij} \rightarrow_p \boldsymbol{\zeta}, \quad \sqrt{n}\left(\widehat{\boldsymbol{\zeta}} - \boldsymbol{\zeta}\right) \rightarrow_d N\left(\mathbf{0}, \Sigma_\zeta\right) \tag{5.27}$$

$$\Sigma_\zeta = 4E\left[\left(E\left(\mathbf{u}_{ij} \mid \mathbf{y}_i\right) - \boldsymbol{\zeta}\right)\left(E\left(\mathbf{u}_{ij} \mid \mathbf{y}_i\right) - \boldsymbol{\zeta}\right)^\top\right], \quad (i,j) \in C_2^n$$

By the Delta method, we obtain the asymptotic distribution of $\widehat{\kappa}$:

$$\sqrt{n}\left(\widehat{\kappa} - \kappa\right) \rightarrow_d N\left(0, \sigma_\kappa^2 = \frac{\partial}{\partial \boldsymbol{\zeta}} f\left(\boldsymbol{\zeta}\right) \Sigma_\zeta \frac{\partial^\top}{\partial \boldsymbol{\zeta}} f\left(\boldsymbol{\zeta}\right)\right) \tag{5.28}$$

Given a consistent estimate of $\widehat{\Sigma}_\zeta$, we immediately obtain a consistent estimate of $\sigma_\kappa^2$ by $\widehat{\sigma}_\kappa^2 = \frac{\partial}{\partial \boldsymbol{\zeta}} f\left(\widehat{\boldsymbol{\zeta}}\right) \widehat{\Sigma}_\zeta \left(\frac{\partial}{\partial \boldsymbol{\zeta}} f\left(\widehat{\boldsymbol{\zeta}}\right)\right)^\top$.

To find a consistent estimate of $\Sigma_\zeta$, note that (see exercise)

$$E\left(\mathbf{u}_{ij} \mid \mathbf{y}_i\right) = \begin{pmatrix} E\left(u_{ij1} \mid \mathbf{y}_i\right) \\ E\left(u_{ij2} \mid \mathbf{y}_i\right) \end{pmatrix} \tag{5.29}$$

$$E\left(u_{ij1} \mid \mathbf{y}_i\right) = \frac{1}{2} \sum_{k=1}^{G} \sum_{m=1}^{G} w_{km} \left(y_{i1k} y_{i2m} - y_{i1k} p_{2m} - p_{1k} y_{i2m} + p_{km}\right)$$

$$E\left(u_{ij2} \mid \mathbf{y}_i\right) = 1 - \frac{1}{2} \sum_{k=1}^{G} \sum_{m=1}^{G} w_{km} \left(y_{i1k} p_{2m} + y_{i2m} p_{1k}\right)$$

Thus, by Slutsky's theorem, we obtain a consistent estimate of $\Sigma_\zeta$:

$$\widehat{\Sigma}_\zeta = \frac{4}{n} \sum_{i=1}^{n} \left[\left(\widehat{E}\left(\mathbf{u}_{ij} \mid \mathbf{y}_i\right) - \widehat{\boldsymbol{\zeta}}\right)\left(\widehat{E}\left(\mathbf{u}_{ij} \mid \mathbf{y}_i\right) - \widehat{\boldsymbol{\zeta}}\right)^\top\right] \tag{5.30}$$

where $\widehat{E}\left(\mathbf{u}_{ij} \mid \mathbf{y}_i\right)$ denotes the estimated $E\left(\mathbf{u}_{ij} \mid \mathbf{y}_i\right)$ in (5.29) obtained by substituting $\widehat{p}_{km}$, $\widehat{p}_{1k}$ and $\widehat{p}_{2m}$ in place of the respective parameters.

In Chapter 4, we introduced the product–moment based intraclass correlation as a measure of overall agreement over multiple raters, but did not discuss its inference because of the lack of theory to determine the joint asymptotic distribution of multiple correlated Pearson correlations. We are now in a position to tackle the complexity in deriving this joint distribution.

**Example 8** (***Product–moment based ICC***). We consider a setting with $n$ subjects and $M$ judges. Let $y_{ik}$ denote the continuous rating from the $k$th rater for the $i$th subject. Let $\rho_{km} = Corr\left(y_{ik}, y_{im}\right)$ denote the product–moment correlation between the $k$th and $m$th judges' ratings $(1 \leq k < m \leq M)$. The product–moment based ICC is defined by averaging all such pairwise correlations:

$$\rho_{\text{PMICC}} = \binom{M}{2}^{-1} \sum_{(k,m)\in C_2^M} \rho_{km} \tag{5.31}$$

For inference about $\rho_{\text{PMICC}}$, we can apply the Delta method once we have the joint distribution of estimates of $\rho_{km}$. For each $\rho_{km}$, we derived the asymptotic distribution of the Pearson estimate in Example 4. Here, we generalize this approach to find the joint asymptotic distribution of such estimates.

Let $\mathbf{z}_{i,km} = \left(y_{ik}, y_{im}\right)^\top$ be a bivariate variable containing the ratings for the $i$th subject from the $k$th and $m$th observers $(1 \leq k < m \leq M)$. Let

$$\begin{aligned}
\mathbf{z}_i &= \left(\mathbf{z}_{i,12}^\top, \mathbf{z}_{i,13}^\top, \ldots, \mathbf{z}_{i,1M}^\top, \ldots, \mathbf{z}_{i,(M-1)M}^\top\right)^\top \\
\boldsymbol{\rho} &= \left(\rho_{12}, \rho_{13}, \ldots, \rho_{1M}, \ldots, \rho_{(M-1)M}\right)^\top \\
\boldsymbol{\theta}_{km} &= \left(Cov\left(y_{ik}, y_{im}\right), Var\left(y_{ik}\right), Var\left(y_{im}\right)\right)^\top \\
\boldsymbol{\theta} &= \left(\boldsymbol{\theta}_{12}^\top, \boldsymbol{\theta}_{13}^\top, \ldots, \boldsymbol{\theta}_{1M}^\top, , \ldots, \boldsymbol{\theta}_{(M-1)M}^\top\right)^\top \\
f\left(\boldsymbol{\theta}_{km}\right) &= \frac{Cov\left(y_{ik}, y_{im}\right)}{\sqrt{Var\left(y_{ik}\right), Var\left(y_{im}\right)}} \\
\mathbf{f}\left(\boldsymbol{\theta}\right) &= \left(f\left(\boldsymbol{\theta}_{12}\right), f\left(\boldsymbol{\theta}_{13}\right), \ldots, f\left(\boldsymbol{\theta}_{1M}\right), \ldots, f\left(\boldsymbol{\theta}_{(M-1)M}\right)\right)^\top
\end{aligned}$$

Then, $\boldsymbol{\rho} = \mathbf{f}\left(\boldsymbol{\theta}\right)$. As Example 4, we now construct U-statistics based estimates $\widehat{\boldsymbol{\theta}}$ of $\boldsymbol{\theta}$ so that $\mathbf{f}\left(\widehat{\boldsymbol{\theta}}\right)$ yields the Pearson correlation for each component $\rho_{km}$ of $\boldsymbol{\rho}$, and determine the asymptotic distribution of $\widehat{\boldsymbol{\theta}}$ using the results in Theorem 1.

Let

$$\begin{aligned}
h_{km}\left(\mathbf{z}_i, \mathbf{z}_j\right) &= \frac{1}{2}\left(y_{ik} - y_{jk}\right)\left(y_{im} - y_{jm}\right), \quad h_k\left(\mathbf{z}_i, \mathbf{z}_j\right) = \frac{1}{2}\left(y_{ik} - y_{jk}\right)^2 \\
\mathbf{h}_{km}\left(\mathbf{z}_i, \mathbf{z}_j\right) &= \left(h_{km}, h_k, h_m\right)^\top \\
\mathbf{h}\left(\mathbf{z}_i, \mathbf{z}_j\right) &= \left(\mathbf{h}_{12}^\top, \mathbf{h}_{13}^\top, \ldots, \mathbf{h}_{1M}^\top, \ldots, \mathbf{h}_{(M-1)M}^\top\right)^\top
\end{aligned}$$

Then, $E[\mathbf{h}(\mathbf{z}_i, \mathbf{z}_j)] = \boldsymbol{\theta}$ and thus the following U-statistic is an unbiased estimate of $\boldsymbol{\theta}$:

$$\widehat{\boldsymbol{\theta}} = \binom{n}{2}^{-1} \sum_{(i,j)\in C_2^n} \mathbf{h}(\mathbf{z}_i, \mathbf{z}_j)$$

By Theorem 1, the asymptotic distribution is given by

$$\sqrt{n}\left(\widehat{\boldsymbol{\theta}} - \boldsymbol{\theta}\right) \to_d N(\mathbf{0}, \Sigma_\theta), \quad \Sigma_\theta = 4Var\left(\widetilde{\mathbf{h}}_1(\mathbf{z}_1)\right) \tag{5.32}$$

To find $\Sigma_\theta$, let us first look at the $km$th component $\widetilde{\mathbf{h}}_{1,km}(\mathbf{z}_1)$ of $\widetilde{\mathbf{h}}_1(\mathbf{z}_1)$. Let

$$\mu_k = E(y_{ik}), \quad \boldsymbol{\mu}_{zkm} = (\mu_k, \mu_m)^\top$$
$$\boldsymbol{\mu}_z = \left(\boldsymbol{\mu}_{z12}^\top, \ldots, \boldsymbol{\mu}_{z1M}^\top, \ldots, \boldsymbol{\mu}_{z(M-1)M}^\top\right)^\top$$

It follows from (5.18) in Example 4 that

$$\widetilde{\mathbf{h}}_{1,km}(\mathbf{z}_1) = \frac{1}{2}\left[\mathbf{p}_{1,km}(\mathbf{z}_1, \boldsymbol{\mu}_z) - \boldsymbol{\theta}_{km}\right] \tag{5.33}$$
$$\mathbf{p}_{i,km}(\mathbf{z}_i, \boldsymbol{\mu}_z) = \left((y_{ik} - \mu_k)(y_{im} - \mu_m), (y_{ik} - \mu_k)^2, (y_{im} - \mu_m)^2\right)^\top$$

Let

$$\widetilde{\mathbf{h}}_1(\mathbf{z}_1) = \left(\widetilde{\mathbf{h}}_{1,12}^\top, \widetilde{\mathbf{h}}_{1,13}^\top, \ldots, \widetilde{\mathbf{h}}_{1,1M}^\top, \ldots, \widetilde{\mathbf{h}}_{1,(M-1)M}^\top\right)^\top$$
$$\mathbf{p}_i(\mathbf{z}_i, \boldsymbol{\mu}_z) = \left(\widetilde{\mathbf{p}}_{i,12}^\top, \widetilde{\mathbf{p}}_{i,13}^\top, \ldots, \widetilde{\mathbf{p}}_{i,1M}^\top, \ldots, \widetilde{\mathbf{p}}_{i,(M-1)M}^\top\right)^\top$$

Then, it follows from (5.33) that the asymptotic variance of $\widehat{\boldsymbol{\theta}}$ in (5.32) is given by, $\Sigma_\theta = 4Var\left(\widetilde{\mathbf{h}}_1(\mathbf{z}_1)\right) = Var(\mathbf{p}_i(\mathbf{z}_i, \boldsymbol{\mu}_z))$. A consistent estimate is given by

$$\widehat{\Sigma}_\theta = \frac{1}{n}\sum_{i=1}^{n}\left(\mathbf{p}_i(\mathbf{z}_i, \widehat{\boldsymbol{\mu}}_z) - \widehat{\boldsymbol{\theta}}\right)\left(\mathbf{p}_i(\mathbf{z}_i, \widehat{\boldsymbol{\mu}}_z) - \widehat{\boldsymbol{\theta}}\right)^\top$$

Note that $\widehat{\boldsymbol{\mu}}_z$ above denotes any consistent estimate of $\boldsymbol{\mu}_z$, such as the sample moment.

From (5.32), we immediately obtain the asymptotic distribution of $\widehat{\boldsymbol{\rho}} = \mathbf{f}\left(\widehat{\boldsymbol{\theta}}\right)$ by applying the Delta method. By expressing the estimate of $\rho_{\text{PMICC}}$

in (5.31) as $\widehat{\rho}_{\mathrm{PMICC}} = \mathbf{1}^{\top}_{\frac{1}{2}M(M-1)}\widehat{\boldsymbol{\rho}}$, we also obtain by the Delta the asymptotic distribution of $\widehat{\rho}_{\mathrm{PMICC}}$ for inference about this product–moment-based ICC.

As seen from the development in Example 8, inference based on the product–moment based ICC is quite complex. Primarily for this reason, the alternative definition based on linear mixed-effects model is more widely used in mental health and psychosocial research. We discussed inference for this model based index within the context of linear mixed-effects model in Chapter 4. As linear mixed-effects models place strong distribution assumptions, such as normality on both the latent effect and model error, parametric inference often yields unreliable results. We now consider a distribution-free approach to provide more robust inference about this model based ICC.

**Example 9** (***Linear mixed-effects model based ICC***). As in Example 8, let $y_{ik}$ denote the rating of the $k$th observer on the $i$th subject $(1 \leq k \leq M,\ 1 \leq i \leq n)$. Consider the following normal based linear mixed-effects model for $y_{ik}$:

$$
\begin{aligned}
& y_{ik} = \mu + \lambda_i + \epsilon_{ik}, \quad \lambda_i \sim \text{i.i.d.} N\left(0, \sigma_\lambda^2\right) \\
& \epsilon_{ik} \sim \text{i.i.d.} N\left(0, \sigma^2\right), \quad 1 \leq i \leq n, \quad 1 \leq k \leq M
\end{aligned}
\tag{5.34}
$$

In the above model, the latent variable $\lambda_i$ measures the between-subject variability and $\epsilon_{im}$ the judges' variability in rating the same subject. Under the model assumptions, we have:

$$
\begin{aligned}
Var\left(y_{ik}\right) = \sigma_\lambda^2 + \sigma^2, \quad & Var\left(\lambda_i\right) = Cov\left(y_{ik}, y_{il}\right) = \sigma_\lambda^2 \\
Corr\left(y_{ik}, y_{im}\right) = \frac{Var\left(\lambda_i\right)}{Var\left(y_{ik}\right)} & = \frac{\sigma_\lambda^2}{\sigma_\lambda^2 + \sigma^2}
\end{aligned}
$$

Thus, $\sigma_\lambda^2$ is interpreted as the between-observer covariance, and the between-rater correlation is a constant and is the ratio of the between-subject variability to the total variability (between-subject and between-judges). This constant between-rater correlation is the model based ICC, $\rho_{\mathrm{LMMICC}} = \frac{\sigma_\lambda^2}{\sigma_\lambda^2+\sigma^2}$. Parametric inference for $\rho_{\mathrm{LMMICC}}$ depends on the normal assumption for both $\lambda_i$ and $\epsilon_{im}$ in (5.34) and as such lacks robustness.

For distribution-free inference, first define a bivariate U-statistics kernel as follows:

$$
\mathbf{h}\left(\mathbf{y}_i, \mathbf{y}_j\right) = \binom{M}{2}^{-1} \sum_{(k,l) \in C_2^M} \frac{1}{2} \begin{pmatrix} \left(y_{ik} - y_{jk}\right)\left(y_{il} - y_{jl}\right) \\ \left(y_{ik} - y_{jk}\right)^2 \end{pmatrix}
\tag{5.35}
$$

Then, it is readily checked that

$$E\left[\left(y_{ik}-y_{jk}\right)\left(y_{il}-y_{jl}\right)\right] = Cov\left(y_{ik}, y_{il}\right) = \sigma_\lambda^2$$
$$E\left(y_{ik}-y_{jk}\right)^2 = Var\left(y_{ik}\right) = \sigma_\lambda^2 + \sigma^2$$

Thus, rather than using the variance and covariance, we can express $\sigma_\lambda^2$ and $\sigma_\lambda^2 + \sigma^2$ in terms of the mean of $\mathbf{h}\left(\mathbf{y}_i, \mathbf{y}_j\right)$, that is,

$$\boldsymbol{\theta} = \begin{pmatrix} \theta_1 \\ \theta_2 \end{pmatrix} = E\left[\mathbf{h}\left(\mathbf{y}_i, \mathbf{y}_j\right)\right] = \begin{pmatrix} \sigma_\lambda^2 \\ \sigma_\lambda^2 + \sigma^2 \end{pmatrix}$$

It follows that we can express the model based ICC as a function of the mean $\boldsymbol{\theta}$ of $\mathbf{h}\left(\mathbf{y}_i, \mathbf{y}_j\right)$, that is, $\rho_{\text{LMMICC}} = g\left(\boldsymbol{\theta}\right) = \frac{\theta_1}{\theta_2}$.

Thus, the following bivariate U-statistic based on $\mathbf{h}\left(\mathbf{y}_i, \mathbf{y}_j\right)$ in (5.35) is a consistent estimate of $\boldsymbol{\theta}$:

$$\widehat{\boldsymbol{\theta}} = \binom{n}{2}^{-1} \sum_{(i,j)\in C_2^n} \mathbf{h}\left(\mathbf{y}_i, \mathbf{y}_j\right)$$
$$= \binom{n}{2}^{-1} \binom{M}{2}^{-1} \sum_{(i,j)\in C_2^n} \sum_{(k,l)\in C_2^M} \frac{1}{2} \begin{pmatrix} \left(y_{ik}-y_{jk}\right)\left(y_{il}-y_{jl}\right) \\ \left(y_{ik}-y_{jk}\right)^2 \end{pmatrix}$$

We estimate $\rho_{\text{LMMICC}}$ by $\widehat{\rho}_{\text{LMMICC}} = g\left(\widehat{\boldsymbol{\theta}}\right)$. By Theorem 1 and the Delta method, the asymptotic distribution of $\widehat{\rho}_{\text{LMMICC}}$ is given by:

$$\sqrt{n}\left(\widehat{\rho}-\rho\right) \to_d N\left(0, \sigma_\rho^2 = \frac{\partial}{\partial\boldsymbol{\theta}} g\left(\boldsymbol{\theta}\right) \Sigma_\theta \left(\frac{\partial}{\partial\boldsymbol{\theta}} g\left(\boldsymbol{\theta}\right)\right)^\top\right)$$

To find $\Sigma_\theta = 4Var\left(\widetilde{\mathbf{h}}_1\left(\mathbf{y}_1\right)\right)$, let

$$\mathbf{q}\left(\mathbf{y}_1, \mu\right) = \binom{M}{2}^{-1} \sum_{(k,l)\in C_2^M} \begin{pmatrix} \left(y_{1k}-\mu\right)\left(y_{1l}-\mu\right) \\ \left(y_{1k}-\mu\right)^2 \end{pmatrix}$$

Then, it is readily checked that (see exercise)

$$\begin{aligned}\widetilde{\mathbf{h}}_1(\mathbf{y}_1) &= E[\mathbf{h}(\mathbf{y}_1,\mathbf{y}_2) \mid \mathbf{y}_1] - \boldsymbol{\theta} \\ &= \binom{M}{2}^{-1} \sum_{(k,l)\in C_2^M} \frac{1}{2} E\left[\begin{pmatrix} (y_{1k}-y_{2k})(y_{1l}-y_{2l}) \\ (y_{1k}-y_{2k})^2 \end{pmatrix} \mid \mathbf{y}_1\right] - \boldsymbol{\theta} \\ &= \frac{1}{2}\binom{M}{2}^{-1} \sum_{(k,l)\in C_2^M} \left[\begin{pmatrix} (y_{1k}-\mu)(y_{1l}-\mu) \\ (y_{1k}-\mu)^2 \end{pmatrix} + \boldsymbol{\theta}\right] - \boldsymbol{\theta} \\ &= \frac{1}{2}[\mathbf{q}(\mathbf{y}_1,\mu) - \boldsymbol{\theta}]\end{aligned} \tag{5.36}$$

Thus, a consistent estimate of $\Sigma_\theta$ is given by

$$\widehat{\Sigma}_\theta = \frac{1}{n}\sum_{i=1}^{n}\left[\left(\mathbf{q}(\mathbf{y}_i,\widehat{\mu}) - \widehat{\boldsymbol{\theta}}\right)\left(\mathbf{q}(\mathbf{y}_i,\widehat{\mu}) - \widehat{\boldsymbol{\theta}}\right)^\top\right]$$

In the above, $\widehat{\mu}$ can be any consistent estimate of $\mu$. A consistent estimate of $\sigma_\rho^2$ is given by: $\widehat{\sigma}_\rho^2 = \frac{\partial}{\partial\boldsymbol{\theta}} g\left(\widehat{\boldsymbol{\theta}}\right)\widehat{\Sigma}_\theta\left(\frac{\partial}{\partial\boldsymbol{\theta}} g\left(\widehat{\boldsymbol{\theta}}\right)\right)^\top$.

As noted in Chapter 4, neither the product–moment nor the model based ICC takes into account raters' bias. For example, if ratings from two judges on six subjects in a hypothetical study are 3, 4, 5, 6, 7, 8 and 5, 6, 7, 8, 9, 10, the LMM in (5.34) does not apply because of the difference in the mean ratings between the two observers. Although the model based ICC derived from a revised LMM by replacing $\mu$ with $\mu_k$ in (5.34) can be applied to the data, it would yield an ICC equal to 1, providing a false or deceptive indication of perfect agreement. The product–moment based ICC has the same problem. A popular index that addresses the limitation of ICC is the concordance correlation coefficient.

**Example 10** (***Concordance correlation coefficient***). Within the context of Example 8, consider only two raters so that $M = 2$. Let

$$\begin{aligned}&\mu_1 = E(y_{i1}), \quad \mu_2 = E(y_{i2}), \quad \sigma_{12} = Cov(y_{i1}, y_{i2}) \\ &\sigma_1^2 = Var(y_{i1}), \quad \sigma_2^2 = Var(y_{i2})\end{aligned}$$

The concordance correlation coefficient (CCC) is defined as follows:

$$\rho_{ccc} = \frac{2\sigma_{12}}{\sigma_1^2 + \sigma_2^2 + (\mu_1 - \mu_2)^2} \tag{5.37}$$

Like the product–moment and LMM based ICC, $\rho_{ccc}$ ranges between $-1$ and 1 and thus can be interpreted as a correlation coefficient. Unlike these popular indices, however, $\rho_{ccc}$ is sensitive to differences in mean rating $\mu_k$ between the observers. For example, $\rho_{ccc} \neq 1$ for the hypothetical data discussed above. Thus, $\rho_{ccc}$ provides a sensible solution to address rater's bias. Note that CCC is closed related to the product–moment correlation;

$$\rho_{ccc} = \rho C$$
$$C = \frac{1}{2}\left[(\sigma_1/\sigma_2)^{-1} + (\sigma_1/\sigma_2) + \left(\frac{\mu_1 - \mu_2}{\sqrt{\sigma_1\sigma_2}}\right)^2\right]^{-1}$$

where $\rho$ is the product–moment correlation and $C$ is a function of $\sigma_1/\sigma_2$ (scale shift) and $\frac{\mu_1-\mu_2}{\sqrt{\sigma_1\sigma_2}}$ (location shift relative to scale).

Let $\boldsymbol{\theta} = (\theta_1, \theta_2)^\top$ with $\theta_k$ defined by

$$\theta_1 = 2\sigma_{12}, \quad \theta_2 = (\mu_1 - \mu_2)^2 + \sigma_1^2 + \sigma_2^2 - 2\sigma_{12} \tag{5.38}$$

Then, $\rho_{ccc} = f(\boldsymbol{\theta}) = \frac{\theta_1}{\theta_1+\theta_1}$. Let

$$\mathbf{h}(\mathbf{y}_i, \mathbf{y}_j) = \begin{pmatrix} h_1(\mathbf{y}_i, \mathbf{y}_j) \\ h_2(\mathbf{y}_i, \mathbf{y}_j) \end{pmatrix} = \begin{pmatrix} (y_{i1} - y_{j1})(y_{i2} - y_{j2}) \\ (y_{i1} - y_{i2})^2 + (y_{j1} - y_{j2})^2 \end{pmatrix} \tag{5.39}$$

Then, we have $E[\mathbf{h}(\mathbf{y}_i, \mathbf{y}_j)] = \boldsymbol{\theta}$ (see exercise). Thus, the U-statistic vector $\widehat{\boldsymbol{\theta}} = \binom{n}{2}^{-1} \sum_{(i,j)\in C_2^n} \mathbf{h}(\mathbf{y}_i, \mathbf{y}_j)$ is an unbiased and asymptotically normal estimate of $\boldsymbol{\theta}$ with the asymptotic variance given by: $\boldsymbol{\Sigma}_\theta = 4Var\left(\widetilde{\mathbf{h}}_1(\mathbf{y}_i)\right)$. By the Delta method, $\widehat{\rho}_{ccc} = f\left(\widehat{\boldsymbol{\theta}}\right)$ is also asymptotically normal with the asymptotic variance $\widehat{\rho}_{ccc} = \frac{\partial}{\partial\boldsymbol{\theta}} f(\boldsymbol{\theta})\, \Sigma_\theta \left(\frac{\partial}{\partial\boldsymbol{\theta}} f(\boldsymbol{\theta})\right)^\top$.

To find the asymptotic variance $\boldsymbol{\Sigma}_\theta$ of $\widehat{\boldsymbol{\theta}}$, note that (see exercise)

$$\begin{aligned} \widetilde{\mathbf{h}}_1(\mathbf{y}_i) &= E(\mathbf{h}(\mathbf{y}_i, \mathbf{y}_j) \mid \mathbf{y}_i) \\ &= \begin{pmatrix} y_{i1}y_{i2} - y_{i2}\mu_1 - y_{i1}\mu_2 + \sigma_{12} + \mu_1\mu_2 \\ \frac{1}{2}\left[(y_{i1} - y_{i2})^2 + \sigma_1^2 + \sigma_2^2 - 2\sigma_{12} + (\mu_1 - \mu_2)^2\right] \end{pmatrix} \\ &= \mathbf{q}(\mathbf{y}_i, \boldsymbol{\zeta}) \end{aligned} \tag{5.40}$$

where $\boldsymbol{\zeta} = \left(\mu_1, \mu_2, \sigma_1^2, \sigma_2^2, \sigma_{12}\right)^\top$. It follows that a consistent estimate of $\boldsymbol{\Sigma}_\theta$ is given by

$$\widehat{\Sigma}_{\boldsymbol{\theta}} = \frac{4}{n}\sum_{i=1}^{n}\left(\mathbf{q}\left(\mathbf{y}_i, \widehat{\boldsymbol{\zeta}}\right) - \widehat{\boldsymbol{\theta}}\right)\left(\mathbf{q}\left(\mathbf{y}_i, \widehat{\boldsymbol{\zeta}}\right) - \widehat{\boldsymbol{\theta}}\right)^\top$$

where $\mathbf{q}\left(\mathbf{y}_i, \widehat{\boldsymbol{\zeta}}\right)$ denotes $\mathbf{q}(\mathbf{y}_i, \boldsymbol{\zeta})$ in (5.40) with $\boldsymbol{\zeta}$ replaced by $\widehat{\boldsymbol{\zeta}}$. We can estimate $\boldsymbol{\zeta}$ by moment estimates for $\mu_k$, $\sigma_k^2$ and $\sigma_{12}$. From the above, we readily obtain a consistent estimate of the asymptotic variance of $\widehat{\rho}_{ccc}$.

In Chapter 4, we considered Cronbach alpha for assessing internal consistency of multiple items within a domain of an instrument or an instrument itself and discussed inference of this index under a multivariate normal assumption for the joint distribution of the items. However, as items in most instruments used in the behavioral and social sciences expect discrete, Likert scale responses, the normal distribution assumption is fundamentally flawed for such item scores. We describe next a distribution-free approach for inference of alpha.

**Example 11 (*Cronbach's coefficient alpha*).** Suppose that a domain of an instrument has $m$ number of individual items. Let $y_{ik}$ denote the response from the $i$th subject to the $k$th item ($1 \leq i \leq n$, $1 \leq k \leq m$). Let

$$\mu_k = E(y_{ik}), \quad \sigma_k^2 = Var(y_{ik}), \quad \sigma_{kl} = Cov(y_{ik}, y_{il})$$
$$\theta_1 = \sum_{k=1}^m \sigma_k^2, \quad \theta_2 = \sum_{k=1}^m \sigma_k^2 + \sum_{k \neq l} \sigma_{kl}$$
$$\mathbf{y}_i = (y_{i1}, \ldots, y_{im})^\top, \quad \boldsymbol{\mu} = (\mu_1, \ldots, \mu_m)^\top$$

Cronbach's coefficient alpha $\alpha$ is defined by:

$$\alpha = \frac{m}{m-1}\left(1 - \frac{\sum_{k=1}^m \sigma_k^2}{\sum_{k=1}^m \sigma_k^2 + \sum_{k \neq l} \sigma_{kl}}\right) = \frac{m}{m-1}\left(1 - \frac{\theta_1}{\theta_2}\right) = f(\boldsymbol{\theta})$$

We estimate $\boldsymbol{\theta}$ and then use the Delta method to make inference about $\alpha = f(\boldsymbol{\theta})$.

It is readily checked that (see exercise)

$$\theta_1 = \sum_{k=1}^m E(y_{ik} - \mu_k)^2 = E\left[(\mathbf{y}_i - \boldsymbol{\mu})^\top (\mathbf{y}_i - \boldsymbol{\mu})\right] \tag{5.41}$$
$$= E\left[\frac{1}{2}(\mathbf{y}_i - \mathbf{y}_j)^\top (\mathbf{y}_i - \mathbf{y}_j)\right]$$
$$\theta_2 = \mathbf{1}^\top E\left[(\mathbf{y}_i - \boldsymbol{\mu})(\mathbf{y}_i - \boldsymbol{\mu})^\top\right]\mathbf{1} = \mathbf{1}^\top E\left[\frac{1}{2}(\mathbf{y}_i - \mathbf{y}_j)(\mathbf{y}_i - \mathbf{y}_j)^\top\right]\mathbf{1}$$

Let

$$\mathbf{h}(\mathbf{y}_i, \mathbf{y}_j) = \frac{1}{2}\begin{pmatrix} (\mathbf{y}_i - \mathbf{y}_j)^\top (\mathbf{y}_i - \mathbf{y}_j) \\ \left[\mathbf{1}^\top (\mathbf{y}_i - \mathbf{y}_j)\right]\left[\mathbf{1}^\top (\mathbf{y}_i - \mathbf{y}_j)\right]^\top \end{pmatrix}$$

Then,

$$\widehat{\boldsymbol{\theta}} = \binom{n}{2}^{-1} \sum_{i,j \in C_2^n} \mathbf{h}(\mathbf{y}_i, \mathbf{y}_j) = \begin{pmatrix} \sum_{k=1}^m s_k^2 \\ \sum_{k=1}^m s_{ki}^2 + \sum_{k<l} s_{kl} \end{pmatrix}$$

$$\widehat{\alpha} = f\left(\widehat{\boldsymbol{\theta}}\right) = \frac{m}{m-1}\left(1 - \frac{\sum_{k=1}^m s_k^2}{\sum_{k=1}^m s_k^2 + \sum_{k \neq l} s_{kl}}\right)$$

where $s_k^2$ and $s_{kl}$ denote the sample variance and covariance of the elements of $\mathbf{y}_i$, respectively. By Theorem 1 and the Delta method, we have

$$\widehat{\alpha} \to_p \alpha, \quad \sqrt{n}\left(\widehat{\alpha} - \alpha\right) \to_d N\left(0, \sigma_\alpha^2 = \frac{\partial}{\partial \boldsymbol{\theta}} f(\boldsymbol{\theta}) \Sigma_\theta \left(\frac{\partial}{\partial \boldsymbol{\theta}} f(\boldsymbol{\theta})\right)^\top\right) \tag{5.42}$$

To find the asymptotic variance $\Sigma_\theta = 4Var\left(\widetilde{\mathbf{h}}_1(\mathbf{y}_1)\right)$, note that

$$\begin{aligned} \mathbf{h}_1(\mathbf{y}_1) &= \left(h_{11}(\mathbf{y}_1), h_{12}(\mathbf{y}_1)\right)^\top = E\left[\mathbf{h}(\mathbf{y}_1, \mathbf{y}_2) \mid \mathbf{y}_1\right] \\ &= E\left[\frac{1}{2}\begin{pmatrix} (\mathbf{y}_1 - \mathbf{y}_2)^\top (\mathbf{y}_1 - \mathbf{y}_2) \\ \left[\mathbf{1}^\top (\mathbf{y}_1 - \mathbf{y}_2)\right]\left[\mathbf{1}^\top (\mathbf{y}_1 - \mathbf{y}_2)\right]^\top \end{pmatrix} \mid \mathbf{y}_1\right] \\ &= \begin{pmatrix} E\left[\frac{1}{2}(\mathbf{y}_1 - \mathbf{y}_2)^\top (\mathbf{y}_1 - \mathbf{y}_2) \mid \mathbf{y}_1\right] \\ \mathbf{1}^\top E\left[\frac{1}{2}(\mathbf{y}_1 - \mathbf{y}_2)(\mathbf{y}_1 - \mathbf{y}_2)^\top \mid \mathbf{y}_1\right]\mathbf{1} \end{pmatrix} \end{aligned} \tag{5.43}$$

Further, it is also readily checked that (see exercise)

$$\begin{aligned} h_{11}(\mathbf{y}_1) &= E\left[\frac{1}{2}(\mathbf{y}_1 - \mathbf{y}_2)^\top (\mathbf{y}_1 - \mathbf{y}_2) \mid \mathbf{y}_1\right] \\ &= \frac{1}{2}\sum_{k=1}^m E\left[(y_{1k} - y_{2k})^2 \mid \mathbf{y}_1\right] = \frac{1}{2}\sum_{k=1}^m \left[(y_{1k} - \mu_k)^2 + \theta_1\right] \\ &= \frac{1}{2}\left[(\mathbf{y}_1 - \boldsymbol{\mu})^\top (\mathbf{y}_1 - \boldsymbol{\mu}) + \theta_1\right] \end{aligned} \tag{5.44}$$

Also, it follows from (5.18) in Example 4 that

$$\begin{aligned} h_{12}(\mathbf{y}_1) &= \mathbf{1}^\top E\left[\frac{1}{2}(\mathbf{y}_1 - \mathbf{y}_2)(\mathbf{y}_1 - \mathbf{y}_2)^\top \mid \mathbf{y}_1\right]\mathbf{1} \\ &= \frac{1}{2}\left[\mathbf{1}^\top \left[(\mathbf{y}_1 - \boldsymbol{\mu})(\mathbf{y}_1 - \boldsymbol{\mu})^\top\right]\mathbf{1} + \mathbf{1}^\top E\left[(\mathbf{y}_i - \boldsymbol{\mu})(\mathbf{y}_i - \boldsymbol{\mu})^\top\right]\mathbf{1}\right] \\ &= \frac{1}{2}\left[\mathbf{1}^\top \left[(\mathbf{y}_1 - \boldsymbol{\mu})(\mathbf{y}_1 - \boldsymbol{\mu})^\top\right]\mathbf{1} + \theta_2\right] \end{aligned} \tag{5.45}$$

By substituting (5.44) and (5.45) into (5.43), we have:

$$\mathbf{h}_1(\mathbf{y}_1) = \frac{1}{2}\left[\begin{pmatrix} (\mathbf{y}_1-\boldsymbol{\mu})^\top(\mathbf{y}_1-\boldsymbol{\mu}) \\ \mathbf{1}^\top(\mathbf{y}_1-\boldsymbol{\mu})(\mathbf{y}_1-\boldsymbol{\mu})^\top\mathbf{1} \end{pmatrix} + \boldsymbol{\theta}\right]$$

$$\widetilde{\mathbf{h}}_1(\mathbf{y}_1) = \mathbf{h}_1(\mathbf{y}_1) - \boldsymbol{\theta} = \frac{1}{2}\left[\mathbf{p}_1(\mathbf{y}_1,\boldsymbol{\mu}) - \boldsymbol{\theta}\right]$$

where

$$\mathbf{p}_i(\mathbf{y}_i,\boldsymbol{\mu}) = \begin{pmatrix} (\mathbf{y}_i-\boldsymbol{\mu})^\top(\mathbf{y}_i-\boldsymbol{\mu}) \\ \mathbf{1}^\top(\mathbf{y}_i-\boldsymbol{\mu})(\mathbf{y}_i-\boldsymbol{\mu})^\top\mathbf{1} \end{pmatrix}$$

A consistent estimate of the asymptotic variance $\Sigma_\theta$ is given by:

$$\widehat{\Sigma}_\theta = \frac{1}{n}\sum_{i=1}^{n}\left(\mathbf{p}_i(\mathbf{y}_i,\widehat{\boldsymbol{\mu}}) - \widehat{\boldsymbol{\theta}}\right)\left(\mathbf{p}_i(\mathbf{y}_i,\widehat{\boldsymbol{\mu}}) - \widehat{\boldsymbol{\theta}}\right)^\top$$

By substituting $\widehat{\Sigma}_\theta$ above and $\frac{\partial}{\partial\boldsymbol{\theta}} f\left(\widehat{\boldsymbol{\theta}}\right)$ in place of $\Sigma_\theta$ and $\frac{\partial}{\partial\boldsymbol{\theta}} f(\boldsymbol{\theta})$ in (5.42), we obtain a consistent estimate of the asymptotic variance of $\sigma_\alpha^2$.

### 5.1.2 General $K$ Sample Multivariate U-Statistics

The generalization of univariate two and $K$ sample U-statistics to a multivariate setting is carried out similar to that of the one-sample case. Consider $K$ i.i.d. samples of random vectors, $\mathbf{y}_{ki}$ $(1 \le i \le n_k,\ 1 \le k \le K)$. Let

$$\mathbf{h}_{11,\dots,1m_1;\dots;K1,\dots,Km_K} = \mathbf{h}(\mathbf{y}_{11},\dots,\mathbf{y}_{1m_1};\dots;\mathbf{y}_{K1},\dots,\mathbf{y}_{Km_K})$$

be some symmetric vector-valued function with respect to input arguments within each sample, that is, $\mathbf{h}$ is constant when $\mathbf{y}_{k1},\dots,\mathbf{y}_{km_k}$ are permuted within each $k$th sample $(1 \le k \le K)$. A $K$-sample $U$-statistic with $m_k$ arguments for the $k$th sample has the general form:

$$\mathbf{U}_n = \left[\prod_{k=1}^{K}\binom{n_k}{m_k}\right]^{-1}\sum_{k=1}^{K}\sum_{\left(i_1,\dots,i_{m_k}\right)\in C_{m_k}^{n_k}} \mathbf{h}_{1i_1,\dots,1i_{m_1};\dots;Ki_1,\dots,Ki_{m_K}} \tag{5.46}$$

As in the one-sample case, $\mathbf{U}_n$ is an unbiased estimate of $\boldsymbol{\theta} = E(\mathbf{h})$.

In Chapter 3, we discussed the importance of examining differences in variances when comparing different treatment groups. For example, if two treatments have the same mean response, but different variances, then the one with the smaller variance may be preferable in terms of treatment generalizability and cost effectiveness, both of which are of great importance in effectiveness research. Also, in applications of the analysis of variance (ANOVA) model to real study data, it is critically important to check variance homogeneity to ensure valid inference, since the F-test in ANOVA for inference about difference in group means are sensitive to this assumption of variance homogeneity. However, testing for differences in variance is beyond the mean based, distribution-free GLM discussed in Chapter 2. In Chapter 3, we discussed a U-statistic based distribution-free model for testing equality of variance between two groups. With the introduction of multivariate U-statistics, we can readily extend this approach to compare both the mean and variance simultaneously between two groups, as we consider in the next example.

**Example 12** (***Model for mean and variance between two groups***). Consider two i.i.d. samples $y_{ki}$ with mean $\mu_k$ and variance $\sigma_k^2$ $(1 \leq i \leq n_k,$ $1 \leq k \leq 2)$. Let

$$\mathbf{h}\left(y_{1i}, y_{1j}; y_{2l}, y_{2m}\right) = \begin{pmatrix} y_{1i} \\ \frac{1}{2}\left(y_{1i} - y_{2j}\right)^2 \end{pmatrix} - \begin{pmatrix} y_{2l} \\ \frac{1}{2}\left(y_{2l} - y_{2m}\right)^2 \end{pmatrix} \tag{5.47}$$

It follows that

$$\boldsymbol{\theta} = E\left[\mathbf{h}\left(y_{1i}, y_{1j}; y_{2l}, y_{2m}\right)\right] = \begin{pmatrix} \mu_1 - \mu_2 \\ \sigma_1^2 - \sigma_2^2 \end{pmatrix}$$

Thus, the two components of $\boldsymbol{\theta}$ represent the difference in the mean and variance between the two groups. If the null $H_0 : \boldsymbol{\theta} = \mathbf{0}$ holds true, then $y_{ki}$ have the same mean and same variance and vice versa.

Since $\mathbf{h}$ is not symmetric (e.g. $\mathbf{h}\left(y_{11}, y_{12}; y_{21}, y_{22}\right) \neq \mathbf{h}\left(y_{12}, y_{11}; y_{21}, y_{22}\right)$), we symmetrize it to obtain

$$\mathbf{g}\left(y_{1i}, y_{1j}; y_{2l}, y_{2m}\right) = \begin{pmatrix} \frac{1}{2}\left(y_{1i} + y_{1j}\right) \\ \frac{1}{2}\left(y_{1i} - y_{1j}\right)^2 \end{pmatrix} - \begin{pmatrix} \frac{1}{2}\left(y_{2l} + y_{2m}\right) \\ \frac{1}{2}\left(y_{2l} - y_{2m}\right)^2 \end{pmatrix}$$

For notational brevity, we denote $\mathbf{g}$ by $\mathbf{h}$. The U-statistic vector based on the above symmetric kernel is

$$\widehat{\boldsymbol{\theta}} = \left[\binom{n_1}{2}\binom{n_2}{2}\right]^{-1} \sum_{(i,j)\in C_2^{n_1}} \sum_{(l,m)\in C_2^{n_2}} E\left[\mathbf{h}\left(y_{1i}, y_{1j}; y_{2l}, y_{2m}\right)\right] \tag{5.48}$$

$$= \begin{pmatrix} \widehat{\mu}_1 - \widehat{\mu}_2 \\ \widehat{\sigma}_1^2 - \widehat{\sigma}_2^2 \end{pmatrix}$$

where $\widehat{\mu}_k = \overline{y}_{n_k} = \frac{1}{n_k}\sum_{i=1}^{n_k} y_{ki}$ and $\widehat{\sigma}_k^2 = s_{n_k}^2 = \frac{1}{n_k-1}\sum_{i=1}^{n_k}\left(y_{ki} - \overline{y}_{n_k}\right)^2$, the usual sample mean and variance of each group $k$ $(= 1, 2)$. Thus, the U-statistic vector in (5.47) generalizes the distribution-free ANOVA for comparing two groups as it can test the equality of both group means and variances simultaneously.

In Chapter 3, we discussed the two-sample Mann–Whitney–Wilcoxon rank sum test for detecting whether two CDFs $F_1$ and $F_2$ that differ by a location shift, $F_1(y) = F_2(y - \mu)$, are identical, that is, $H_0 : \mu = 0$ or $F_1(y) = F_2(y)$. By using multivariate $K$-sample U-statistics, we can generalize this classic test to comparing the equality of multiple CDFs if they differ by a location shift. We consider such an extension for the special case of three samples ($K = 3$) in the example below.

**Example 13 (*K-sample Mann–Whitney–Wilcoxon test*).** Let $y_{gi}$ denote three i.i.d. samples from three continuous CDFs that differ by a location shift, that is, $F_g(y) = F\left(y - \mu_g\right)$, for some $\mu_g \in R$ ($1 \le i \le n_g$, $1 \le g \le 3$). Let

$$\mathbf{h}\left(y_{1i}; y_{2j}; y_{3k}\right) = \begin{pmatrix} h_1\left(y_{1i}; y_{2j}\right) \\ h_2\left(y_{1i}; y_{3k}\right) \end{pmatrix} = \begin{pmatrix} I_{\{y_{1i}-y_{2j}\le 0\}} \\ I_{\{y_{1i}-y_{3k}\le 0\}} \end{pmatrix}$$

Consider the bivariate U-statistic,

$$\mathbf{U}_n = \left(n_1 n_2 n_3\right)^{-1} \sum_{i=1}^{n_1}\sum_{j=1}^{n_2}\sum_{k=1}^{n_3} \mathbf{h}\left(y_{1i}; y_{2j}; y_{3k}\right) \tag{5.49}$$

$$= \left(n_1 n_2 n_3\right)^{-1} \sum_{i=1}^{n_1}\sum_{j=1}^{n_2}\sum_{k=1}^{n_3} \begin{pmatrix} I_{\{y_{1i}-y_{2j}\le 0\}} \\ I_{\{y_{1i}-y_{3k}\le 0\}} \end{pmatrix}$$

The mean of $\mathbf{U}_n$ is $\boldsymbol{\theta} = E(\mathbf{h}) = \left(E\left(I_{\{y_{1i}-y_{2j}\le 0\}}\right), E\left(I_{\{y_{1i}-y_{3k}\le 0\}}\right)\right)^{\top}$. As discussed in Chapter 3 for comparing two such CDFs, the null of no location shift, $H_0 : \mu_g = 0$ $(1 \leq g \leq 3)$, corresponds to the identity:

$$\theta_1 = \theta_2 = E\left(I_{\{y_{1i}-y_{2j}\leq 0\}}\right) = E\left(I_{\{y_{1i}-y_{3k}\leq 0\}}\right) = \frac{1}{2}$$

Thus, in terms of the parameter vector $\boldsymbol{\theta}$ of the mean of the U-statistic $\mathbf{U}_n$, the null of no location shift is: $H_0 : \boldsymbol{\theta} = \left(\frac{1}{2}, \frac{1}{2}\right)^\top$. Thus, we can ascertain whether there is any location shift across three such CDFs by testing this null $H_0$ regarding the parameters of the U-statistic defined in (5.49).

The asymptotic distribution of the $K$-sample U-statistic vector in (5.46) is established in a similar fashion as in the one-sample case by utilizing the projection of the U-statistic and its asymptotic equivalence to the distribution of the U-statistic counterpart. More specifically, define the projection of $\mathbf{U}_n$ as follows:

$$\widehat{\mathbf{U}}_n = \sum_{k=1}^{K}\sum_{i=1}^{n_k} E\left(\mathbf{U}_n \mid \mathbf{y}_{ki}\right) - \boldsymbol{\theta}\left(\sum_{k=1}^{K} n_k - 1\right) \tag{5.50}$$
$$= \sum_{k=1}^{K}\sum_{i=1}^{n_k} \left[E\left(\mathbf{U}_n \mid \mathbf{y}_{ki}\right) - \boldsymbol{\theta}\right] + \boldsymbol{\theta}$$

where $\boldsymbol{\theta} = E\left(\mathbf{U}_n\right)$. Then, it is readily checked that the centered-projection is given by (see exercise):

$$\widehat{\mathbf{U}}_n - \boldsymbol{\theta} = \sum_{k=1}^{K} \frac{m_k}{n_k} \sum_{i=1}^{n_k} \left(\mathbf{h}_{k1}\left(\mathbf{y}_{ki}\right) - \boldsymbol{\theta}\right) = \sum_{k=1}^{K} \frac{m_k}{n_k} \sum_{i=1}^{n_k} \widetilde{\mathbf{h}}_{k1}\left(\mathbf{y}_{ki}\right) \tag{5.51}$$

where for $1 \leq k \leq K$,

$$\mathbf{h}_{k1}\left(\mathbf{y}_{k1}\right) = E\left[\mathbf{h}\left(\mathbf{y}_{11}, .., \mathbf{y}_{1m_1}; \ldots; \mathbf{y}_{K1}, .., \mathbf{y}_{Km_K}\right) \mid \mathbf{y}_{k1}\right] \tag{5.52}$$

As in the one-sample case, by applying CLT to (5.51), we immediately obtain the asymptotic distribution for the projection $\widehat{\mathbf{U}}_n$. The asymptotic distribution of the U-statistic vector $\mathbf{U}_n$ defined in (5.46) is then obtained by establishing the fact that $\widehat{\mathbf{U}}_n$ and $\mathbf{U}_n$ have the same asymptotic distribution. Such an asymptotic equivalence between $\widehat{\mathbf{U}}_n$ and $\mathbf{U}_n$ is proved by showing (as in the proof of Theorem 1) that the difference sequence, $\mathbf{e}_n = \sqrt{n}\left(\mathbf{U}_n - \widehat{\mathbf{U}}_n\right)$, converges to 0 in square mean, that is, $E\left(\mathbf{e}_n^\top \mathbf{e}_n\right) \to 0$ (see exercise). We summarize the asymptotic properties of the multivariate $K$-sample U-statistic in the following theorem.

**Theorem 2** (Multivariate $K$-sample). Let $n = \sum_{k=1}^{K} n_k$ and assume that $\lim_{n\to\infty} \frac{n}{n_k} = \rho_k^2 < \infty$ $(1 \leq k \leq K)$. Let

$$\mathbf{h}_{k1}(\mathbf{y}_{k1}) = E(\mathbf{h} \mid \mathbf{y}_{k1}), \quad \widetilde{\mathbf{h}}_{k1}(\mathbf{y}_{k1}) = \mathbf{h}_{k1}(\mathbf{y}_{k1}) - \boldsymbol{\theta}$$

$$\Sigma_{h_k} = Var\left[\widetilde{\mathbf{h}}_{k1}(\mathbf{y}_{k1})\right] = E\left[\widetilde{\mathbf{h}}_{k1}(\mathbf{y}_{k1})\right]^2, \quad 1 \leq k \leq K$$

Under mild regularity conditions, we have:

$$\mathbf{U}_n \to_p \boldsymbol{\theta}, \quad \sqrt{n}(\mathbf{U}_n - \boldsymbol{\theta}) \to_d N\left(\mathbf{0}, \Sigma_U = \sum_{k=1}^{K} \rho_k^2 m_k^2 \Sigma_{h_k}\right) \tag{5.53}$$

Note that as in the proof of Theorem 1 for the one-sample case, the asymptotic distribution in (5.53) is actually derived based on the projection of $\mathbf{U}_n$ by directly applying CLT to the sum of i.i.d. random vectors of the centered projection of $\mathbf{U}_n$ in (5.51) (see exercise). Also, as noted in the univariate $K$-sample U-statistics case, $n$ may be defined differently and the assumption $\lim_{n\to\infty} \frac{n}{n_k} = \rho_k^2 < \infty$ is to ensure that $n_k$ increase at similar rates so that $\frac{n_k}{n_l} \to c_{kl} \in (0, \infty)$ for any $n_l$ and $n_k$ as $n \to \infty$ $(1 \leq k, l \leq K)$. Further, the asymptotic distribution of $\mathbf{U}_n$ retains the same form as given in (5.53) regardless of how $n$ is defined.

**Example 14**. Consider again the U-statistic vector for simultaneously comparing the mean and variance between two independent samples discussed in Example 12. In this application, $K = 2$, $m_1 = m_2 = 2$. In addition, we have

$$\mathbf{h}_{11}(y_{11}) = E(\mathbf{h}(y_{11}, y_{12}, y_{21}, y_{22}) \mid y_{11}) = \begin{pmatrix} \frac{1}{2}y_{11} + \frac{1}{2}\mu_1 - \mu_2 \\ \frac{1}{2}(y_{11} - \mu_1)^2 + \frac{1}{2}\sigma_1^2 - \sigma_2^2 \end{pmatrix}$$

It follows that

$$\widetilde{\mathbf{h}}_{11}(y_{11}) = \mathbf{h}_{11}(y_{11}) - \boldsymbol{\theta} = \begin{pmatrix} \frac{1}{2}(y_{11} - \mu_1) \\ \frac{1}{2}\left[(y_{11} - \mu_1)^2 - \sigma_1^2\right] \end{pmatrix}$$

Let $\mu_{k4} = E(y_{k1} - \mu_k)^4$, the fourth-moment of $y_{ki}$. Since $E(y_{11} - \mu_1)^2 = \sigma_1^2$, it follows that

$$\Sigma_{h_1} = E\left[\widetilde{\mathbf{h}}_{11}(y_{11})\right]^2 = \frac{1}{4}\begin{pmatrix} \sigma_1^2 & Cov\left(y_{11}, (y_{11} - \mu_1)^2\right) \\ & \mu_{14} - \sigma_1^4 \end{pmatrix}$$

Similarly, we have

$$\Sigma_{h_2} = E\left[\widetilde{\mathbf{h}}_{21}(y_{21})\right]^2 = \frac{1}{4}\begin{pmatrix} \sigma_2^2 & Cov\left(y_{21}, (y_{21}-\mu_2)^2\right) \\ & \mu_{24}-\sigma_2^4 \end{pmatrix}$$

Thus, the asymptotic distribution of the estimate $\widehat{\boldsymbol{\theta}}$ in (5.48) is given by:

$$\sqrt{n}\left(\widehat{\boldsymbol{\theta}}-\boldsymbol{\theta}\right) \to_d N\left(\mathbf{0}, \Sigma_\theta = 4\left(\rho_1^2\Sigma_{h_1}+\rho_2^2\Sigma_{h_2}\right)\right) \tag{5.54}$$

Given a real data set, we can readily use the above expressions to estimate the asymptotic variance $\Sigma_{h_k}$ $(k=1,2)$. For example, we may estimate the parameters in $\Sigma_{h_k}$ by

$$\widehat{\sigma}_k^2 = \frac{1}{n_k}\sum_{i=1}^{n_k}(y_{ki}-\widehat{\mu}_k)^2, \quad \widehat{\mu}_{k4} = \frac{1}{n_k}\sum_{i=1}^{n_k}(y_{ki}-\widehat{\mu}_k)^4$$

$$\widehat{Cov}\left(y_{k1}, (y_{k1}-\mu_k)^2\right) = \frac{1}{n_k}\sum_{i=1}^{n_k}(y_{ki}-\widehat{\mu}_k)\left[(y_{ki}-\widehat{\mu}_k)^2-\widehat{\mu}_{k4}\right]$$

We estimate $\rho_k$ by $\widehat{\rho}_k = \frac{n_1+n_2}{n_k}$ $(k=1,2)$. Denote the estimates of $\Sigma_{h_k}$ by $\widehat{\Sigma}_{h_k}$. Then for large sample sizes $n_1$ and $n_2$, $\widehat{\boldsymbol{\theta}}$ follows approximately the following normal distribution:

$$\widehat{\boldsymbol{\theta}} \sim N\left(\boldsymbol{\theta}, \widehat{\Sigma}_\theta = \frac{4}{n}\left(\widehat{\rho}_1^2\widehat{\Sigma}_{h_1}+\widehat{\rho}_2^2\widehat{\Sigma}_{h_2}\right)\right) \tag{5.55}$$

As noted above, we may define $n$ differently such as $\widetilde{n} = \min(n_1, n_2)$. Then $\widehat{\rho}_k^2$ are estimated by the same expression, $\widehat{\rho}_k^2 = \frac{\widetilde{n}}{n_k}$, albeit a different value $\widetilde{n}$. By substituting the different definition of $n$ and $\widehat{\rho}_k^2$ into (5.55), we immediately obtain

$$\begin{aligned}\widehat{\Sigma}_\theta &= \frac{4}{\widetilde{n}}\left(\frac{\widetilde{n}}{n_1}\widehat{\Sigma}_{h_1}+\frac{\widetilde{n}}{n_1}\widehat{\Sigma}_{h_2}\right) = 4\left(\frac{1}{n_1}\widehat{\Sigma}_{h_1}+\frac{1}{n_1}\widehat{\Sigma}_{h_2}\right) \\ &= \frac{4}{n}\left(\frac{n}{n_1}\widehat{\Sigma}_{h_1}+\frac{n}{n_1}\widehat{\Sigma}_{h_2}\right)\end{aligned}$$

The above confirms the prior assertion that the asymptotic distribution of $\widehat{\boldsymbol{\theta}}$ is independent of how $n$ is defined.

In the above example, we estimated the asymptotic variance of $\widehat{\boldsymbol{\theta}}$ by first obtaining a closed form expression for each $\Sigma_{h_k}$. Alternatively, as in the one-sample case, we can construct a U-statistic to estimate each $\Sigma_{h_k}$ without attempting to evaluate these variance matrices. To illustrate, consider estimating $\Sigma_{h_1}$, which can be expressed as:

$$\begin{aligned}\Sigma_{h_1} &= E\left[E\left(\mathbf{h}\left(y_{11}, y_{12}, y_{21}, y_{22}\right) \mid y_{11}\right)\right]^2 - \boldsymbol{\theta}\boldsymbol{\theta}^\top \\ &= \boldsymbol{\Phi}_{h_1} - \boldsymbol{\theta}\boldsymbol{\theta}^\top\end{aligned}$$

We can estimate $\boldsymbol{\theta}\boldsymbol{\theta}^\top$ by $\widehat{\boldsymbol{\theta}}\widehat{\boldsymbol{\theta}}^\top$. To estimate $\boldsymbol{\Phi}_{h_1}$, note that:

$$\begin{aligned}\boldsymbol{\Phi}_{h_1} &= E\left[E\left(\mathbf{h}\left(y_{11}, y_{12}, y_{21}, y_{22}\right) \mid y_{11}\right) E\left(\mathbf{h}\left(y_{11}, y_{13}, y_{23}, y_{24}\right) \mid y_{11}\right)\right] \\ &= E\left\{E\left[\mathbf{h}\left(y_{11}, y_{12}, y_{21}, y_{22}\right) \mathbf{h}\left(y_{11}, y_{13}, y_{23}, y_{24}\right) \mid y_{11}\right]\right\} \\ &= E\left[\mathbf{h}\left(y_{11}, y_{12}, y_{21}, y_{22}\right) \mathbf{h}\left(y_{11}, y_{13}, y_{23}, y_{24}\right)\right]\end{aligned}$$

Let $\mathbf{g}_{123;1234} = \mathbf{g}\left(y_{11}, y_{12}, y_{13}, y_{21}, y_{22}, y_{23}, y_{24}\right)$ be a symmetric kernel of

$$\mathbf{h}\left(y_{11}, y_{12}, y_{21}, y_{22}\right) \mathbf{h}\left(y_{11}, y_{13}, y_{23}, y_{24}\right) \tag{5.56}$$

Then, the U-statistic matrix,

$$\widehat{\boldsymbol{\Phi}}_{h_1} = \left[\binom{n_1}{3}\binom{n_2}{4}\right]^{-1} \sum_{(i_1,i_2,i_3)\in C_3^{n_1}} \sum_{(j_1,j_2,j_3,j_4)\in C_4^{n_2}} \mathbf{g}_{i_1 i_2 i_3; j_1 j_2 j_3 j_4}$$

is a consistent estimate of $\boldsymbol{\Phi}_{h_1}$. This approach is especially effective when the asymptotic variance cannot be evaluated analytically.

## 5.2 Models for Longitudinal Study Designs

The longitudinal models discussed in Chapter 4 are mostly generalizations of the models for cross-sectional study designs developed in Chapter 2 and are primarily used for modeling the conditional mean of a single response given a set of independent variables. As seen in Chapter 3, many important statistics, such as the classic two-sample Mann–Whitney–Wilcoxon rank sum test, cannot be framed as regression models and are best studied by the theory of U-statistics. Thus, in parallel to the development in Chapter 4, we consider generalizing these U-statistics based models to a longitudinal data setting in this section.

In Section 5.1, we discussed models that require multivariate U-statistics because of the number of parameters involved in defining and estimating the quantities of interest. In recent years, longitudinal studies have not only become more popular in biomedical and psychosocial research but also more complex in their designs. In such applications, we have multiple outcomes from the same subject due to repeated assessments. Interest for such longitudinal studies usually centers around patterns of response over time. Thus, even for parameters defined by a single response, studying the change patterns of such parameters over time requires considering multiple responses collected together at various assessment times, giving rise to multivariate outcomes. However, because of the opportunity for missing responses, the methods discussed in Section 5.1 no longer apply to data from such study designs. One approach is to simply use subjects with no missing response at any of the assessment points, or *complete data*. However, such an approach at best does not utilize all available data and thereby lacks efficiency in inference of model parameters. Further, as we discussed for regression analysis in Chapter 4, such a complete-data approach generally gives rise to biased estimates if missing data follows the missing at random (MAR) model. In this Section, we generalize the results in the preceding section to address missing data. As in Chapter 4, we focus on the missing completely at random (MCAR) and MAR mechanisms.

### 5.2.1 Inference in the Absence of Missing Data

The multivariate U-statistics discussed in Section 5.1 arise from the need to construct multidimensional statistic vectors to estimate the parameters of interest. Thus, even for cross-sectional study designs, such problems require multivariate U-statistics. Another common type of problem that requires multivariate statistics is analysis of longitudinal study data. As discussed in Chapter 4, longitudinal study designs offer many benefits and address the fundamental limitation of cross-sectional study — the inability to provide causal inference. On the other hand, longitudinal study designs yield multiple correlated responses, which turn otherwise simple univariate statistical analyses into complex models involving multivariate distributions. To differentiate the two types of multivariate statistics, we refer to the multivariate U-statistics that are required to express the parameters of interest as *endogenous* while those resulting from longitudinal study designs as *exogenous* type. Thus, although the theory of multivariate U-statistics can be applied to *exogenous* type of multivariate problems, we must address the inherent missing data issue. In this section, we assume no missing data and illustrate how the theory of multivariate U-statistics developed for *endogenous* problems can be applied to address *exogenous* multivariate outcomes.

**Example 1** (***Model for variance over time***). Consider a longitudinal study with $n$ subjects and $m$ assessment times. Let $y_{it}$ denote the response from the $i$th subject at time $t$ ($1 \leq i \leq n$, $1 \leq t \leq m$). We are interested in modeling the variance of the response $y_{it}$ over time. Let

$$\sigma_t^2 = Var(y_{it}), \quad \mathbf{y}_i = (y_{i1}, y_{i2}, \ldots, y_{im})^\top, \quad \boldsymbol{\theta} = \left(\sigma_1^2, \sigma_2^2, \ldots, \sigma_m^2\right)^\top$$

A primary area of interest in longitudinal studies is trend over time. Within the current context, we are interested in testing variance homogeneity over time, that is

$$\begin{aligned} H_0 &: \sigma_t^2 = \sigma^2, \quad \text{for all } 1 \le t \le m \quad \text{versus} \\ H_a &: \sigma_t^2 \ne \sigma_s^2, \quad \text{for some } 1 \le s < t \le m \end{aligned}$$

In this example, the quantity of interest at each assessment time is $\sigma_t^2$ and the associated univariate U-statistic based estimate has been discussed in Chapter 3. However, testing the above hypothesis requires jointly modeling all these univariate statistics simultaneously.

To this end, consider the following multivariate U-statistic vector:

$$\widehat{\boldsymbol{\theta}} = \binom{n}{2}^{-1} \sum_{(i,j) \in C_2^n} \mathbf{h}(\mathbf{y}_i, \mathbf{y}_j) = \binom{n}{2}^{-1} \sum_{(i,j) \in C_2^n} \frac{1}{2} \begin{pmatrix} (y_{i1} - y_{j1})^2 \\ \vdots \\ (y_{im} - y_{jm})^2 \end{pmatrix} \tag{5.57}$$

Since $E\left(\widehat{\boldsymbol{\theta}}\right) = E\left[\mathbf{h}(\mathbf{y}_i, \mathbf{y}_j)\right] = \boldsymbol{\theta}$, $\widehat{\boldsymbol{\theta}}$ is an unbiased estimate of $\boldsymbol{\theta}$. The asymptotic distribution of $\widehat{\boldsymbol{\theta}}$ is readily obtained using the development in Section 5.1. In particular, the asymptotic variance is given by (see exercise):

$$\Sigma_\theta = Var(\mathbf{p}_1(\mathbf{y}_1)), \quad \mathbf{p}_i(\mathbf{y}_i) = \left((y_{i1} - \mu_1)^2, \ldots, (y_{im} - \mu_m)^2\right)^\top \tag{5.58}$$

With such a joint distribution, we can readily test the hypothesis of variance homogeneity using linear contrasts.

**Example 2** (***Mann–Whitney–Wilcoxon for longitudinal data***). Suppose that we have two groups of subjects who are followed over time in a longitudinal study with $m$ assessment times. Let $y_{kit}$ denote the response from the $i$th subject within the $k$th group at time $t$ ($1 \le i \le n_k$, $k = 1, 2$, $1 \le t \le m$). At each time $t$, we can compare the two groups using the Mann–Whitney–Wilcoxon rank sum test as discussed in Chapter 3. In this example, we generalize this classic test for no location shift between the two CDFs of $y_{kit}$ at a single time to a longitudinal data setting involving multiple assessment times.

For each time $t$, let $\theta_t = \Pr[y_{1it} \leq y_{2it}]$ $(1 \leq t \leq m)$. Then, it follows from the discussion of the two-sample Mann–Whitney–Wilcoxon test in Chapter 3 that the hypothesis of no location shift can be expressed as:

$$H_0 : \theta_t = \frac{1}{2}, \quad \text{for all } 1 \leq t \leq m \quad \text{versus} \tag{5.59}$$
$$H_a : \theta_t \neq \frac{1}{2}, \quad \text{for some } 1 \leq t \leq m$$

As in Example 1, the quantity of interest is univariate, but the longitudinal study design requires that we model $\theta_t$ simultaneously over all $t$.

Let

$$\mathbf{y}_{ki} = (y_{ki1}, y_{ki2}, \ldots, y_{kim})^\top, \quad 1 \leq i \leq n_k, \quad k = 1, 2$$
$$\mathbf{h}(\mathbf{y}_{1i}, \mathbf{y}_{2j}) = \left(I_{\{y_{1i1} \leq y_{2j1}\}}, \ldots, I_{\{y_{1im} \leq y_{2jm}\}}\right)^\top, \quad \boldsymbol{\theta} = E[\mathbf{h}(\mathbf{y}_{1i}, \mathbf{y}_{2j})]$$

Then, the $U$-statistic vector,

$$\widehat{\boldsymbol{\theta}} = \binom{n_1}{2}^{-1} \binom{n_2}{2}^{-1} \sum_{i=1}^{n_1} \sum_{j=1}^{n_2} \mathbf{h}(\mathbf{y}_{1i}, \mathbf{y}_{2j}) \tag{5.60}$$

is an unbiased estimate of $\boldsymbol{\theta}$. The asymptotic distribution of $\widehat{\boldsymbol{\theta}}$ is readily computed using the theory of multivariate U-statistics in Section 5.1 to provide inference for testing the hypothesis in (5.59) (see exercise).

One major limitation of the product–moment correlation between two continuous variables $x$ and $y$ is the assumption of a linear relationship between $x$ and $y$. In Chapter 3, we discussed some popular alternatives that require no such assumption. One such alternative is Kendall's $\tau$, which assumes no functional relationship between the variables. In Chapter 3, we discussed inference for $\tau$ using the theory of univariate U-statistics. In Example 3, we consider extending this approach to data from a longitudinal study design.

**Example 3 (*Kendall's $\tau$ for continuous outcome*).** Consider a longitudinal study with $n$ subjects and $m$ assessment times. Let $\mathbf{z}_{it} = (x_{it}, y_{it})^\top$ be an i.i.d. sample from a continuous bivariate distribution $(1 \leq t \leq m$, $1 \leq i \leq n)$. As discussed in Chapter 3, for each time $t$, we can develop a univariate U-statistic based on the kernel, $h(\mathbf{z}_{it}, \mathbf{z}_{jt}) = I_{\{(x_{it}-x_{jt})(y_{it}-y_{jt})>0\}}$, to test whether there is association between $x_i$ and $y_i$.

To generalize this univariate test to the current longitudinal data setting,

let

$$h_t(\mathbf{z}_{it}, \mathbf{z}_{jt}) = I_{\{(x_{it}-x_{jt})(y_{it}-y_{jt})>0\}}, \quad \mathbf{z}_i = \left(\mathbf{z}_{i1}^\top, \ldots, \mathbf{z}_{im}^\top\right)^\top \tag{5.61}$$
$$\mathbf{h}(\mathbf{z}_i, \mathbf{z}_j) = (h_1, \ldots, h_m)^\top, \quad 1 \le i < j \le n$$

Then, $\boldsymbol{\tau} = \mathbf{E}[\mathbf{h}(\mathbf{z}_i, \mathbf{z}_j)]$ is the vector containing Kendall's $\tau$s from all the $m$ assessments. The U-statistic vector, $\widehat{\boldsymbol{\tau}} = \binom{n}{2}^{-1} \sum_{i=1}^{n} \mathbf{h}(\mathbf{z}_i, \mathbf{z}_j)$, is an unbiased estimate of $\boldsymbol{\tau}$. The asymptotic distribution again is readily derived using the theory in Section 5.1 and can be used to test the null of no association between $x_{it}$ and $y_{it}$ over the study period, $H_0 : \boldsymbol{\tau} = \mathbf{0}$ (see exercise). If this *overall* test is rejected, we may apply the procedure to a subvector of $\boldsymbol{\tau}$ to carry out follow-up tests to determine the assessment times over which $x_{it}$ and $y_{it}$ are associated. In addition, we may want to perform follow-up tests to compare magnitudes of association across different assessment times. For example, we can test whether there is any change in association between times 1 and 2 by testing the linear contrast, $H_0 : \tau_1 = \tau_2$.

In Example 3 of Section 5.1.1, we also discussed inference for Kendall's $\tau_b$, a version of $\tau$ for association of between two ordinal outcomes. The example below extends this index to a longitudinal data setting.

**Example 4 (*Kendall's $\tau_b$ for ordinal outcome*).** In Example 3, let $\mathbf{z}_{it} = (u_{it}, v_{it})^\top$ and $\mathbf{z}_i = \left(\mathbf{z}_{i1}^\top, \ldots, \mathbf{z}_{im}^\top\right)^\top$ be an i.i.d. sample of bivariate ordinal outcomes. Assume that $u$ has $K$ and $v$ has $M$ levels indexed by $k$ and $m$, respectively.

To generalize the kernel function for $\tau_b$ developed in Section 5.1.1 to the current longitudinal context, let

$$h_{1t}(\mathbf{z}_{it}, \mathbf{z}_{jt}) = I_{\{(u_{it}-u_{jt})(v_{it}-v_{jt})>0\}} - I_{\{(u_{it}-u_{jt})(v_{it}-v_{jt})<0\}}$$
$$h_{2t}(\mathbf{z}_{it}, \mathbf{z}_{jt}) = I_{\{u_{it}=u_{jt}\}}, \quad h_{3t}(\mathbf{z}_{it}, \mathbf{z}_{jt}) = I_{\{v_{it}=v_{jt}\}}$$
$$\mathbf{h}_t(\mathbf{z}_{it}, \mathbf{z}_{jt}) = (h_{1t}, h_{2t}, h_{3t})^\top, \quad \mathbf{h}(\mathbf{z}_i, \mathbf{z}_j) = \left(\mathbf{h}_1^\top, \ldots, \mathbf{h}_m^\top\right)^\top$$

Then, the U-statistic vector:

$$\widehat{\boldsymbol{\theta}} = \binom{n}{2}^{-1} \sum_{(i,j)\in C_2^n} \mathbf{h}(\mathbf{z}_i, \mathbf{z}_j) = \binom{n}{2}^{-1} \sum_{(i,j)\in C_2^n} \begin{pmatrix} \mathbf{h}_1(\mathbf{z}_{i1}, \mathbf{z}_{j1}) \\ \vdots \\ \mathbf{h}_m(\mathbf{z}_{im}, \mathbf{z}_{jm}) \end{pmatrix} \tag{5.62}$$

is an unbiased estimate of $\boldsymbol{\theta} = E(\mathbf{h}(\mathbf{z}_i, \mathbf{z}_j))$. We can readily determine the asymptotic distribution of $\widehat{\boldsymbol{\theta}}$ by applying Theorem 2 of Section 5.1.

Now let

$$\tau_{bt} = f_t(\boldsymbol{\theta}_t) = \frac{\theta_{1t}}{\sqrt{(1-\theta_{2t})(1-\theta_{3t})}}, \quad \boldsymbol{\tau}_b = (\tau_{b1}, \ldots, \tau_{bm})^\top \tag{5.63}$$
$$\mathbf{f} = (f_1, \ldots, f_m)^\top$$

Then, $\widehat{\boldsymbol{\tau}}_b = \mathbf{f}\left(\widehat{\boldsymbol{\theta}}\right)$ is a consistent and asymptotic normal estimate of $\boldsymbol{\tau}_b$. The asymptotic variance of $\widehat{\boldsymbol{\tau}}_b$ is readily obtained from the Delta method (see exercise). As in Example 4, we can use this asymptotic distribution to test hypotheses concerning $\boldsymbol{\tau}_b$ or a subset of $\boldsymbol{\tau}_b$.

### 5.2.2 Inference Under MCAR

In all the above examples, the multivariate $U$-statistics are required because of the repeated assessments in longitudinal study designs. In such studies, missing data is a common and persistent problem, and as a result, the approaches discussed can only be applied to subjects with complete data. However, under *random missingness*, such an approach still provides valid inference. As discussed in Chapter 4, a formal definition of random missingness is the missing completely at random (MCAR) assumption. Under MCAR, the occurrence of missing response is unrelated to or independent of the response and thus estimates based on the observed data are still consistent. Although such a complete-data approach still provides valid inference, it may be quite inefficient. For example, if the occurrence of missing data is random over time and across subjects, the number of subjects with complete data may be substantially smaller than the original sample. As a result, such a strategy of removing subjects with missing data may seriously affect power. An alternative and indeed a more efficient way is to use all observed data.

As discussed in Chapter 4, missing data arising in many longitudinal trials is the result of treatment response patterns and thus is often predicted by observed responses. Such a *response-dependent* dropout process or missing at random (MAR) mechanism invalidates the MCAR assumption. Consequently, estimates based on the complete-data subsample become biased or inconsistent. As pointed out in Chapter 4, such a dependence of occurrence of missing data on the observed response must be specifically modeled and explicitly accounted for when making inferences for mean based distribution-free generalized linear and related regression models. This MAR mechanism or response-dependent dropout process exerts a similar effect on U-statistics based distribution-free models discussed in the above sections and therefore must be carefully addressed. In this section, we first generalize the missing data techniques employed for the distribution-free regression models in Chapter 4 to develop approaches for inference for U-statistics based models under the relatively stronger, but more easily addressed MCAR assumption.

**Example 5** (***Model for variance over time***). Consider the model for variance in Example 1 of Section 5.2.1. Under the complete-data approach, only subjects with observed $y_{it}$ over all assessment times $t$ contribute to inference ($1 \leq t \leq m$). Let $n_{obs}$ denote the number of such subjects. In this case, we estimate $\boldsymbol{\theta}$ in Example 1 by applying the procedure to only these subjects. Although unbiased, such estimates are based on an effective sample size, $n_{obs}$. If missing data occurs sporadically over time and across subjects, $n_{obs}$ may severely undercut the original sample size $n$.

Now consider the more efficient approach that uses all observed data. For each time $t$, we can estimate $\sigma_t^2$ based on all available responses. Although it is not difficult to compute such estimates, it is notationally complex to express the asymptotic distribution of these joint estimates. For example, consider only two time points, $s$ and $t$ ($1 \leq s < t \leq m$). Let $n_s$ and $n_t$ denote the sample size at each of the times and let $\{i_1, \ldots, i_{n_s}\}$ and $\{j_1, \ldots, j_{n_t}\}$ denote the subsets of the indices for the subjects with observed responses at times $s$ and $t$, respectively. Then, we can express the two estimates of $\sigma_s^2$ and $\sigma_t^2$ as follows:

$$\widehat{\sigma}_s^2 = \binom{n_s}{2}^{-1} \sum_{(k,l)\in C_2^{n_s}} \frac{1}{2}\left(y_{ks} - y_{ls}\right)^2 \tag{5.64}$$

$$\widehat{\sigma}_t^2 = \binom{n_t}{2}^{-1} \sum_{(k,l)\in C_2^{n_t}} \frac{1}{2}\left(y_{kt} - y_{lt}\right)^2, \quad 1 \leq s, t \leq m$$

The difficulty lies in comparing $\sigma_s^2$ and $\sigma_t^2$, since $\widehat{\sigma}_s^2$ and $\widehat{\sigma}_t^2$ are correlated and the joint asymptotic distribution is needed to make inference concerning these parameters. It is at best cumbersome to derive this joint distribution directly using the expressions in (5.64). The major problem is tracking the indices for the subjects with observed data at each of the two times points when deriving the asymptotic distribution. This problem becomes more challenging when considering more time points.

We now describe an alternative approach based on a similar development for addressing missing data for the distribution-free generalized linear models in Chapter 4. This approach allows us to effectively deal with the notational complexity discussed above and model the effect of missing data in a systematic fashion. This approach employs missing data indicators to track the occurrence of missing response and defines estimates as a function of such missing data indicators, thus alleviating the algebraic complexity in expressing observed responses using subsets of subject and time indices. In the following example, we describe how to adapt this approach for U-statistics-based models.

**Example 6.** Consider again the model for variance in Example 1. Rather than employing subject and time indices to express the observed data, we can do so alternatively using binary indicators. For each subject $i$, define a set of missing (or rather, observed) data indicators as follows:

$$r_{it} = \begin{cases} 1 \text{ if } y_{it} \text{ is observed} \\ 0 \text{ if } y_{it} \text{ is missing} \end{cases}, \quad \mathbf{r}_i = (r_{i1}, ..., r_{im})^\top \tag{5.65}$$

$$R_i = \operatorname{diag}(\mathbf{r}_i) = \begin{pmatrix} r_{i1} & \cdots & 0 \\ \vdots & \ddots & \vdots \\ 0 & \cdots & r_{im} \end{pmatrix}, \quad 1 \le i \le n$$

By using $\mathbf{r}_i$, we can readily revise the U-statistic estimate $\widehat{\boldsymbol{\theta}}$ defined in (5.57) for this problem to reflect its dependence on observed data. For example, consider the $t$th component $h_t(y_{it}, y_{jt})$ of the kernel $\mathbf{h}(\mathbf{y}_i, \mathbf{y}_j)$ for this U-statistic estimate $\widehat{\boldsymbol{\theta}}$ in (5.57). Although $h_t(y_{it}, y_{jt})$ is not defined when one or both of $y_{it}$ and $y_{jt}$ is missing, the following revised kernel $g_t(y_{it}, r_{it}, y_{jt}, r_{jt})$ is well defined

$$h_t(y_{it}, y_{jt}) = \frac{1}{2}(y_{it} - y_{jt})^2 \tag{5.66}$$

$$g_t(y_{it}, r_{it}, y_{jt}, r_{jt}) = r_{it} r_{jt} h_t(y_{it}, y_{jt}) = \frac{1}{2} r_{it} r_{jt} (y_{it} - y_{jt})^2$$

Unlike $h_t(y_{it}, y_{jt})$, $g_t(y_{it}, r_{it}, y_{jt}, r_{jt})$ is also a function of $r_{it}$ and $r_{jt}$. It is readily checked that the estimate $\widehat{\sigma}_t^2$ in (5.64) can be expressed in terms of this kernel function as follows (see exercise):

$$\widehat{\sigma}_t^2 = \frac{\binom{n}{2}^{-1} \sum_{(i,j)\in C_2^n} g_t(y_{it}, r_{it}, y_{jt}, r_{jt})}{\binom{n}{2}^{-1} \sum_{(i,j)\in C_2^n} r_{it} r_{jt}} \tag{5.67}$$

However, unlike (5.64), the above does not involve the use of a subset of indices of subject and time for tracking observed data, making it particularly easy to derive the joint asymptotic distribution of $\widehat{\sigma}_t^2$ as we develop below.

Let $\mathbf{z}_i = (\mathbf{y}_i, \mathbf{r}_i)^\top$ and $\boldsymbol{\theta} = \left(\sigma_1^2, \sigma_2^2, \ldots, \sigma_m^2\right)^\top$. By expressing (5.66) in the form of a vector, we have

$$\begin{aligned}
\mathbf{g}(\mathbf{z}_i, \mathbf{z}_j) &= R_i R_j \mathbf{h}(\mathbf{y}_i, \mathbf{y}_j) \\
&= R_i R_j \frac{1}{2} \left( (y_{i1} - y_{j1})^2, \ldots, (y_{im} - y_{jm})^2 \right)^\top \\
&= \left( r_{i1} r_{j1} \frac{1}{2} (y_{i1} - y_{j1})^2, \ldots, r_{im} r_{jm} \frac{1}{2} (y_{im} - y_{jm})^2 \right)^\top
\end{aligned}$$

It is readily checked that $\mathbf{g}(\mathbf{z}_i, \mathbf{z}_j)$ is a symmetric function with respect to $\mathbf{z}_i$ and $\mathbf{z}_j$. Thus, the U-statistic

$$\mathbf{U}_n = \binom{n}{2}^{-1} \sum_{(i,j) \in C_2^n} \mathbf{g}(\mathbf{z}_i, \mathbf{z}_j) = \binom{n}{2}^{-1} \sum_{(i,j) \in C_2^n} R_i R_j \mathbf{h}(\mathbf{y}_i, \mathbf{y}_j)$$

is an unbiased estimate of $E(\mathbf{g}(\mathbf{z}_i, \mathbf{z}_j))$. Under MCAR, $R_i$ and $\mathbf{y}_i$ are independent. It follows that

$$E(\mathbf{g}(\mathbf{z}_i, \mathbf{z}_j)) = E^2(R_i) \left(\sigma_1^2, \sigma_2^2, \ldots, \sigma_m^2\right)^\top = E^2(R_i) \boldsymbol{\theta}$$

where $E^2(R_i) = [E(R_i)]^2$. If $E(R_i) = \mathbf{I}_m$, that is, there is no missing data, $E(\mathbf{g}(\mathbf{z}_i, \mathbf{z}_j)) = \boldsymbol{\theta}$ and $\mathbf{U}_n$ is an unbiased estimate of $\boldsymbol{\theta}$. Otherwise, missing data occurs with a positive probability and $\mathbf{U}_n$ is not an estimate of $\boldsymbol{\theta}$. However, we can construct a consistent estimate of $\boldsymbol{\theta}$ by generalizing the univariate estimate $\widehat{\sigma}_t^2$ in (5.67).

Let

$$G = \begin{pmatrix} n_1 & \cdots & 0 \\ \vdots & \ddots & \vdots \\ 0 & \cdots & n_m \end{pmatrix} = \sum_{(i,j) \in C_2^n} R_i R_j = \sum_{(i,j) \in C_2^n} \begin{pmatrix} r_{i1} r_{j1} & \cdots & 0 \\ \vdots & \ddots & \vdots \\ 0 & \cdots & r_{im} r_{jm} \end{pmatrix}$$

Now define an estimate of $\boldsymbol{\theta}$ by

$$\widehat{\boldsymbol{\theta}} = G^{-1} \sum_{(i,j) \in C_2^n} \mathbf{g}(\mathbf{z}_i, \mathbf{z}_j) \tag{5.68}$$

It is readily checked that (see exercise)

$$\begin{aligned}
\widehat{\theta}_t &= \frac{1}{\sum_{(i,j)\in C_2^n} r_{it}r_{jt}} \sum_{(i,j)\in C_2^n} r_{it}r_{jt}\frac{1}{2}\left(y_{it}-y_{jt}\right)^2 \qquad (5.69)\\
&= \frac{1}{\binom{n}{2}^{-1}\sum_{(i,j)\in C_2^n} r_{it}r_{jt}} \binom{n}{2}^{-1} \sum_{(i,j)\in C_2^n} \frac{1}{2}r_{it}r_{jt}\left(y_{it}-y_{jt}\right)^2\\
&\to_p \frac{E\left[\frac{1}{2}r_{it}r_{jt}\left(y_{it}-y_{jt}\right)^2\right]}{E^2\left(r_{it}\right)} = \frac{E^2\left(r_{it}\right)\sigma_t^2}{E^2\left(r_{it}\right)}\\
&= \sigma_t^2
\end{aligned}$$

where the second to the last equality follows from the independence between $R_i$ and $\mathbf{y}_i$ under the MCAR assumption. Thus, $\widehat{\boldsymbol{\theta}}$ in (5.67) is a consistent estimate of $\boldsymbol{\theta}$.

Since $\boldsymbol{\theta}$ can be expressed as $\boldsymbol{\theta} = G^{-1}\sum_{(i,j)\in C_2^n} R_iR_j\boldsymbol{\theta}$, it follows that (see exercise)

$$\begin{aligned}
\sqrt{n}\left(\widehat{\boldsymbol{\theta}}-\boldsymbol{\theta}\right) &= \sqrt{n}\left[G^{-1}\sum_{(i,j)\in C_2^n}\mathbf{g}\left(\mathbf{z}_i,\mathbf{z}_j\right) - G^{-1}\sum_{(i,j)\in C_2^n} R_iR_j\boldsymbol{\theta}\right] \qquad (5.70)\\
&= \sqrt{n}G^{-1}\sum_{(i,j)\in C_2^n} R_iR_j\left[\mathbf{h}\left(\mathbf{y}_i,\mathbf{y}_j\right)-\boldsymbol{\theta}\right]\\
&= \binom{n}{2}G^{-1}\sqrt{n}\binom{n}{2}^{-1}\sum_{(i,j)\in C_2^n} R_iR_j\left[\mathbf{h}\left(\mathbf{y}_i,\mathbf{y}_j\right)-\boldsymbol{\theta}\right]\\
&= \binom{n}{2}G^{-1}\sqrt{n}\binom{n}{2}^{-1}\sum_{(i,j)\in C_2^n}\mathbf{v}\left(\mathbf{z}_i,\mathbf{z}_j\right)\\
&= \binom{n}{2}G^{-1}\sqrt{n}\mathbf{V}_n
\end{aligned}$$

where $\mathbf{v}\left(\mathbf{z}_i,\mathbf{z}_j\right) = R_iR_j\left[\mathbf{h}\left(\mathbf{y}_i,\mathbf{y}_j\right)-\boldsymbol{\theta}\right]$ and $\mathbf{V}_n = \binom{n}{2}^{-1}\sum_{(i,j)\in C_2^n}\mathbf{v}\left(\mathbf{z}_i,\mathbf{z}_j\right)$. As $\mathbf{v}\left(\mathbf{z}_i,\mathbf{z}_j\right)$ is symmetric, $\mathbf{V}_n$ is a U-statistic-like random vector, that is, $\mathbf{V}_n$ is a U-statistic if $\boldsymbol{\theta}$ is known. By Theorem 1 of Section 5.1, we have:

$$\binom{n}{2}G^{-1} \to_p E^{-2}\left(R_1\right), \quad \sqrt{n}\mathbf{V}_n \to_d N\left(\mathbf{0},\Sigma_V = 4Var\left(\widetilde{\mathbf{v}}_1\left(\mathbf{z}_1\right)\right)\right) \qquad (5.71)$$

The asymptotic normality of $\widehat{\boldsymbol{\theta}}$ follows by applying Slutsky's theorem to (5.71):

$$\sqrt{n}\left(\widehat{\boldsymbol{\theta}}-\boldsymbol{\theta}\right) \to_p N\left(\mathbf{0},\Sigma_\theta = E^{-2}\left(R_1\right)\Sigma_V E^{-2}\left(R_1\right)\right) \qquad (5.72)$$

To estimate the asymptotic variance $\Sigma_\theta$ in (5.72), we need to find $\Sigma_V$. Consider the $t$th component $\widetilde{v}_t(\mathbf{z}_1)$ of $\widetilde{\mathbf{v}}_1(\mathbf{z}_1)$ $(1 \leq t \leq m)$. It is readily checked that

$$\begin{aligned} \widetilde{v}_t(\mathbf{z}_1) &= E\left[r_{1t}r_{2t}\left(h_t(\mathbf{y}_1, \mathbf{y}_2) - \theta_t\right) \mid \mathbf{z}_1\right] \\ &= E\left[r_{1t}r_{2t}\left(h_t(\mathbf{y}_1, \mathbf{y}_2) - \theta_t\right) \mid y_{1t}, r_{1t}\right] \\ &= \frac{1}{2}r_{1t}E(r_{1t})\left[(y_{1t} - \mu_t)^2 - \sigma_t^2\right] \end{aligned} \tag{5.73}$$

where $\mu_t = E(y_{it})$. Let $\mathbf{p}_i(\mathbf{y}_i, \boldsymbol{\mu}) = \left((y_{i1} - \mu_1)^2, \ldots, (y_{im} - \mu_m)^2\right)^\top$. By expressing $\widetilde{v}_t(\mathbf{z}_1)$ in (5.73) in terms of the vector $\widetilde{\mathbf{v}}_1(\mathbf{z}_1)$, we have

$$\widetilde{\mathbf{v}}_1(\mathbf{z}_1) = \frac{1}{2}E(R_1)R_1\left[\mathbf{p}_1(\mathbf{y}_1, \boldsymbol{\mu}) - \boldsymbol{\theta}\right] \tag{5.74}$$

Thus, we have (see exercise):

$$\begin{aligned} \Sigma_V &= 4Var\left(\widetilde{\mathbf{v}}_1(\mathbf{z}_1)\right) = 4E\left[\widetilde{\mathbf{v}}_1(\mathbf{z}_1)\widetilde{\mathbf{v}}_1^\top(\mathbf{z}_1)\right] \\ &= E(R_1)Var\left(R_1\mathbf{p}_1(\mathbf{y}_1, \boldsymbol{\mu})\right)E(R_1) \end{aligned} \tag{5.75}$$

Thus, it follows from (5.75) that

$$\begin{aligned} \Sigma_\theta &= E^{-2}(R_1)\Sigma_V E^{-2}(R_1) \\ &= E^{-1}(R_1)Var\left(R_1\mathbf{p}_1(\mathbf{y}_1, \boldsymbol{\mu})\right)E^{-1}(R_1) \end{aligned} \tag{5.76}$$

When expressed in terms of the elements of $\Sigma_\theta = [\sigma_{st}]$, we have:

$$\sigma_{st} = E^{-1}(r_{1s})E^{-1}(r_{1t})E(r_{1s}r_{1t})Cov(p_{1s}, p_{1t}) \tag{5.77}$$

A consistent estimate of $\Sigma_\theta$ is obtained by substituting consistent estimates for $E(r_{1t})$, $E(r_{1s}r_{1t})$ and $Cov(p_{1s}, p_{1t})$. For example, we can estimate these quantities by the following moment estimates:

$$\widehat{E}(r_{1t}) = n^{-1}\sum_{i=1}^n r_{it}, \quad \widehat{E}(r_{1s}r_{1t}) = n^{-1}\sum_{i=1}^n r_{is}r_{it}, \quad \widehat{p}_{it} = (y_{it} - \widehat{\mu}_t)^2$$

$$\widehat{Cov}(p_{1s}, p_{1t}) = \left(\sum_{i=1}^n r_{is}r_{it}\right)^{-1}\sum_{i=1}^n r_{is}r_{it}\left(\widehat{p}_{is} - \widehat{\theta}_s\right)\left(\widehat{p}_{it} - \widehat{\theta}_t\right)$$

In Examples 5 and 6, the U-statisics for the parameters of interest are univariate and the multivariate $U$-statistics are created by the repeated assessments as dictated by longitudinal study designs. The above considerations are similarly applied to *endogenous* multivariate U-statistics within a longitudinal data setting as illustrated by the next example.

**Example 7** (***Product–moment correlation for longitudinal data***). Consider the product–moment correlation problem in Example 1 of Section 5.1.1 within a longitudinal study setting. Even within a cross-sectional study design, this problem requires consideration of a three-component multivariate U-statistic. Here, we extend the development to a longitudinal data setting with $m$ assessments.

Let $\mathbf{z}_{it} = (x_{it}, y_{it})^\top$ be an i.i.d. sample of bivariate variables $m$ assessments. Let

$$\begin{aligned}
\sigma_{xyt} &= Cov\left(x_{it}, y_{it}\right), \quad \sigma_{xt}^2 = Var\left(x_{it}\right), \quad \sigma_{yt}^2 = Var\left(y_{it}\right) \qquad (5.78)\\
\boldsymbol{\theta}_t &= \left(\sigma_{xyt}, \sigma_{xt}^2, \sigma_{yt}^2\right)^\top, \quad \boldsymbol{\theta} = \left(\boldsymbol{\theta}_1, \ldots, \boldsymbol{\theta}_m\right)^\top \\
\rho_t &= \frac{\sigma_{xyt}}{\sqrt{\sigma_{xt}^2 \sigma_{yt}^2}} = f_t\left(\boldsymbol{\theta}_t\right), \quad \boldsymbol{\rho} = \mathbf{f}\left(\boldsymbol{\theta}\right) = \left(f_1\left(\boldsymbol{\theta}_1\right), , \ldots, f_m\left(\boldsymbol{\theta}_m\right)\right)^\top
\end{aligned}$$

We are interested in inference for $\boldsymbol{\rho}$ as well as hypotheses concerning linear combinations $\boldsymbol{\rho}$.

First, consider the problem without missing data. For each time $t$, let $h_{kt}\left(\mathbf{z}_{it}, \mathbf{z}_{jt}\right)$ be defined as in (5.2) of Example 1 in Section 5.1 for $\mathbf{z}_{it}$ and let

$$\begin{aligned}
\mathbf{h}_t\left(\mathbf{z}_{it}, \mathbf{z}_{jt}\right) &= \left(h_{1t}\left(\mathbf{z}_{it}, \mathbf{z}_{jt}\right), h_{2t}\left(\mathbf{z}_{it}, \mathbf{z}_{jt}\right), h_{3t}\left(\mathbf{z}_{it}, \mathbf{z}_{jt}\right)\right)^\top \\
\mathbf{h}\left(\mathbf{z}_i, \mathbf{z}_j\right) &= \left(\mathbf{h}_1^\top\left(\mathbf{z}_{i1}, \mathbf{z}_{j1}\right), \ldots, \mathbf{h}_m^\top\left(\mathbf{z}_{im}, \mathbf{z}_{jm}\right)\right)^\top \\
\widehat{\boldsymbol{\theta}} &= \binom{n}{2}^{-1} \sum_{(i,j)\in C_2^n} \mathbf{h}\left(\mathbf{z}_i, \mathbf{z}_j\right) \\
\widehat{\rho}_t &= \frac{\sum_{(i,j)\in C_2^n} h_{1t}\left(\mathbf{z}_{it}, \mathbf{z}_{jt}\right)}{\sqrt{\sum_{(i,j)\in C_2^n} h_{2t}\left(\mathbf{z}_{it}, \mathbf{z}_{jt}\right) \sum_{(i,j)\in C_2^n} h_{3t}\left(\mathbf{z}_{it}, \mathbf{z}_{jt}\right)}} = f_t\left(\widehat{\boldsymbol{\theta}}_t\right) \\
\widehat{\boldsymbol{\rho}} &= \mathbf{f}\left(\widehat{\boldsymbol{\theta}}\right) = \left(f_1\left(\widehat{\boldsymbol{\theta}}_1\right), , \ldots, f_m\left(\widehat{\boldsymbol{\theta}}_m\right)\right)^\top
\end{aligned}$$

By applying the results in Example 4 of Section 5.1.1 to each $\widehat{\rho}_t$, it follows that $\widehat{\boldsymbol{\rho}}$ is a consistent estimate of $\boldsymbol{\rho}$. Also, by the Delta method, we immediately obtain the asymptotic distribution of $\widehat{\boldsymbol{\rho}}$:

$$\begin{aligned}
\sqrt{n}\left(\widehat{\boldsymbol{\rho}} - \boldsymbol{\rho}\right) &\to_d N\left(\mathbf{0}, \Sigma_\rho = \frac{\partial}{\partial \boldsymbol{\theta}} \mathbf{f}\left(\boldsymbol{\theta}\right) \Sigma_\theta \left(\frac{\partial}{\partial \boldsymbol{\theta}} \mathbf{f}\left(\boldsymbol{\theta}\right)\right)^\top\right) \\
\Sigma_\theta &= 4Var\left(\widetilde{\mathbf{h}}_1\left(\mathbf{z}_1\right)\right),
\end{aligned}$$

where $\widetilde{\mathbf{h}}_1\left(\mathbf{z}_1\right) = E\left(\mathbf{h}\left(\mathbf{z}_1, \mathbf{z}_2\right) \mid \mathbf{z}_1\right) - \boldsymbol{\theta}$.

To estimate $\Sigma_\theta$, let $\boldsymbol{\mu}_x = E(\mathbf{x}_i)$ and $\boldsymbol{\mu}_y = E(\mathbf{y}_i)$. Then, for each $t$, it follows from Example 4 that

$$\begin{aligned} \widetilde{\mathbf{h}}_{1t}(\mathbf{z}_1) &= E[\mathbf{h}_t(\mathbf{z}_1, \mathbf{z}_2) \mid \mathbf{z}_1] - \boldsymbol{\theta}_t \\ &= \frac{1}{2}\left[\mathbf{p}_{1t}\left(\mathbf{z}_i, \mu_{xt}, \mu_{yt}\right) - \boldsymbol{\theta}_t\right] \\ \mathbf{p}_{it}\left(\mathbf{z}_i, \mu_{xt}, \mu_{yt}\right) &= \left(\left(x_{it} - \mu_{xt}\right)\left(y_{it} - \mu_{yt}\right), \left(x_{it} - \mu_{xt}\right)^2, \left(y_{it} - \mu_{yt}\right)^2\right)^\top \end{aligned} \tag{5.79}$$

Thus, it follows from (5.79) that

$$\Sigma_\theta = 4Var\left(\widetilde{\mathbf{h}}_1(\mathbf{z}_1)\right) = Var\left(\mathbf{p}_i\left(\mathbf{z}_i, \boldsymbol{\mu}_x, \boldsymbol{\mu}_y\right)\right)$$

where $\mathbf{p}_i\left(\mathbf{z}_i, \boldsymbol{\mu}_x, \boldsymbol{\mu}_y\right) = \left(\widetilde{\mathbf{p}}_{i1}^\top, \ldots, \widetilde{\mathbf{p}}_{im}^\top\right)^\top$. A consistent estimate of $\Sigma_\theta$ is given by

$$\widehat{\Sigma}_\theta = \frac{1}{n}\sum_{i=1}^n \left(\mathbf{p}_i\left(\mathbf{z}_i, \widehat{\boldsymbol{\mu}}_x, \widehat{\boldsymbol{\mu}}_y\right) - \widehat{\boldsymbol{\theta}}\right)\left(\mathbf{p}_i\left(\mathbf{z}_i, \widehat{\boldsymbol{\mu}}_x, \widehat{\boldsymbol{\mu}}_y\right) - \widehat{\boldsymbol{\theta}}\right)^\top$$

Now consider the missing data case. Let $r_{it}$ be defined as in (5.65), except that $y_{it}$ is replaced by $\mathbf{z}_{it} = (x_{it}, y_{it})^\top$. Thus, we assume that $x_{it}$ and $y_{it}$ are either observed or missing together as a pair, that is

$$r_{it} = \begin{cases} 1 & \text{if the pair } (x_{it}, y_{it}) \text{ is observed} \\ 0 & \text{if otherwise} \end{cases} \tag{5.80}$$

Let $\mathbf{r}_i$ and $R_i$ be defined as in (5.65). For real study data, this assumption is plausible if missing data is the result of missed study visits. We relax this assumption to allow $x_{it}$ and $y_{it}$ to have their own missing data patterns and discuss this more general case in Section 5.2.3.

In the presence of missing data, we define the Pearson estimate at each assessment time $t$ as follows:

$$\begin{aligned} \widehat{\rho}_t &= \frac{\sum_{(i,j)\in C_2^n} r_{it}r_{jt}\left(x_{it} - x_{jt}\right)\left(y_{it} - y_{jt}\right)}{\sqrt{\sum_{(i,j)\in C_2^n} r_{it}r_{jt}\left(x_{it} - x_{jt}\right)^2 \sum_{(i,j)\in C_2^n} r_{it}r_{jt}\left(y_{it} - y_{jt}\right)^2}} \\ &= \frac{\sum_{(i,j)\in C_2^n} g_{1t}\left(\mathbf{z}_{it}, \mathbf{z}_{jt}, r_{it}, r_{jt}\right)}{\sqrt{\sum_{(i,j)\in C_2^n} g_{2t}\left(\mathbf{z}_{it}, \mathbf{z}_{jt}, r_{it}, r_{jt}\right) \sum_{(i,j)\in C_2^n} g_{3t}\left(\mathbf{z}_{it}, \mathbf{z}_{jt}, r_{it}r_{jt}\right)}} \end{aligned} \tag{5.81}$$

where

$$
\begin{aligned}
g_{1t}\left(\mathbf{z}_{it},\mathbf{z}_{jt},r_{it},r_{jt}\right) &= r_{it}r_{jt}h_{1t}\left(\mathbf{z}_{it},\mathbf{z}_{jt}\right) \\
&= \frac{1}{2}r_{it}r_{jt}\left(x_{it}-x_{jt}\right)\left(y_{it}-y_{jt}\right) \\
g_{2t}\left(\mathbf{z}_{it},\mathbf{z}_{jt},r_{it},r_{jt}\right) &= r_{it}r_{jt}h_{2t}\left(\mathbf{z}_{it},\mathbf{z}_{jt}\right) = \frac{1}{2}r_{it}r_{jt}\left(x_{it}-x_{jt}\right)^2 \\
g_{3t}\left(\mathbf{z}_{it},\mathbf{z}_{jt},r_{it}r_{jt}\right) &= r_{it}r_{jt}h_{3t}\left(\mathbf{z}_{it},\mathbf{z}_{jt}\right) = \frac{1}{2}r_{it}r_{jt}\left(y_{it}-y_{jt}\right)^2
\end{aligned}
\tag{5.82}
$$

Thus, $\widehat{\rho}_t$ not only depends on $\mathbf{z}_{it}$ but on $r_{it}$ and $r_{jt}$ as well ($1 \leq i \leq n$). Let $\widehat{\boldsymbol{\rho}} = \left(\widehat{\rho}_1, \widehat{\rho}_2, \ldots, \widehat{\rho}_m\right)^\top$.

Before embarking on the task to find the asymptotic distribution of $\widehat{\boldsymbol{\rho}}$, let us first establish the consistency of $\widehat{\boldsymbol{\rho}}$ as an estimate of $\boldsymbol{\rho}$. To this end, let

$$
R_{it} = r_{it}\mathbf{I}_3, \quad \mathbf{g}_t\left(\mathbf{z}_{it},\mathbf{z}_{jt},r_{it}r_{jt}\right) = \left(g_{1t},g_{2t},g_{3t}\right)^\top, \quad G_t = \sum_{(i,j)\in C_2^n} R_{it}R_{jt}
$$

where $\mathbf{I}_3$ denotes the $3\times 3$ identity matrix. For each $t$, define an estimate of $\boldsymbol{\theta}_t$ as follows:

$$
\begin{aligned}
\widehat{\boldsymbol{\theta}}_t &= G_t^{-1}\sum_{(i,j)\in C_2^n} \mathbf{g}_t\left(\mathbf{z}_{it},\mathbf{z}_{jt},r_{it}r_{jt}\right) \\
&= \binom{n}{2}G_t^{-1}\binom{n}{2}^{-1}\sum_{(i,j)\in C_2^n} \mathbf{g}_t\left(\mathbf{z}_{it},\mathbf{z}_{jt},r_{it}r_{jt}\right) \\
&= \binom{n}{2}G_t^{-1}\mathbf{U}_{nt}
\end{aligned}
\tag{5.83}
$$

where $\mathbf{U}_{nt} = \binom{n}{2}^{-1}\sum_{(i,j)\in C_2^n} \mathbf{g}_t\left(\mathbf{z}_{it},\mathbf{z}_{jt},r_{it}r_{jt}\right)$ is a three-dimensional U-statistic vector. Thus, although the estimate $\widehat{\boldsymbol{\theta}}_t$ above is not a U-statistic as in the complete data case, it is a function of such statistics since $\binom{n}{2}^{-1}G_t$ is also a U-statistic. Further, under MCAR, it is readily shown that $\widehat{\boldsymbol{\theta}}_t \rightarrow_p \boldsymbol{\theta}_t$ ($1 \leq t \leq m$) (see exercise). Thus, $\widehat{\boldsymbol{\rho}} = \mathbf{f}\left(\widehat{\boldsymbol{\theta}}\right) \rightarrow_p \mathbf{f}\left(\boldsymbol{\theta}\right) = \boldsymbol{\rho}$ and $\widehat{\boldsymbol{\rho}}$ defined in (5.81) is a consistent estimate of $\boldsymbol{\rho}$.

Since $\widehat{\boldsymbol{\theta}}$ defined in (5.83) is consistent, as in the complete data case, we can use the Delta method to find the asymptotic variance of $\widehat{\boldsymbol{\rho}}$ once we have the asymptotic distribution of $\widehat{\boldsymbol{\theta}}$. To find the latter, let

$$\mathbf{g} = \left(\mathbf{g}_1^\top, \ldots, \mathbf{g}_m^\top\right)^\top, \quad R_i = diag_t\left(R_{it}\right), \quad G = \sum_{(i,j)\in C_2^n} R_i R_j$$
$$\widehat{\boldsymbol{\theta}} = \left(\widehat{\boldsymbol{\theta}}_1^\top, \ldots, \widehat{\boldsymbol{\theta}}_m^\top\right)^\top, \quad \mathbf{f}\left(\widehat{\boldsymbol{\theta}}\right) = \left(f\left(\widehat{\boldsymbol{\theta}}_1\right), \ldots, f\left(\widehat{\boldsymbol{\theta}}_m\right)\right)^\top$$

where $diag_t\left(R_{it}\right)$ denotes the $3m \times 3m$ block diagonal matrix with $R_{it}$ forming the $t$th block diagonal. Then, we have (see exercise):

$$\begin{aligned}
\sqrt{n}\left(\widehat{\boldsymbol{\theta}} - \boldsymbol{\theta}\right) &= \sqrt{n}G^{-1} \sum_{(i,j)\in C_2^n} \left[\mathbf{g}\left(\mathbf{z}_i, \mathbf{z}_j, R_i, R_j\right) - R_i R_j \boldsymbol{\theta}\right] \qquad (5.84)\\
&= \binom{n}{2} G^{-1} \sqrt{n} \binom{n}{2}^{-1} \sum_{(i,j)\in C_2^n} R_i R_j \left[\mathbf{h}\left(\mathbf{z}_i, \mathbf{z}_j\right) - \boldsymbol{\theta}\right]\\
&= \binom{n}{2} G^{-1} \sqrt{n} \mathbf{U}_n
\end{aligned}$$

where

$$\mathbf{U}_n = \binom{n}{2}^{-1} \sum_{(i,j)\in C_2^n} R_i R_j \left[\mathbf{h}\left(\mathbf{z}_i, \mathbf{z}_j\right) - \boldsymbol{\theta}\right] = \binom{n}{2}^{-1} \sum_{(i,j)\in C_2^n} \mathbf{u}\left(\mathbf{z}_i, \mathbf{z}_j, \mathbf{r}_i, \mathbf{r}_j\right)$$

is a U-statistic-like vector with mean $\mathbf{0}$ as in Example 6. Thus, $\mathbf{U}_n$ is asymptotically normal:

$$\sqrt{n}\mathbf{U}_n \to_d N\left(\mathbf{0}, \boldsymbol{\Sigma}_U = 4Var\left(\widetilde{\mathbf{u}}_1\left(\mathbf{z}_1, \mathbf{r}_1\right)\right)\right) \qquad (5.85)$$

It follows from (5.84), (5.85) and Slutsky's theorem that

$$\sqrt{n}\left(\widehat{\boldsymbol{\theta}} - \boldsymbol{\theta}\right) \to_d N\left(\mathbf{0}, \boldsymbol{\Sigma}_\theta = 4E^{-2}\left(R_1\right) \boldsymbol{\Sigma}_U E^{-2}\left(R_1\right)\right)$$

To find $\boldsymbol{\Sigma}_U$ in the asymptotic variance $\boldsymbol{\Sigma}_\theta$ above, it is readily checked that (see exercise)

$$\begin{aligned}
\widetilde{\mathbf{u}}_1\left(\mathbf{z}_1, \mathbf{r}_1\right) &= E\left[R_1 R_2 \left(\mathbf{h}\left(\mathbf{z}_1, \mathbf{z}_2\right) - \boldsymbol{\theta}\right) \mid \mathbf{z}_1, \mathbf{r}_1\right] \qquad (5.86)\\
&= \frac{1}{2} R_1 E\left(R_1\right) \left[\mathbf{p}_1\left(\mathbf{z}_1, \boldsymbol{\mu}_x, \boldsymbol{\mu}_y\right) - \boldsymbol{\theta}\right]
\end{aligned}$$

Thus,

$$Var\left(\widetilde{\mathbf{u}}_1\left(\mathbf{z}_1, \mathbf{r}_1\right)\right) = \frac{1}{4} E\left(R_1\right) Var\left(R_1 \mathbf{p}_1\left(\mathbf{z}_1, \boldsymbol{\mu}_x, \boldsymbol{\mu}_y\right)\right) E\left(R_1\right) \qquad (5.87)$$

from which it follows that

$$\boldsymbol{\Sigma}_\theta = E^{-1}(R_1)\,Var\left(R_1\mathbf{p}_1\left(\mathbf{z}_1,\boldsymbol{\mu}_x,\boldsymbol{\mu}_y\right)\right)E^{-1}(R_1) \tag{5.88}$$

Note that (5.87) and (5.88) are quite similar to (5.75) and (5.76), except for the difference in $\mathbf{p}_1\left(\mathbf{z}_1,\boldsymbol{\mu}_x,\boldsymbol{\mu}_y\right)$ and $R_1$, as these are all $3m\times 3m$ matrices in the current example. A consistent estimate of $\boldsymbol{\Sigma}_\theta$ is obtained by substituting consistent estimates for the respective parameters, such as the ones given below:

$$\widehat{E}(R_{1t}) = n^{-1}\sum_{i=1}^{n} R_{it}, \quad \widehat{E}(R_{1s}R_{1t}) = n^{-1}\sum_{i=1}^{n} R_{is}R_{it}$$

$$\widehat{\mathbf{p}}_{it} = \mathbf{p}_{it}\left(\mathbf{z}_i,\widehat{\mu}_{xt},\widehat{\mu}_{yt}\right)$$

$$\widehat{Cov}(\mathbf{p}_{1s},\mathbf{p}_{1t}) = \left(\sum_{i=1}^{n} R_{is}R_{it}\right)^{-1}\sum_{i=1}^{n} R_{is}R_{it}\left(\widehat{\mathbf{p}}_{is}-\widehat{\boldsymbol{\theta}}_s\right)\left(\widehat{\mathbf{p}}_{it}-\widehat{\boldsymbol{\theta}}_t\right)^{\top}$$

### 5.2.3 Inference Under MAR

In Chapter 4, we discussed inference for distribution-free generalized linear and other related models under response-dependent missingness or MAR within the context of generalized estimating equation (GEE). The idea is to use weight function to statistically "recover" missing data for study dropouts and include them in estimation of model parameters. Although we do not consider GEE for inference in this section, the general strategy is equally applicable to address selection bias when missing data follows the MAR model.

Consider the model for variance within a longitudinal data setting introduced in Example 1 of Section 5.2.1. We developed an approach that employs missing data indicators to facilitate the construction of a consistent estimate $\widehat{\boldsymbol{\theta}}$ of $\boldsymbol{\theta} = \left(\sigma_1^2,\ldots,\sigma_m^2\right)^{\top}$ as well as the derivation of the asymptotic distribution of $\widehat{\boldsymbol{\theta}}$ under MCAR in Example 6 of Section 5.2.2. Under MAR, however, $\widehat{\boldsymbol{\theta}}$ is generally not consistent and a new estimate must be introduced.

To see why in general $\widehat{\boldsymbol{\theta}}$ does not consistently estimate $\boldsymbol{\theta}$, let us revisit the steps to show the consistency of $\widehat{\sigma}_t^2$ in (5.69). The crucial step is the following identity:

$$\frac{E\left[\frac{1}{2}r_{it}r_{jt}\left(y_{it}-y_{jt}\right)^2\right]}{E^2(r_{it})} = \frac{E^2(r_{it})\,\sigma_t^2}{E^2(r_{it})} \tag{5.89}$$

Under MCAR, $r_{it}$ and $y_{it}$ are independent and thus

$$E\left[\frac{1}{2}r_{it}r_{jt}\left(y_{it}-y_{jt}\right)^2\right]=E\left(r_{it}r_{jt}\right)E\left[\frac{1}{2}\left(y_{it}-y_{jt}\right)^2\right]=E^2\left(r_{it}\right)\sigma_t^2$$

Without the independence between $r_{it}$ and $y_{it}$, such as under MAR, the identity in (5.89) generally does not hold true, and as a result $\widehat{\boldsymbol{\theta}}$ constructed in Example 6 of Section 5.2.2 may not be consistent estimate of $\boldsymbol{\theta}$. In the following example, we discuss how to construct a consistent estimate of $\boldsymbol{\theta}$ for this particular model as a way of introducing a general approach to address MAR in our context.

**Example 8 (*Model for variance over time*).** Within the context of Example 6 of Section 5.2.2, assume that missingness depends on the response vector $\mathbf{y}_i$. Let $\pi_{it}=\Pr\left(r_{it}=1\mid \mathbf{y}_i\right)$ denote the probability of observing the response $y_{it}$ given $\mathbf{y}_i$. For the moment, assume that $\pi_{it}$ is known ($1\leq i\leq n$, $1\leq t\leq m$).

Let

$$h_t\left(y_{it},y_{jt}\right)=\frac{1}{2}\left(y_{it}-y_{jt}\right)^2,\quad g_t\left(y_{it},y_{jt},r_{it},r_{jt}\right)=\frac{r_{it}r_{jt}}{\pi_{it}\pi_{jt}}h_t\left(y_{it},y_{jt}\right)$$

Consider the following estimate of $\sigma_t^2$:

$$\begin{aligned}\widehat{\sigma}_t^2&=\binom{n}{2}^{-1}\sum_{(i,j)\in C_2^n}g_t\left(y_{it},y_{jt},r_{it},r_{jt}\right)\\&=\binom{n}{2}^{-1}\sum_{(i,j)\in C_2^n}\pi_{it}^{-1}\pi_{jt}^{-1}r_{it}r_{jt}h_t\left(y_{it},y_{jt}\right)\end{aligned}\tag{5.90}$$

It follows from the law of iterated conditional expectation that (see exercise)

$$\begin{aligned}E\left(\widehat{\sigma}_t^2\right)&=E\left[g_t\left(y_{it},y_{jt},r_{it},r_{jt}\right)\right]\\&=E\left[E\left(\pi_{it}^{-1}\pi_{jt}^{-1}r_{it}r_{jt}h_t\left(y_{it},y_{jt}\right)\mid y_{it},y_{jt}\right)\right]\\&=E\left[\pi_{it}^{-1}\pi_{jt}^{-1}h_t\left(y_{it},y_{jt}\right)E\left(r_{it}r_{jt}\mid y_{it},y_{jt}\right)\right]\\&=E\left[\pi_{it}^{-1}\pi_{jt}^{-1}h_t\left(y_{it},y_{jt}\right)\pi_{it}\pi_{jt}\right]\\&=\sigma_t^2\end{aligned}\tag{5.91}$$

Thus, $\sigma_t^2$ is an unbiased estimate of $\sigma_t^2$. By the consistency of U-statistics, (5.90) defines a consistent estimate of $\sigma_t^2$ and $\widehat{\boldsymbol{\theta}}=\left(\widehat{\sigma}_1^2,\widehat{\sigma}_2^2,\ldots,\widehat{\sigma}_m^2\right)^\top$ is a consistent estimate of $\boldsymbol{\theta}=\left(\sigma_1^2,\sigma_2^2,\ldots,\sigma_m^2\right)^\top$.

Note that under MCAR, since $r_{it}$ is independent of $\mathbf{y}_i$, (5.90) reduces to

$$\begin{aligned}\widehat{\sigma}_t^2 &= \binom{n}{2}^{-1} \sum_{(i,j)\in C_2^n} \pi_t^{-1}\pi_t^{-1} r_{it} r_{jt} h_t(y_{it}, y_{jt}) \\ &= \pi_t^{-2} \binom{n}{2}^{-1} \sum_{(i,j)\in C_2^n} r_{it} r_{jt} h_t(y_{it}, y_{jt})\end{aligned} \tag{5.92}$$

On the other hand, it follows from the consistency of U-statistics that

$$\binom{n}{2}^{-1} \sum_{(i,j)\in C_2^n} r_{it} r_{jt} \to E(r_{it} r_{jt}) = \pi_t^2 \tag{5.93}$$

Thus, it follows from Slutsky's theorem that $\widehat{\sigma}_t^2$ defined in (5.90) has the same asymptotic distribution as the estimate of $\sigma_t^2$ definition in (5.67) in Example 6 under MCAR. The difference between the two estimates is that the version in (5.90) requires a known or an estimate of $\pi_{it}$, while the estimate in (5.67) does not.

To find the asymptotic distribution of $\widehat{\boldsymbol{\theta}}$, let

$$G_i = diag_t\left(\pi_{it}^{-1} r_{it}\right), \quad \mathbf{h}(\mathbf{y}_i, \mathbf{y}_j) = (h_1, h_2, \ldots, h_m)^\top$$
$$\mathbf{g}(\mathbf{y}_i, \mathbf{y}_j, \mathbf{r}_i, \mathbf{r}_j) = G_i G_j \mathbf{h}(\mathbf{y}_i, \mathbf{y}_j), \quad \widehat{\boldsymbol{\theta}} = \binom{n}{2}^{-1} \sum_{(i,j)\in C_2^n} \mathbf{g}(\mathbf{y}_i, \mathbf{y}_j, \mathbf{r}_i, \mathbf{r}_j)$$

Then, it follows from the theory of multivariate U-statistics in Section 5.1 that the asymptotic distribution of $\widehat{\boldsymbol{\theta}}$ is given by:

$$\sqrt{n}\left(\widehat{\boldsymbol{\theta}} - \boldsymbol{\theta}\right) \to_d N\left(\mathbf{0}, \Sigma_\theta = 4Var\left(\widetilde{\mathbf{g}}_1(\mathbf{y}_1, \mathbf{r}_1)\right)\right) \tag{5.94}$$

The asymptotic variance $\Sigma_\theta$ in (5.94) is evaluated as in Example 6 (see exercise). Likewise, we can readily construct a consistent estimate of $\Sigma_\theta$ by following the same steps in that example (see exercise).

In Example 8, we showed how to incorporate weight functions into estimates to address selection bias due to missing data under MAR. Such an approach has been used for regression analysis within the context of GEE in Chapter 4. As discussed there, it is critical that $\pi_{it} > 0$ for all $1 \le i \le n$ and $1 \le t \le m$ so that the observed responses can serve as representatives for their missing counterparts. Otherwise, we will not be able to statistically recover all missing data to include them in estimation to ensure consistent parameter estimates. Further, we often assume $\pi_{it} \geq c > 0$, that is, $\pi_{it}$ is bounded away from 0, to obtain stable estimates of parameters of interest $(1 \leq i \leq n,\ 1 \leq t \leq m)$.

While $\pi_{it}$ may be known in some study designs (e.g., deliberately discontinuing some subjects in a multiphase clinical trial), the relationship between $\mathbf{r}_i$ and $\mathbf{y}_i$ is unknown in most applications. Under MAR, $\pi_{it}$ only depend on the observed responses. As discussed in Chapter 4, it is generally difficult to model $\pi_{it}$ without imposing the monotone missing data pattern (MMDP) assumption because of the large number and complexity of missing data patterns. The structured patterns under MMDP eliminate a potentially large number of missing data patterns and make it practical to model and estimate $\pi_{it}$ under MAR.

**Example 9**. In Example 8, assume MAR, that is, $\pi_{it}$ is a function of observed $y_{it}$. Further, we assume no missing data at baseline $t = 1$. Under MMDP, $\pi_{it}$ only depends on the observed responses and thus can be expressed as:

$$\begin{aligned} \pi_{it} &= \Pr\left(r_{it} = 1 \mid \mathbf{y}_i\right) = \Pr\left(r_{it} = 1 \mid \widetilde{\mathbf{y}}_{it}\right) \\ \widetilde{\mathbf{y}}_{it} &= \left(y_{i1}, ..., y_{i(t-1)}\right)^\top, \quad 2 \leq t \leq m \end{aligned} \tag{5.95}$$

where $\widetilde{\mathbf{y}}_{it} = \left(y_{i1}, \ldots, y_{i(t-1)}\right)^\top$, the sub-vector of $\mathbf{y}_i$ containing $y$ variables up to and including time $t-1$. As discussed in Chapter 4, we can readily model $\pi_{it}$ in (5.95) under MAR and MMDP using logistic regression. More specifically, we first model the one-step transition probability from observing the response at $t-1$ to $t$ using a logistic regression as follows:

$$\begin{aligned} &\operatorname{logit}\ (p_{it}) = \operatorname{logit}\left[\Pr\left(r_{it} = 1 \mid r_{i(t-1)} = 1, \widetilde{\mathbf{y}}_{it}\right)\right] = \alpha_t + \boldsymbol{\beta}_t^\top \widetilde{\mathbf{y}}_{it} \\ &\qquad 2 \leq t \leq m \end{aligned} \tag{5.96}$$

Let $\boldsymbol{\gamma}_t = \left(\alpha_t, \boldsymbol{\beta}_t^\top\right)^\top$ denote the parameters of the logistic model above. Then, by invoking MMDP, we obtain

$$\begin{aligned} \pi_{it}(\boldsymbol{\gamma}) &= \Pr\left[r_{is} = 1, r_{it} = 1 \mid \widetilde{\mathbf{y}}_{is}, \widetilde{\mathbf{y}}_{it}\right] = \Pr\left[R_{it} = 1 \mid \widetilde{\mathbf{y}}_{it}\right] \\ &= \Pr\left[r_{it} = 1 \mid r_{i(t-1)} = 1, \widetilde{\mathbf{y}}_{it}\right] \Pr\left[r_{i(t-1)} = 1 \mid \widetilde{\mathbf{y}}_{i(t-1)}\right] \\ &= \prod_{s=2}^{t} p_{is}\left(\boldsymbol{\gamma}_s\right), \quad 2 \leq t \leq m, \quad 1 \leq i \leq n \end{aligned} \tag{5.97}$$

where $\boldsymbol{\gamma} = \left(\boldsymbol{\gamma}_1^\top, \ldots, \boldsymbol{\gamma}_m^\top\right)^\top$. We can estimate $\pi_{it}(\boldsymbol{\gamma})$ by $\pi_{it}(\widehat{\boldsymbol{\gamma}})$ with $\widehat{\boldsymbol{\gamma}}$ obtained from the logistic regression for one-step transition probability in (5.96).

Note that when $\pi_{it}$ is estimated by (5.97), the asymptotic variance in (5.94) is likely to underestimate the variability of $\widehat{\boldsymbol{\theta}}$. In Chapter 4, we discussed how to account for the variability of the estimate $\widehat{\gamma}$ in the model for $\pi_{it}(\gamma)$. If $\widehat{\gamma}$ is obtained from GEE or the maximum likelihood method, we can readily find the asymptotic variance of $\widehat{\boldsymbol{\theta}}$ adjusted for the variability of $\widehat{\gamma}$.

Consider a one-sample U-statistic vector with $q$ arguments of the form $\mathbf{U}_n(\boldsymbol{\gamma})$, where $\boldsymbol{\gamma}$ is the parameter vector describing the missing data model as in the above example. It follows from the theory of multivariate statistics that

$$\sqrt{n}\left(\mathbf{U}_n(\boldsymbol{\gamma})-\boldsymbol{\theta}\right)=q\frac{\sqrt{n}}{n}\sum_{i=1}^{n}\widetilde{\mathbf{h}}_1(\mathbf{y}_i,\mathbf{r}_i,\gamma)+\mathbf{o}_p(1) \tag{5.98}$$

As in Chapter 4, let $\widehat{\gamma}$ be the solution to the estimating equation: $\mathbf{w}_n(\gamma)=\frac{1}{n}\sum_{i=1}^{n}\mathbf{w}_{ni}(\gamma)=\mathbf{0}$, where $\mathbf{w}_{ni}$ is the score (for maximum likelihood estimation) or score-like vector (for GEE estimation) for the $i$th subject ($1\leq i\leq n$). By a Taylor series expansion, we have

$$\begin{aligned}\sqrt{n}\left(\widehat{\gamma}-\gamma\right)&=\left(-\frac{\partial}{\partial\gamma}\mathbf{w}_n\right)^{-\top}\sqrt{n}\mathbf{w}_n+\mathbf{o}_p(1) \\ &=-H^{-1}\frac{\sqrt{n}}{n}\sum_{i=1}^{n}\mathbf{w}_{ni}+\mathbf{o}_p(1)\end{aligned} \tag{5.99}$$

where $H=E\left(\frac{\partial}{\partial\gamma}\mathbf{w}_{ni}\right)^{\top}$. It follows from (5.98) and (5.99) that (see exercise):

$$\begin{aligned}\sqrt{n}\left(\mathbf{U}_n(\widehat{\gamma})-\boldsymbol{\theta}\right)&=\sqrt{n}\left[\mathbf{U}_n(\widehat{\gamma})-\mathbf{U}_n(\gamma)+\mathbf{U}_n(\gamma)-\boldsymbol{\theta}\right] \\ &=\sqrt{n}\left(\mathbf{U}_n(\gamma)-\boldsymbol{\theta}\right)+\left(\frac{\partial}{\partial\gamma}\mathbf{U}_n(\gamma)\right)^{\top}\sqrt{n}\left(\widehat{\gamma}-\gamma\right)+\mathbf{o}_p(1) \\ &=q\frac{\sqrt{n}}{n}\sum_{i=1}^{n}\left(\widetilde{\mathbf{h}}_1(\mathbf{y}_i,\mathbf{r}_i,\gamma)-CH^{-1}\mathbf{w}_{ni}\right)+\mathbf{o}_p(1)\end{aligned} \tag{5.100}$$

where $C=E\left(\frac{\partial}{\partial\gamma}\widetilde{\mathbf{h}}_1(\mathbf{y}_i,\mathbf{r}_i,\gamma)\right)^{\top}$. By applying CLT to (5.100), we see that $\widehat{\boldsymbol{\theta}}$ is asymptotically normal with the asymptotic variance given by (see exercise):

$$\begin{aligned}\Sigma_\theta&=4\left[Var\left(\widetilde{\mathbf{h}}_1(\mathbf{y}_i,\gamma)\right)+\Phi\right] \\ \Phi&=CH^{-1}Var(\mathbf{w}_{ni})H^{-\top}C^{\top}-2E\left(\widetilde{\mathbf{h}}_1(\mathbf{y}_i,\mathbf{r}_i,\gamma)\mathbf{w}_{ni}^{\top}H^{-\top}C^{\top}\right)\end{aligned} \tag{5.101}$$

Thus, the extra term $\Phi$ in (5.101) accounts for the additional variability due to $\widehat{\gamma}$. A consistent estimate of $\Phi$ is readily obtained by substituting consistent estimates for the respective quantities in (5.101).

Now consider extending the considerations for the product–moment correlation discussed in Example 7 of Section 5.2.2 to account for response-dependent missingness under MAR. Although we can apply the same approach in Example 9 to the product–moment correlation, we consider a more general problem of inference about the *cross-variable, cross-time* product–moment correlation $\rho_{xyst}$ between $x_{is}$ and $y_{it}$, as well as by allowing $x_{it}$ and $y_{it}$ to have their own missing data patterns and in the process introduce a version of MMDP for bivariate responses.

**Example 10 (*Product–moment correlation*).** Let $z_{it} = (x_{it}, y_{it})^{\top}$ be an i.i.d. sample of bivariate outcomes over time $t$ ($1 \leq t \leq m$, $1 \leq i \leq n$). Here, we generalize the consideration in Example 7 to discuss inference about the cross-variable, cross-time product–moment correlation $\rho_{xyst}$ between $x_{is}$ and $y_{it}$ ($1 \leq s, t \leq m$). Further, we relax the missing data assumption considered earlier to allow $x_{it}$ and $y_{it}$ to have their own missing data patterns, that is, $x_{it}$ and $y_{it}$ may be missing at different time points.

We define two missing data indicators, one for each of the outcomes as follows:

$$r_{iut} = \begin{cases} 1 \text{ if } u_{it} \text{ is observed} \\ 0 \text{ if } u_{it} \text{ is missing} \end{cases}, \quad u = x, y \tag{5.102}$$

Let

$$\pi_{iuvst} = \Pr\left(r_{ius} = 1, r_{ivt} = 1 \mid \mathbf{z}_i\right), \quad u = x, y \tag{5.103}$$

The above define three sets of probabilities, $\pi_{ixyst}$, $\pi_{ixst}$ and $\pi_{iyst}$. For notational brevity, we adopt the convention that $\pi_{iuvst} = \pi_{iust}$ if $u = v$ and $s \leq t$ so that we can use $\pi_{iuvst}$ to represent all three. Note that unlike $\pi_{iuvst}$, $\pi_{iust} = \pi_{iuts}$ for $u = x, y$ and thus we only consider $\pi_{iust}$ for $s \leq t$.

While our interest lies in estimating $\rho_{st}$, for convenience, we discuss inference for the covariance $\sigma_{xyst}$ between $x_{is}$ and $y_{it}$ ($1 \leq s, t \leq m$). It is straightforward to generalize the consideration to develop a similar procedure for inference about $\rho_{st}$. As before, we first assume that $\pi_{iuvst}$ is known and derive a consistent estimate of $\sigma_{xyst}$ and its asymptotic distribution. We then discuss how to model and estimate $\pi_{iuvst}$ under MAR. Note that for our consideration of $\sigma_{xyst}$, we only need $\pi_{ixyst}$. However, when considering inference about $\rho_{st}$, we also need $\pi_{ixst}$ and $\pi_{iyst}$ for constructing estimates of $\sigma_{xst}$ and $\sigma_{yst}$ (see exercise). Thus, we consider modeling the general $\pi_{iuvst}$ below.

Let

$$h_{st}(\mathbf{z}_i,\mathbf{z}_j)=\frac{1}{2}(x_{is}-x_{js})(y_{it}-y_{jt}) \tag{5.104}$$
$$g_{st}(\mathbf{z}_i,\mathbf{z}_j,\mathbf{r}_i,\mathbf{r}_j)=\pi_{ixyst}^{-1}\pi_{jxyst}^{-1}r_{ixs}r_{iyt}r_{jxs}r_{jyt}h_{st}(\mathbf{z}_i,\mathbf{z}_j)$$

The above generalizes (5.82) to account for dependence of the occurrence of missing data on $\mathbf{z}_i$. It follows from the law of iterated conditional expectation that (see exercise)

$$\begin{aligned}E[g_{st}(\mathbf{z}_i,\mathbf{z}_j,\mathbf{r}_i,\mathbf{r}_j)] &= E\left[E\left(\frac{r_{ixs}r_{iyt}r_{jxs}r_{jyt}}{\pi_{ixyst}\pi_{jxyst}}h_{st}(\mathbf{z}_i,\mathbf{z}_j)\mid \mathbf{z}_i,\mathbf{z}_j\right)\right] \\ &= E\left[h_{st}(\mathbf{z}_i,\mathbf{z}_j)\pi_{ixyst}^{-1}\pi_{jxyst}^{-1}E(r_{ixs}r_{iyt}r_{jxs}r_{jyt}\mid \mathbf{z}_i,\mathbf{z}_j)\right] \\ &= E[h_{st}(\mathbf{z}_i,\mathbf{z}_j)]=\sigma_{xyst}\end{aligned} \tag{5.105}$$

Thus, by the theory of U-statistics, the statistic,

$$\widehat{\sigma}_{xyst}=\binom{n}{2}^{-1}\sum_{(i,j)\in C_2^n} g_{st}(\mathbf{z}_i,\mathbf{z}_j,\mathbf{r}_i,\mathbf{r}_j)$$

is an unbiased and consistent estimate of $\sigma_{xyst}$. Further, it follows from Theorem 2 of Section 5.2 that the asymptotic distribution of $\widehat{\sigma}_{xyst}$ is given by

$$\sqrt{n}(\widehat{\sigma}_{xyst}-\sigma_{xyst})\rightarrow_d N\left(0,\ \phi_{st}^2=4Var(\widetilde{g}_{st1}(\mathbf{z}_1,\mathbf{r}_1))\right) \tag{5.106}$$

Since

$$Var(\widetilde{g}_{st1}(\mathbf{z}_1,\mathbf{r}_1))=E[g_{st1}(\mathbf{z}_1,\mathbf{z}_2,\mathbf{r}_1,\mathbf{r}_2)\,g_{st1}(\mathbf{z}_1,\mathbf{z}_3,\mathbf{r}_1,\mathbf{r}_3)]$$

a consistent estimate of $\phi_{st}^2$ is readily constructed using U-statistics.

In the above, we only considered inference for a single $\sigma_{xyst}$. However, we can readily extend the consideration to the vector statistic $\mathbf{g}=(g_{11},\ldots,g_{1m},g_{21},\ldots,g_{mm})^\top$ so that we can make joint inference of $\boldsymbol{\sigma}_{xy}=(\sigma_{xy11},\ldots,\sigma_{xy1m},\sigma_{xy21},\ldots,\sigma_{xymm})^\top$.

As in Example 9. $\pi_{ixyst}$ are rarely known in practice. Under MCAR, $r_{iut}$ are stochastically independent of $\mathbf{z}_i$. As a result, $\pi_{iuvst}=E(r_{ius}r_{ivt})$, which are readily estimated by sample moments of the form: $\frac{1}{n}\sum_{i=1}^n r_{ius}r_{ivt}$. Further, if $r_{ixt}=r_{iyt}=r_{it}$, that is, $x_i$ and $y_i$ are observed or missing together, $\pi_{iuvst}=E(r_{is}r_{it})=\pi_{st}$ and are again readily estimated.

When MCAR fails, $\pi_{iuvst}$ becomes dependent on $\mathbf{z}_i$. Note that $\pi_{iuvst}$ is different from $\pi_{it}$ discussed in Example 9 as it is a function of two outcomes,

$x_{is}$ and $y_{it}$, and the approach for modeling $\pi_{it}$ does not apply to $\pi_{iuvst}$. As in the case of $\pi_{it}$, it is difficult to model $\pi_{iuvst}$ as a function of $\mathbf{z}_i$ in general without imposing strong assumptions on the relationship between the occurrence of missing data and outcomes. Under MAR, $\pi_{iuvst}$ only depends on the observed $\mathbf{z}_i$. Unfortunately, it is still quite complex to model the occurrence of missing data and a common approach is to further impose the monotone missing data pattern (MMDP) assumption as we discussed earlier in Example 9. For the current bivariate response setting, the single-response version of MMDP can still create a potentially large number of distinct patterns. For example, if $m = 5$, $x_{it}$ may be observed at all times, while $y_{it}$ may be last observed at time $t = 3$, 4 or 5. Although possible, it is difficult to formulate a model for $\pi_{iuvst}$ given the large number of missing data patterns. Thus, it is necessary to limit our consideration to some special cases. To this end, we first introduce an index to categorize the different MMDPs within a bivariate response setting.

For each subject $i$, let $l_i$ denote the absolute difference between the times of the last observed $x_{it}$ and $y_{it}$. This lag time can vary between 0 and $m-1$. For example, if $x_{it}$ is last observed at $t = 1$ (baseline), while $y_{it}$ is observed at all times, then $l_i = m - 1$. Let $k = \max\{l_i; 1 \le i \le n\}$ denote the largest difference in lag time across all subjects in the study. This index $k$ characterizes the lag time of bivariate MMDP (BMMDP) for the study sample. We focus on the special cases when $k = 0$ and $k = 1$ and discuss modeling $\pi_{iuvst}$ under the assumption that $r_{iut}$ only depend on $\widetilde{\mathbf{z}}_{it} = \{\mathbf{z}_{is}; 1 \le s \le t-1\}$, the history of responses observed prior to time $t$.

For BMMDP with lag 0, $r_{ixt} = r_{iyt} = r_{it}$. In this case, $x$ and $y$ are either observed or missed together. By BMMDP, we have

$$\pi_{ist} = \Pr\left(r_{is} = 1, r_{it} = 1 \mid \widetilde{\mathbf{z}}_{is}, \widetilde{\mathbf{z}}_{it}\right) = \Pr\left(r_{it} = 1 \mid \widetilde{\mathbf{z}}_{it}\right) = \pi_{it}, \quad 1 \le s \le t \le m$$

We can estimate $\pi_{ist}$ through a process similar to the steps defined by (5.96) and (5.97) for modeling $\pi_{it}$ in Example 9, except that the predictor of the logistic regression in (5.96) is defined by the bivariate outcomes $\widetilde{\mathbf{z}}_{it} = (x_{it}, y_{it})^{\top}$, that is

$$\text{logit } (p_{it}) = \text{logit}\left(\Pr\left(r_{it} = 1 \mid r_{i(t-1)} = 1, \widetilde{\mathbf{z}}_{it}\right)\right) = \alpha_t + \boldsymbol{\beta}_t^{\top} \widetilde{\mathbf{z}}_{it}$$
$$2 \le t \le m$$

If $k = 1$, the lag times between the last observed responses of $x_{it}$ and $y_{it}$ can differ at most by 1 across all subjects. Given that $(x_{is}, y_{is})$ is observed at time $t-1$, four possible missing data patterns emerge at time $t$, as illustrated

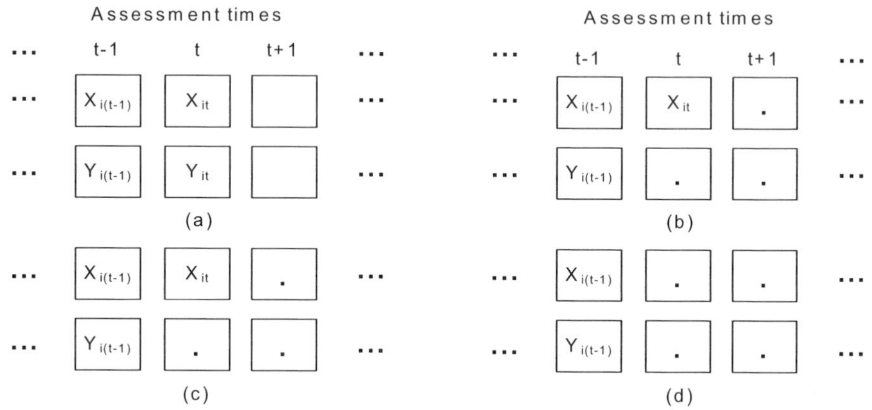


Figure 5.1. Four possible missing data patterns at time $t$ given observed pair $\left(x_{i(t-1)}, y_{i(t-1)}\right)$ at time $t-1$.

in Figure 5.1, with Patterns (a) and (d) correspond to the two missing data patterns for the BMMDP case with $k=0$. Denote the four patterns by

$$
\begin{array}{llll}
\text{(a)} & \omega_{it1}=\{r_{ixt}=1, r_{iyt}=1\}, & \text{(b)} & \omega_{it2}=\{r_{ixt}=0, r_{iyt}=1\} \\
\text{(c)} & \omega_{it3}=\{r_{ixt}=1, r_{iyt}=0\}, & \text{(d)} & \omega_{it4}=\{r_{ixt}=0, r_{iyt}=0\}
\end{array}
$$

Let $I_{itl}=1$ if $\omega_{itl}$ is observed and 0 if otherwise $(1 \leq l \leq 4)$. As discussed in Chapter 2, a popular model for such nominal response is the generalized logit or multinomial response model:

$$
p_{itl}=E\left(I_{itl}=1 \mid I_{i(t-1)1}=1, \widetilde{\mathbf{z}}_{it}\right)=\frac{\exp\left(\alpha_{tl}+\boldsymbol{\beta}_{tl}^{\top}\widetilde{\mathbf{z}}_{it}\right)}{\Phi\left(\widetilde{\mathbf{z}}_{it}\right)} \tag{5.107}
$$

$$
\Phi\left(\widetilde{\mathbf{z}}_{it}\right)=1+\sum_{l=2}^{4}\exp\left(\alpha_{tl}+\boldsymbol{\beta}_{tl}^{\top}\widetilde{\mathbf{z}}_{it}\right), \quad \alpha_{t1}=0, \quad \boldsymbol{\beta}_{t1}=\mathbf{0}, \quad \gamma_{t1}=\mathbf{0}
$$

$$
\boldsymbol{\beta}_{tl}=\left(\beta_{tl1},\ldots,\beta_{tl(t-1)}\right)^{\top}, \quad \gamma_{tl}=\left(\gamma_{tl1},\ldots,\gamma_{tl(t-1)}\right)^{\top}
$$
$$
1 \leq l \leq 4, \quad 2 \leq l \leq m
$$

Under the assumptions, $\mathbf{z}_{it}$ are observed for all subjects at baseline $t=1$ and so $p_{i1l}=1$ if $l=1$ and 0 if otherwise. Thus, starting at $t=2$, we apply (5.107) successively to model the occurrence of missing data patterns at each time $t$. We then estimate $\pi_{iuvst}$ using a relationship between $p_{itl}$ and $\pi_{iuvst}$ similar to (5.97) for the univariate $\pi_{it}$.

By applying BMMDP, we have for $s = t$:

$$\begin{aligned}\pi_{ixytt} &= \Pr\left(I_{it1} = 1 \mid \widetilde{\mathbf{z}}_{it}\right) \\ &= \Pr\left(I_{it1} = 1 \mid I_{i(t-1)1} = 1, \widetilde{\mathbf{z}}_{it}\right) \Pr\left(I_{i(t-1)1} = 1 \mid \widetilde{\mathbf{z}}_{i(t-1)}\right) \\ &= p_{it1} \Pr\left(I_{i(t-1)1} = 1 \mid \widetilde{\mathbf{z}}_{i(t-1)}\right) \\ &= \prod_{k=2}^{t} p_{ik1}\end{aligned}$$

for $s < t$:

$$\begin{aligned}\pi_{ixyst} &= \Pr\left(r_{ixs} = 1, r_{iyt} = 1 \mid \widetilde{\mathbf{z}}_{it}\right) \\ &= \Pr\left(r_{ixt} = 1, r_{iyt} = 1 \mid \widetilde{\mathbf{z}}_{it}\right) + \Pr\left(r_{ixt} = 0, r_{iyt} = 1 \mid \widetilde{\mathbf{z}}_{it}\right) \\ &= \Pr\left(I_{it1} = 1 \mid \widetilde{\mathbf{z}}_{it}\right) + \\ &+ \Pr\left(r_{ixt} = 0, r_{iyt} = 1 \mid I_{i(t-1)1} = 1, \widetilde{\mathbf{z}}_{it}\right) \Pr\left(I_{i(t-1)1} = 1 \mid \widetilde{\mathbf{z}}_{i(t-1)}\right) \\ &= \left(p_{it1} + p_{it2}\right) \prod_{k=2}^{t-1} p_{ik1}\end{aligned}$$

and for $s > t$,

$$\pi_{ixyst} = \left(p_{is1} + p_{is3}\right) \prod_{k=2}^{s-1} p_{ik1}$$

Similarly, for $u = v$, we have

$$\pi_{ixst} = \left(p_{it1} + p_{it3}\right) \prod_{k=2}^{t-1} p_{ik1}, \quad \pi_{iyst} = \left(p_{it1} + p_{it2}\right) \prod_{k=2}^{t-1} p_{ik1}, \quad s \leq t$$

In many applications, $p_{itl}$ may only depend on the most recently observed responses, $\widetilde{\mathbf{z}}_{it} = \mathbf{z}_{i(t-1)}$, and the predictor of the model (5.107) under this Markov condition is simply: $\alpha_{tl} + \boldsymbol{\beta}_{tl}^{\top} \mathbf{z}_{i(t-1)}$.

By assuming BMMDP in the above example, the total number of missing data patterns reduces from $m^2$ to $m$ for $k = 0$ and to $3m - 2$ for $k = 1$ (see exercise). For BMMDPs with lag time $k$ larger than 1, the number of potential missing data patterns at each time $t$ will increase at the rate of $(k+1)^2$ (see exercise). Although possible, it is difficult to model and estimate parameters given insufficient replications of missing data patterns in most applications. On the other hand, BMMDPs with large lag times are unlikely to occur under MAR in real study applications. Since MAR is often posited for modeling subject dropout due to deteriorated/improved health and related conditions, it is unlikely that one outcome will be continuously observed while the other is not. Thus, the BMMDPs with lag 0 and 1 provide good approximations for modeling the occurrence of missing data under MAR in most applications.

As in Example 9, we may want to adjust the asymptotic variance of $\widehat{\sigma}_{xyst}$ in (5.106) to account for the variability of $\widehat{\gamma}$ in the estimated $\pi_{ixyst}$. The discussion parallels that for Example 9 and in particular, the formula in (5.101) can be applied to obtained the adjusted asymptotic variance.

## 5.3 Exercises

**Section 5.1**

1. Consider a study with $n$ subjects and let $\mathbf{z}_i = (x_i, y_i)^\top$ be a pair of variables from the $i$th subjects $(1 \leq i \leq n)$. Consider a simple linear regression:

$$y_i = \beta_0 + \beta_1 x_i + \epsilon_i, \quad \epsilon_i \sim \text{i.i.d.} \left(0, \sigma^2\right), \quad 1 \leq i \leq n$$

(a) Find the least-square estimate $\widehat{\beta}_1$ of $\beta_1$.

(b) Show that $\widehat{\beta}_1 = \frac{s_{ny}}{s_{nx}} \widehat{\rho}$, where $s_{nz} = \left[\sum_{i=1}^n (z_i - \overline{z}_n)^2 / (n-1)\right]^{\frac{1}{2}}$ and $\widehat{\rho}$ is the Pearson correlation given in (5.1).

2. Consider the model in (5.3).

(a) Define an GEE II of the form $\mathbf{w}_n(\boldsymbol{\theta}) = \sum_{i=1}^n G_i S_i$ to estimate $\boldsymbol{\zeta} = \left(\mu_x, \mu_y, \sigma_x^2, \sigma_y^2, \sigma_{xy}\right)^\top$.

(b) Select an appropriate $G_i$ and $S_i$ so that the GEE II in (a) yields the sample mean, variance and covariance for the respective components of $\boldsymbol{\zeta}$.

3. Let $\mathbf{e}_n$ be defined in the proof of Theorem 1 of Section 5.1.1. Show $E\left(\mathbf{e}_n^\top \mathbf{e}_n\right) \to 0$ implies that $\mathbf{e}_n \to_p 0$.

4. For the one-group U-statistic vector $\mathbf{U}_n$ defined in (5.4), show

$$E\left(\mathbf{U}_n \mid \mathbf{y}_i\right) = \frac{m}{n}\mathbf{h}_1(\mathbf{y}_i) + \frac{n-m}{n}\boldsymbol{\theta} = \frac{m}{n}\widetilde{\mathbf{h}}_1(\mathbf{y}_i) + \boldsymbol{\theta}$$

5. Consider the product–moment correlation in Example 4 of Section 5.1.1. Verify (5.19).

6. In Problem 5, assume that $\mathbf{z}_i = (x_i, y_i)^\top \sim N(\boldsymbol{\mu}, \boldsymbol{\Sigma})$, a bivariate normal with mean $\boldsymbol{\mu}$ and variance $\boldsymbol{\Sigma} = [\sigma_{kl}]$.

(a) Find $\Sigma_\theta = Var\left(\widetilde{\mathbf{h}}_1(\mathbf{z}_1)\right)$ and express it in terms of $\sigma_{kl}$.

(b) Show that the asymptotic variance of $\widehat{\rho}$ in (5.19) reduces to $\sigma_\rho^2 = \left(1 - \rho^2\right)^2$.

7. Consider the Concordance correlation coefficient $\rho_{ccc}$ defined in (5.37).

(a) Show $E(\mathbf{h}(\mathbf{y}_i, \mathbf{y}_j)) = \boldsymbol{\theta}$, where $\mathbf{h}(\mathbf{y}_i, \mathbf{y}_j)$ is defined in (5.39) and $\boldsymbol{\theta}$ in (5.38).

(b) Find a bivariate U-statistic vector $\mathbf{U}_n = (U_{n1}, U_{n2})^\top$ such that

$$E(U_{n1}) = 2\sigma_{12}, \quad E(U_{n2}) = \sigma_1^2 - 2\sigma_{12} + \sigma_2^2 + (\mu_1 - \mu_2)^2$$

(c) Find the asymptotic distribution of $\mathbf{U}_n$.

(d) Find the asymptotic distribution of the estimated CCC, $\widehat{\rho}_{ccc} = \frac{U_{n1}}{U_{n2}}$.

8. Verify (5.26), (5.29), (5.36), (5.40), (5.41), and (5.44).

9. For the kernel function $\mathbf{h}(y_{1i}, y_{1j}; y_{2l}, y_{2m})$ defined in (5.47), find a symmetric version of this kernel.

10. For the $K$-group U-statistic vector defined in (5.46).

(a) Show that the projection of the U-statistic can be expressed as:

$$\widehat{\mathbf{U}}_n = \sum_{k=1}^{K} \frac{m_k}{n_k} \sum_{i=1}^{n_k} \widetilde{\mathbf{h}}_{k1}(\mathbf{y}_{ki}) + \boldsymbol{\theta}$$

where $\mathbf{h}_{k1}(y_{k1})$ is defined in (5.52).

(b) Use the result in (a) to verify (5.51).

(c) By applying the central limit theorem to the centered projection $\widehat{\mathbf{U}}_n - \boldsymbol{\theta}$ in (5.51), show that $\widehat{\mathbf{U}}_n$ has the asymptotic distribution given by (5.53).

11. Let $\mathbf{U}_n$ and $\widehat{\mathbf{U}}_n$ be defined in (5.46) and (5.51). Let $\mathbf{e}_n = \sqrt{n}\left(\mathbf{U}_n - \widehat{\mathbf{U}}_n\right)$. Show that $E\left(\mathbf{e}_n^\top \mathbf{e}_n\right) \to 0$ as $n \to \infty$.

12. Find a symmetric version of the kernel function in (5.56).

**Section 5.2**

1. Consider the estimate $\widehat{\boldsymbol{\theta}}$ defined by a multivariate U-statistic in (5.57). Show that the asymptotic variance of $\widehat{\boldsymbol{\theta}}$ can be expressed in (5.58).

2. Consider the U-statistic vector $\widehat{\boldsymbol{\theta}}$ in (5.60) for testing the hypothesis of no location shift between two CDFs in a longitudinal data setting in (5.59).

(a) Find the asymptotic variance $\Sigma_\theta$ of $\widehat{\boldsymbol{\theta}}$.

(b) Construct a consistent estimate of $\Sigma_\theta$ in (a).

3. Consider the U-statistic vector $\widehat{\boldsymbol{\tau}} = \binom{n}{2}^{-1} \sum_{i=1}^{n} \mathbf{h}(\mathbf{z}_i, \mathbf{z}_j)$ defined by the kernel $\mathbf{h}(\mathbf{z}_i, \mathbf{z}_j)$ in (5.61) for Kendall's $\tau$.

(a) Find the asymptotic distribution of $\widehat{\boldsymbol{\tau}}$.

(b) Use the asymptotic distribution of $\widehat{\boldsymbol{\tau}}$ in (a) to develop an asymptotic test for the null of no association between $x_{it}$ and $y_{it}$ over time, $H_0 : \boldsymbol{\tau} = \mathbf{0}$.

4. Consider the U-statistic vector $\widehat{\boldsymbol{\theta}}$ defined in (5.62) for inference about Kendall's $\tau_b$ within a longitudinal data setting.

(a) Find the asymptotic distribution of $\widehat{\boldsymbol{\theta}}$.

(b) Find the asymptotic distribution of $\widehat{\boldsymbol{\tau}}_b = \mathbf{f}\left(\widehat{\boldsymbol{\theta}}\right)$, where $\mathbf{f}(\cdot)$ is defined in (5.63).

5. Show that (5.64) and (5.67) define the same estimate of $\widehat{\sigma}_t^2$ based on the observed data.

6. Verify (5.69), (5.70), (5.73), (5.75), (5.76) and (5.77). [*Hint*: $E(r_{1t}) = E(r_{2t})$ in (5.73).]

7. Consider the estimate $\widehat{\boldsymbol{\theta}}$ defined in (5.83).

(a) Show that $\widehat{\boldsymbol{\theta}}$ is consistent, that is, $\widehat{\boldsymbol{\theta}} \to_p \boldsymbol{\theta}$, where $\boldsymbol{\theta}$ is defined in (5.78).

(b) Verify (5.84) and (5.86). [*Hint*: $E(R_2) = E(R_1)$ in (5.86).]

8. Consider the estimate $\widehat{\boldsymbol{\theta}} = \left(\widehat{\sigma}_1^2, \widehat{\sigma}_2^2, \ldots, \widehat{\sigma}_m^2\right)^\top$ with $\widehat{\sigma}_t^2$ defined in (5.90).

(a) By verifying (5.91), show that $\widehat{\boldsymbol{\theta}}$ is an unbiased estimate of $\boldsymbol{\theta} = \left(\sigma_1^2, \sigma_2^2, \ldots, \sigma_m^2\right)^\top$ with $\sigma_t^2$ denoting the variance of $y_{it}$.

(b) Show that $\widehat{\sigma}_t^2$ has the same asymptotic distribution as the estimate of $\widehat{\sigma}_t^2$ definition in (5.67) by applying the results in (5.92), (5.93), and Slutsky theorem.

(c) Evaluate the asymptotic variance $\Sigma_\theta$ of $\widehat{\boldsymbol{\theta}}$ given in (5.94).

(d) Use the result in (c) to construct a consistent estimate of $\Sigma_\theta$.

9. Consider the estimate $\widehat{\boldsymbol{\theta}} = \left(\widehat{\sigma}_1^2, \widehat{\sigma}_2^2, \ldots, \widehat{\sigma}_m^2\right)^\top$ defined in (5.90). Let $\pi_{it}(\boldsymbol{\gamma})$ be modeled by (5.96) and (5.97).

(a) Assume $\boldsymbol{\gamma}$ is known and evaluate its asymptotic variance $\Sigma_\theta = 4Var\left(\widetilde{\mathbf{g}}_1(\mathbf{y}_1, \mathbf{r}_1)\right)$ given in (5.94).

(b) Evaluate $\Sigma_\theta = 4\left[Var\left(\widetilde{\mathbf{h}}_1(\mathbf{y}_1, \mathbf{r}_1, \boldsymbol{\gamma})\right) + \Phi\right]$ with $\Phi$ given in (5.101) if $\boldsymbol{\gamma}$ is estimated by $\widehat{\boldsymbol{\gamma}}$ satisfying (5.99).

10. Verify (5.100), (5.101), and (5.105).

11. In Example 10, let

$$
\begin{aligned}
g_{2st}(\mathbf{z}_i, \mathbf{z}_j, \mathbf{r}_i, \mathbf{r}_j) &= \pi_{ixs}^{-1}\pi_{jxs}^{-1} r_{ixt} r_{jxt} h_{2st}(\mathbf{z}_i, \mathbf{z}_j) \\
g_{3st}(\mathbf{z}_i, \mathbf{z}_j, \mathbf{r}_i, \mathbf{r}_j) &= \pi_{iys}^{-1}\pi_{jys}^{-1} r_{iyt} r_{jyt} h_{3st}(\mathbf{z}_i, \mathbf{z}_j) \\
h_{2st}(\mathbf{z}_i, \mathbf{z}_j) &= \frac{1}{2}(x_{is} - x_{js})(x_{it} - x_{jt}) \\
h_{3st}(\mathbf{z}_i, \mathbf{z}_j) &= \frac{1}{2}(y_{is} - y_{js})(y_{it} - y_{jt})
\end{aligned}
$$

Assume $\pi_{ixyst}$ are known.

(a) Show that

$$E\left(g_{2st}\left(\mathbf{z}_i,\mathbf{z}_j,\mathbf{r}_i,\mathbf{r}_j\right)\right)=Cov\left(x_{is},x_{it}\right),\quad E\left(g_{3st}\left(\mathbf{z}_i,\mathbf{z}_j,\mathbf{r}_i,\mathbf{r}_j\right)\right)=\sigma_{yst}$$

(b) Let

$$\widehat{\rho}_{xyst}=\frac{\sum_{(i,j)\in C_2^n} g_{st}\left(\mathbf{z}_i,\mathbf{z}_j,\mathbf{r}_i,\mathbf{r}_j\right)}{\sqrt{\sum_{(i,j)\in C_2^n} g_{2ss}\left(\mathbf{z}_i,\mathbf{z}_j,\mathbf{r}_i,\mathbf{r}_j\right)\sum_{(i,j)\in C_2^n} g_{2tt}\left(\mathbf{z}_i,\mathbf{z}_j,\mathbf{r}_i,\mathbf{r}_j\right)}}$$

where $g_{st}\left(\mathbf{z}_i,\mathbf{z}_j,\mathbf{r}_i,\mathbf{r}_j\right)$ is defined in (5.104). Show that $\widehat{\rho}_{xyst}$ is a consistent and asymptotically normal estimate of $\rho_{xyst}$.

(c) Find the asymptotic variance of $\widehat{\rho}_{xyst}$.

(d) Let $\widehat{\boldsymbol{\rho}}_{xy}=\left(\widehat{\rho}_{xy11},\ldots,\widehat{\rho}_{xy1m},\widehat{\rho}_{xy21},\ldots,\widehat{\rho}_{xymm}\right)^{\top}$. Find the asymptotic distribution of $\widehat{\boldsymbol{\rho}}_{xy}$.

12. In Problem 11, assume that $\pi_{ixyst}$ are unknown and modeled using the approach described in Example 10. Let $\gamma$ denote the parameter vector in the model for $\pi_{ixyst}$.

(a) Find the asymptotic variance of $\widehat{\rho}_{xyst}$.

(b) Find the asymptotic variance of $\widehat{\boldsymbol{\rho}}_{xy}$.

13. Consider a study with bivariate outcomes $z_{it}=\left(x_{it},y_{it}\right)^{\top}$ collected over $m$ assessment times ($1\leq t\leq m$, $1\leq i\leq n$). Let $k$ denote the lag time defined in Example 10 to characterize BMMDPs. Show

(a) The total number of missing data patterns is $2^m$.

(b) The total number of missing data patterns under MMDP on each $x_{it}$ and $y_{it}$ is $m^2$.

(c) The total number of missing data patterns under BMMDP is $m$ if lag time $k=0$ and $3m-2$ if lag time $k=1$.

(d) The total number of missing data patterns at each time $t$ is $(k+1)^2$ ($1\leq t\leq m$).

# Chapter 6

# Functional Response Models

In Chapter 5, we introduced multivariate U-statistics, developed an inference theory for such vector-valued statistics for longitudinal data analysis and discussed applications of this new theory to address an array of important statistical problems involving modeling second order moments and statistics defined by responses from multiple subjects. From a methodologic standpoint, the theory extended the many classic univariate U-statistics based models and tests to a longitudinal data setting and addressed missing data under the general MAR assumption so that these models and tests can provide more powerful applications in modern-day research studies populated by longitudinal study designs. From a practical application perspective, Chapter 5 demonstrated the widespread utility of multivariate U-statistics by providing effective solutions to problems that involve second order moments, such as in modeling the variance and correlation, and functions of multiple subjects' responses in the fields of biomedical and psychosocial research.

A distinctive feature of the new methods is the paradigm shift from classic, univariate- and continuous-variable based U-statistics to applications of discrete outcomes within the context of longitudinal study data. This methodological breakthrough opens the door to many new applications that we continue to expand upon in this chapter. At the same time, the new theoretical development presented in Chapter 5 to address new applications of U-statistics also highlighted some critical limitations — in particular, the lack of a systematic approach to modeling complex high order moments and functions of responses from multiple subjects. Typically, in such cases, models are developed to address specific problems in an ad hoc fashion with no formal framework for either model estimation or inference. Additionally, the inability to accommodate covariates within such a modeling setting seriously limits the potential for the broad application of such methods.

In this chapter, we introduce a new class of distribution-free or semi-parametric regression models to simultaneously address all these limitations. By generalizing traditional regression models through defining the response variable as a general function of several responses from multiple subjects, this class of functional response models (FRM) subsumes linear, generalized linear and even nonlinear regression models as special cases. In addition, FRM provides a general platform to integrate many classic nonparametric statistics and popular distribution-free models, as discussed in Chapters 2–5, within a single, unified modeling framework to systematically address the various limitations of these statistics and models. For example, by viewing the classic, nonparametric, two-sample Mann–Whitney–Wilcoxon rank sum statistic as a regression under FRM, we are readily able to generalize it to account for multiple groups and clustered data as in a longitudinal study setting. Such extensions are especially important for effectiveness studies which by their design introduce variability, as well as for microarray studies in which additional heterogeneity in expression may be introduced simply from the type of sample used, such as biopsies from skin tissue. With FRM, we are also able to formally address the limitations of the classic mean based, distribution-free regression model for inference about second and higher order moments, such as the random effects parameters in the linear mixed-effects model, and causal-effect parameters in structural equation models, both of which are popular in psychosocial research and more recently, in cancer vaccine trials with molecular endpoints. To further achieve the over arching scope of this book for analyses of modern day applications, we also systematically address MCAR and MAR assumptions.

In Section 6.1, we highlight the limitations of the classic distribution-free regression model discussed in Chapter 2 and 4 for modeling second and higher order moments and how the classic models can be extended to overcome this major difficulty by using functional response. In Section 6.2, we formally define FRM and discuss their applications in integrating classic non-parametric statistics and distribution-free models discussed in Chapters 2–5. In Section 6.3, we discuss inference for FRM by developing a new class of U-statistics based estimating equations. In Section 6.4, we address missing data. We discuss the impact of missing data on model inference under different missing data mechanisms and develop procedures for consistent estimation of model parameters under the two most common MCAR and MAR assumptions.

## 6.1 Limitations of Linear Response Models

In this section, we illustrate some of the key limitations associated with the classic, mean-based distribution-free regression models discussed in Chapter 2 for modeling second and higher order moments as well as statistics defined by multisubject responses by way of examples. We start with a relative simple analysis of variance model (ANOVA) model.

**Example 1** (***ANOVA***). Consider the ANOVA model discussed in Chapter 2 for comparing $K$ treatment conditions. Let $y_{ki}$ denote some continuous response of interest from the $i$th subject within the $k$th group for $1 \leq i \leq n_k$, where $n_k$ denotes the sample size of group $k$ $(1 \leq k \leq K)$. Under the distribution-free ANOVA setup discussed in Chapter 2, we model the mean of $y_{kj}$ as follows:

$$E(y_{ki}) = \mu_k, \quad 1 \leq i \leq n_k, \quad 1 \leq k \leq K \tag{6.1}$$

where $E(\cdot)$ denotes mathematical expectation. The above model is then used to test hypotheses concerning the mean responses, $\mu_k$, across the different treatment groups. As noted in Chapter 3, for many real data applications, especially in the area of effectiveness research, there may also exist difference in variance in addition to the mean, in which case, a comparison of such second order variability among the different groups is also of interest. For example, if two treatments have the same mean response, but different variances, the treatment associated with the smaller variance in response may be preferable in terms of its generalizability and cost effectiveness — two important issues in effectiveness research. However, a test of difference in variance across the treatment conditions is beyond the capability of the mean response based model in (6.1), since it is restricted to modeling only the mean of $y_{ki}$.

In Chapter 3, we discussed a U-statistics based approach for comparing the variance between two treatment groups. By setting $K = 2$ within the context of Example 1 and denoting the variance of $y_{ki}$ by $\sigma_k^2$ $(k = 1, 2)$, we can use the following $U$-statistic to compare the variance between the two groups:

$$U = \binom{n_1}{2}^{-1} \binom{n_2}{2}^{-1} \sum_{i,j \in C_2^{n_1}} \sum_{l,m \in C_2^{n_2}} \frac{1}{2} \left[ (y_{1i} - y_{1j})^2 - (y_{2l} - y_{2m})^2 \right]$$

Since $\theta = E(U) = \sigma_1^2 - \sigma_2^2$, we can use this U-statistic to test the null hypothesis of equal variance between the two groups, $H_0 : \sigma_1^2 = \sigma_2^2$. Although one may generalize this statistic to compare the variance for more than two groups using multivariate U-statistics as illustrated in Chapter 5 for extending the two-sample Mann–Whitney–Wilcoxon rank sum test to a multi-group setting, such an approach is ad hoc. Alternatively, a regression model provides a general framework to systematically address this and other similar problems involving higher order moments.

**Example 2** (***Generalized ANOVA for comparing variance***). Consider a generalization of the ANOVA in Example 1 to compare variances across the $K$ groups. To this end, let $\sigma_k^2 = Var(y_{ki})$ denote the variance of $y_{ki}$ $(1 \leq k \leq K)$.

Let

$$f(y_{ki}, y_{kj}) = \frac{1}{2}(y_{ki} - y_{kj})^2, \quad h(\sigma_k^2) = \sigma_k^2, \quad 1 \leq i < j \leq n_k, \; 1 \leq k \leq K$$

Then, the mean of the two-subject response based functional $f(y_{ki}, y_{kj})$ is given by

$$E[f(y_{ki}, y_{kj})] = h(\sigma_k^2) = \sigma_k^2, \quad 1 \leq i < j \leq n_k, \; 1 \leq k \leq K \qquad (6.2)$$

The above model has a similar form as the mean-based distribution-free ANOVA in (6.1). If we want to test the hypothesis of variance homogeneity, we can simply test the following linear contrast:

$$\begin{aligned} H_0 &: \sigma_l^2 = \sigma_m^2, \quad \text{for all } 1 \leq l, m \leq K \quad \text{versus} \\ H_a &: \sigma_l^2 \neq \sigma_m^2, \quad \text{for some } 1 \leq l \neq m \leq K \end{aligned}$$

The mean based ANOVA model in (6.1) differs from the one in (6.2) in several respects. First, the classic mean based ANOVA only involves a single response $y_{ki}$, while the latter model involves a pair of responses, $y_{ki}$ and $y_{kj}$, from two different subjects (within the same group). Second, the dependent variable in the mean based ANOVA is $y_{ki}$, whereas the dependent variable for the model in (6.2) is defined by a more complex, quadratic function of two individual responses $y_{ki}$ and $y_{kj}$. It is these extensions in the dependent variable that allow the distribution-free model in (6.2) to overcome the fundamental limitation of the mean based ANOVA and provide inference for hypotheses concerning variance or second order moment of the response.

**Example 3** (***Random-factor ANOVA***). A classic one-way ANOVA with random factor levels is defined by

$$\begin{aligned} & y_{ki} = \mu + \lambda_k + \epsilon_{ki}, \quad \lambda_k \sim \text{i.i.d.} N(0, \sigma_\lambda^2) \qquad (6.3) \\ & \epsilon_{ki} \sim \text{i.i.d.} N(0, \sigma^2), \quad \lambda_k \perp \epsilon_{ki}, \quad 1 \leq i \leq n, \quad 1 \leq k \leq K \end{aligned}$$

where $\mu$, $\sigma_\lambda^2$ and $\sigma^2$ are parameters, $N(\mu, \sigma^2)$ denotes a normal distribution with mean $\mu$ and variance $\sigma^2$ and $\perp$ stochastic independence. Unlike the fixed-factor ANOVA in Example 1, the random factor level effect, $\lambda_k$, is now a latent variable and as a result, $y_{ki}$ are not independent across the subjects within each cluster $k$, though $y_{ki}$ are independent across the different clusters $k$ $(1 \leq k \leq K)$. As discussed in Chapters 4, the random-factor ANOVA in (6.3) is a special case of the linear mixed-effects model and widely used to model rater agreement and instrument reliability in biomedical and psychosocial research.

As shown in Chapter 4, the variance of $y_{ki}$ and covariance between any pair of within-cluster responses, $y_{ki}$ and $y_{kj}$, are given by

$$\begin{aligned} Var(y_{ki}) &= Var(\lambda_k) + Var(\epsilon_{ki}) = \sigma_\lambda^2 + \sigma^2 \\ Cov(y_{ki}, y_{kj}) &= Var(\lambda_k) = \sigma_\lambda^2, \quad 1 \leq i \leq n, \ 1 \leq k \leq K \end{aligned} \tag{6.4}$$

Thus, the correlation between two within-cluster responses $y_{ki}$ and $y_{kj}$ is given by

$$\rho = Corr(y_{ki}, y_{kj}) = \frac{Cov(y_{ki}, y_{kj})}{Var(y_{ki})} = \frac{\sigma_\lambda^2}{\sigma_\lambda^2 + \sigma^2} \tag{6.5}$$

Under the model assumption, this within-cluster or intraclass correlation (ICC) is a constant and a ratio of the between-cluster variability $\sigma_\lambda^2$ to the total variability $\sigma_\lambda^2 + \sigma^2$.

A major drawback of the model in (6.3) is the normal–normal distribution assumption for the response and random factor. In Chapter 5, we discussed a U-statistics based approach to facilitate distribution-free inference about $\rho$. Here, we develop a regression model to provide distribution-free inference for the between-cluster variance $\sigma_\lambda^2$. In many applications, we may want to test the null hypothesis, $H_0 : \sigma_\lambda^2 = 0$. If this null is not rejected, we can then ignore the between-cluster variability and use a simpler model for $y_{ki}$ without any random factor.

To model $\sigma_\lambda^2$ directly, first observe that under the model assumptions in (6.3), if $(i, j) \neq (i', j')$, that is, both $i$ and $j$ are different from $i'$ and $j'$, we have

$$\begin{aligned} E\left[f\left(y_{ki}, y_{ki'}; y_{lj}, y_{lj'}\right)\right] &= E\left[\frac{1}{2}\left(y_{ki} - y_{lj}\right)\left(y_{ki'} - y_{lj'}\right)\right] \\ &= h\left(\sigma_\lambda^2\right) = \sigma_\lambda^2 \\ (i, j) \in C_2^{n_k}, &\quad (i', j') \in C_2^{n_l}, \quad 1 \leq k, l \leq K \end{aligned} \tag{6.6}$$

In other words, the parameter of interest $\sigma_\lambda^2$ is the mean of a functional of response of the form: $f\left(y_{ki}, y_{ki'}; y_{lj}, y_{lj'}\right) = \frac{1}{2}\left(y_{ki} - y_{lj}\right)\left(y_{ki'} - y_{lj'}\right)$.

To express (6.6) in a vector form, let $\mathbf{y}_k = (y_{k1}, \ldots, y_{kn})^\top$. Then, $\mathbf{y}_k$ are independent across $k$, though the components within $\mathbf{y}_k$ are correlated. Let

$$\mathbf{f}_{ij,i'j'}(\mathbf{y}_k, \mathbf{y}_l) = \frac{1}{2}(y_{ki} - y_{lj})(y_{ki'} - y_{lj'})$$

$$\mathbf{f}(\mathbf{y}_k, \mathbf{y}_l) = \left(f_{12,34}, \ldots, f_{12,(n-1)n}, f_{13,24}, \ldots, f_{(n-3)(n-2),(n-1)n}\right)^\top$$

$$\mathbf{h}(\sigma_\lambda^2) = \sigma_\lambda^2 \mathbf{1}_{3C_4^n}$$

where $\mathbf{1}_{3C_4^n}$ denotes the $3C_4^n \times 1$ column vector of 1's. The number $3C_4^n$ is the dimension of the vector $\mathbf{f}(\mathbf{y}_k, \mathbf{y}_l)$, which is also the total number of combinations of $\{i,j\}$ and $\{i',j'\}$, $\frac{1}{2}\, C_4^n C_2^4$ (see exercise). The factor $\frac{1}{2}$ accounts for the fact that the order of $\{i,j\}$ and $\{i',j'\}$ does not matter. Then, we can express (6.6) succinctly as:

$$E[\mathbf{f}(\mathbf{y}_k, \mathbf{y}_l)] = \mathbf{h}(\sigma_\lambda^2), \quad 1 \le k < l \le K \tag{6.7}$$

Note that unlike the original parametric model in (6.3), the nuisance parameter, $\sigma^2$, no longer appears in the distribution-free model in (6.7). Note also that the above model differs from the one in (6.1) for Example 1 in that it involves a vector-valued functional response. This difference reflects the need to group correlated responses into independent outcomes and is analogous to modeling clustered data such as arises from longitudinal study designs.

In Chapter 4, we introduced the notion of internal consistency for grouping item scores in an instrument to form dimensional measures of latent constructs in psychosocial research and discussed a normal distribution based model for Cronbach's alpha coefficient for assessing such consistency. As noted there, the major difficulty in applying such a model to real data, particularly in psychosocial research, is the distribution assumption which is fundamentally at odds with the data distribution of item scores for most instruments. We addressed this problem using theory of multivariate U-statistics in Chapter 5. In the next example, we show how to model this index using a regression-like formulation.

**Example 4 (*Model for Cronbach's alpha*).** Consider an instrument (or questionnaire) with $m$ individual items and let $y_{ik}$ denote the response from the $i$th subject to the $k$th item for a sample of $n$ subjects. The Cronbach's coefficient alpha for assessing internal consistency of questionnaire with $m$ items is defined as follows:

$$\alpha = \frac{m}{m-1}\left(1 - \frac{\gamma}{\beta}\right), \quad \gamma = \sum_{k=1}^{m} \sigma_k^2, \quad \beta = \sum_{k=1}^{m} \sigma_k^2 + \sum_{k \neq l} \sigma_{kl} \tag{6.8}$$

where $\sigma_k^2 = Var(y_{ik})$ and $\sigma_{kl} = Cov(y_{ik}, y_{il})$ denote the variance and covariance of the item responses, respectively. In (6.8), $\alpha$ is expressed as a function of two parameters $\gamma$ and $\beta$ and thus we need to define two response functionals to model this parameter of interest.

To this end, let

$$\begin{aligned} f_1(\mathbf{y}_i, \mathbf{y}_j) &= \frac{1}{2}(\mathbf{y}_i - \mathbf{y}_j)^\top (\mathbf{y}_i - \mathbf{y}_j) \\ f_2(\mathbf{y}_i, \mathbf{y}_j) &= \mathbf{1}^\top (\mathbf{y}_i - \boldsymbol{\mu})(\mathbf{y}_i - \boldsymbol{\mu})^\top \mathbf{1} \\ h_1(\alpha, \beta) &= \beta\left(1 - \frac{m-1}{m}\alpha\right), \quad h_2(\alpha, \beta) = \beta \end{aligned} \tag{6.9}$$

Then, we have:

$$E[f_1(\mathbf{y}_i, \mathbf{y}_j)] = h_1(\alpha, \beta), \quad E[f_2(\mathbf{y}_i, \mathbf{y}_j)] = h_2(\alpha, \beta)$$

or simply,

$$E[\mathbf{f}(\mathbf{y}_i, \mathbf{y}_j)] = \mathbf{h}(\alpha, \beta), \quad \mathbf{f} = (f_1, f_2)^\top, \quad \mathbf{h} = (h_1, h_2)^\top \tag{6.10}$$

Thus, by defining two functionals of response in (6.9), we are able to model a complex function of second order moments using an ANOVA-like model in (6.10). As in the two previous examples, (6.10) are fundamentally different from the mean based ANOVA by the form of the dependent variable. In this example, the mean response, $\mathbf{h}(\alpha, \beta)$, is a more complex nonlinear function of the parameters.

The three examples above demonstrate the limitations of existing regression models for modeling complex functions of second and high order moments. Although the examples are contrasted with the relatively simple, distribution-free ANOVA model, the comparisons nonetheless illustrate an inherent weakness with the classic distribution-free regression model. For example, the class of distribution-free generalized linear model (GLM) for continuous, binary and count response is defined by

$$E(y_i \mid \mathbf{x}_i) = h\left(\mathbf{x}_i^\top \boldsymbol{\beta}\right), \quad 1 \leq i \leq n \tag{6.11}$$

where $y_i$ denotes a response and $\mathbf{x}_i$ a $p \times 1$ vector of independent variables from the $i$th individual of a sample with $n$ subjects, $E(y_i \mid \mathbf{x}_i)$ the conditional mean of $y_i$ given $\mathbf{x}_i$, $\boldsymbol{\beta}$ a vector of parameters and $g(\cdot)$ a link function. Although (6.11) generalizes the multiple linear model to accommodate more complex types of response variables such as binary, the left side retains the same form as in the mean-based ANOVA, in that it is restricted to a single subject response $y_i$. Although GEE II discussed in Chapter 4 is capable of modeling second order moments, its applications are limited to relatively simple problems. Further, GEE II does not address many popular statistics that are complex functions of response from multiple subjects, such as the Mann–Whitney–Wilcoxon test.

Note that the distribution-free GLM in (6.11) has a slightly different form than the definition in Chapter 2 in that the conditional mean of $y_i$ given $\mathbf{x}_i$ is expressed as a function of the linear predictor $\eta_i = \mathbf{x}_i^\top \boldsymbol{\beta}$ using the inverse of the link function $g(\cdot)$ for the convenience of discussion in this chapter. But, other than this difference in appearance, (6.11) defines exactly the same distribution-free GLM as in Chapter 2.

**Example 5** (***Wilcoxon signed rank test***). Consider the one-sample Mann–Whitney–Wilcoxon signed rank test. Let $y_i$ denote the continuous response of interest from an i.i.d. sample of $n$ subjects $(1 \leq i \leq n)$. Let

$$f(y_i, y_j) = I_{\{y_i + y_j \leq 0\}}, \quad h(\mu) = \mu, \quad 1 \leq i < j \leq n$$

where $I_{\{\cdot\}}$ denotes a set indicator. Then, the mean of the functional response $f(y_i, y_j)$ is given by

$$E[f(y_i, y_j)] = E\left(I_{\{y_i + y_j \leq 0\}}\right) = h(\mu) = \mu \tag{6.12}$$

As discussed in Chapter 3, the parameter of interest $\mu$ in the above model can be used to test whether $y_i$ are symmetrically distributed around 0.

The dependent variable, $f(y_i, y_j)$, in the model defined in (6.12) is again a function of multiple subjects' responses as in the previous examples. However, what is different from the other examples considered earlier is that the parameter of interest $\mu$ in this model cannot be expressed as a function of moments. For such an *intrinsic* or *endogenous* multisubject based functional, moment-based methods such as GEE II do not apply.

## 6.2 Models with Functional Responses

The models discussed in the examples in the previous section differ from the distribution-free GLMs discussed in Chapters 2 and 4 in that the response variable can be a complex functional involving multiple subjects' responses rather than a linear function of a single subject response. This generalization not only widens the class of existing regression models to accommodate new challenges in modeling real data applications, but also provides a general framework to generalize classic nonparametric methods such as the Mann–Whitney–Wilcoxon rank sum test and unify them under the rubric of regression analysis.

Let $\mathbf{y}_i$ be a vector of response and $\mathbf{x}_i$ a vector of independent variables from an i.i.d. sample $(1 \leq i \leq n)$. By broadening the response variable as a general non-linear functional of several responses from multiple subjects, we define the general functional response model as follows:

$$E\left[f\left(\mathbf{y}_{i_1}, \ldots, \mathbf{y}_{i_q}\right) \mid \mathbf{x}_{i_1}, \ldots, \mathbf{x}_{i_q}\right] = h\left(\mathbf{x}_{i_1}, \ldots, \mathbf{x}_{i_q}; \boldsymbol{\beta}\right)$$
$$\left(i_1, \ldots, i_q\right) \in C_q^n$$

where $f(\cdot)$ is some function, $h(\cdot)$ some smooth function (with continuous second order derivatives), $C_q^n$ denotes the set of $\binom{n}{q}$ combinations of $q$ distinct elements $(i_1, \ldots, i_q)$ from the integer set $\{1, \ldots, n\}$ and $\boldsymbol{\beta}$ a vector of parameters. By generalizing the response variable in such a way, this new class of FRM provides a single distribution-free framework for modeling first and higher order moments, such as mean and variance, and endogenous multisubject based statistics, such as the Mann–Whitney–Wilcoxon test. More importantly, we can readily extend such models to clustered and longitudinal data settings and address the inherent missing data issue under a unified paradigm.

As noted earlier, the distribution-free regression models discussed in Chapters 2 and 4 are all defined based on a single-subject response. The generalized linear model extends linear regression to accommodate more complex types of response such as binary and count variable. Yet, the fact remains that GLM is still defined by a single subject response. The extension only occurs to the right side of the model specification by generalizing the linear predictor $\eta_i = \mathbf{x}_i^\top \boldsymbol{\beta}$ in linear regression to a more general function of $\mathbf{x}_i^\top \boldsymbol{\beta}$ in GLM. But the left side of the model definition remains identical to the original linear regression model.

For ease of exposition, we define the FRM and discuss its applications for comparisons of multiple samples (or treatment groups) and regression analyses separately. We start with models for comparing multiple samples involving inference based on $K$ $(\geq 2)$ groups.

### 6.2.1 Models for Group Comparisons

Consider the following class of models for comparing $K$ independent groups:

$$E\left[\mathbf{f}\left(\mathbf{y}_{1i_1}, \ldots, \mathbf{y}_{1i_q}; \ldots; \mathbf{y}_{Kj_1}, \ldots, \mathbf{y}_{Kj_q}\right)\right] = \mathbf{h}(\boldsymbol{\theta}) \tag{6.13}$$
$$\left(i_1, \ldots, i_q\right) \in C_q^{n_k}, \quad 1 \leq q$$

where $\mathbf{y}_{kj}$ denotes a vector of response from the $j$th subject within the $k$th group, $\mathbf{f}$ some vector-valued function, $\mathbf{h}(\boldsymbol{\theta})$ some vector-valued smooth function (with continuous derivatives up to the second order), $\boldsymbol{\theta}$ a vector of parameters of interest, $q$ some positive integer, $n_k$ the sample size of group $k$ and $C_q^{n_k}$ the set of $\binom{n_k}{q}$ combinations of $q$ distinct elements $(i_1, \ldots, i_q)$ from the integer set $\{1, \ldots, n_k\}$ $(1 \leq k \leq K)$. Since no parametric distribution is assumed, (6.13) defines a class of distribution-free GLM-like models for inference about the $K$ groups. Note that $\mathbf{y}_{kj}$ are assumed to be stochastically independent across both $k$ and $j$. The response $\mathbf{f}$ of the FRM above is a function of multiple individual responses, $\mathbf{y}_{ki_1}, \ldots, \mathbf{y}_{ki_q}$, from within each of the $K$ groups.

**Example 1** (***Generalized ANOVA for modeling variance***). Consider the ANOVA like variance model in (6.2). Based on the definition above, this model is an FRM with $K$ independent groups and

$$q = 2, \quad f(y_{ki}, y_{kj}) = \frac{1}{2}(y_{ki} - y_{kj})^2, \quad h(\sigma_k^2) = \sigma_k^2$$
$$1 \leq i < j \leq n_k, \quad 1 \leq k \leq K$$

**Example 2** (***Random-factor ANOVA***). We can also express the distribution-free ANOVA with random factor levels in (6.6) as an FRM:

$$\mathbf{f}(\mathbf{y}_k, \mathbf{y}_l) = \left(f_{12}, \ldots, f_{1n}, f_{23}, \ldots, f_{2n}, \ldots, f_{(n-1)n}\right)^\top$$
$$\mathbf{h}(\sigma_\lambda^2) = \sigma_\lambda^2 \mathbf{1}_{\frac{1}{2}n(n-1)}, \quad 1 \leq k < l \leq K$$

Note that although similar in appearance, the random-effect ANOVA is different from its fixed-effect counterpart. For the fixed-effect ANOVA, the responses $y_{ki}$ within the $k$th group are independent across $i$, while for the random-effect version, these responses are dependent within each $k$th group. Thus, for the random-effect ANOVA, we must group the responses within each random factor to form independent vectors $\mathbf{y}_k$ when defining the FRM for this model.

The one- and two-sample Mann–Whitney–Wilcoxon rank based tests are often used as a robust alternative to parametric tests for comparing mean differences. We now define the FRMs that give rise to these test statistics.

**Example 3** (***Wilcoxon signed rank test***). Let $y_i$ denote a continuous response of interest from an i.i.d. sample of size $n$. Let

$$K = 1, \quad q = 2$$
$$f(y_i, y_j) = I_{\{y_i + y_j \leq 0\}}, \quad h(\mu) = \mu, \quad 1 \leq i < j \leq n$$

Then, $\mu = E\left[\, f\left(y_i, y_j\right)\right] = \Pr\left(y_i + y_j \leq 0\right)$. Thus, the null hypothesis that $y_i$ are symmetrically distributed around zero can be expressed in terms of the model parameter as $H_0 : \mu = \frac{1}{2}$. Note that the above FRM has the same form as the model in (6.12) considered in the previous section for the discussion of limitations of the classic regression paradigm.

**Example 4** (***Mann–Whitney–Wilcoxon rank sum test***). Let $y_{ki}$ denote some continuous response from two i.i.d. samples ($1 \leq i \leq n_k$, $k = 1, 2$). Define an FRM as follows:

$$K = 2, \quad q = 1 \tag{6.14}$$
$$f\left(y_{1i}, y_{2j}\right) = I_{\{y_{1i} \leq y_{2j}\}}, \quad h\left(\mu\right) = \mu, \quad 1 \leq i \leq n_1, \quad 1 \leq j \leq n_2$$

As discussed in detail in Chapters 3, the model parameter $\mu = \Pr\left(y_{1i} \leq y_{2j}\right)$ can be used to express the null hypothesis of no location shift when comparing the CDFs of the two samples, that is, $H_0 : \mu = \frac{1}{2}$. Thus, by using a functional response, we can frame this classic test within a regression-like model in (6.14).

**Example 5** (***ROC analysis***). Consider the problem of assessing the quality of a diagnostic test for detecting some medical or mental health condition of interest, such as an infectious disease or depression. Let $y_{1i}$ and $y_{2j}$ denote the continuous test outcomes from the diseased and nondiseased samples ($1 \leq i \leq n_1, 1 \leq j \leq n_2$). As discussed in Chapter 3, the area under the receiver operating characteristics curve (AUC), $\theta = \Pr\left(y_{1i} \leq y_{2j}\right)$, is widely used to evaluate the performance of a diagnostic test. By identifying $y_{ki}$ as the two samples in Example 4, we can see that this parameter of interest $\theta$ is precisely the model parameter $\mu$ in (6.14). Thus, we can define the same FRM as in the previous example to model the AUC of the diagnostic test.

Although the two FRMs are the same, interest is no longer centered around testing the null, $H_0 : \theta = \frac{1}{2}$, within the current context. In fact, $\theta$ is well interpreted in its entire range $(0, 1)$. If the diagnostic test determines disease status purely in a random fashion, that is, with a probability 0.5 for issuing disease and nondisease outcomes, the ROC curve, the plot of $\left(F_1\left(c\right), F_2\left(c\right)\right)$ with $F_k\left(c\right)$ denoting the CDF of $y_{ki}$, is a straight line with a $45°$ angle between 0 and 1 and thus the corresponding AUC is 0.5. On the other hand, if the test has some diagnostic value, that is, detecting disease status with a probability higher than 0.5, then $\mu > 0.5$. Thus, within the current context, it is of interest to test null hypotheses of the form: $H_0 : \mu = a$, where $a$ is some known constant between 0.5 and 1.

The real advantage of expressing these popular models in the form of an FRM is to generalize them. For example, we generalized the two-sample

Mann–Whitney–Wilcoxon rank sum test to compare three samples in Chapter 5. Within the framework of FRM, we can readily generalize this classic test to a general setting involving any number of groups without running into the notational complications found with multivariate U-statistics in Chapter 4, as illustrated in the next example.

**Example 6** (***K-sample Mann–Whitney–Wilcoxon rank sum test***). Let $y_{ki}$ denote some continuous response of interest from $K$ i.i.d. samples $(1 \leq i \leq n_k,\ 1 \leq k \leq K)$. Let $L = \binom{K}{2}$ and $q = 1$. Consider the following FRM:

$$\begin{aligned} f(y_{ki}, y_{hj}) = I_{\{y_{ki} \leq y_{hj}\}}, \quad h(\boldsymbol{\theta}) = \theta_{kh} \\ \boldsymbol{\theta} = \left(\theta_{12}, \ldots, \theta_{1K}, \theta_{23}, \ldots, \theta_{2K}, \ldots, \theta_{(K-1)K}\right)^\top \\ 1 \leq i \leq n_k, \quad 1 \leq j \leq n_h, \quad 1 \leq k < h \leq K \end{aligned}$$

It follows that

$$E\left[f(y_{ki}, y_{hj})\right] = h(\boldsymbol{\theta}) = \theta_{kh}, \quad 1 \leq i \leq n_k,\ 1 \leq j \leq n_h,\ 1 \leq k < h \leq K$$

If $y_{kj}$ have the same distribution across the $K$ samples, then $\theta_{kh} = \frac{1}{2}$ for all $1 \leq k < h \leq K$ and vice versa. Thus, we can test the null hypothesis, $H_0 : \theta_{kh} = \frac{1}{2}$ for all $1 \leq k < h \leq K$, to determine whether there is any location shift in CDF across the $K$ independent samples.

In Chapter 5, we developed a U-statistics based approach for simultaneously comparing the mean and variance between two groups. By combining the classic mean-based ANOVA with the generalized ANOVA for variance discussed earlier, we can readily generalize this approach by developing an FRM to jointly compare the mean and variance of a response across multiple samples, as illustrated below.

**Example 7** (***Model for K-sample mean and variance***). Consider $K$ i.i.d. samples and let $y_{ki}$ denote the response of interest from the $i$th subject within the $k$th group with mean $\mu_k$ and variance $\sigma_k^2$ $(1 \leq i \leq n_k,$ $1 \leq k \leq K)$. Let

$$\mathbf{f}(y_{ki}, y_{kj}) = \begin{pmatrix} y_{ki} \\ \frac{1}{2}\left(y_{ki} - y_{kj}\right)^2 \end{pmatrix}, \quad \mathbf{h}(\boldsymbol{\theta}) = \begin{pmatrix} \mu_k \\ \sigma_k^2 \end{pmatrix} = \boldsymbol{\theta}_k \tag{6.15}$$

$$\boldsymbol{\theta}_k = \left(\mu_k, \sigma_k^2\right)^\top, \quad \boldsymbol{\theta} = \left(\boldsymbol{\theta}_1^\top, \ldots, \boldsymbol{\theta}_K^\top\right)^\top, \quad (i,j) \in C_2^{n_k}, \quad 1 \leq k \leq K$$

Then, $\mathbf{f}(y_{ki}, y_{kj}) = \mathbf{h}(\boldsymbol{\theta}) = \boldsymbol{\theta}_k$ and the parameter vector $\boldsymbol{\theta}_k$ contains both the mean and variance of the $k$th sample. If $H_0 : \boldsymbol{\theta}_k = \boldsymbol{\theta}$ for all $1 \leq k \leq K$, then $y_{ki}$ not only have the same mean, but the same variance across the $K$ groups as well.

As discussed in Chapter 4, we can also perform joint inference about the mean and variance by using GEE II, which extends the mean-based distribution-free GLM to include additional models and estimating equations for the variance parameters. The difference is that the model for the variance component in GEE II is not really in the form of a regression since the dependent variable on the left side of the model specification also involves parameters. For example, consider again the problem for comparing the mean and variance of $y_{ki}$ across $K$ groups in Example 7. If we use GEE II to model both the mean and variance simultaneously, we have

$$E(y_{ki}) = \mu_k, \quad E(s_{ki}) = \sigma_k^2, \quad 1 \leq i \leq n_k, \ 1 \leq k \leq K \tag{6.16}$$

where $s_{ki} = (y_{ki} - \mu_k)^2$. Since the dependent variable $s_{ki}$ for the variance component in (6.16) is a function of the group mean $\mu_k$, (6.16) is not a regression in the usual sense. In contrast, the dependent vector $\mathbf{f}(y_{ki}, y_{kj})$ of the FRM in (6.15) does not contain any unknown parameter and thus conforms to the usual notion of regression analysis.

Count response arises quite often in biomedical and psychosocial research. The standard approach for modeling such a response is the Poisson distribution. Under the Poisson model, the mean and variance are constrained to be the same. However, in many applications, the variance is often larger than the mean, a phenomenon known as overdispersion. As discussed in Chapter 2, a common cause for overdispersion is data clustering and a popular approach for modeling this type of overdispersed data is the Negative Binomial (NB) model. In practical applications, overdispersion can be assessed by the deviance and Pearson's chi-square statistics (see Chapter 2). The appropriateness of an NB as a parametric model to address overdispersion can also be formally assessed by the Vuong likelihood-ratio test (Vuong, 1989). Within the framework of FRM, we can readily develop a distribution-free model to relax the parametric assumption in NB and to assess the appropriateness of fitting such an NB-like model.

**Example 8** (***Negative binomial model***). Consider $K$ i.i.d. samples and let $y_{ki}$ be some count response of interest from the $i$th subject within the $k$th group $(1 \leq i \leq n_k, 1 \leq k \leq K)$. Following Chapter 2, the standard parametric approach for modeling such a response type is the Poisson model. By removing the random component of GLM, we obtain a distribution-free, Poisson-like log-linear model as follows:

$$E(y_{ki}) = \exp(\mu_k), \quad 1 \leq i \leq n, 1 \leq k \leq K \tag{6.17}$$

If $y_{ki}$ does follow the Poisson distribution for each $k$, then we have

$$E\left[\frac{1}{2}(y_{ki}-y_{kj})^2\right]=\exp(\mu_k), \quad 1\leq i<j\leq n, 1\leq k\leq K$$

When overdispersion occurs, the above will not hold true. If $y_{ki}$ follows the NB model, we have:

$$E\left[\frac{1}{2}(y_{ki}-y_{kj})^2\right]=\exp(\mu_k)(1+\alpha_k\exp(\mu_k)) \qquad (6.18)$$
$$1\leq i<j\leq n, 1\leq k\leq K$$

where $\alpha_k$ is a parameter indicating the degree of overdispersion within each sample $(1 \leq k \leq K)$.

By combining (6.17) and (6.18), we arrive at an FRM that addresses overdispersion without imposing any parametric model on the data distribution:

$$E[\mathbf{f}(y_{ki},y_{kj})]=\mathbf{h}(\boldsymbol{\theta}), \quad 1\leq i<j\leq n_k, \quad 1\leq k\leq K \qquad (6.19)$$
$$\mathbf{f}(y_{ki},y_{kj})=\begin{pmatrix} y_{ki} \\ \frac{1}{2}(y_{ki}-y_{kj})^2 \end{pmatrix}$$
$$\mathbf{h}(\boldsymbol{\theta})=\begin{pmatrix} \exp(\mu_k) \\ \exp(\mu_k)(1+\alpha_k\exp(\mu_k)) \end{pmatrix}$$
$$\boldsymbol{\theta}_k=(\mu_k,\alpha_k)^\top, \quad \boldsymbol{\theta}=\left(\boldsymbol{\theta}_1^\top,\ldots,\boldsymbol{\theta}_K^\top\right)^\top$$

As in the parametric model case, we can assess overdisperson under this distribution-free, NB-like FRM (6.19) by testing the hypothesis:

$$H_0:\alpha_k=0 \quad \text{versus} \quad H_a:\alpha_k>0, \quad 1\leq k\leq K$$

Note that in the parametric NB model $NB(\mu,\alpha)$, $\alpha=0$ does not correspond to a valid NB distribution since the PDF of $NB(\mu,\alpha)$ becomes undefined in this case. In addition, $\alpha>0$. Thus, the constraint $\alpha>0$ must be imposed when fitting the parametric NB model. In contrast, this *boundary* issue does not arise for the FRM defined in (6.19) since $\alpha_k$ can be negative as long as the right side in (6.18) stays positive. Note also that the Poisson model is not a member of the NB family, though it can be viewed as the limiting case of the NB distribution when $\alpha \to 0$. As a result, assessing the appropriateness of Poisson within the NB context is complicated since the Poisson model corresponds to the boundary value 0 in the NB family and regular maximum likelihood theory may not apply. However, for the distribution-free FRM defined in (6.19), this boundary issue is again irrelevant since $\alpha$ can be 0 or even negative, in which case, there is under-dispersion in the data.

As discussed in Chapter 2, count outcomes arise in many studies in biomedical and behavioral research that do not follow either the Poisson or NB model because of structural zeros in the data distribution. In such cases, the frequency of zeros in the data exceeds their expected frequency by either the Poisson or NB model, also causing overdispersion. The zero-inflated Poisson (ZIP) is widely used to address this type of overdispersion. We discussed the parametric ZIP and the difficulty in developing a distribution-free version of the model under the distribution-free GLM in Chapter 2. The major problem in the latter is the general lack of information to identify model parameters from a mixture-distribution based model such as ZIP. By including a component for modeling the second order moment, we will be able to define an FRM to identify the parameters of ZIP in a distribution-free setting, as we discuss in the next example.

**Example 9** (***Zero-inflated Poisson***). Within the context of Example 8, now consider modeling $y_{ki}$ in the presence of structural zeros using the ZIP model. This ZIP approach models the structural zeros separately using a dedicated point mass at zero so that the remaining amount of sampling zeros, together with the positive responses, can be modeled by the Poisson.

Let $f_0(y)$ denote the PDF of a degenerate distribution centered at zero and $f_{P,k}(y)$ the PDF of a Poisson with mean $\mu_k$. Then, the PDF of the parametric ZIP for $y_{ki}$ is defined by the following mixture of two component distributions, $f_0(y)$ and $f_{P,k}(y)$,

$$f_{ZIP,k}(y_{ki}) = (1 - \rho_k) f_{P,k}(y_{ki}) + \rho_k f_0(y_{ki}) \qquad (6.20)$$
$$y_{ki} = 0, 1, \ldots, \quad 1 \le k \le K$$

In the above, $f_0(\cdot)$ absorbs the structural zeros, while $f_{P,k}(\cdot)$ models the remaining observations to follow the Poisson distribution.

Under the distribution-free GLM, only the conditional mean response given the $k$th sample is specified, that is,

$$E(y_{ki}) = (1 - \rho_k) \exp(\mu_k), \quad 1 \le k \le K \qquad (6.21)$$

The above specification will not yield unique estimates of $\rho$ and $\mu$ since the mean-based GLM in (6.21) is overparameterized. One approach to address this issue of model identifiability without reverting back to the parametric ZIP defined in (6.20) is to add a component for modeling the variance of $y_{ki}$. Under this approach, we supplement the mean response with a component for the variance to obtain

$$E\left[\frac{1}{2}(y_{ki}-y_{kj})^2\right] = \exp(\mu_k)(1-\rho_k)[1+\rho_k\exp(\mu_k)] \qquad (6.22)$$

By combining (6.21) and (6.22), we obtain a distribution-free, ZIP-like model that rectifies the identifiability issue with the distribution-free GLM in (6.21):

$$E[\mathbf{f}(y_{ki},y_{kj})] = \mathbf{h}(\boldsymbol{\theta}), \quad 1 \le i < j \le n_k, \quad 1 \le k \le K \qquad (6.23)$$

$$\mathbf{f}(y_{ki},y_{kj}) = \begin{pmatrix} y_{ki} \\ \frac{1}{2}(y_{ki}-y_{kj})^2 \end{pmatrix}$$

$$\mathbf{h}(\boldsymbol{\theta}) = \begin{pmatrix} (1-\rho_k)\exp(\mu_k) \\ \exp(\mu_k)(1-\rho_k)[1+\rho_k\exp(\mu_k)] \end{pmatrix}$$

$$\boldsymbol{\theta}_k = (\mu_k,\rho_k)^\top, \quad \boldsymbol{\theta} = \left(\boldsymbol{\theta}_1^\top,\ldots,\boldsymbol{\theta}_K^\top\right)^\top$$

The functional response $\mathbf{f}(y_{ki},y_{kj})$ of this FRM is the same as the one defined in (6.19) for the Poisson-like log-linear FRM. However, the FRM in (6.23) differs from the one in (6.19) in the mean response to reflect the assumptions between the NB and ZIP models.

All the examples above concern cross-sectional data. As in the case of distribution-free GLM, FRM also applies to longitudinal study designs. In Chapter 4, we discussed models for longitudinal study data and contrasted the two popular approaches, the generalized linear mixed-effects model (GLMM) and the generalized distribution-free GLM (or GEE), in terms of their relative strengths and weaknesses. In the next example, we develop an FRM to integrate the strengths of the linear mixed-effects model (LMM) with the GEE.

**Example 10** (***Linear mixed-effects model***). Consider a longitudinal study with $K$ independent treatment groups and $m$ assessment times. Let $y_{kit}$ denote some response of interest from the $i$th subject at time $t$ within the $k$th group ($1 \le i \le n_k$, $1 \le t \le m$, $1 \le k \le K$). We are interested in modeling the growth-curve over time and comparing such curves across the different treatment conditions. For convenience, we assume a linear time trend for each group and allow the slope to vary across the treatment groups.

The normal-based LMM has the following form:

$$\begin{aligned} y_{kit} &= \beta_{k0} + t\beta_{k1} + b_{k0i} + tb_{k1i} + \epsilon_{kit} \\ \mathbf{b}_{ki} &= (b_{k0i}, b_{k1i})^\top \sim \text{i.i.d.} N(0, D_k) \\ \boldsymbol{\epsilon}_{ki} &= (\epsilon_{ki1}, \ldots, \epsilon_{kim})^\top \sim \text{i.i.d.} N\left(0, G = diag\left(\sigma_t^2\right)\right) \end{aligned} \tag{6.24}$$

where $diag\left(\sigma_t^2\right)$ denotes a diagonal matrix with $\sigma_t^2$ on the $t$th diagonal. In this model, $\beta_{k0}+\beta_{k1}t$ represents the mean response of $y_{kit}$ for the $k$th group and $b_{k0i}+tb_{ki}$ the deviation of the $i$th individual from the group mean. Thus, the fixed-effects parameters, $(\beta_{k0}, \beta_{k1})$, represent the intercept and slope for the group response, while the random-effects parameters, $D_k$, measure the variability of individual deviations from the group mean within each sample. The model error $\epsilon_{kit}$ captures the variation in response over time within each individual.

Inference for parametric LMM under the normal distribution assumption has been discussed in Chapter 4. We now consider distribution-free inference about the parameters of interest by developing an appropriate FRM.

First, it is readily checked that

$$\begin{aligned} E\left(y_{kit}\right) &= \beta_{k0} + t\beta_{k1} \\ E\left[\frac{1}{2}\left(y_{kis} - y_{kjs}\right)\left(y_{kit} - y_{kjt}\right)\right] &= \left[\sigma_{kb0}^2 + (s+t)\,\sigma_{kb01} + st\sigma_{kb1}^2\right] + \sigma_t^2\delta_{st} \end{aligned} \tag{6.25}$$

where $\sigma_{kb0}^2$, $\sigma_{kb1}^2$ and $\sigma_{kb01}$ denote the diagonal and off-diagonal elements of $D_k$ and $\delta_{st} = 1$ if $s = t$ and 0 if otherwise. By (6.25), we define the vectors of functional and mean response as follows:

$$\begin{aligned} f_{k1t}\left(\mathbf{y}_{ki}\right) &= y_{kit}, \quad f_{k2st}\left(\mathbf{y}_{ki}, \mathbf{y}_{kj}\right) = \frac{1}{2}\left(y_{kis} - y_{kjs}\right)\left(y_{kit} - y_{kjt}\right) \\ \mathbf{f}_{k1} &= \left(f_{k11}, \ldots, f_{k1m}\right)^\top, \quad \mathbf{f}_{k2} = \left(f_{k211}, f_{k212}, \ldots, f_{k2(m-1)m}, f_{k2mm}\right)^\top \\ \mathbf{f} &= \left(\mathbf{f}_{11}^\top, \ldots, \mathbf{f}_{K1}^\top, \mathbf{f}_{21}^\top, \ldots, \mathbf{f}_{K2}^\top\right)^\top \end{aligned}$$

and

$$\begin{aligned} h_{k1t}(\boldsymbol{\theta}) &= \beta_{k0} + t\beta_{k1}, \quad \mathbf{h}_{k1} = \left(h_{k11}, \ldots, h_{k1m}\right)^\top \\ h_{k2st}(\boldsymbol{\theta}) &= \left[\sigma_{kb0}^2 + (s+t)\,\sigma_{kb01} + st\sigma_{kb1}^2\right] + \sigma_t^2\delta_{st} \\ \mathbf{h}_{k2} &= \left(h_{k211}, h_{k212}, \ldots, h_{k2(m-1)m}, h_{k2mm}\right)^\top \\ \mathbf{h} &= \left(\mathbf{h}_{11}^\top, \ldots, \mathbf{h}_{K1}^\top, \mathbf{h}_{21}^\top, \ldots, \mathbf{h}_{K2}^\top\right)^\top \end{aligned}$$

where $\boldsymbol{\theta}$ denotes the collection of all the model parameters, $\beta_{k0}$, $\beta_{k1}$, $D_k$ and $G$ $(1 \leq k \leq K)$. It follows from (6.25) that the FRM for distribution-free inference of the normal-based LMM in (6.24) is given by

$$E\left[f_{k1t}\left(\mathbf{y}_{ki}\right)\right] = h_{k1t}\left(\boldsymbol{\theta}\right), \quad E\left[f_{k2st}\left(\mathbf{y}_{ki}, \mathbf{y}_{kj}\right)\right] = h_{k2st}\left(\boldsymbol{\theta}\right)$$

With this FRM, we are able to not only test hypotheses concerning the parameters for the mean response, $\beta_{k0}$ and $\beta_{k1}$, but also those pertaining to the variability of individual subjects, $D_k$.

### 6.2.2 Models for Regression Analysis

In this section, we discuss FRM for regression analysis. Consider a class of distribution-free regression models defined by

$$E\left[\mathbf{f}\left(\mathbf{y}_{i_1}, \ldots, \mathbf{y}_{i_q}\right) \mid \mathbf{x}_{i_1}, \ldots, \mathbf{x}_{i_q}\right] = \mathbf{h}\left(\mathbf{x}_{i_1}, \ldots, \mathbf{x}_{i_q}; \boldsymbol{\theta}\right) \tag{6.26}$$
$$\left(i_1, \ldots, i_q\right) \in C_q^n, \quad 1 \leq q, \ 1 \leq i \leq n$$

where $\mathbf{f}$, $\mathbf{h}$, $\boldsymbol{\theta}$, $n$ and $q$ all have the same interpretation as before. Note that as in the case of group comparison, the vectors of individual response, $\mathbf{y}_i, \ldots, \mathbf{y}_n$, are assumed to be stochastically independent.

The FRM defined in (6.26) subsumes all regression models discussed in Chapters 2 and 4 as special cases. For example, by setting:

$$q = 1, \quad f\left(y_i\right) = y_i, \quad h_i = h\left(\mathbf{x}_i; \boldsymbol{\beta}\right)$$

where $h(\cdot;\cdot)$ denotes some nonlinear function, we obtain from (6.26) the distribution-free, nonlinear regression model:

$$E\left(y_i \mid \mathbf{x}_i\right) = h\left(\mathbf{x}_i; \boldsymbol{\beta}\right), \quad 1 \leq i \leq n \tag{6.27}$$

If $h\left(\mathbf{x}_i; \boldsymbol{\beta}\right) = h\left(\mathbf{x}_i^\top \boldsymbol{\beta}\right)$, we obtain the distribution-free GLM. Of course, if $h\left(\mathbf{x}_i; \boldsymbol{\beta}\right) = \mathbf{x}_i^\top \boldsymbol{\beta}$, we obtain the distribution-free linear model.

The general FRM in (6.26) also applies to clustered and longitudinal study data. For example, consider a longitudinal study with $m$ assessment times. Let $y_{it}$ and $\mathbf{x}_{it}$ denote the response and vector of independent variables of interest from the $i$th subject at the $t$th assessment $(1 \leq i \leq n$, $1 \leq t \leq m)$. Let

$$q = 1, \quad \mathbf{y}_i = \left(y_{i1}, \ldots, y_{im}\right)^\top, \quad h_{it} = h_t\left(\mathbf{x}_{it}; \boldsymbol{\beta}\right)$$
$$\mathbf{h} = \left(h_{i1}, \ldots, h_{im}\right)^\top, \quad f_t\left(\mathbf{y}_i\right) = y_{it}, \quad \mathbf{f}\left(\mathbf{y}_i\right) = \mathbf{y}_i$$

We then immediately obtain from (6.26) the generalization of the class of models in (6.27) for repeated measures data arising from longitudinal study designs:

$$E\left[f_t\left(\mathbf{y}_i\right) \mid \mathbf{x}_{it}\right] = E\left(y_{it} \mid \mathbf{x}_{it}\right) = h_t\left(\mathbf{x}_{it}; \boldsymbol{\beta}\right), \quad 1 \leq i \leq n, \ 1 \leq t \leq m \quad (6.28)$$

In particular, if $h_t\left(\mathbf{x}_{it}; \boldsymbol{\beta}\right) = h_t\left(\mathbf{x}_{it}^{\top} \boldsymbol{\beta}\right)$, (6.28) reduces to the distribution-free GLM for continuous, binary and count response studied in Chapter 4.

The classic mean-based distribution GLM only models the conditional mean of the response given the independent variables. As a result, parametric models that differ in second and higher order moments such as the Poisson and negative binomial log-linear models are not distinguishable when modeled under this paradigm. It is this insensitivity to parametric distribution assumptions that enables the distribution-free GLM to provide robust estimates and valid inference under a wider class of data distributions. A potential drawback of this classic approach is that unlike its parametric counterpart it may not provide asymptotically efficient estimates when the data at hand does support the parametric assumption. This limitation is exacerbated by our interest in modeling second and higher order moments in many studies. We can address this weakness with an ad-hoc approach similar to the one discussed in Chapters 4: by adding to (6.28) a component for modeling the conditional variance of the mean given the independent variables. As noted in Example 6, such a GEE II model does not constitute a formal extension of (6.28) since the dependent variable for the variance component involves unknown parameters and thus GEE II is not really a regression in the usual sense. Within the general FRM framework, we can readily accommodate modeling components for variance or higher order moments without such a conceptual difficulty, as illustrated by the next example.

**Example 11** (***Model for overdispersion***). Consider a cross-sectional study with $n$ subjects. Let $y_i$ and $\mathbf{x}_i$ denote some count response and vector of independent variables of interest from the $i$th subject. The classic distribution-free Poisson-like log-linear model is given by

$$E\left(y_i \mid \mathbf{x}_i\right) = \mu_i = \exp\left(\mathbf{x}_i^{\top} \boldsymbol{\beta}\right), \quad 1 \leq i \leq n \quad (6.29)$$

As discussed in Chapter 2, (6.29) provides valid inference under a wide class of distributions including Poisson and negative binomial. If we are interested in formally testing over- or underdispersion, we may use a GEE II by including a component for modeling the conditional variance of $y_i$ given $\mathbf{x}_i$ as follows:

$$E(s_i \mid \mathbf{x}_i) = \lambda^2 \mu_i, \quad s_i = (y_i - \mu_i)^2, \quad 1 \leq i \leq n \tag{6.30}$$

This GEE II defined by (6.29) and (6.30) allows us to make simultaneous inferences about $\boldsymbol{\beta}$ and $\lambda^2$. Thus, we can test for over- or underdispersion by considering the following hypotheses:

$$\begin{aligned} \text{Overdispersion} &: H_0 : \lambda^2 = 1 \quad \text{versus} \quad H_a : \lambda^2 > 1 \\ \text{Underdispersion} &: H_0 : \lambda^2 = 1 \quad \text{versus} \quad H_a : \lambda^2 < 1 \end{aligned}$$

The dependent variable $s_i$ depends on $\mu_i$ which is a function of the parameter vector $\boldsymbol{\beta}$, and (6.30) is not a regression model. To define an FRM as a formal regression alternative to GEE II, let

$$\begin{aligned} f_1(y_i, y_j) &= y_i + y_j, \quad f_2(y_i, y_j) = (y_i - y_j)^2 \\ h_1(\mathbf{x}_i, \mathbf{x}_j; \boldsymbol{\beta}, \lambda^2) &= \mu_i + \mu_j, \quad h_2(\mathbf{x}_i, \mathbf{x}_j; \boldsymbol{\beta}, \lambda^2) = \lambda^2(\mu_i + \mu_j) + (\mu_i - \mu_j)^2 \\ \mathbf{f} &= (f_1, f_2)^\top, \quad \mathbf{h} = (h_1, h_2)^\top \end{aligned}$$

Then, it is readily checked that the FRM below defines the same model as GEE II above (see exercise):

$$E(\mathbf{f}(y_i, y_j) \mid \mathbf{x}_i, \mathbf{x}_j) = \mathbf{h}(\mathbf{x}_i, \mathbf{x}_j; \boldsymbol{\beta}, \lambda^2), \quad 1 \leq i < j \leq n \tag{6.31}$$

However, unlike the GEE II, the FRM in (6.31) is a generalization of regression model.

One of the most popular models for regression analysis of longitudinal data with continuous response is the linear mixed-effects models (LMM). Consider a longitudinal study with $n$ subjects and $m$ assessment points. Let $y_{it}$ denote some response, and $\mathbf{u}_{it}$ and $\mathbf{v}_{it}$ vectors of independent variables from the $i$th subject at the $t$th assessment. Note that $\mathbf{u}_{it}$ and $\mathbf{v}_{it}$ may overlap and share common variables. Let $\mathbf{x}_{it} = \left(\mathbf{u}_{it}^\top, \mathbf{v}_{it}^\top\right)^\top$. The LMM is specified as follows:

$$\begin{aligned} &y_{it} = \mathbf{u}_{it}^\top \boldsymbol{\beta} + \mathbf{v}_{it}^\top \mathbf{b}_i + \epsilon_{it}, \quad \mathbf{b}_i \sim \text{i.i.d.} N(\mathbf{0}, D) \\ &\boldsymbol{\epsilon}_i \sim \text{i.i.d.} N\left(\mathbf{0}, G = diag\left(\sigma_t^2\right)\right), \quad \mathbf{b}_i \perp \boldsymbol{\epsilon}_i, \quad 1 \leq i \leq n, 1 \leq t \leq m \end{aligned} \tag{6.32}$$

where $\mathbf{b}_i = (b_{i1}, \ldots, b_{ik})^\top$, $\boldsymbol{\epsilon}_i = (\epsilon_{i1}, \ldots, \epsilon_{im})^\top$, $\mathbf{u}_{it}$ ($\mathbf{v}_{it}$) denote the vectors of independent variables for the fixed (random) effect, $diag\left(\sigma_t^2\right)$ a $m \times m$

diagonal matrix with $\sigma_t^2$ on the $t$th diagonal and $D$ the variance matrix of $\mathbf{b}_i$. In (6.32), the parameter vector $\boldsymbol{\beta}$ describes the population or fixed effects, the elements of the matrix $D$ measure individual deviations from the popular mean or random effects and the error term $\epsilon_{it}$ accounts for the within-individual variation in the response over time. Under the model assumptions, inference for $\boldsymbol{\beta}$ and $D$ based on the likelihood theory has been discussed in Chapter 4.

A primary limitation of the normal-based LMM in (6.32) is the strong distribution assumption on both the response $\mathbf{y}_i$ and the random effects $\mathbf{b}_i$. For robust inference, we can consider the following distribution-free GLM as an alternative to model the longitudinal data:

$$E\left(y_{it} \mid \mathbf{u}_{it}\right) = \mathbf{u}_{it}^{\top}\boldsymbol{\beta}, \quad 1 \leq i \leq n, 1 \leq t \leq m \tag{6.33}$$

However, the above model can only be used to make inference about the fixed effect $\boldsymbol{\beta}$. In many applications, we are also interested in inference about the random effect parameters $D$. The classic mean-based distribution LMM in (6.33) does not contain information for inference about this parameter matrix. By capitalizing on the strength of FRM for modeling second order moments, we can readily develop an FRM model to address this issue.

**Example 12** (***Linear mixed-effects model***). To derive a distribution-free FRM for the LMM in (6.32), first define the functional response as follows:

$$\begin{aligned} f_{1t}\left(\mathbf{y}_i, \mathbf{y}_j \mid \mathbf{x}_i, \mathbf{x}_j\right) &= y_{it} + y_{jt}, \quad 1 \leq t \leq m \\ f_{2st}\left(\mathbf{y}_i, \mathbf{y}_j \mid \mathbf{x}_i, \mathbf{x}_j\right) &= \left(y_{is} - y_{js}\right)\left(y_{it} - y_{jt}\right), \quad 1 \leq s, t \leq m \\ \mathbf{f}_1 &= \left(f_{11}, \ldots, f_{1m}\right)^{\top} \\ \mathbf{f}_2 &= \left(f_{211}, f_{212}, \ldots, f_{2(m-1)m}, f_{2mm}\right)^{\top} \\ \mathbf{f} &= \left(\mathbf{f}_1^{\top}, \mathbf{f}_2^{\top}\right)^{\top} \end{aligned} \tag{6.34}$$

Next, we define the mean of functional response by:

$$\begin{aligned} h_{1t}(\boldsymbol{\theta}) &= \left(\mathbf{u}_{it} + \mathbf{u}_{jt}\right)^{\top}\boldsymbol{\beta} \\ h_{2st}(\boldsymbol{\theta}) &= \boldsymbol{\beta}^{\top}\left(\mathbf{u}_{is} - \mathbf{u}_{js}\right)\left(\mathbf{u}_{it} - \mathbf{u}_{jt}\right)^{\top}\boldsymbol{\beta} + \mathbf{v}_{is}^{\top}D\mathbf{v}_{it} + \mathbf{v}_{js}^{\top}D\mathbf{v}_{jt} + 2\sigma_t^2\delta_{st} \\ \mathbf{h}_1 &= \left(h_{11}, \ldots, h_{1m}\right)^{\top}, \quad \mathbf{h}_2 = \left(h_{211}, h_{212}, \ldots, h_{2(m-1)m}, h_{2mm}\right)^{\top} \\ \mathbf{h} &= \left(\mathbf{h}_1^{\top}, \mathbf{h}_2^{\top}\right)^{\top} \end{aligned} \tag{6.35}$$

where $\boldsymbol{\theta}$ includes all the parameters, $\boldsymbol{\beta}$, $D$ and $\boldsymbol{\sigma} = \left(\sigma_1^2, \ldots, \sigma_m^2\right)$. Then, it is readily checked that under (6.32) $\mathbf{f}$ and $\mathbf{h}$ satisfy the following (see exercise):

$$E\left[\mathbf{f}_1\left(\mathbf{y}_i, \mathbf{y}_j\right) \mid \mathbf{x}_i, \mathbf{x}_j\right] = \mathbf{h}_1\left(\boldsymbol{\theta}\right), \quad E\left[\mathbf{f}_2\left(\mathbf{y}_i, \mathbf{y}_j\right) \mid \mathbf{x}_i, \mathbf{x}_j\right] = \mathbf{h}_2\left(\boldsymbol{\theta}\right) \qquad (6.36)$$

Structural equations models (SEM) are widely used in psychosocial research. One important application of SEM in this area is modeling the mediational process. We discussed SEM and its applications in mediation analysis in Chapter 4. We also discussed inference for SEM using both the classic likelihood theory and modern GEE II. As many applications of SEM involve a large number of variables, likelihood methods based on parametric assumptions (often multivariate normal) do not provide reliable estimates. Within the framework of FRM, we can readily develop a distribution-free approach to address this key limitation. In the next example, we illustrate how such a development is carried out for a relatively simple SEM for mediation analysis.

**Example 13** (***Structural equation model***). The primary hypothesis of interest in a mediation analysis is to test whether the effect of a factor $x_0$ (e.g., an intervention) on the response $y$ can be mediated by a change in the mediating variable $z$. In a full mediation process, the effect of $x_0$ on $y$ is 100% mediated by $z$, that is, in the presence of $z$, the pathway connecting $x_0$ and $y$ is completely broken so that $x_0$ has no direct effect on $y$. In most applications, however, partial mediation is more common, in which case $z$ only mediates part of the effect of $x_0$ on $y$, that is, $x_0$ has some residual direct effect after $z$ is introduced into the model.

Consider an i.i.d. sample with $n$ subjects. In addition to the response $y_i$, predictor $x_{i0}$ and mediator $z_i$, let $x_{i1}, \ldots, x_{ip}$ denote a set of covariates from the $i$th subject $(1 \leq i \leq n)$. When modeled using SEM, the mediation process has the following form:

$$\begin{aligned} y_i &= \gamma_{10} x_{i0} + \gamma_{11} x_{i1} + \cdots + \gamma_{1p} x_{ip} + \beta z_i + \epsilon_{yi} \qquad (6.37) \\ z_i &= \gamma_{20} x_{i0} + \gamma_{21} x_{i1} + \cdots + \gamma_{2p} x_{ip} + \epsilon_{zi} \\ E\left(\epsilon_{yi}\right) = 0, &\quad Cov\left(z_i, \epsilon_{yi}\right) = 0, \quad E\left(\epsilon_{zi}\right) = 0 \\ Cov\left(x_{ik}, \epsilon_{yi}\right) = 0, &\quad Cov\left(x_{ik}, \epsilon_{zi}\right) = 0, \quad 0 \leq k \leq p, \ 1 \leq i \leq n \end{aligned}$$

Let

$$\mathbf{x}_i = (x_{i0}, x_{i1}, \ldots, x_{ip})^\top, \quad \mathbf{y}_i = (y_{i1}, y_{i2})^\top = (y_i, z_i)^\top \tag{6.38}$$

$$B = \begin{pmatrix} 1 & -\beta \\ 0 & 1 \end{pmatrix}, \quad \Gamma = \begin{pmatrix} \gamma_{10} & \gamma_{11} & \cdots & \gamma_{1p} \\ \gamma_{20} & \gamma_{2p} & \cdots & \gamma_{2p} \end{pmatrix}$$

Under the normal distribution assumption, we can express (6.37) in matrix form as follows:

$$\mathbf{y}_i = B^{-1}(\Gamma \mathbf{x}_i + \boldsymbol{\epsilon}_i), \quad \boldsymbol{\epsilon}_i = (\epsilon_{yi}, \epsilon_{zi})^\top \sim \text{i.i.d.} N(0, \Sigma_\epsilon) \tag{6.39}$$

where $B$ and $\Gamma$ are parameter matrices of interest and the matrix $\Sigma_\epsilon$ consists of nuisance parameters.

To derive an FRM for distribution-free inference, define the functional response as follows:

$$f_{1t}(\mathbf{y}_i) = y_{it}, \quad f_{2st}(\mathbf{y}_i, \mathbf{y}_j) = \frac{1}{2}(y_{is} - y_{js})(y_{it} - y_{jt})$$

$$\mathbf{f}_1 = (f_{11}, f_{12})^\top, \quad \mathbf{f}_2 = (f_{211}, f_{212}, f_{222})^\top, \quad \mathbf{f} = \left(\mathbf{f}_1^\top, \mathbf{f}_2^\top\right)^\top$$

Similarly, define the mean of the functional response as:

$$h_{1t}(\boldsymbol{\theta}) = \mathbf{b}_t^\top \Gamma \mathbf{x}_i, \quad h_{2st}(\boldsymbol{\theta}) = \mathbf{b}_s^\top \Gamma (\mathbf{x}_i - \mathbf{x}_j)(\mathbf{x}_i - \mathbf{x}_j)^\top \Gamma^\top \mathbf{b}_t + \mathbf{b}_s^\top \Sigma_\epsilon \mathbf{b}_t$$

$$\mathbf{h}_1 = (h_{11}, h_{12})^\top, \quad \mathbf{h}_2 = (h_{211}, h_{212}, h_{222})^\top, \quad \mathbf{h} = \left(\mathbf{h}_1^\top, \mathbf{h}_2^\top\right)^\top$$

where $\mathbf{b}_t^\top$ denotes the $t$th row of $B^{-1}$. If $z$ is a mediator, then $\gamma_{20} \neq 0$ and $\beta \neq 0$. The hypothesis of a full mediation process is supported if the condition $\gamma_{20} \neq 0$ and $\beta = 0$ holds true and vice versa.

In Section 6.2, we discussed FRM based negative binomial (NB) and zero-inflated Poisson (ZIP) models for count response for comparing multiple groups when there is over- or underdispersion in the data. We next discuss similar FRM-based models for regression analysis.

**Example 14** (***Negative binomial model***). Consider a sample of $n$ subjects and let $y_i$ be a count response and $\mathbf{x}_i$ a set of independent variables. As in the group comparison case, if there is overdispersion due to data clustering, the Poisson log-linear model is inappropriate and the NB log-linear model is often used to address this issue. The mean-based distribution-free Poisson- or NB-like log-linear model has the form:

$$E(y_i \mid \mathbf{x}_i) = \mu_i, \quad \log(\mu_i) = \mathbf{x}_i^\top \boldsymbol{\beta}, \quad 1 \leq i \leq n \tag{6.40}$$

Since NB has the same mean as the Poisson, the above specification does not distinguish between the two models. As in the group comparison case, we can add a component for modeling the variance to tease out such a second order difference between the two models. Under the NB model, the conditional variance of $y_i$ given $\mathbf{x}_i$ is

$$Var\left(y_i \mid \mathbf{x}_i\right) = \mu_i\left(1 + \alpha\mu_i\right), \quad 1 \leq i \leq n \tag{6.41}$$

Thus, we define an FRM by adding the following component to the mean-based model in (6.40):

$$E\left[\left(y_i - y_j\right)^2 \mid \mathbf{x}_i, \mathbf{x}_j\right] = \mu_i + \mu_j + \alpha\left(\mu_i^2 + \mu_j^2\right) + \left(\mu_i - \mu_j\right)^2 \tag{6.42}$$
$$1 \leq i < j \leq n$$

where $\alpha$ is a parameter indicating the degree of overdispersion.

Note that by adding the above variance component, we have implicitly imposed another constraint upon modeling the relationship between $y_i$ and $\mathbf{x}_i$, that is, in addition to satisfying the specification for the mean response (6.40), the conditional second order moment must also follow a particular relationship as defined in (6.42). If the latter relationship holds true, then the FRM increases the efficiency for inference about $\boldsymbol{\beta}$. Otherwise, inference based on the FRM model may be incorrect. For example, if $y_i$ does follow a negative binomial distribution conditional on $\mathbf{x}_i$, then (6.42) is true and inference based on the FRM is more efficient than the mean-based model in (6.40). However, if the conditional variance of $y_i$ given $\mathbf{x}_i$ does not satisfy (6.40), inference about $\boldsymbol{\beta}$ based on the FRM may be incorrect.

Under the FRM, we can assess whether there is overdispersion by testing the following hypothesis,

$$H_0 : \alpha = 0 \quad \text{versus} \quad H_a : \alpha > 0$$

As in the discussion of the NB model within the context of comparing multiple samples in Section 6.2.1, $\alpha$ can be negative in the FRM model. Thus, the boundary issue for $\alpha$ in the parametric NB model does not exist and the FRM defined by (6.40) and (6.42) can be used to assess either over- or underdispersion in the data distribution.

**Example 15** (***Zero-inflated Poisson***). Within the context of Example 14, if there are structural zeros in the distribution of $y_i$, then the FRM developed is not appropriate and ZIP must be used to address this type of overdispersion.

As discussed in Chapter 2, the parametric ZIP is based on the following two-component mixture model:

$$f_{ZIP}(y_i \mid \mathbf{x}_i) = \rho(\mathbf{x}_i) f_0(y_i) + [1 - \rho(\mathbf{x}_i)] f_P(y_i \mid \mathbf{x}_i) \quad (6.43)$$
$$y_i = 0, 1, \ldots$$

where $f_0(y_i)$ denotes a degenerate distribution at 0 to account for the structural zeros, $f_P(y_i \mid \mathbf{x}_i)$ the Poisson distribution to model the remaining observations and $\rho(\mathbf{x}_i)$ a mixing proportion of the mixture model. The two components $\rho(\mathbf{x}_i)$ and $f_P(y_i \mid \mathbf{x}_i)$ are typically modeled to depend on two different subsets of independent variables, $\mathbf{u}_i$ and $\mathbf{v}_i$, which may overlap and share some common variables. The Poisson mean of $y_i$ and the mixing proportion $\rho(\mathbf{u}_i)$ are modeled as a function of $\mathbf{u}_i$ and $\mathbf{v}_i$, respectively, as follows:

$$\operatorname{logit}(\rho(\mathbf{u}_i)) = \log\left(\frac{\rho(\mathbf{u}_i)}{1-\rho(\mathbf{u}_i)}\right) = \mathbf{u}_i^\top \boldsymbol{\beta}_u, \quad \log(\mu(\mathbf{v}_i)) = \mathbf{v}_i^\top \boldsymbol{\beta}_v$$

Inference under the parametric assumption has been discussed in Chapter 2.

For distribution-free inference under FRM, we first specify the mean response of $y_i$ based on the mixture model as follows:

$$E(f_1(y_i) \mid \mathbf{x}_i) = E(y_i \mid \mathbf{x}_i) = [1 - \rho(\mathbf{u}_i)] \mu(\mathbf{v}_i) = (1 - \rho_i) \mu_i \quad (6.44)$$

As in the group comparison case discussed in Section 6.2.1, the above mean specification is not sufficient to identify the parameters, $\boldsymbol{\beta}_u$ and $\boldsymbol{\beta}_v$. Now let $f_2(y_i, y_j) = (y_i - y_j)^2$. It is readily checked that under the parametric ZIP assumption, we have

$$\begin{aligned} E[f_2(y_i, y_j) \mid \mathbf{x}_i, \mathbf{x}_j] = {} & \mu_i (1 - \rho_i)(1 + \mu_i \rho_i) + \\ & + \mu_j (1 - \rho_j)(1 + \mu_j \rho_j) + \\ & + \left[(1 - \rho_i) \mu_i - (1 - \rho_j) \mu_j\right]^2 \end{aligned} \quad (6.45)$$

Thus, by combining (6.44) and (6.45), we obtain an FRM for distribution-free inference about the parameters of interest, $\boldsymbol{\beta} = \left(\boldsymbol{\beta}_u^\top, \boldsymbol{\beta}_v^\top\right)^\top$.

## 6.3 Model Estimation

In Chapters 2 and 4, we discussed inference for regression models under the two dueling likelihood theory and estimating equation based paradigms.

Regardless of which approach is used, an assumption fundamental to both is independence in the response variable across the sampling subjects. For example, consider a cross-sectional study consisting of $n$ subjects and let $y_i$ and $\mathbf{x}_i$ denote some continuous response and $p \times 1$ vector of independent variables of interest. Suppose that we model the relation between $y_i$ and $\mathbf{x}_i$ using a generalized linear model:

$$E(y_i \mid \mathbf{x}_i) = h\left(\mathbf{x}_i^\top \boldsymbol{\beta}\right) = h_i, \quad 1 \leq i \leq n \tag{6.46}$$

where $\boldsymbol{\beta}$ is the $p \times 1$ vector of parameters of interest. Then, as discussed in Chapter 2, irrespective of the inference approach used, estimates of $\boldsymbol{\beta}$ are obtained as solutions to the following equation:

$$\mathbf{w}_n(\boldsymbol{\beta}) = \sum_{i=1}^{n} \mathbf{w}_{ni}(\boldsymbol{\beta}) = \sum_{i=1}^{n} G_i(\mathbf{x}_i)(y_i - h_i) = \mathbf{0} \tag{6.47}$$

where $G_i(\mathbf{x}_i)$ is some $p \times 1$ function of $\mathbf{x}_i$ (not $y_i$). In (6.47), $\mathbf{w}_n(\boldsymbol{\beta})$ is a sum of i.i.d. random quantities $G_i(\mathbf{x}_i) S_i$. As seen in Chapters 2, this particular form of $\mathbf{w}_n(\boldsymbol{\beta})$ allows us to apply standard asymptotic techniques, such as the central limit theorem to show large sample properties of estimates of $\boldsymbol{\beta}$ obtained from (6.47).

Although exceptions occur when responses become clustered such as those arising from repeated assessments on the same subject in longitudinal studies, the basic form of the estimating equation does not change. For example, as discussed in Chapter 4, for longitudinal data analysis, estimates for distribution-free inference are defined by equations identical in form to (6.47). For maximum likelihood inference, the score vector equation is also a sum of i.i.d. random quantities, although the summand may not have an explicit partition involving the theoretical residual $(y_i - h_i)$ as does the estimating equation for distribution-free inference.

For an FRM, the dependent variable is defined by a function of responses from multiple subjects and as a result, estimating equations of the form for the single response based dependent variable given in (6.47) do not apply to such a class of models. To illustrate this important difference as well as to motivate the development of a new class of estimation equations for FRM, consider the generalized ANOVA for modeling variance heterogeneity discussed in Section 6.1. For convenience, consider a single sample so that $K = 1$. The FRM in this case is given by

$$E[f(y_i, y_j)] = E\left[\frac{1}{2}(y_i - y_j)^2\right] = \sigma^2, \quad 1 \leq i < j \leq n \tag{6.48}$$

Clearly, the estimating equation for the distribution-free GLM defined in (6.47) does not apply to the above model, since the dependent variables $f(y_i, y_j)$ are defined by a function of two individual subjects' responses, $y_i$ and $y_j$, and thus are not stochastically independent across $i$ and $j$.

For this simple problem, we know that the sample variance of $y_i$ is an unbiased estimate of $\sigma^2$. Further, from the consideration of this model in Chapter 3, we know that we can express the unbiased estimate of $\sigma^2$ as a U-statistic as follows:

$$\widehat{\sigma}^2 = \binom{n}{2}^{-1} \sum_{(i,j)\in C_2^n} \frac{1}{2}(y_i - y_j)^2 = \binom{n}{2}^{-1} \sum_{(i,j)\in C_2^n} f(y_i, y_j) \tag{6.49}$$

Now by applying some simple algebra, we can readily show that $\widehat{\sigma}^2$ is the solution to the following U-statistics defined equation

$$U_n(\sigma^2) = \sum_{(i,j)\in C_2^n} U_{nij} = \sum_{(i,j)\in C_2^n} \left[f(y_i, y_j) - \sigma^2\right] = 0 \tag{6.50}$$

where $U_{nij} = f(y_i, y_j) - \sigma^2$ denotes the theoretical residual for the FRM in (6.48). Thus, by identifying itself as an equation for $\sigma^2$, the *estimating equation* in (6.50) suggests a generalization of (6.47) to accommodate dependent variables defined by functions of responses from multiple individuals. In addition, since $U_{nij}$ is symmetric with respect to $y_i$ and $y_j$, that is, $U_{ij} = U_{ji}$, $U_n$ defined in (6.50) is a U-statistic-like quantity ($U_n$ is not a statistic as it contains the unknown parameter $\sigma^2$). Thus, the theories of U-statistics in Chapters 3 and 5 can be utilized to provide the basis to develop estimating equations as well as to study the asymptotic properties of estimates obtained from such estimating equations for FRM. As in Section 6.2, we consider models for group comparison and regression analysis separately.

### 6.3.1 Inference for Models for Group Comparison

Consider the class of FRM for comparing $K$ independent samples defined in 6.13 of Section 6.2.1. Let

$$\begin{aligned} S_{\mathbf{i}_1,\ldots,\mathbf{i}_K} &= \mathbf{f}_{\mathbf{i}_1,\ldots,\mathbf{i}_K} - \mathbf{h}, \quad \mathbf{i}_k = (i_{k1}, \ldots, i_{kq}), \\ \mathbf{f}_{\mathbf{i}_1,\ldots,\mathbf{i}_K} &= \mathbf{f}\left(\mathbf{y}_{1i_{11}}, \ldots, \mathbf{y}_{1i_{1q}}; \ldots; \mathbf{y}_{Ki_{K1}}, \ldots, \mathbf{y}_{Ki_{Kq}}\right) \end{aligned}$$

where $(i_{k1}, \ldots, i_{kq}) \in C_q^{n_k}$ $(1 \le k \le K)$. For notational brevity, assume that $\mathbf{f}_{\mathbf{i}_1,\ldots,\mathbf{i}_K}$ is symmetric with respect to each $\mathbf{i}_k$, that is, $\mathbf{f}_{\mathbf{i}_1,\ldots,\mathbf{i}_k,\ldots,\mathbf{i}_K} =$

$\mathbf{f}_{\mathbf{i}_1,\ldots,\mathbf{i}'_k,\ldots,\mathbf{i}_K}$ for any permutation $\mathbf{i}'_k$ of $\mathbf{i}_k$ $(1 \leq k \leq K)$. For nonsymmetric $\mathbf{f}$, a symmetric version is readily constructed (see Chapters 3 and 5).

Define the U-statistics based generalized estimating equation (UGEE) as follows:

$$\mathbf{U}_n(\boldsymbol{\theta}) = \sum_{\mathbf{i}_1 \in C_q^{n1},\ldots,\mathbf{i}_K \in C_q^{nK}} \mathbf{U}_{n\mathbf{i}_1,\ldots,\mathbf{i}_K} = \sum_{\mathbf{i}_1 \in C_q^{n1},\ldots,\mathbf{i}_K \in C_q^{nK}} GS_{\mathbf{i}_1,\ldots,\mathbf{i}_K} = \mathbf{0} \tag{6.51}$$

where $G$ is some known $p \times l$ matrix function of $\boldsymbol{\theta}$ with $p$ and $l$ denoting the dimension of $\boldsymbol{\theta}$ and $S_{\mathbf{i}_1,\ldots,\mathbf{i}_K}$. The above UGEE generalizes the estimating equation in (6.47) and GEEs (GEE I and II) for single response based distribution-free models discussed in Chapter 4 to accommodate dependent variables defined by nonlinear functions of multiple individual responses. The Newton–Raphson algorithm is readily applied to solve the UGEE in (6.51) to obtain estimates of $\boldsymbol{\theta}$. If $E(S_{\mathbf{i}_1,\ldots,\mathbf{i}_K}) = \mathbf{0}$, it follows that (see exercise):

$$E(\mathbf{U}_n(\boldsymbol{\theta})) = \sum_{\mathbf{i}_1 \in C_q^{n1},\ldots,\mathbf{i}_K \in C_q^{nK}} E(GS_{\mathbf{i}_1,\ldots,\mathbf{i}_K}) = \mathbf{0} \tag{6.52}$$

Thus, as in the case of GEEs, the estimating equation defined in (6.51) is *unbiased* provided that the FRM is specified correctly.

Like GEEs, the choice of $G$ is not unique. In most applications, $G = DV^{-1} = \left(\frac{\partial}{\partial \boldsymbol{\theta}}\mathbf{h}\right) V^{-1}$, where $V$ denotes some known $l \times l$ matrix. As in the case of GEEs, $V(\boldsymbol{\alpha})$ may be parameterized by some vector $\boldsymbol{\alpha}$. If $\boldsymbol{\alpha}$ is unknown, it must be estimated before the UGEE in (6.51) can be solved for $\boldsymbol{\theta}$. For GEEs, we showed in Chapter 4 that consistency of GEE estimate is independent of how $\boldsymbol{\alpha}$ is estimated, while asymptotic normality of such estimate is guaranteed only when $\sqrt{n}$-consistent estimates of $\boldsymbol{\alpha}$ are used. As we show shortly, these nice asymptotic properties carry directly over to UGEE estimates.

**Example 1 (*Generalized ANOVA for variance*).** Consider the FRM for the one-group variance ANOVA in (6.48) again. In this case,

$$\mathbf{i} = (i_1, i_2), \quad \mathbf{y_i} = (y_{i_1}, y_{i_2})^\top, \quad f_{\mathbf{i}} = \frac{1}{2}(y_{i_1} - y_{i_2})^2$$
$$h = \theta = \sigma^2, \quad U_{\mathbf{i}} = f_{\mathbf{i}} - \theta = f_{\mathbf{i}} - \sigma^2$$

Set $G = D = \frac{\partial h}{\partial \boldsymbol{\theta}}$. Since $\frac{\partial h}{\partial \boldsymbol{\theta}} = \frac{\partial h}{\partial \sigma^2} = 1$, the estimating equation for $\theta = \sigma^2$

is given by

$$U_n(\theta) = \sum_{\mathbf{i} \in C_2^n} GS_{\mathbf{i}} = \sum_{\mathbf{i} \in C_2^n} \left[ \frac{1}{2} (y_{i_1} - y_{i_2})^2 - \sigma^2 \right] = 0 \tag{6.53}$$

The above is the same estimating equation as given in (6.48). By solving for $\sigma^2$, we immediately obtain (6.49).

**Example 2 (*Generalized ANOVA for mean and variance*).** In Example 1, now consider adding a component for the mean of $y_i$ such that the FRM has the form:

$$\mathbf{f}(y_{i_1}, y_{i_2}) = \begin{pmatrix} \frac{1}{2}(y_{i_1} + y_{i_2}) \\ \frac{1}{2}(y_{i_1} - y_{i_2})^2 \end{pmatrix}, \quad \mathbf{h}(\boldsymbol{\theta}) = \begin{pmatrix} \mu \\ \sigma^2 \end{pmatrix} = \boldsymbol{\theta}, \quad 1 \le i < j \le n$$

Let

$$\mathbf{i} = (i_1, i_2), \quad \mathbf{y_i} = (y_{i_1}, y_{i_2})^\top, \quad \mathbf{f_i} = \begin{pmatrix} \frac{1}{2}(y_{i_1} + y_{i_2}) \\ \frac{1}{2}(y_{i_1} - y_{i_2})^2 \end{pmatrix}$$

$$\mathbf{h} = \boldsymbol{\theta} = (\mu, \sigma^2)^\top, \quad \mathbf{U_i} = \mathbf{f_i} - \boldsymbol{\theta}$$

Set $G = D = \frac{\partial \mathbf{h}}{\partial \boldsymbol{\theta}}$. Since $\frac{\partial \mathbf{h}}{\partial \boldsymbol{\theta}} = \mathbf{I}_2$ with $\mathbf{I}_2$ denoting the $2 \times 2$ identity matrix, the UGEE in (6.51) for $\boldsymbol{\theta}$ is given by

$$\mathbf{U}_n(\boldsymbol{\theta}) = \sum_{\mathbf{i} \in C_2^n} GS_{\mathbf{i}} = \sum_{\mathbf{i} \in C_2^n} \left[ \begin{pmatrix} \frac{1}{2}(y_{i_1} + y_{i_2}) \\ \frac{1}{2}(y_{i_1} - y_{i_2})^2 \end{pmatrix} - \begin{pmatrix} \mu \\ \sigma^2 \end{pmatrix} \right] = 0 \tag{6.54}$$

By solving for $\boldsymbol{\theta}$, we obtain $\widehat{\boldsymbol{\theta}} = (\overline{y}_n, s_n^2)^\top$, where $\overline{y}_n$ and $s_n^2$ denote the sample mean and variance of $y_i$.

The UGEE estimate $\widehat{\boldsymbol{\theta}}$ defined by (6.51) has asymptotic properties that parallel those of GEE studied in Chapter 4. The following theorem summarizes these nice asymptotic results.

**Theorem 1** (UGEE estimates for group comparison). Let

$$\Sigma_k = Var\left[ E\left( U_{n\mathbf{i}_1,\ldots,\mathbf{i}_K} \mid \mathbf{y}_{i_{k1}} \right) \right], \quad B = E\left( GD^\top \right), \quad D = \frac{\partial}{\partial \boldsymbol{\theta}} \mathbf{h} \tag{6.55}$$

$$n = \sum_{k=1}^{K} n_k, \quad \rho_k^2 = \lim_{n \to \infty} \frac{n}{n_k} < \infty, \quad (1 \le k \le K)$$

Then, under mild regularity conditions, we have

1. $\widehat{\boldsymbol{\theta}}$ is consistent.
2. If $\sqrt{n}\left(\widehat{\boldsymbol{\alpha}}-\boldsymbol{\alpha}\right)=\mathbf{O}_p(1)$, $\widehat{\boldsymbol{\theta}}$ is asymptotically normal:

$$\sqrt{n}\left(\widehat{\boldsymbol{\theta}}-\boldsymbol{\theta}\right)\to_d N\left(\mathbf{0},\Sigma_\theta=q^2\sum_{k=1}^{K}\rho_k B^{-1}\Sigma_k B^{-\top}\right)$$

3. For $\mathbf{i}_k=(i_{k1},i_{k2},\ldots,i_{kq})$, let

$$\begin{aligned}\widetilde{\mathbf{i}}_k(1)&=\left(i_{k1},i_{k(q+1)},\ldots,i_{k(2q-1)}\right)\\ \widetilde{\mathbf{i}}_k(0)&=\left(i_{k(q+1)},i_{k(q+2)},\ldots,i_{k(2q-1)},i_{k(2q)}\right)\end{aligned} \tag{6.56}$$

For each $1\le k\le K$, let

$$\mathbf{j}_r=\begin{cases}\mathbf{i}_r\cup\widetilde{\mathbf{i}}_r(1)=\left(i_{k1},\ldots,i_{k(2q-1)}\right) & \text{if } r=k\\ \mathbf{i}_r\cup\widetilde{\mathbf{i}}_r(0)=\left(i_{k1},\ldots,i_{k(2q-1)},i_{k(2q)}\right) & \text{if } r\neq k\end{cases} \tag{6.57}$$

Let $\Phi_{k,\mathbf{j}_1,\ldots,\mathbf{j}_K}$ be some symmetric kernel of $U_{\mathbf{i}_1,\ldots,\mathbf{i}_k,\ldots,\mathbf{i}_K}U^{\top}_{\widetilde{\mathbf{i}}_1(0),\ldots,\widetilde{\mathbf{i}}_k(1),\ldots,\widetilde{\mathbf{i}}_K(0)}$, with respect to each $\mathbf{j}_k$ $(1\le r\le K)$. Then, a consistent estimate of $\Sigma_\theta$ is given by

$$\begin{aligned}\widehat{\Sigma}_\theta&=q^2\sum_{k=1}^{K}\frac{n}{n_k}\widehat{B}^{-1}\widehat{\Sigma}_k\widehat{B}^{-\top}\\ \widehat{\Sigma}_k&=\binom{n_k}{2q-1}^{-1}\left[\prod_{r\neq k}\binom{n_k}{2q}\right]^{-1}\sum_{\mathbf{j}_k\in C^{n_k}_{2q-1}}\sum_{r\neq k}\sum_{\mathbf{j}_r\in C^{n_r}_{2q}}\widehat{\Phi}_{k,\mathbf{j}_1,\ldots,\mathbf{j}_K}\end{aligned} \tag{6.58}$$

where $\widehat{A}$ denotes the estimated $A$ obtained by substituting estimates $\widehat{\boldsymbol{\theta}}$ and $\widehat{\boldsymbol{\alpha}}$ in place of their respective parameters $\boldsymbol{\theta}$ and $\boldsymbol{\alpha}$.

Proof. For convenience, we assume $\boldsymbol{\alpha}$ is known. The case with $\boldsymbol{\alpha}$ substituted by a $\sqrt{n}$-consistent estimate is similarly considered (see exercise).

1. Since $\mathbf{U}_{n\mathbf{i}_1,\ldots,\mathbf{i}_K}$ is symmetric with respect to $\mathbf{i}_k$, it follows that the normalized $\prod_{k=1}^{K}\binom{n_k}{m_k}^{-1}\mathbf{U}_n$ is a $K$-sample U-statistic-like quantity. For notational brevity, denote this normalized vector by $\mathbf{U}_n$. By applying the theory of multivariate U-statistics in Chapter 5, we have

$$\sqrt{n}\mathbf{U}_n\to_d N\left(\mathbf{0},\Sigma_U=q^2\sum_{k=1}^{K}\rho_k\Sigma_k\right) \tag{6.59}$$

where $\Sigma_k$ and $\rho_k$ are defined in (6.55). Given that $\mathbf{U}_n$ is asymptotically normal with mean 0, consistency of $\widehat{\boldsymbol{\theta}}$ follows from an argument similar to the one employed in proving consistency of maximum likelihood estimate in Chapter 2 (see exercise).

2. It follows from a Taylor series expansion that

$$\sqrt{n}\left(\widehat{\boldsymbol{\theta}}-\boldsymbol{\theta}\right)=\left(-\frac{\partial}{\partial\boldsymbol{\theta}}\mathbf{U}_n\right)^{-\top}\sqrt{n}\mathbf{U}_n+\mathbf{o}_p\left(1\right) \tag{6.60}$$

where $\mathbf{o}_p(\cdot)$ denotes the stochastic $\mathbf{o}(\cdot)$ (see Chapter 1) and

$$\frac{\partial}{\partial\boldsymbol{\theta}}\mathbf{U}_n=\left[\prod_{k=1}^{K}\binom{n_k}{q}\right]^{-1}\sum_{\mathbf{i}_1\in C_q^{n1},\ldots,\mathbf{i}_K\in C_q^{nK}}\frac{\partial}{\partial\boldsymbol{\theta}}\mathbf{U}_{n\mathbf{i}_1,\ldots,\mathbf{i}_K} \tag{6.61}$$

Since $\mathbf{U}_{n\mathbf{i}_1,\ldots,\mathbf{i}_K}$ is symmetric with respect to $\mathbf{i}_k$, it follows that $\frac{\partial}{\partial\boldsymbol{\theta}}\mathbf{U}_n$ is also a $K$-sample U-statistic-like quantity. Thus, by applying the theory of multivariate U-statistics to (6.61), we have

$$\frac{\partial}{\partial\boldsymbol{\theta}}\mathbf{U}_n\rightarrow_p E\left(\frac{\partial}{\partial\boldsymbol{\theta}}S_{\mathbf{i}_1,\ldots,\mathbf{i}_K}G^\top\right)=-E\left(DG^\top\right)=-B^\top \tag{6.62}$$

where $B$ is defined in (6.55). The conclusion follows from (6.60), (6.62) and an application of the Delta method.

3. Since $E\left[E\left(\mathbf{U}_{n\mathbf{i}_1,\ldots,\mathbf{i}_K}\mid\mathbf{y}_{i_{k1}}\right)\right]=E\left[GE\left(S_{\mathbf{i}_1,\ldots,\mathbf{i}_K}\mid\mathbf{y}_{i_{k1}}\right)\right]=\mathbf{0}$, it follows that

$$\begin{aligned}\Sigma_k&=Var\left(E\left(\mathbf{U}_{n\mathbf{i}_1,\ldots,\mathbf{i}_K}\mid\mathbf{y}_{i_{k1}}\right)\right)\\&=E\left[E\left(\mathbf{U}_{n\mathbf{i}_1,\ldots,\mathbf{i}_K}\mid\mathbf{y}_{i_{k1}}\right)E\left(\mathbf{U}_{n\mathbf{i}_1,\ldots,\mathbf{i}_K}^\top\mid\mathbf{y}_{i_{k1}}\right)\right]\\&=E\left[E\left(\mathbf{U}_{n\mathbf{i}_1,\ldots,\mathbf{i}_k\ldots,\mathbf{i}_K}\mathbf{U}_{n\widetilde{\mathbf{i}}_1(0),\ldots,\widetilde{\mathbf{i}}_k(1),\ldots,\widetilde{\mathbf{i}}_K(0)}^\top\mid\mathbf{y}_{k1}\right)\right]\\&=E\left(U_{n\mathbf{i}_1,\ldots,\mathbf{i}_k\ldots,\mathbf{i}_K}\mathbf{U}_{n\widetilde{\mathbf{i}}_1(0),\ldots,\widetilde{\mathbf{i}}_k(1),\ldots,\widetilde{\mathbf{i}}_K(0)}^\top\right)\end{aligned}$$

where $\widetilde{\mathbf{i}}_k(1)$ and $\widetilde{\mathbf{i}}_k(0)$ are defined in (6.56). Let $\mathbf{j}_r$ be defined in (6.57). Then, it is readily checked that $\mathbf{V}_{n\mathbf{j}_1,\ldots,\mathbf{j}_K}=\mathbf{U}_{n\mathbf{i}_1,\ldots,\mathbf{i}_k\ldots,\mathbf{i}_K}\mathbf{U}_{n\widetilde{\mathbf{i}}_1(0),\ldots,\widetilde{\mathbf{i}}_k(1),\ldots,\widetilde{\mathbf{i}}_K(0)}^\top$ depends on $\mathbf{i}_k$, $\widetilde{\mathbf{i}}_k(0)$ and $\widetilde{\mathbf{i}}_k(1)$ only through $\mathbf{j}_1,\ldots,\mathbf{j}_K$ (see exercise). Thus, if $\Phi_{\mathbf{j}_1,\ldots,\mathbf{j}_K}$ is a symmetric kernel of $\mathbf{V}_{n\mathbf{j}_1,\ldots,\mathbf{j}_K}$, then it follows from the theory of multivariate U-statistics and Slutsky's theorem that $\widehat{\Sigma}_k$ in (6.58) is a consistent estimate of $\Sigma_k=Var\left(E\left(\mathbf{U}_{n\mathbf{i}_1,\ldots,\mathbf{i}_K}\mid\mathbf{y}_{i_{k1}}\right)\right)$. The conclusion follows by applying this result and the Delta method.

**Example 3 (*Generalized ANOVA for variance*).** Consider again the FRM in Example 1 for the one-group variance ANOVA. Here, $q = 2$ and the estimating equation for $\sigma^2$ is given by (6.53). Thus, $G = D = 1$ and

$$B = E\left(GD^\top\right) = E\left(DD^\top\right) = 1$$

It follows from Theorem 1 that the asymptotic variance $\phi^2$ of $\widehat{\sigma}^2$ is

$$\phi^2 = 4BVar\left[E\left(U_{\mathbf{i}} \mid y_{i_1}\right)\right]B = 4Var\left[E\left(f_{\mathbf{i}} - \sigma^2 \mid y_{i_1}\right)\right]$$

However,

$$E\left(f_{\mathbf{i}} - \sigma^2 \mid y_{i_1}\right) = E\left[\frac{1}{2}\left(y_{i_1} - y_{i_2}\right)^2 \mid y_{i_1}\right] - \sigma^2 = \frac{1}{2}\left(y_{i_1} - \mu\right)^2 - \sigma^2$$

Thus,

$$\phi^2 = 4Var\left(\frac{1}{2}\left(y_i - \mu\right)^2\right) = \mu_4 - \sigma^4$$

where $\mu_4 = E\left(y_i^4\right)$ denotes the 4th-order moment of $y_i$. A consistent estimate of $\phi^2$ is obtained by substituting consistent estimates of $\mu_4$ and $\sigma^2$ in the above such as the sample moments.

Note that we can also use Theorem 1/3 to construct a U-statistics based estimate of the asymptotic variance $\phi^2$ (see exercise). But, in this simple example, it is necessary to do so since we can evaluate $Var\left[E\left(U_{\mathbf{i}} \mid y_{i_1}\right)\right]$ in closed form.

**Example 4 (*Generalized ANOVA for mean and variance*).** Consider the FRM in Example 2. The UGEE for $\boldsymbol{\theta} = \left(\mu, \sigma^2\right)^\top$ is given by (6.54) with $q = 2$. Since $G = D = \frac{\partial}{\partial \boldsymbol{\theta}}\mathbf{h} = \mathbf{I}_2$, it follows that

$$B = E\left(GD^\top\right) = E\left(DD^\top\right) = \mathbf{I}_2$$

By Theorem 1, the asymptotic variance of $\widehat{\boldsymbol{\theta}} = \left(\overline{y}_n, s_n^2\right)^\top$ is given by

$$\Sigma_\theta = 4BVar\left[E\left(\mathbf{U}_{n\mathbf{i}} \mid y_{i_1}\right)\right]B = 4Var\left[E\left(\mathbf{f}_{\mathbf{i}} - \boldsymbol{\theta} \mid y_{i_1}\right)\right]$$

Since

$$\begin{aligned} E\left(\mathbf{f}_{\mathbf{i}} - \boldsymbol{\theta} \mid \mathbf{y}_i\right) &= \begin{pmatrix} E\left[\frac{1}{2}\left(y_{i_1} + y_{i_2}\right) \mid y_{i_1}\right] - \mu \\ E\left[\frac{1}{2}\left(y_{i_1} - y_{i_2}\right)^2 \mid y_{i_1}\right] - \sigma^2 \end{pmatrix} \\ &= \frac{1}{2}\begin{pmatrix} y_i - \mu \\ \left(y_i - \mu\right)^2 - \sigma^2 \end{pmatrix} \end{aligned}$$

it follows that

$$\Sigma_\theta = 4Var\left[\frac{1}{2}\begin{pmatrix} y_i \\ (y_i-\mu)^2 \end{pmatrix}\right] = \left[\begin{pmatrix} \sigma^2 & \mu_3 \\ & \mu_4-\sigma^4 \end{pmatrix}\right]$$

**Example 5** (***Generalized K-sample ANOVA for mean and variance***). We have extended the FRM in Example 2 to a more general setting with three samples in Example 7 of Section 6.2.1. Here, we consider inference for this three-sample FRM for simultaneously modeling the mean and variance.

Let $y_{ki}$ denote the response of interest from three i.i.d. samples ($1 \le i \le n_k$, $1 \le k \le 3$). Also, let

$$\begin{aligned}
\mathbf{i}_k &= (i_{k1}, i_{k2}), \quad \mathbf{y}_{\mathbf{i}_k} = (y_{ki_{k1}}, y_{ki_{k2}})^\top \\
\mathbf{f}_{\mathbf{i}_k} &= \left(\frac{1}{2}(y_{ki_{k1}} + y_{ki_{k2}}), \frac{1}{2}(y_{ki_{k1}} - y_{ki_{k2}})^2\right)^\top \\
\mathbf{h}_k &= \boldsymbol{\theta}_k = (\mu_k, \sigma_k^2)^\top, \quad \mathbf{f}_{\mathbf{i}_1,\mathbf{i}_2,\mathbf{i}_3} = \left(\mathbf{f}_{\mathbf{i}_1}^\top, \mathbf{f}_{\mathbf{i}_2}^\top, \mathbf{f}_{\mathbf{i}_3}^\top\right)^\top \\
\mathbf{h} &= \left(\mathbf{h}_1^\top, \mathbf{h}_2^\top, \mathbf{h}_3^\top\right)^\top, \quad \mathbf{U}_{n\mathbf{i}_1,\mathbf{i}_2,\mathbf{i}_3} = \mathbf{f}_{\mathbf{i}_1,\mathbf{i}_2,\mathbf{i}_3} - \boldsymbol{\theta}
\end{aligned}$$

It follows from (6.51) that the UGEE for this three-sample FRM is given by

$$\begin{aligned}
\mathbf{U}_n(\boldsymbol{\theta}) &= \sum_{\mathbf{i}_1 \in C_2^{n_1}} \sum_{\mathbf{i}_2 \in C_2^{n_2}} \sum_{\mathbf{i}_3 \in C_2^{n_3}} \mathbf{S}_{\mathbf{i}_1,\mathbf{i}_2,\mathbf{i}_3} \\
&= \sum_{\mathbf{i}_1 \in C_2^{n_1}} \sum_{\mathbf{i}_2 \in C_2^{n_2}} \sum_{\mathbf{i}_3 \in C_2^{n_3}} (\mathbf{f}_{\mathbf{i}_1,\mathbf{i}_2,\mathbf{i}_3} - \boldsymbol{\theta}) = \mathbf{0}
\end{aligned} \tag{6.63}$$

The solution to the above equation is in closed form

$$\widehat{\boldsymbol{\theta}}_k = (\overline{y}_{kn}, s_{kn}^2)^\top, \quad \widehat{\boldsymbol{\theta}} = \left(\widehat{\boldsymbol{\theta}}_1^\top, \widehat{\boldsymbol{\theta}}_2^\top, \widehat{\boldsymbol{\theta}}_3^\top\right)^\top \tag{6.64}$$

where $\overline{y}_{kn}$ and $s_{kn}^2$ denote the sample mean and variance of the $k$th sample ($1 \le k \le 3$).

It follows from Theorem 1 that $\widehat{\boldsymbol{\theta}}$ is asymptotically normal with the asymptotic variance given by

$$\Sigma_k = Var\left[E\left(\mathbf{U}_{n\mathbf{i}_1,\mathbf{i}_2,\mathbf{i}_3} \mid y_{ki_{k1}}\right)\right], \quad \Sigma_\theta = 4\sum_{k=1}^{3} \rho_k \Sigma_k$$

where $\rho_k = \lim_{n\to\infty} \frac{n}{n_k}$ with $n = \sum_{k=1}^{3} n_k$. To find $\Sigma_k$, note first that

$$E\left(\mathbf{f}_{\mathbf{i}_k} - \mathbf{h}_k \mid y_{ki_{k1}}\right) = \begin{pmatrix} E\left[\frac{1}{2}\left(y_{ki_{k1}} + y_{ki_{k2}}\right) \mid y_{ki_{k1}}\right] - \mu_k \\ E\left[\frac{1}{2}\left(y_{ki_{k1}} - y_{ki_{k2}}\right)^2 \mid y_{ki_{k1}}\right] - \sigma_k^2 \end{pmatrix}^\top = \frac{1}{2}\begin{pmatrix} y_{ki_{k1}} - \mu_k \\ \left(y_{ki_{k1}} - \mu_k\right)^2 - \sigma_k^2 \end{pmatrix}^\top$$

and

$$E\left(\mathbf{f}_{\mathbf{i}_k} - \mathbf{h}_k \mid y_{mi_{m1}}\right) = \mathbf{0}, \quad \text{if } m \neq k$$

Thus, it follows that

$$\begin{aligned} E\left(\mathbf{U}_{\mathbf{i}_1,\mathbf{i}_2,\mathbf{i}_3} \mid y_{1i}\right) &= E\left(\mathbf{f}_{\mathbf{i}_1,\mathbf{i}_2,\mathbf{i}_3} \mid y_{1i}\right) - \mathbf{h} \\ &= \frac{1}{2}\left(\left(y_{1i} - \mu_1\right), \left(y_{1i} - \mu_1\right)^2 - \sigma_1^2, 0, 0, 0, 0\right)^\top \end{aligned} \tag{6.65}$$

Similarly,

$$E\left(\mathbf{U}_{\mathbf{i}_1,\mathbf{i}_2,\mathbf{i}_3} \mid y_{2i}\right) = \frac{1}{2}\left(0, 0, \left(y_{2i} - \mu_2\right), \left(y_{2i} - \mu_2\right)^2 - \sigma_2^2, 0, 0\right)^\top$$
$$E\left(\mathbf{U}_{\mathbf{i}_1,\mathbf{i}_2,\mathbf{i}_3} \mid y_{3i}\right) = \frac{1}{2}\left(0, 0, 0, 0, \left(y_{3i} - \mu_3\right), \left(y_{3i} - \mu_3\right)^2 - \sigma_3^2\right)^\top$$

Thus, we have

$$\Sigma_1 = \frac{1}{4}\begin{pmatrix} \Phi_1 & 0 & 0 \\ & 0 & 0 \\ & & 0 \end{pmatrix}, \quad \Sigma_2 = \frac{1}{4}\begin{pmatrix} 0 & 0 & 0 \\ & \Phi_2 & 0 \\ & & 0 \end{pmatrix}, \quad \Sigma_3 = \frac{1}{4}\begin{pmatrix} 0 & 0 & 0 \\ & 0 & 0 \\ & & \Phi_3 \end{pmatrix}$$
$$\Phi_k = \begin{pmatrix} \sigma_k^2 & \mu_{k3} \\ & \mu_{k4} - \sigma_k^4 \end{pmatrix}$$

where $\sigma_k^2$, $\mu_{k3}$ and $\mu_{k4}$ denote the variance, the third and forth moment of $y_{ki}$, respectively. It follows that the asymptotic variance of $\widehat{\boldsymbol{\theta}}$ is given by

$$\Sigma_\theta = 4\sum_{k=1}^{3} \rho_k \Sigma_k = \begin{pmatrix} \rho_1\Phi_1 & 0 & 0 \\ & \rho_2\Phi_2 & 0 \\ & & \rho_3\Phi_3 \end{pmatrix} \tag{6.66}$$

Note that we may also obtain the above asymptotic results by applying Slutsky's theorem. For each $k$, it follows from Example 4 that

$$\sqrt{n_k}\left(\widehat{\boldsymbol{\theta}}_k - \boldsymbol{\theta}_k\right) \to_d N\left(\mathbf{0}, \Phi_k\right), \quad 1 \leq k \leq 3.$$

It follows from Slutsky's theorem that

$$\sqrt{n}\left(\widehat{\boldsymbol{\theta}} - \boldsymbol{\theta}\right) = \begin{pmatrix} \sqrt{\frac{n}{n_1}} & 0 & 0 \\ 0 & \sqrt{\frac{n}{n_2}} & 0 \\ 0 & 0 & \sqrt{\frac{n}{n_3}} \end{pmatrix} \begin{pmatrix} \sqrt{n_1}\left(\widehat{\boldsymbol{\theta}}_1 - \boldsymbol{\theta}_1\right) \\ \sqrt{n_2}\left(\widehat{\boldsymbol{\theta}}_2 - \boldsymbol{\theta}_2\right) \\ \sqrt{n_3}\left(\widehat{\boldsymbol{\theta}}_3 - \boldsymbol{\theta}_3\right) \end{pmatrix} \to_d N\left(\mathbf{0}, \Sigma_\theta\right)$$

yielding the same asymptotic variance as in (6.66).

**Example 6** (***Wilcoxon signed rank test***). Consider the FRM for the Wilcoxon signed-rank test introduced in Example 5 of Section 6.2.1:

$$f\left(y_{i_1}, y_{i_2}\right) = I_{\left\{y_{i_1}+y_{i_2} \leq 0\right\}}, \quad h\left(\theta\right) = \theta, \quad 1 \leq i_1 < i_2 \leq n$$

By setting $G = \frac{\partial}{\partial\theta}h = 1$, the UGEE for $\theta$ is given by

$$U_n\left(\theta\right) = \binom{n}{2}^{-1} \sum_{\mathbf{i}\in C_2^n} S_{\mathbf{i}} = \binom{n}{2}^{-1} \sum_{(i_1,i_2)\in C_2^n} \left(I_{\left\{y_{i_1}+y_{i_2}\leq 0\right\}} - \theta\right) = 0$$

The above is readily solved, yielding $\widehat{\theta} = \binom{n}{2}^{-1} \sum_{(i_1,i_2)\in C_2^n} I_{\left\{y_{i_1}+y_{i_2}\leq 0\right\}}$. This UGEE estimate is precisely the Wilcoxon one-sample signed rank statistic. By Theorem 1, $\widehat{\theta}$ is asymptotically normal. Since $B = 1$, it follows that the asymptotic variance of $\widehat{\theta}$ is given by

$$\sigma_\theta^2 = 4Var\left[E\left(f_{\mathbf{i}} \mid y_{i_1}\right)\right] = 4Var\left[E\left(I_{\left\{y_{i_1}+y_{i_2}\leq 0\right\}} \mid y_{i_1}\right)\right] \tag{6.67}$$

Unlike the previous examples, we cannot evaluate the above in closed form. We can use Theorem 1/3 to construct a U-statistics based estimate of $\sigma_\theta^2$ (see exercise).

As discussed in Chapter 3, the Wilcoxon signed-rank statistic is used to test whether $y_i$ is distributed symmetrically around 0. Under the null, $\theta = \frac{1}{2}$ and the asymptotic variance $\sigma_\theta^2$ can be evaluated in closed form $\sigma_\theta^2 = \frac{1}{3}$.

The Mann–Whitney–Wilcoxon signed rank test provides an alternative to the paired t test for comparing treatment difference in a pre-post study design. As discussed in Chapter 4, the pre-post comparison is a special case of the longitudinal study design with two assessment times. The classic application of pre-post study design is in evaluating the effect of a treatment of interest by examining the change or difference of the outcome from pre- to posttreatment assessment. In this case, the study subjects are assessed at pre- and posttreatment times, yielding outcomes of interest before and after the administration of the intervention. We can then assess treatment effect by testing whether the mean change of the outcome is significantly different from zero using the paired t test. Alternatively, we can use the signed-rank test for this purpose. If we let $y_i$ be the change score between pre- and posttreatment assessment, then no treatment effect corresponds to the null that $y_i$ is distributed symmetrically about 0, that is, $H_0 : \theta = \frac{1}{2}$ for the parameter $\theta$ defined by the Wilcoxon signed-rank statistic.

**Example 7** (***Generalized*** $K$***-sample Mann–Whitney–Wilcoxon rank sum test***). Consider the generalized $K$-sample Mann–Whitney–Wilcoxon test introduced in Section 6.2.1. For notational brevity, consider the three group case, that is, $K = 3$. Let

$$\mathbf{i}_k = i_k, \quad \mathbf{y}_{\mathbf{i}_k} = y_{ki_k}, \quad \mathbf{h} = \boldsymbol{\theta} = (\theta_{12}, \theta_{13}, \theta_{23})^\top, \quad 1 \leq k \leq 3$$
$$\mathbf{f}_{i_1,i_2,i_3} = \left( I_{\{y_{1i_1} \leq y_{2i_2}\}}, I_{\{y_{1i_1} \leq y_{3i_3}\}}, I_{\{y_{2i_2} \leq y_{3i_3}\}} \right)^\top$$

Set $G = D = \frac{\partial}{\partial \boldsymbol{\theta}} \mathbf{h}$. Then, $D = \frac{\partial}{\partial \boldsymbol{\theta}} \mathbf{h} = \mathbf{I}_3$ and $B = E\left(GD^\top\right) = \mathbf{I}_3$. The UGEE for this model is given by

$$\mathbf{U}_n(\boldsymbol{\theta}) = \sum_{i_1 \in C_1^{n_1}} \sum_{i_2 \in C_1^{n_2}} \sum_{i_3 \in C_1^{n_3}} \left(\mathbf{f}_{i_1,i_2,i_3} - \boldsymbol{\theta}\right) = \mathbf{0}$$

Let $\widehat{\boldsymbol{\theta}}$ denote the estimate obtained from solving the above equation. Then, $\widehat{\boldsymbol{\theta}}$ is asymptotically normal, with the asymptotic variance given by

$$\Sigma_\theta = 4 \sum_{k=1}^{3} \rho_k \Sigma_k, \quad \Sigma_k = Var\left[E\left(\mathbf{U}_{ni_1,i_2,i_3} \mid y_{ki_k}\right)\right]$$

To find $\Sigma_k$, first consider $\Sigma_1$. We have

$$\Sigma_1 = E\left(\mathbf{U}_{ni_1,i_2,i_3} \mid y_{1i_1}\right) = E\left(\mathbf{f}_{i_1,i_2,i_3} \mid y_{1i_1}\right) - \boldsymbol{\theta} \tag{6.68}$$

$$= \begin{pmatrix} E\left(E\left(I_{\{y_{1i_1} \leq y_{2i_2}\}} \mid y_{1i_1}\right)\right) \\ E\left(I_{\{y_{1i_1} \leq y_{3i_3}\}} \mid y_{1i_1}\right), \\ E\left(I_{\{y_{2i_2} \leq y_{3i_3}\}} \mid y_{1i_1}\right) \end{pmatrix} = \begin{pmatrix} 1 - F_2(y_{1i_1}) - \theta_1 \\ 1 - F_3(y_{1i_1}) - \theta_2 \\ E\left(I_{\{y_{i_2} \leq y_{i_3}\}}\right) - \theta_3 \end{pmatrix}$$

$$= \begin{pmatrix} 1 - F_2(y_{1i_1}) - \theta_1 \\ 1 - F_3(y_{1i_1}) - \theta_2 \\ 0 \end{pmatrix}$$

where $F_k$ denotes the CDF of $y_{ki}$ $(1 \leq k \leq 3)$. To further evaluate $\Sigma_1$, we consider two case scenarios.

First, consider testing the null hypothesis:

$$H_0 : F_1\left(y_{1i_1}\right) = F_2\left(y_{2i_2}\right) = F_3\left(y_{3i_3}\right). \tag{6.69}$$

Under $H_0$, $F_k$ are all uniformly distributed between 0 and 1. It follows that (see exercise):

$$\Sigma_1 = Var\left[E\left(\mathbf{U}_{i_1,i_2,i_3} \mid y_{1i_1}\right)\right] = \begin{pmatrix} \frac{1}{12} & \frac{1}{12} & 0 \\ & \frac{1}{12} & 0 \\ & & 0 \end{pmatrix} \tag{6.70}$$

Similarly, we have for $\Sigma_2$ and $\Sigma_3$:

$$\Sigma_2 = \begin{pmatrix} \frac{1}{12} & 0 & \frac{1}{12} \\ & 0 & 0 \\ & & \frac{1}{12} \end{pmatrix}, \quad \Sigma_3 = \begin{pmatrix} 0 & 0 & 0 \\ & \frac{1}{12} & \frac{1}{12} \\ & & \frac{1}{12} \end{pmatrix} \tag{6.71}$$

Thus, it follows from (6.70) and (6.71) that

$$\Sigma_\theta = 4\sum_{k=1}^{3} \rho_k \Sigma_k = \begin{pmatrix} \rho_1 & 0 & 0 \\ 0 & \rho_2 & 0 \\ 0 & 0 & \rho_3 \end{pmatrix} \begin{pmatrix} \frac{2}{3} & \frac{1}{3} & \frac{1}{3} \\ \frac{1}{3} & \frac{2}{3} & \frac{1}{3} \\ \frac{1}{3} & \frac{1}{3} & \frac{2}{3} \end{pmatrix} \tag{6.72}$$

The asymptotic variance in (6.72) is derived under the null hypothesis in (6.69) concerning the equality of distribution functions of the response. Now, consider an alternative approach that does not require such a strong assumption.

First, note that

$$\begin{aligned}\Sigma_1 &= Var\left[E\left(\mathbf{U}_{ni_1,i_2,i_3} \mid y_{1i_1}\right)\right] = E\left[E\left(\mathbf{U}_{ni_1,i_2,i_3} \mid y_{1i_1}\right) E\left(\mathbf{U}^{\top}_{ni_1,i_2,i_3} \mid y_{1i_1}\right)\right] \\ &= E\left[E\left(\mathbf{U}_{ni_1,i_2,i_3}\mathbf{U}^{\top}_{ni_1,j_2,j_3} \mid y_{1i_1}\right)\right] = E\left(\mathbf{U}_{ni_1,i_2,i_3}\mathbf{U}^{\top}_{ni_1,j_2,j_3}\right)\end{aligned}$$

Since

$$\mathbf{U}_{ni_1,i_2,i_3} = \left(I_{\{y_{1i_1} \le y_{2i_2}\}} - \theta_{12}, I_{\{y_{1i_1} \le y_{3i_3}\}} - \theta_{13}, 0\right)^{\top}$$

it follows that

$$\begin{aligned}\Sigma_1 &= E\begin{pmatrix} d_{12}\left(y_{1i_1}, y_{2i_2}\right) d_{12}\left(y_{1i_1}, y_{2j_2}\right) & d_{12}\left(y_{1i_1}, y_{2i_2}\right) d_{13}\left(y_{1i_1}, y_{3j_3}\right) & 0 \\ & d_{13}\left(y_{1i_1}, y_{3i_3}\right) d_{13}\left(y_{1i_1}, y_{3j_3}\right) & 0 \\ & 0 & 0 \end{pmatrix} \\ &= E\left[G_1\left(y_{1i_1}, y_{2i_2}, y_{2j_2}, y_{3i_3}, y_{3j_3}\right)\right] \qquad (6.73)\end{aligned}$$

where $d_{kl}\left(y_{ki_k}, y_{li_l}\right) = I_{\{y_{ki_k} \le y_{li_l}\}} - \theta_{kl}$ $(1 \le k < l \le 3)$.

Similarly, we obtain:

$$\begin{aligned}\Sigma_2 &= E\begin{pmatrix} d_{12}\left(y_{1i_1}, y_{2i_2}\right) d_{12}\left(y_{1j_1}, y_{2i_2}\right) & 0 & d_{12}\left(y_{1i_1}, y_{2i_2}\right) d_{23}\left(y_{2i_2}, y_{3i_3}\right) \\ 0 & 0 & 0 \\ & & d_{23}\left(y_{2i_2}, y_{3i_3}\right) d_{23}\left(y_{2i_2}, y_{3j_3}\right) \end{pmatrix} \\ &= E\left[G_2\left(y_{1i_1}, y_{1j_1}, y_{2i_2}, y_{3i_3}, y_{3j_3}\right)\right] \qquad (6.74)\end{aligned}$$

and

$$\begin{aligned}\Sigma_3 &= E\begin{pmatrix} 0 & 0 & 0 \\ 0 & d_{13}\left(y_{1i_1}, y_{3i_3}\right) d_{13}\left(y_{1j_1}, y_{3i_3}\right) & d_{13}\left(y_{1i_1}, y_{3i_3}\right) d_{23}\left(y_{2j_2}, y_{3i_3}\right) \\ 0 & & d_{23}\left(y_{2i_2}, y_{3i_3}\right) d_{23}\left(y_{2j_2}, y_{3i_3}\right) \end{pmatrix} \\ &= E\left[G_3\left(y_{1i_1}, y_{1j_1}, y_{2i_2}, y_{2j_2}, y_{3i_3}\right)\right]. \qquad (6.75)\end{aligned}$$

Let $H_k$ be a symmetric version of $G_k$. For example, $H_1$ below is a symmetric kernel of $G_k$ (see exercise):

$$H_1 = \frac{1}{2}\left[G_1\left(y_{1i_1}, y_{2i_2}, y_{2j_2}, y_{3i_3}, y_{3j_3}\right) + G_1\left(y_{1j_1}, y_{2i_2}, y_{2j_2}, y_{3i_3}, y_{3j_3}\right)\right] \quad (6.76)$$

Then, the following U-statistic matrix is a consistent estimate of $\Sigma_\theta$:

$$\widehat{\Sigma}_\theta = 4\left[\binom{n_1}{2}\binom{n_2}{2}\binom{n_3}{2}\right]^{-1} \sum_{\mathbf{i}_1 \in C_2^{n_1}} \sum_{\mathbf{i}_2 \in C_2^{n_2}} \sum_{\mathbf{i}_3 \in C_2^{n_3}} \left(\widehat{H}_1 + \widehat{H}_2 + \widehat{H}_3\right)$$

where $\mathbf{i}_k = (i_k, j_k)$ and $\widehat{H}_k$ denotes the estimated $H_k$ obtained by substituting $\widehat{\theta}_{kl}$ in place of $\theta_{kl}$ $(1 \le k < l \le 3)$.

### 6.3.2 Inference for Models for Regression Analysis

Now consider the FRM for regression analysis defined in (6.26) of Section 6.2.2. Let

$$\mathbf{i} = (i_1, \ldots, i_q), \quad \mathbf{h}_\mathbf{i} = \mathbf{h}\left(\mathbf{x}_{i_1}, \ldots, \mathbf{x}_{i_q}; \boldsymbol{\theta}\right), \quad \mathbf{f}_\mathbf{i} = \mathbf{f}\left(\mathbf{y}_{i_1}, \ldots, \mathbf{y}_{i_q}\right)$$
$$S_\mathbf{i} = \mathbf{f}_\mathbf{i} - \mathbf{h}_\mathbf{i}$$

where $\mathbf{i} = (i_1, \ldots, i_q) \in C_q^n$. As in Section 6.3.1, assume that $\mathbf{f}_\mathbf{i}$ is symmetric with respect to $\mathbf{i}$, that is, $\mathbf{f}_\mathbf{i} = \mathbf{f}_\mathbf{j}$ for any permutation $\mathbf{j}$ of $\mathbf{i}$.

Define the U-statistic based estimating equation for $\boldsymbol{\theta}$ as follows:

$$\mathbf{U}_n(\boldsymbol{\theta}) = \sum_{\mathbf{i} \in C_q^n} \mathbf{U}_{n\mathbf{i}} = \sum_{\mathbf{i} \in C_q^n} G_\mathbf{i}(\mathbf{x}_\mathbf{i}) S_\mathbf{i} = 0 \qquad (6.77)$$

where $G_\mathbf{i}(\mathbf{x}_\mathbf{i})$ is some known $p \times l$ matrix function of $\mathbf{x}_\mathbf{i}$ with $p$ and $l$ denoting the dimension of $\boldsymbol{\theta}$ and $S_\mathbf{i}$. As in the group comparison case, this new class of U-statistics based estimating equations is a generalization of the GEEs discussed in Chapter 4 for distribution-free inference of single-response based regression models to the current context of FRM where dependent variables are defined by general nonlinear functions of multisubject responses. Again, the Newton–Raphson algorithm is readily applied to solve the UGEE in (6.77) to obtain estimates of $\boldsymbol{\theta}$. Also, it is again readily shown that the UGEE is unbiased, that is, $E(\mathbf{U}_n(\boldsymbol{\theta})) = \mathbf{0}$, if the FRM is specified correctly (see exercise).

As in the case of UGEE for group comparison discussed in Section 6.3.1, the choice of $G_\mathbf{i}(\mathbf{x}_\mathbf{i})$ is not unique and consistency of the UGEE estimate defined in (6.77) is independent of the selection of $G_\mathbf{i}(\mathbf{x}_\mathbf{i})$. In most applications, we set $G_\mathbf{i}(\mathbf{x}_\mathbf{i}) = D_\mathbf{i} V_\mathbf{i}^{-1}$, where $D_\mathbf{i} = \frac{\partial}{\partial \boldsymbol{\theta}} \mathbf{h}_\mathbf{i}$ and $V_\mathbf{i}$ is some $l \times l$ matrix. The matrix $V_\mathbf{i}(\boldsymbol{\alpha})$ may depend on some parameter vector $\boldsymbol{\alpha}$. As in Theorem 1 for the group comparison case, the UGEE estimate obtained from (6.77) is guaranteed to be asymptotically normal if $\boldsymbol{\alpha}$ is substituted by some $\sqrt{n}$-consistent. These asymptotic properties are summarized by the following theorem.

**Theorem 2** (UGEE estimate for regression analysis). Let

$$\Sigma_U = Var\left(E\left(\mathbf{U}_{n\mathbf{i}} \mid \mathbf{y}_{i_1}, \mathbf{x}_{i_1}\right)\right), \quad B = E\left(G_{\mathbf{i}} D_{\mathbf{i}}^{\top}\right)$$

Then, under mild regularity conditions, we have

1. $\widehat{\boldsymbol{\theta}}$ is consistent.
2. If $\sqrt{n}\left(\widehat{\boldsymbol{\alpha}} - \boldsymbol{\alpha}\right) = \mathbf{O}_p(1)$, $\widehat{\boldsymbol{\theta}}$ is asymptotically normal:

$$\sqrt{n}\left(\widehat{\boldsymbol{\theta}} - \boldsymbol{\theta}\right) \to_d N\left(\mathbf{0}, \Sigma_\theta = q^2 B^{-1} \Sigma_U B^{-\top}\right) \tag{6.78}$$

3. For $\mathbf{i} = (i_1, i_2, \ldots, i_q)$, let

$$\widetilde{\mathbf{i}} = \left(i_1, i_{q+1}, \ldots, i_{2q-1}\right), \quad \mathbf{j} = \mathbf{i} \cup \widetilde{\mathbf{i}} = \left(i_1, \ldots, i_q, i_{q+1}, \ldots, i_{2q-1}\right) \tag{6.79}$$

Let $\Phi_{\mathbf{j}}$ denote some symmetric kernel of $\mathbf{U}_{n\mathbf{i}} \mathbf{U}_{n\widetilde{\mathbf{i}}}^{\top}$ with respect to $\mathbf{j}$. Then, a consistent estimate of $\Sigma_{\boldsymbol{\theta}}$ is given by

$$\widehat{\Sigma}_{\boldsymbol{\theta}} = q^2 \binom{n}{2q-1}^{-1} \widehat{B}^{-1} \left(\sum_{\mathbf{j} \in C_{2q-1}^n} \widehat{\Phi}_{\mathbf{j}}\right) \widehat{B}^{-\top}$$

where $\widehat{A}$ denotes the estimated $A$ obtained by substituting estimates $\widehat{\boldsymbol{\theta}}$ and $\widehat{\boldsymbol{\alpha}}$ in place of the respective parameters $\boldsymbol{\theta}$ and $\widehat{\boldsymbol{\alpha}}$.

Proof. Again, for convenience, we only consider the case with known $\boldsymbol{\alpha}$. The general case with $\boldsymbol{\alpha}$ substituted by a $\sqrt{n}$-consistent estimate is similarly established (see exercise).

1. As in the proof of Theorem 1, we first show that the normalized U-statistic-like vector $\binom{n}{q}^{-1} \mathbf{U}_n(\boldsymbol{\theta})$ in (6.77) is asymptotically normal with mean $\mathbf{0}$. For notational brevity, we denote the normalized vector by $\mathbf{U}_n(\boldsymbol{\theta})$.

The asymptotic normality of $\mathbf{U}_n(\boldsymbol{\theta})$ is established by following the same argument as used to prove this property for the similar U-statistic-like vector $\mathbf{U}_n(\boldsymbol{\theta})$ in Theorem 1 (see exercise). The consistency of $\widehat{\boldsymbol{\theta}}$ then follows by employing an argument similar to the one used in proving consistency of maximum likelihood estimate in Chapter 2 (see exercise).

2. Similar to the proof of Theorem 1/2 in Section 6.3.1, we obtain from a Taylor series expansion the following:

$$\sqrt{n}\left(\widehat{\boldsymbol{\theta}} - \boldsymbol{\theta}\right) = \left(-\frac{\partial}{\partial \boldsymbol{\theta}} \mathbf{U}_n\right)^{-\top} \sqrt{n} \mathbf{U}_n + \mathbf{o}_p(1)$$

Since $\mathbf{U}_{n\mathbf{i}}$ is symmetric with respect to $\mathbf{i}$, it follows that $\frac{\partial}{\partial\boldsymbol{\theta}}\mathbf{U}_n$ is a U-statistic-like quantity. Thus, by convergence results for multivariate U-statistics in Chapter 5, we have

$$\frac{\partial}{\partial\boldsymbol{\theta}}\mathbf{U}_n \to_p E\left[\left(\frac{\partial}{\partial\boldsymbol{\theta}}S_{\mathbf{i}}\right)G_{\mathbf{i}}^{\top}\right] = E\left[-\left(\frac{\partial}{\partial\boldsymbol{\theta}}h_{\mathbf{i}}\right)G_{\mathbf{i}}^{\top}\right] = -B^{\top} \tag{6.80}$$

where $\mathbf{i} = (i_1, \ldots, i_q) \in C_q^n$. The conclusion follows from an application of the Delta method.

3. Since $E\left[E\left(\mathbf{U}_{n\mathbf{i}} \mid \mathbf{y}_{i_1}, \mathbf{x}_{i_1}\right)\right] = \mathbf{0}$, it follows from an argument similar to the proof of Theorem 1/3 that

$$\Sigma_U = q^2 Var\left(E\left(\mathbf{U}_{n\mathbf{i}} \mid \mathbf{y}_{i_1}, \mathbf{x}_{i_1}\right)\right) = q^2 E\left(\mathbf{U}_{n\mathbf{i}}\mathbf{U}_{n\widetilde{\mathbf{i}}}^{\top}\right)$$

where $\mathbf{i} = (i_1, i_2, \ldots, i_q)$ and $\widetilde{\mathbf{i}} = (i_1, i_{q+1}, \ldots, i_{2q-1})$. As in the proof of Theorem 1/3, it is readily checked that any symmetric kernel of $\mathbf{U}_{n\mathbf{i}}\mathbf{U}_{n\widetilde{\mathbf{i}}}^{\top}$ depends on $\mathbf{i}$ and $\widetilde{\mathbf{i}}$ through $\mathbf{j}$, where $\mathbf{j}$ is defined in (6.79). Thus, if $\Phi_{\mathbf{j}}$ is some symmetric kernel of $\mathbf{U}_{n\mathbf{i}}\mathbf{U}_{n\widetilde{\mathbf{i}}}^{\top}$, then

$$\widehat{\Sigma}_U = q^2 \binom{n}{2q-1}^{-1} \sum_{\mathbf{j} \in C_{2q-1}^n} \widehat{\Phi}_{\mathbf{j}} \tag{6.81}$$

is a U-statistic-like vector and thus a consistent estimate of the asymptotic variance $\Sigma_U = q^2 E\left(\mathbf{U}_{n\mathbf{i}}\mathbf{U}_{n\widetilde{\mathbf{i}}}^{\top}\right)$ of $\mathbf{U}_n$. The conclusion follows from (6.78), (6.81) and the Delta method.

**Example 8 (*Negative binomial FRM*).** Consider the NB-like FRM for regression analysis with clustered data discussed in Example 14 of Section 6.2.2:

$$\begin{aligned} f_1(y_i, y_j) &= y_i + y_j, \quad f_2(y_i, y_j) = (y_i - y_j)^2, \quad \log(\mu_i) = \mathbf{x}_i^{\top}\boldsymbol{\beta} \\ h_1(\mathbf{x}_i, \mathbf{x}_i; \boldsymbol{\theta}) &= \mu_i + \mu_j \\ h_2(\mathbf{x}_i, \mathbf{x}_i; \boldsymbol{\theta}) &= \mu_i + \mu_j + \alpha\left(\mu_i^2 + \mu_j^2\right) + \left(\mu_i - \mu_j\right)^2 \\ \mathbf{f} = (f_1, f_2)^{\top}, &\quad \mathbf{h} = (h_1, h_2)^{\top}, \quad \boldsymbol{\theta} = \left(\boldsymbol{\beta}^{\top}, \alpha\right)^{\top} \end{aligned}$$

Also, let

$$S_{\mathbf{i}} = \mathbf{f}_{\mathbf{i}} - \mathbf{h}_{\mathbf{i}}, \quad G_{\mathbf{i}} = D_{\mathbf{i}} = \frac{\partial}{\partial\boldsymbol{\theta}}\mathbf{h}_{\mathbf{i}}, \quad \mathbf{i} = (i, j)$$

Then, the estimating equation for this FRM has the form:

$$\begin{aligned}\mathbf{U}_n(\boldsymbol{\theta}) &= \sum_{(i,j)\in C_2^n} D_{\mathbf{i}} \begin{pmatrix} f_1(y_i, y_j) - h_1(\mathbf{x}_i, \mathbf{x}_i; \boldsymbol{\theta}) \\ f_2(y_i, y_j) - h_2(\mathbf{x}_i, \mathbf{x}_i; \boldsymbol{\theta}) \end{pmatrix} \qquad (6.82)\\ &= \sum_{(i,j)\in C_2^n} D_{\mathbf{i}} \begin{pmatrix} (y_i + y_j) - (\mu_i + \mu_j) \\ (y_i - y_j)^2 - \left[\mu_i + \mu_j + \alpha\left(\mu_i^2 + \mu_j^2\right) + (\mu_i - \mu_j)^2\right] \end{pmatrix} \\ &= \mathbf{0}\end{aligned}$$

Unlike the UGEEs in the examples for group comparisons discussed in Section 6.3.1, $G_{\mathbf{i}}$ is not an identity matrix. In addition, the estimating equation in (6.82) does not have a closed form solution, but estimates of $\boldsymbol{\theta}$ are readily computed numerically by using the Newton–Raphson method. Let $\widehat{\boldsymbol{\theta}}$ denote the solution to the equation. It follows from Theorem 2 that $\widehat{\boldsymbol{\theta}}$ is asymptotically normal with the asymptotic variance given by

$$\Sigma_\theta = 4B^{-1}\Sigma_U B^{-1} = 4B^{-1}Var\left(E\left(U_{n\mathbf{i}} \mid \mathbf{y}_i, \mathbf{x}_i\right)\right)B^{-1} \qquad (6.83)$$

where $B = E\left(D_{\mathbf{i}}D_{\mathbf{i}}^{\top}\right)$. Although $\Sigma_\theta$ cannot be evaluated analytically, a consistent estimate is readily obtained as we illustrate below.

First, we estimate $B$ by

$$\widehat{B} = \binom{n}{2}^{-1} \sum_{\mathbf{i}\in C_2^n} \widehat{D}_{\mathbf{i}}\widehat{D}_{\mathbf{i}}^{\top}. \qquad (6.84)$$

The above is a U-statistic-like quantity and thus consistency of $\widehat{B}$ as an estimate of $B$ follows from the theory of multivariate U-statistics and Slutsky's theorem. To find a consistent estimate of $\Sigma_U = Var\left(E\left(U_{n\mathbf{i}} \mid \mathbf{y}_i, \mathbf{x}_i\right)\right)$, define a symmetric kernel of $U_{(i,j)}U_{(i,k)}$ by

$$\Phi_{(i,j,k)} = \frac{1}{3}\left(U_{(i,j)}U_{(i,k)} + U_{(j,k)}U_{(j,k)} + U_{(k,i)}U_{(k,j)}\right)$$

Then, by applying Theorem 2, we obtain a consistent $\widehat{\Sigma}_U$ given by

$$\widehat{\Sigma}_U = 4\binom{n}{3}^{-1} \sum_{(i,j,k)\in C_3^n} \widehat{\Phi}_{(i,j,k)} \qquad (6.85)$$

where $\widehat{\Phi}_{(i,j,k)}$ denotes the estimated $\Phi_{(i,j,k)}$ with $\boldsymbol{\theta}$ replaced by $\widehat{\boldsymbol{\theta}}$. By combining (6.83), (6.84) and (6.85), we obtain a consistent estimate of $\Sigma_\theta$.

Note that the choice of $G_{\mathbf{i}} = D_{\mathbf{i}} = \frac{\partial}{\partial \boldsymbol{\theta}} \mathbf{h}_{\mathbf{i}}$ may not be optimal. As in the case of GEEs for single-response based distribution-free regression models, we may use $G_{\mathbf{i}} = D_{\mathbf{i}} V_{\mathbf{i}}^{-1}$ to improve efficiency, where the matrix $V_{\mathbf{i}}$ may be parameterized by an unknown vector $\boldsymbol{\alpha}$. If $\boldsymbol{\alpha}$ is substituted by some $\sqrt{n}$-consistent estimate, the UGEE estimate $\widehat{\boldsymbol{\theta}}$ obtained from (6.82) is still asymptotically normal with the same asymptotic variance $\Sigma_\theta$ given in (6.83). We can still use $\widehat{\Sigma}_U$ in (6.85) to estimate $\Sigma_U$. Of course, $\widehat{\Phi}_{(i,j,k)}$ in this case is not only a function of $\widehat{\boldsymbol{\theta}}$, but $\widehat{\boldsymbol{\alpha}}$ as well.

**Example 9** (***Linear mixed effects model***). Consider the FRM for distribution-free inference about the linear mixed-effects model introduced in Example 12 of Section 6.2.2:

$$E\left[f_{1t}\left(\mathbf{y}_i, \mathbf{y}_j\right) \mid \mathbf{x}_i, \mathbf{x}_j\right] = h_{1t}\left(\boldsymbol{\theta}\right) \qquad (6.86)$$
$$E\left[f_{2st}\left(\mathbf{y}_i, \mathbf{y}_j\right) \mid \mathbf{x}_i, \mathbf{x}_j\right] = h_{2st}\left(\boldsymbol{\theta}\right)$$

where $\boldsymbol{\theta}$ denotes the collection of all the distinct parameters in $\boldsymbol{\beta}$ and $D$ and:

$$f_{1t}\left(\mathbf{y}_i, \mathbf{y}_j\right) = y_{it} + y_{jt}, \quad h_{1t}\left(\mathbf{x}_i, \mathbf{x}_j; \boldsymbol{\theta}\right) = \left(\mathbf{u}_{it} + \mathbf{u}_{jt}\right)^\top \boldsymbol{\beta} \qquad (6.87)$$
$$f_{2st}\left(\mathbf{y}_i, \mathbf{y}_j\right) = \left(y_{is} - y_{js}\right)\left(y_{it} - y_{jt}\right)$$
$$h_{2st}\left(\mathbf{x}_i, \mathbf{x}_j; \boldsymbol{\theta}\right) = \boldsymbol{\beta}^\top \left(\mathbf{u}_{is} - \mathbf{u}_{js}\right)\left(\mathbf{u}_{it} - \mathbf{u}_{jt}\right)^\top \boldsymbol{\beta} + \mathbf{v}_{is}^\top D \mathbf{v}_{it} + \mathbf{v}_{js}^\top D \mathbf{v}_{jt} + 2\sigma_t^2 \delta_{st}$$

where $\mathbf{u}_{it}$ and $\mathbf{v}_{it}$ are subsets of $\mathbf{x}_{it}$ $(1 \le t \le m,\ 1 \le i \le n)$.

Let

$$\mathbf{f}_1 = \left(f_{11}, \ldots, f_{1m}\right)^\top, \quad \mathbf{h}_1 = \left(h_{11}, \ldots, h_{1m}\right)^\top \qquad (6.88)$$
$$\mathbf{f}_2 = \left(f_{211}, f_{212}, \ldots, f_{21m}, f_{222}, \ldots, f_{22m}, \ldots, f_{2(m-1)m}, f_{2mm}\right)^\top$$
$$\mathbf{h}_2 = \left(h_{211}, h_{212}, \ldots, h_{21m}, h_{222}, \ldots, h_{22m}, \ldots, h_{2(m-1)m}, h_{2mm}\right)^\top$$
$$\mathbf{h} = \left(\mathbf{h}_1^\top, \mathbf{h}_2^\top\right)^\top, \quad \mathbf{f} = \left(\mathbf{f}_1^\top, \mathbf{f}_2^\top\right)^\top$$
$$S_{\mathbf{i}} = \mathbf{f}_{\mathbf{i}} - \mathbf{h}_{\mathbf{i}}, \quad G_{\mathbf{i}} = D_{\mathbf{i}} = \frac{\partial}{\partial \boldsymbol{\theta}} \mathbf{h}_{\mathbf{i}}, \quad \mathbf{i} = (i, j)$$

Then, we estimate $\boldsymbol{\theta}$ by solving the following equation:

$$U_n\left(\boldsymbol{\theta}\right) = \sum_{\mathbf{i} \in C_2^n} G_{\mathbf{i}} S_{\mathbf{i}} = \sum_{(i,j) \in C_2^n} D_{\mathbf{i}} \left[\mathbf{f}\left(\mathbf{y}_i, \mathbf{y}_j\right) - \mathbf{h}\left(\mathbf{x}_i, \mathbf{x}_j; \boldsymbol{\theta}\right)\right] = \mathbf{0} \qquad (6.89)$$

Again, the UGEE estimate $\widehat{\boldsymbol{\theta}}$ is readily obtained numerically. The asymptotic variance of $\widehat{\boldsymbol{\theta}}$ and a consistent estimate are computed similarly to the previous example. Also, we may set $G_{\mathbf{i}} = D_{\mathbf{i}} V_{\mathbf{i}}^{-1}$ and select $V_{\mathbf{i}}^{-1}$ based on the variance of $\mathbf{f}$. For example, since the variance of $\mathbf{f}_1$ is readily obtained, we can set $V_{\mathbf{i}}$ to a block diagonal matrix and use the variance of $\mathbf{f}_1$ and the identity matrix to form the first and second block diagonal of $V_{\mathbf{i}}$. The consideration is similar to the selection of $G$ for GEE II discussed in Chapter 4.

**Example 10** (***Zero-inflated Poisson***). Consider the FRM for the distribution-free ZIP model discussed in Example 15 of Section 6.2.2:

$$
\begin{aligned}
E\left(y_i \mid \mathbf{x}_i\right) &= \left(1-\rho\left(\mathbf{u}_i\right)\right) \mu\left(\mathbf{v}_i\right) = \left(1-\rho_i\right) \mu_i \qquad (6.90)\\
E\left[\left(y_i-y_j\right)^2 \mid \mathbf{x}_i, \mathbf{x}_j\right] &= \mu_i\left(1-\rho_i\right)\left(1+\mu_i \rho_i\right)+\mu_j\left(1-\rho_j\right)\left(1+\mu_j \rho_j\right)+ \\
&\quad +\left[\left(1-\rho_i\right) \mu_i-\left(1-\rho_j\right) \mu_j\right]^2 \\
\operatorname{logit}\left(\rho\left(\mathbf{u}_i\right)\right) &= \mathbf{u}_{1i}^{\top} \boldsymbol{\beta}_u, \quad \log\left(\mu\left(\mathbf{v}_i\right)\right)=\mathbf{v}_i^{\top} \boldsymbol{\beta}_v, \quad \boldsymbol{\theta}=\left(\boldsymbol{\beta}_1^{\top}, \boldsymbol{\beta}_2^{\top}\right)^{\top}
\end{aligned}
$$

where $\mathbf{u}_i$ and $\mathbf{v}_i$ are subsets of $\mathbf{x}_i$. Let

$$
\begin{aligned}
f_1\left(y_i, y_j\right) &= y_i+y_j, \quad f_2\left(y_i, y_j\right)=\left(y_i-y_j\right)^2, \quad h_1=\left(1-\rho_i\right) \mu_i \qquad (6.91)\\
h_2\left(\mathbf{x}_i, \mathbf{x}_j ; \boldsymbol{\theta}\right) &= \mu_i\left(1-\rho_i\right)\left(1+\mu_i \rho_i\right)+\mu_j\left(1-\rho_j\right)\left(1+\mu_j \rho_j\right)+ \\
&\quad +\left[\left(1-\rho_i\right) \mu_i-\left(1-\rho_j\right) \mu_j\right]^2 \\
\mathbf{h} &= \left(h_1, h_2\right)^{\top}, \quad \mathbf{f}=\left(f_1, f_2\right)^{\top}, \quad S_{\mathbf{i}}=\mathbf{f}_{\mathbf{i}}-\mathbf{h}_{\mathbf{i}}, \quad G_{\mathbf{i}}=\frac{\partial}{\partial \boldsymbol{\theta}} \mathbf{h}_{\mathbf{i}}
\end{aligned}
$$

The UGEE for $\boldsymbol{\theta}$ has the same form as the one given in (6.89). The estimate of $\boldsymbol{\theta}$ and its asymptotic distribution are computed similarly to the previous example (see exercise).

## 6.4 Inference for Longitudinal Data

When analyzing data from longitudinal studies, missing data is a common issue. In Chapter 4, we discussed inference for single-response based, distribution-free regression models, including the popular class of generalized linear models under missing data. In Chapter 5, we applied and extended such missing data considerations for inference about multivariate U-statistics under the two most popular missing data mechanisms — the MCAR and the more general MAR assumption. In this section, we extend the discussions in Chapters 4 and 5 to address missing data when making inference for FRM. We first consider missing data following the MCAR mechanism.

### 6.4.1 Inference Under MCAR

Under MCAR, the probability of occurrence of missing response does not depend on the response of interest. As discussed in Chapters 4 and 5, the important issue is how to use all available data so that more power is achieved for inference of model parameters since the traditional approach based only on the subjects with complete data does not utilize all available data and thus may be quite inefficient. We discussed in these chapters a missing-data indicator approach to facilitate the development of asymptotic distribution for inference of model parameters based on all observed data within the context of each chapter. Such an approach is readily extended to the current FRM setting.

**Example 1** (***Generalized ANOVA for variance***). Consider the one-group variance ANOVA in Example 1 of Section 6.3.1 within the setting of a longitudinal study with $n$ subjects and two assessments at pre- and posttreatment times indexed by $t$ ($t = 1, 2$). Let $\mathbf{y}_i = (y_{i1}, y_{i2})^\top$ denote the bivariate response from the $i$th subject with $y_{i1}$ and $y_{i2}$ denoting the outcome at the pre- and posttreatment assessment points ($1 \leq i \leq n$). Assume that missing data only occurs at posttreatment so that $y_{i2}$ is partially observed.

Let

$$\sigma_t^2 = Var(y_{it}), \quad \boldsymbol{\theta} = \left(\sigma_1^2, \sigma_2^2\right)^\top, \quad \mathbf{h}(\boldsymbol{\theta}) = \boldsymbol{\theta} = \left(\sigma_1^2, \sigma_2^2\right)^\top \tag{6.92}$$
$$\mathbf{i} = (i_1, i_2), \quad \mathbf{f}(\mathbf{y_i}, \mathbf{y_j}) = \left(\frac{1}{2}(y_{i1} - y_{j1})^2, \frac{1}{2}(y_{i2} - y_{j2})^2\right)^\top$$

Then, the FRM defined by

$$E\left[\mathbf{f}(\mathbf{y_i}, \mathbf{y_j})\right] = \mathbf{h}(\boldsymbol{\theta}) \tag{6.93}$$

can be used for inference about the variance parameters $\sigma_k^2$ or their difference between the pre- and posttreatment assessments. Setting $G = D = \frac{\partial}{\partial \boldsymbol{\theta}} \mathbf{h}(\boldsymbol{\theta}) = \mathbf{I}_2$, the UGEE given by

$$\begin{aligned} \mathbf{U}_n &= \sum_{(i,j) \in C_2^n} G\left(\mathbf{f}(\mathbf{y_i}, \mathbf{y_j}) - \mathbf{h}\right) \\ &= \sum_{(i,j) \in C_2^n} \left[\begin{pmatrix} \frac{1}{2}(y_{i1} - y_{j1})^2 \\ \frac{1}{2}(y_{i2} - y_{j2})^2 \end{pmatrix} - \begin{pmatrix} \sigma_1^2 \\ \sigma_2^2 \end{pmatrix}\right] = \mathbf{0} \end{aligned} \tag{6.94}$$

can be used to provide inference about $\boldsymbol{\theta}$ in the absence of missing data.

Solving the above for $\boldsymbol{\theta}$ yields

$$\widehat{\sigma}_t^2 = \binom{n}{2}^{-1} \sum_{(i,j)\in C_2^n} \frac{1}{2}\left(y_{it} - y_{jt}\right)^2, \quad t = 1, 2 \tag{6.95}$$

which is the familiar sample variance estimate at each time $t$ $(t = 1, 2)$.

When $y_{i2}$ is partially observed, the UGEE in (6.94) can only be applied to subjects with complete observations at the pre- and posttreatment assessment times, that is, those with both $y_{i1}$ and $y_{i2}$ observed. As discussed in Chapters 4 and 5, although such a complete-data approach still yields consistent estimates of $\boldsymbol{\theta}$ under MCAR, it may severely undercut power, especially when many subjects have intermittent missing responses. Within the context of the example, the loss of power or efficiency depends on the number of subjects who have missing data at posttreatment.

To utilize all available data to estimate $\boldsymbol{\theta}$, let $r_{it} = 1$ if $y_{it}$ is observed and $r_{it} = 0$ if otherwise $(1 \leq i \leq n,\ 1 \leq t \leq 2)$. Since the response is always observed at pre-treatment, $r_{i1} \equiv 1$. Now consider the following revised estimating equation:

$$\begin{aligned}
\mathbf{U}_n &= \sum_{(i,j)\in C_2^n} G \begin{pmatrix} 1 & 0 \\ 0 & r_{i2}r_{j2} \end{pmatrix} (\mathbf{f}_{ij} - \mathbf{h}) \qquad (6.96) \\
&= \sum_{(i,j)\in C_2^n} \begin{pmatrix} 1 & 0 \\ 0 & r_{i2}r_{j2} \end{pmatrix} \left[ \begin{pmatrix} \frac{1}{2}\left(y_{i1} - y_{j1}\right)^2 \\ \frac{1}{2}\left(y_{i2} - y_{j2}\right)^2 \end{pmatrix} - \begin{pmatrix} \sigma_1^2 \\ \sigma_2^2 \end{pmatrix} \right] \\
&= \sum_{(i,j)\in C_2^n} \left[ \begin{pmatrix} \frac{1}{2}\left(y_{i1} - y_{j1}\right)^2 \\ \frac{1}{2} r_{i2}r_{j2}\left(y_{i2} - y_{j2}\right)^2 \end{pmatrix} - \begin{pmatrix} \sigma_1^2 \\ \sigma_2^2 \end{pmatrix} \right] = \mathbf{0}
\end{aligned}$$

Equation (6.96) is well defined for all subjects regardless of their missing response status. For example, if $y_{i2}$ or $y_{j2}$ or both are missing, then $r_{i2}r_{j2} = 0$ and in this case $\frac{1}{2} r_{i2}r_{j2}\left(y_{i2} - y_{j2}\right)^2 = 0$, a well-defined quantity.

The equation is readily solved to yield

$$\begin{aligned}
\widehat{\sigma}_1^2 &= \binom{n}{2}^{-1} \sum_{(i,j)\in C_2^n} \frac{1}{2}\left(y_{i1} - y_{i1}\right)^2 \qquad (6.97) \\
\widehat{\sigma}_2^2 &= \frac{1}{\sum_{(i,j)\in C_2^n} r_{i2}r_{j2}} \sum_{(i,j)\in C_2^n} \frac{1}{2} r_{i2}r_{j2}\left(y_{i2} - y_{j2}\right)^2
\end{aligned}$$

The estimate $\widehat{\boldsymbol{\theta}} = \left(\widehat{\sigma}_1^2, \widehat{\sigma}_2^2\right)^\top$ in (6.97) has been derived for the same problem using a similar missing data indicator approach within the context of multivariate U-statistics in Chapter 5. It was shown there that $\widehat{\boldsymbol{\theta}}$ is both consistent and asymptotically normal. It is also readily checked that $\widehat{\sigma}_2^2$ is the usual sample variance estimate calculated based on the observed $y_{i2}$.

We can also readily show the consistency and asymptotic normality of $\widehat{\boldsymbol{\theta}}$ by using Theorem 1 of Section 6.3. To see this, let

$$\mathbf{i} = (i, j), \quad R_{\mathbf{i}} = \begin{pmatrix} 1 & 0 \\ 0 & r_{i2} r_{j2} \end{pmatrix} \tag{6.98}$$

Then, the estimating equation in (6.96) can be expressed in the following form:

$$\mathbf{U}_n = \sum_{\mathbf{i} \in C_2^n} \mathbf{U}_{n\mathbf{i}} = \sum_{\mathbf{i} \in C_2^n} R_{\mathbf{i}} \left(\mathbf{f}_{\mathbf{i}} - \mathbf{h}\right) \tag{6.99}$$

It is readily checked that $E\left(\mathbf{U}_n\right) = \mathbf{0}$ and thus the above equation is unbiased (see exercise). Further, $\mathbf{U}_n$ is a U-statistic-like quantity and thus it follows from Theorem 1 of Section 6.3 that (6.99) defines consistent and asymptotically normal estimates of $\boldsymbol{\theta}$.

**Example 2** (***Generalized ANOVA for mean and variance***). In Example 1, now consider adding a component for the mean response at each assessment to simultaneously model the mean and variance of $y_{it}$. Let

$$\mu_t = E\left(y_{it}\right), \quad \boldsymbol{\theta}_t = \left(\mu_t, \sigma_t^2\right)^\top, \quad \boldsymbol{\theta} = \left(\boldsymbol{\theta}_1^\top, \boldsymbol{\theta}_2^\top\right)^\top, \quad \mathbf{h}_t\left(\boldsymbol{\theta}\right) = \left(\mu_t, \sigma_t^2\right)^\top = \boldsymbol{\theta}_t$$

$$\mathbf{f}_t = \left(\frac{1}{2}\left(y_{it} + y_{jt}\right), \frac{1}{2}\left(y_{it} - y_{jt}\right)^2\right)^\top, \quad \mathbf{h} = \left(\mathbf{h}_1^\top, \mathbf{h}_2^\top\right)^\top, \quad \mathbf{f} = \left(\mathbf{f}_1^\top, \mathbf{f}_2^\top\right)^\top$$

The FRM has the same general form as in the previous example, except for the added component for the mean response. Again, set $G = \frac{\partial}{\partial \boldsymbol{\theta}} \mathbf{h}\left(\boldsymbol{\theta}\right) = \mathbf{I}_4$. Then, in the absence of missing data, the UGEE is given by

$$\mathbf{U}_n = \sum_{(i,j) \in C_2^n} \left[ \begin{pmatrix} \frac{1}{2}\left(y_{i1} + y_{j1}\right) \\ \frac{1}{2}\left(y_{i1} - y_{j1}\right)^2 \\ \frac{1}{2}\left(y_{i2} + y_{j2}\right) \\ \frac{1}{2}\left(y_{i2} - y_{j2}\right)^2 \end{pmatrix} - \begin{pmatrix} \mu_1 \\ \sigma_1^2 \\ \mu_2 \\ \sigma_2^2 \end{pmatrix} \right] = \mathbf{0} \tag{6.100}$$

We obtain from (6.100) the usual sample mean and variance $y_{it}$ as estimates of $\mu_t$ and $\sigma_t^2$.

In the presence of missing data, we can readily generalize the approach in Example 1 to accommodate the extra component for the mean response. Let $R_{ij2} = r_{i2}r_{j2}\mathbf{I}_2$ and define the estimating equation as follows:

$$\mathbf{U}_n = \sum_{(i,j)\in C_2^n} \begin{pmatrix} I_2 & 0 \\ 0 & R_{ij2} \end{pmatrix} \left[ \begin{pmatrix} \frac{1}{2}(y_{i1}+y_{j1}) \\ \frac{1}{2}(y_{i1}-y_{j1})^2 \\ \frac{1}{2}(y_{i2}+y_{j2}) \\ \frac{1}{2}(y_{i2}-y_{j2})^2 \end{pmatrix} - \begin{pmatrix} \mu_1 \\ \sigma_1^2 \\ \mu_2 \\ \sigma_2^2 \end{pmatrix} \right] = \mathbf{0} \qquad (6.101)$$

As in Example 1, the estimating equation above is well defined. Further, it is readily checked that (6.101) yields the same estimates of $\sigma_1^2$ and $\sigma_2^2$ as those in Example 1 (see exercise). The estimates of $\mu_1$ and $\mu_2$ are given by (see exercise)

$$\widehat{\mu}_1 = \frac{1}{n} \sum_{(i,j)\in C_2^n} y_{i1}, \quad \widehat{\mu}_2 = \frac{1}{\sum_{i=1}^n r_{i2}} \sum_{i=1}^n r_{i2} y_{i2} \qquad (6.102)$$

As in Example 1, it is readily shown that the estimate $\widehat{\boldsymbol{\theta}}$ obtained from (6.101) with $\widehat{\mu}_t$ and $\widehat{\sigma}_t^2$ given in (6.102) and (6.97) is consistent and asymptotically normal (see exercise).

**Example 3** (***Product–moment correlation***). Consider again the product–moment correlation for longitudinal study data discussed in Chapter 5. Let $\mathbf{z}_{it} = (x_{it}, y_{it})^\top$ denote the continuous bivariate outcome from the $i$th subject at time $t$ from a longitudinal study with $n$ subjects and $m$ assessment times $(1 \le i \le n,\ 1 \le t \le m)$. Let $\rho_t = Corr(x_{it}, y_{it})$ denote the product–moment correlation between $x_{it}$ and $y_{it}$ and $\boldsymbol{\rho} = (\rho_1, , \ldots, \rho_m)^\top$. We discussed inference about $\boldsymbol{\rho}$ under both MCAR and MAR within the context of multivariate U-statistics in Chapter 5. Alternatively, we can also model $\boldsymbol{\rho}$ using an FRM.

Let

$$f_{kt}(\mathbf{z}_{it}, \mathbf{z}_{it}) = \begin{cases} \frac{1}{2}(x_{it}-x_{jt})(y_{it}-y_{jt}) \\ \frac{1}{2}(x_{it}-x_{it})^2 \\ \frac{1}{2}(y_{it}-y_{it})^2 \end{cases}, \quad h_{kt} = \begin{cases} \sqrt{\sigma_{xt}^2\sigma_{yt}^2}\rho_t & k=1 \\ \sigma_{xt}^2 & k=2 \\ \sigma_{yt}^2 & k=3 \end{cases}$$

$$\mathbf{f}_t = (f_{1t}, f_{2t}, f_{3t})^\top, \quad \mathbf{h}_t = (h_{1t}, h_{2t}, h_{3t})^\top \qquad (6.103)$$

Also, let

$$\boldsymbol{\theta}_t = \left(\rho_t, \sigma_{xt}^2, \sigma_{yt}^2\right)^\top, \quad \boldsymbol{\theta} = \left(\boldsymbol{\theta}_1, \ldots, \boldsymbol{\theta}_m\right)^\top$$

Define the FRM as follows:

$$E\left(\mathbf{f}_t\left(\mathbf{z}_{it}, \mathbf{z}_{it}\right)\right) = \mathbf{h}_t\left(\boldsymbol{\theta}\right), \quad 1 \leq t \leq m \tag{6.104}$$

In the FRM model above, $\boldsymbol{\rho}$ is part of the model parameter vector $\boldsymbol{\theta}$. In Chapter 5, we first employed multivariate U-statistics to model the covariance $\sigma_{xyt} = Cov\left(x_{it}, y_{it}\right)$ together with the variance components $\sigma_{xt}^2$ and $\sigma_{yt}^2$ and then obtained inference about $\boldsymbol{\rho}$ by applying the Delta method, as $\rho_t$ is a function of these parameters. In contrast, the FRM in (6.104) directly models $\boldsymbol{\rho}$ as part of its parameters.

Now let

$$\mathbf{f} = \left(\mathbf{f}_1^\top, \ldots, \mathbf{f}_m^\top\right)^\top, \quad \mathbf{h} = \left(\mathbf{h}_1^\top, \ldots, \mathbf{h}_m^\top\right)^\top, \quad \mathbf{z}_i = \left(\mathbf{z}_{i1}^\top, \ldots, \mathbf{z}_{im}^\top\right)^\top$$

Then, by setting $G = D = \frac{\partial}{\partial \boldsymbol{\theta}} \mathbf{h}\left(\boldsymbol{\theta}\right) = \mathbf{I}_{3\times m}$, the estimating equation for $\boldsymbol{\theta}$ in the absence of missing data is given by

$$\mathbf{U}_n\left(\boldsymbol{\theta}\right) = \sum_{i,j \in C_2^n} DS_{ij} = \sum_{i,j \in C_2^n} \left[\mathbf{f}\left(\mathbf{z}_i, \mathbf{z}_i\right) - \mathbf{h}\right] = \mathbf{0} \tag{6.105}$$

The above UGEE can be solved in closed form to obtain the estimate $\widehat{\boldsymbol{\theta}}$ of $\boldsymbol{\theta}$. In addition, it is readily checked that $\widehat{\rho}_t$ are the Pearson correlation estimates and $\widehat{\sigma}_{xt}^2$ and $\widehat{\sigma}_{yt}^2$ are the sample variances of $x_{it}$ and $y_{it}$, respectively (see exercise).

When there is missing data, let $r_{it} = 1$ if $\mathbf{z}_{it}$ is observed, that is, if both $x_{it}$ and $y_{it}$ are observed. Also, let

$$R_{ijt} = r_{it} r_{jt} I_3, \quad R_{ij} = \text{diag}_t\left(R_{ijt}\right) = \begin{pmatrix} R_{ij1} & \cdots & 0 \\ \vdots & \ddots & \vdots \\ 0 & \cdots & R_{ijm} \end{pmatrix}, \quad 1 \leq i, j \leq n$$

Define the estimating equation in the presence of missing data by

$$U_n\left(\boldsymbol{\theta}\right) = \sum_{(i,j) \in C_2^n} R_{ij}\left(\mathbf{f}\left(\mathbf{z}_i, \mathbf{z}_i\right) - \mathbf{h}\left(\boldsymbol{\theta}\right)\right) = \mathbf{0} \tag{6.106}$$

Like the estimating equations for Examples 1 and 2 discussed above, (6.106) is well defined. It is again readily checked that the estimate obtained as the solution to (6.106) is $\widehat{\boldsymbol{\theta}} = \left(\widehat{\rho}_t, \widehat{\sigma}^2_{xt}, \widehat{\sigma}^2_{yt}\right)^\top$, where $\widehat{\rho}_t$ is the Pearson correlation between $x_{it}$ and $y_{it}$, while $\widehat{\sigma}^2_{xt}$ and $\widehat{\sigma}^2_{yt}$ are the respective sample variance estimates of $x_{it}$ and $y_{it}$ calculated based on the observed data (see exercises), that is,

$$\widehat{\sigma}^2_{xt} = \frac{1}{n_t}\sum_{i=1}^{n} r_{it}\left(x_{it} - \widehat{\mu}_{xt}\right)^2, \quad \widehat{\mu}_{xt} = \frac{1}{n_t}\sum_{i=1}^{n} r_{it}x_{it}, \quad 1 \le t \le m \tag{6.107}$$

$$\widehat{\sigma}^2_{yt} = \frac{1}{n_t}\sum_{i=1}^{n} r_{it}\left(y_{it} - \widehat{\mu}_{yt}\right)^2, \quad \widehat{\mu}_{yt} = \frac{1}{n_t}\sum_{i=1}^{n} r_{it}y_{it}$$

$$\widehat{\rho}_t = \frac{1}{n_t}\sum_{i=1}^{n} r_{it}\frac{\left(x_{it} - \widehat{\mu}_{xt}\right)\left(x_{it} - y_{j2t}\right)}{\sqrt{\widehat{\sigma}^2_{xt}\widehat{\sigma}^2_{yt}}}, \quad n_t = \frac{1}{n}\sum_{i=1}^{n} r_{it}$$

The consistency and asymptotic normality of the estimate $\widehat{\boldsymbol{\theta}}$ also follows from an argument similar to the one used in Example 2.

Note that in the missing data considerations above, we assumed that both $x_{it}$ and $y_{it}$ are either observed or missing together. As in the case of the multivariate U-statistics approach discussed in Chapter 5 for the product–moment correlation, we can relax this assumption and allow $x_{it}$ and $y_{it}$ to have different missing data patterns. The above considerations are readily generalized to this more general situation (see exercise).

**Example 4 (*Linear mixed-effects model*).** Consider the FRM defined by (6.86) and (6.87) in Example 9 of Section 6.3.2 for distribution-free inference of the linear mixed-effects model. Let $\mathbf{f}_1$, $\mathbf{f}_2$, $\mathbf{h}_1$, $\mathbf{h}_2$, $\mathbf{f}_\mathbf{i}$, $\mathbf{h}_\mathbf{i}$, $S_\mathbf{i}$, and $\mathbf{i}$ be defined in (6.88). As discussed in that example, the UGEE in (6.89) defines consistent estimates of the parameter vector of interest $\boldsymbol{\theta}$ under complete data. In the presence of missing data, this estimating equation can be applied to the subset of the data formed by the subjects with complete response $y_{it}$ over all assessment times to provide valid inference about $\boldsymbol{\theta}$. However, as noted earlier, such an approach generally leads to less efficient estimates.

To define an appropriate estimating equation to take advantage of all available data, let $r_{it} = 1$ if $y_{it}$ is observed and 0 if otherwise. We assume no missing data at $t = 1$ so that $r_{i1} \equiv 1$. Also, let

$$\mathbf{r}_i = \left(r_{i1}, r_{i2}, \ldots, r_{im}\right)^\top, \quad \mathbf{r}_{1ij} = \left(r_{i1}r_{j1}, r_{i2}r_{j2}, \ldots, r_{im}r_{jm}\right)^\top \tag{6.108}$$

$$\mathbf{r}_{2ij} = \left(r_{i1}r_{j1}, r_{i1}r_{j2}, \ldots, r_{i1}r_{jm}, r_{i2}r_{j2}, \ldots, r_{i(m-1)}r_{jm}, r_{im}r_{jm}\right)^\top$$

$$\mathbf{r}_{ij} = \left(\mathbf{r}^\top_{1ij}, \mathbf{r}^\top_{2ij}\right)^\top$$

In the above, $\mathbf{r}_{1ij}$ is an $m \times 1$ vector, while $\mathbf{r}_{2ij}$ is an $\left(m + \frac{1}{2}m(m+1)\right) \times 1$ vector. Thus, $\mathbf{r}_{ij}$ is a $v \times 1$ vector, where $v = 2m + \frac{1}{2}m(m+1)$. Let $\mathbf{i} = (i, j)$ and $R_{\mathbf{i}}$ be a $v \times v$ diagonal matrix with $\mathbf{r}_{\mathbf{i}}$ forming the diagonal of the matrix.

Define the estimating equation for $\boldsymbol{\theta}$ as follows:

$$\mathbf{U}_n(\boldsymbol{\theta}) = \sum_{\mathbf{i} \in C_2^n} \mathbf{U}_{n\mathbf{i}}(\boldsymbol{\theta}) = \sum_{\mathbf{i} \in C_2^n} G_{\mathbf{i}} R_{\mathbf{i}} S_{\mathbf{i}} = \mathbf{0} \tag{6.109}$$

To see if the above estimating equation is well defined, consider the term $R_{\mathbf{i}} S_{\mathbf{i}}$. This term is a $v \times 1$ column vector. The $l$th component $\Delta_{\mathbf{i}l}$ of $R_{\mathbf{i}} S_{\mathbf{i}}$ has the following form:

$$\Delta_{\mathbf{i}l} = \begin{cases} r_{it} r_{jt} \left[f_{1t}(\mathbf{y}_i, \mathbf{y}_j) - h_{1t}\right] & \text{if } 1 \le l \le m \\ r_{is} r_{it} r_{js} r_{jt} \left[f_{2st}(\mathbf{y}_i, \mathbf{y}_j) - h_{2t}\right] & \text{if } m+1 \le l \le v \end{cases}$$

For $1 \le l \le m$, $\Delta_{\mathbf{i}l} = r_{it} r_{jt} \left[f_{1t}(\mathbf{y}_i, \mathbf{y}_j) - h_{1t}\right]$, with $1 \le t \le m$. If $y_{it}$ or $y_{jt}$ is missing, then one of $r_{it}$ and $r_{jt}$ will be 0. So, $\Delta_{\mathbf{i}l} = 0$ regardless of what value is assigned to the missing response. For $m + 1 \le l \le v$, $\Delta_{\mathbf{i}l} = r_{is} r_{it} r_{js} r_{jt} \left[f_{2st}(\mathbf{y}_i, \mathbf{y}_j) - h_{2t}\right]$, with $1 \le s < t \le m$. If one of $y_{is}$, $y_{js}$, $y_{it}$ and $y_{jt}$ is missing, then one of $r_{is}$, $r_{js}$, $r_{it}$ and $r_{jt}$ is 0 and $E(\Delta_{\mathbf{i}l}) = 0$. Thus, the estimating equation for $\boldsymbol{\theta}$ in (6.109) is well defined. In addition, it is quite similar to the one defined in (6.89) for estimating $\boldsymbol{\theta}$ in the absence of missing data, with the only difference being replacing $S_{\mathbf{i}}$ with the revised residual term $R_{\mathbf{i}} S_{\mathbf{i}}$.

Under MCAR, the missing data indicator vector $\mathbf{r}_i$ is stochastically independent of $\mathbf{y}_i$. Thus, it follows from the iterated conditional expectation that for $1 \le l \le m$,

$$\begin{aligned} E(\Delta_{\mathbf{i}l} \mid \mathbf{x}_i, \mathbf{x}_j) &= E\left[r_{it} r_{jt} \left(f_{1t}(\mathbf{y}_i, \mathbf{y}_j) - h_{1t}\right) \mid \mathbf{x}_i, \mathbf{x}_j\right] \\ &= E(r_{it} r_{jt}) E\left[\left(f_{1t}(\mathbf{y}_i, \mathbf{y}_j) - h_{1t}\right) \mid \mathbf{x}_i, \mathbf{x}_j\right] \\ &= 0 \end{aligned}$$

Similarly, for $m + 1 \le l \le v$, we have

$$\begin{aligned} E(\Delta_{\mathbf{i}l} \mid \mathbf{x}_i, \mathbf{x}_j) &= E\left[r_{is} r_{it} r_{js} r_{jt} \left(f_{2st}(\mathbf{y}_i, \mathbf{y}_j) - h_{2t}\right) \mid \mathbf{x}_i, \mathbf{x}_j\right] \\ &= E(r_{is} r_{it} r_{js} r_{jt}) E\left[\left(f_{2st}(\mathbf{y}_i, \mathbf{y}_j) - h_{2t}\right) \mid \mathbf{x}_i, \mathbf{x}_j\right] \\ &= 0 \end{aligned}$$

It follows that

$$E(\mathbf{U}_n(\boldsymbol{\theta})) = \sum_{\mathbf{i} \in C_2^n} E(\mathbf{U}_{n\mathbf{i}}(\boldsymbol{\theta})) = \sum_{\mathbf{i} \in C_2^n} E\left[G_{\mathbf{i}} R_{\mathbf{i}} E(S_{\mathbf{i}} \mid \mathbf{x}_i, \mathbf{x}_j)\right] = \mathbf{0}$$

Thus, the estimating equation defined in (6.109) is unbiased. In addition, as $\mathbf{U}_{n\mathbf{i}}(\boldsymbol{\theta})$ are U-statistic like quantities, it follows from Theorem 2 that the estimate of $\boldsymbol{\theta}$ obtained from (6.109) is consistent and asymptotically normal.

### 6.4.2 Inference Under MAR

When missing data are MAR, the missingness will depend on observed responses. Under such response-dependent missingness, the estimating equations discussed in the preceding section no longer guarantee consistent estimates for the models considered. In Chapter 4, we discussed how to deal with this issue for the mean-based distribution-free regression models. In Chapter 5, we generalized the approaches in Chapter 4 to address MAR for multivariate $U$-statistics-based models. These considerations can also be applied to address MAR for the functional response model.

**Example 5** (***Generalized ANOVA for variance***). Consider again the bivariate response model in Example 1 of Section 6.4.1 arising from a pre-post treatment study design. As before, assume that missing data only occurs at posttreatment. But, now suppose that the occurrence of missing response $y_{i2}$ at posttreatment depends on pretreatment response $y_{i1}$. Under this response-dependent missingness or MAR, the consistency of the estimate of $\boldsymbol{\theta}$, or more specifically $\sigma_2^2$, obtained from the estimating equation in (6.94) can no longer be assured and a new estimating equation must be developed.

As seen in the discussion of Example 1, we can estimate $\boldsymbol{\theta}$ from a modified UGEE given by

$$\mathbf{U}_n = \sum_{\mathbf{i}\in C_2^n} \mathbf{U}_{n\mathbf{i}} = \sum_{\mathbf{i}\in C_2^n} GR_{\mathbf{i}}S_{\mathbf{i}} = \sum_{\mathbf{i}\in C_2^n} GR_{\mathbf{i}}\left(\mathbf{f}_{\mathbf{i}} - \mathbf{h}\right) = \mathbf{0} \tag{6.110}$$

where $R_{\mathbf{i}}$ is given in (6.98), $\mathbf{i} = (i_1, i_2) \in C_2^n$ and $G$ can be parameterized by a vector $\boldsymbol{\alpha}$. We can see that the diagonal elements, $r_{\mathbf{i}t} = r_{i_1t}r_{i_2t}$, are binary variables indicating whether the $t$th component $f_{\mathbf{i}t}$ of the functional response $\mathbf{f}_{\mathbf{i}} = (f_{\mathbf{i}1}, f_{\mathbf{i}2})^\top$ is missing. Thus, $r_{\mathbf{i}t}$ is a generalization of the single-response missing data indicator $r_{it}$ to the functional response of FRM. Now we can generalize the weighted generalized estimating equation (WGEE) approach for single-response based regression models discussed in Chapter 4 to address MAR for inference about FRM within our current setting.

To this end, let $\pi_{\mathbf{i}t} = E\left(r_{\mathbf{i}t} \mid \mathbf{y}_{\mathbf{i}}\right)$ and assume that for some $c > 0$, $\pi_{\mathbf{i}t} \geq c$ for all $\mathbf{i} \in C_2^n$ and $1 \leq t \leq 2$, that is, $\pi_{\mathbf{i}t}$ are bounded away from 0. This assumption is a generalization of a similar condition discussed in Chapters

4 and 5 for weighting function employed for single-response based regression models and multivariate U-statistics to ensure consistent estimation and stable estimates of the model parameter vector $\boldsymbol{\theta}$ when $\pi_{\mathbf{i}t}$ are used as weights to statistically "recover" missing *functional* responses. Following the discussion in these chapters, we first assume that $\pi_{\mathbf{i}t}$ are known and consider estimation of $\boldsymbol{\theta} = \left(\sigma_1^2, \sigma_2^2\right)^\top$. We then discuss the more practical situation when $\pi_{\mathbf{i}t}$ are unknown and modeled by observed study data.

Let

$$\Delta_{\mathbf{i}} = \begin{pmatrix} \pi_{\mathbf{i}1}^{-1} r_{\mathbf{i}1} & 0 \\ 0 & \pi_{\mathbf{i}2}^{-1} r_{\mathbf{i}2} \end{pmatrix} = \begin{pmatrix} 1 & 0 \\ 0 & \pi_{i_1 2}^{-1} \pi_{i_2 2}^{-1} r_{i_1 2} r_{i_2 2} \end{pmatrix}$$

Note that the second equality above follows since $y_{i1}$ is always observed and

$$\pi_{\mathbf{i}t} = E\left(r_{i_1 t} \mid \mathbf{y}_{i_1}\right) E\left(r_{i_2 t} \mid \mathbf{y}_{i_1}\right) = \pi_{i_1 t} \pi_{i_2 t}, \quad \mathbf{i} = (i_1, i_2) \in C_2^n \tag{6.111}$$

Consider the following U-statistics-based weighted generalized estimating equation (UWGEE):

$$\mathbf{U}_n = \sum_{\mathbf{i} \in C_2^n} \mathbf{U}_{n\mathbf{i}} = \sum_{\mathbf{i} \in C_2^n} G \Delta_{\mathbf{i}} S_{\mathbf{i}} = \mathbf{0} \tag{6.112}$$

The UWGEE is quite similar to the unweighted UGEE in (6.110) for the same model defined by (6.92) and (6.93). The only difference is that $\Delta_{\mathbf{i}}$ in (6.112) depends not only on the missing data indicators $r_{\mathbf{i}t}$, but also on their conditional probabilities $\pi_{\mathbf{i}t}$.

Now consider the special case with $G = D = \frac{\partial}{\partial \boldsymbol{\theta}} \mathbf{h}(\boldsymbol{\theta}) = \mathbf{I}_2$. The UWGEE in (6.112) becomes:

$$\sum_{(i,j) \in C_2^n} \begin{pmatrix} 1 & 0 \\ 0 & \pi_{i2}^{-1} \pi_{j2}^{-1} r_{i2} r_{j2} \end{pmatrix} \left[ \begin{pmatrix} \frac{1}{2}\left(y_{i1} - y_{j1}\right)^2 \\ \frac{1}{2}\left(y_{i2} - y_{j2}\right)^2 \end{pmatrix} - \begin{pmatrix} \sigma_1^2 \\ \sigma_2^2 \end{pmatrix} \right] = \mathbf{0} \tag{6.113}$$

As in Example 1 of Section 6.4.1, it is readily checked that the above equation is well defined (see exercise). Further, it follows from the iterated conditional expectation that

$$\begin{aligned} E\left(\mathbf{U}_{n\mathbf{i}2}\right) &= E\left[E\left(\pi_{i2}^{-1} \pi_{j2}^{-1} r_{i2} r_{j2} \left(f_{ij2} - h_2\right) \mid \mathbf{y}_i, \mathbf{y}_j\right)\right] \\ &= E\left[\pi_{i2}^{-1} \pi_{j2}^{-1} E\left(r_{i2} r_{j2} \left(f_{ij2} - h_2\right) \mid \mathbf{y}_i, \mathbf{y}_j\right)\right] \\ &= E\left[\pi_{i2}^{-1} \pi_{j2}^{-1} \left(f_{ij2} - h_2\right) E\left(r_{i2} r_{j2} \mid y_{i1}, y_{j1}\right)\right] \\ &= E\left[\pi_{i2}^{-1} \pi_{j2}^{-1} \left(f_{ij2} - h_2\right) E\left(r_{i2} \mid y_{i1}\right) E\left(r_{j2} \mid y_{j1}\right)\right] \\ &= E\left(f_{ij2} - h_2\right) = 0 \end{aligned} \tag{6.114}$$

Thus, $E(\mathbf{U}_n) = 0$ and the UWGEE in (6.113) is unbiased. Since $\mathbf{U}_n$ is a U-statistic-like random vector, it follows from Theorem 1 of Section 6.3.1 that the estimate of $\boldsymbol{\theta}$ obtained from the estimating equation in (6.113) is consistent and asymptotically normal.

Note that for notational brevity, we expressed $\mathbf{i}$ as $\mathbf{i} = (i, j)$ instead of $\mathbf{i} = (i_1, i_2)$ in equation (6.114). We will continue to employ this notation whenever convenient provided that no confusion arises.

The UWGEE in this simple example can be solved in closed form:

$$\widehat{\sigma}_1^2 = \binom{n}{2}^{-1} \sum_{(i,j)\in C_2^n} \frac{1}{2}(y_{i1} - y_{j1})^2 \tag{6.115}$$

$$\widehat{\sigma}_2^2 = \frac{1}{\sum_{(i,j)\in C_2^n} \pi_{i2}^{-1}\pi_{j2}^{-1} r_{i2} r_{j2}} \sum_{(i,j)\in C_2^n} \frac{1}{2} \frac{r_{i2} r_{j2}}{\pi_{i2}\pi_{j2}} (y_{i2} - y_{j2})^2$$

In comparison to the estimate in (6.97) obtained under MCAR, it is seen that $\widehat{\sigma}_2^2$ has the additional weight $\pi_{i2}\pi_{j2}$. As a special case, if $\pi_{i2}$ is independent of $y_{i1}$, that is, missing data is MCAR, then $\pi_{i2} = \pi_2$ and (6.115) reduces to (6.97). Thus, the UWGEE estimate $\widehat{\boldsymbol{\theta}}$ in (6.115) is a generalization of the unweighted version in (6.97) to account for response-dependent missingness under MAR.

We can also directly verify that $\widehat{\boldsymbol{\theta}}$ is consistent. The consistency of $\widehat{\sigma}_1^2$ follows immediately from the univariate U-statistics theory. To show that $\widehat{\sigma}_2^2$ is consistent, note that

$$\begin{aligned}
\widehat{\sigma}_2^2 &= \frac{1}{\sum_{(i,j)\in C_2^n} \pi_{i2}^{-1}\pi_{j2}^{-1} r_{i2} r_{j2}} \sum_{(i,j)\in C_2^n} \frac{r_{i2} r_{j2}}{\pi_{i2}\pi_{j2}} (y_{i2} - y_{j2})^2 \\
&= \frac{1}{\binom{n}{2} \sum_{(i,j)\in C_2^n} \pi_{i2}^{-1}\pi_{j2}^{-1} r_{i2} r_{j2}} \binom{n}{2} \sum_{(i,j)\in C_2^n} \frac{r_{i2} r_{j2}}{\pi_{i2}\pi_{j2}} f_{ij2}
\end{aligned}$$

Since both the numerator and denominator are univariate U-statistics, it follows from the consistency of U-statistics and Slutsky's theorem that

$$\widehat{\sigma}_2^2 \rightarrow_p \frac{E\left(\pi_{i2}^{-1} r_{i2} \pi_{j2}^{-1} r_{j2} f_{ij2}\right)}{E\left(\pi_{i2}^{-1} r_{i2}\right) E\left(\pi_{j2}^{-1} r_{j2}\right)} = \frac{\sigma_2^2}{\pi_{i2}^{-2} E^2(r_{i2})} = \sigma_2^2$$

By applying Theorem 1 of Section 6.3, it is readily shown that the asymptotic variance of the UWGEE estimate $\widehat{\boldsymbol{\theta}}$ in (6.115) is given by (see exercise):

$$\Sigma_\theta = 4\Sigma_U = 4Var\left(E\left(\Delta_{ij} S_{ij} \mid \mathbf{y}_i, \mathbf{r}_i\right)\right) \tag{6.116}$$

A consistent estimate of the asymptotic variance $\Sigma_\theta$ is readily constructed following Theorem 1/3 (see exercise).

In practical applications, $\pi_{\mathbf{i}t}$, or rather $\pi_{\mathbf{i}2}$, are unknown and some estimates of this weight function must be substituted into (6.113) before estimates of $\boldsymbol{\theta}$ can be computed. Given the relationship in (6.111), we can model bivariate $\pi_{\mathbf{i}2}$ through single-response based $\pi_{i2}$. Under MAR, we can readily model $\pi_{i2}$ using a logistic regression as follows:

$$\operatorname{logit}\ (\pi_{i2}) = \operatorname{logit}\left[\Pr\left(r_{i2} = 1 \mid y_{i1}\right)\right] = \alpha + \beta y_{i1} \tag{6.117}$$

We can estimate $\pi_{\mathbf{i}2}$ for each pair of subjects $\mathbf{i}$ by the fitted $\pi_{\mathbf{i}2}(\widehat{\gamma}) = \pi_{i_1 2}(\widehat{\gamma})\,\pi_{i_2 2}(\widehat{\gamma})$ based on estimates of the parameter vector $\gamma = (\alpha, \beta)^\top$ of the above logistic regression and then substitute them into (6.113) for computing estimates of $\boldsymbol{\theta}$. We can also use (6.116) to estimate the asymptotic variance of the resulting UGEE estimate $\widehat{\boldsymbol{\theta}}$. However, as discussed in Chapters 4 and 5, $\Sigma_\theta$ in (6.116) may underestimate the variability of $\widehat{\boldsymbol{\theta}}$. By following the discussion in these chapters, we can readily derive the asymptotic variance that accounts for this extra variability.

As noted in Chapters 4 and 5, we can estimate the parameter vector $\gamma = (\alpha, \beta)^\top$ of the logistic model using either parametric or distribution-free procedures. Regardless of the approach taken, $\widehat{\gamma}$ can be expressed as the solution to an estimating equation of the form:

$$\mathbf{w}_n(\gamma) = \frac{1}{n}\sum_{i=1}^{n} \mathbf{w}_{ni}(\gamma) = \mathbf{0} \tag{6.118}$$

where $\mathbf{w}_{ni}$ is the score (for maximum likelihood estimation) or score-like vector (for GEE estimation) for the $i$th subject $(1 \le i \le n)$. By a Taylor series expansion, we have

$$\begin{aligned} \sqrt{n}\,(\widehat{\gamma} - \gamma) &= \left(-\frac{\partial}{\partial \gamma}\mathbf{w}_n\right)^{-\top} \sqrt{n}\mathbf{w}_n + \mathbf{o}_p(1) \\ &= -H^{-1}\frac{\sqrt{n}}{n}\sum_{i=1}^{n}\mathbf{w}_{ni} + \mathbf{o}_p(1) \end{aligned} \tag{6.119}$$

where $H = E\left(\frac{\partial}{\partial \gamma}\mathbf{w}_{ni}\right)^\top$. By slightly modifying the asymptotic expansion of $\mathbf{U}_n$ in (6.60) in the proof of Theorem 1 to account for variability in the estimated $\widehat{\gamma}$, we have (see exercise)

$$\sqrt{n}\left(\widehat{\boldsymbol{\theta}}-\boldsymbol{\theta}\right) = B\sqrt{n}\left[\mathbf{U}_n + \left(\frac{\partial}{\partial\gamma}\mathbf{U}_n\right)^{-\top}(\widehat{\gamma}-\gamma)\right] + \mathbf{o}_p(1) \qquad (6.120)$$

$$= 2\frac{\sqrt{n}}{n}B\sum_{i=1}^{n}\left(\widetilde{\mathbf{h}}_1(\mathbf{y}_i,\mathbf{r}_i,\gamma) - CH^{-1}\mathbf{w}_{ni}\right) + \mathbf{o}_p(1)$$

where $C = E\left(\frac{\partial}{\partial\gamma}\widetilde{\mathbf{h}}_1(\mathbf{y}_i,\mathbf{r}_i,\gamma)\right)^{\top}$ and $\widetilde{\mathbf{h}}_1(\mathbf{y}_1,\mathbf{r}_1,\gamma) = E(\mathbf{U}_{n12} \mid \mathbf{y}_1,\mathbf{r}_1)$. Note that the coefficient 2 in the last equality of (6.120) is from the asymptotic expansion of $\mathbf{U}_n$ around its projection. By applying CLT to (6.120), we obtain the asymptotic variance of $\widehat{\boldsymbol{\theta}}$ (see exercise):

$$\Sigma_\theta = 4B(\Sigma_U + \Phi)B \qquad (6.121)$$

$$\Phi = CH^{-1}Var(\mathbf{w}_{ni})H^{-\top}C^{\top} - 2E\left(\widetilde{\mathbf{h}}_1(\mathbf{y}_i,\mathbf{r}_i,\gamma)\mathbf{w}_{ni}^{\top}H^{-\top}C^{\top}\right)$$

where $\Sigma_U$ is defined in (6.116). Thus, the extra term $\Phi$ in (6.121) accounts for the additional variability due to $\widehat{\gamma}$. A consistent estimate of $\Phi$ is readily obtained by substituting consistent estimates for the respective quantities in (6.121) (see exercise).

Note that the above consideration only applied to the special case with $G = D = \frac{\partial}{\partial\boldsymbol{\theta}}\mathbf{h}(\boldsymbol{\theta})$. If $G$ is also parameterized by a vector $\boldsymbol{\alpha}$, the UWGEE estimate $\widehat{\boldsymbol{\theta}}$ from (6.112) is still consistent when $\boldsymbol{\alpha}$ is substituted by some estimate $\widehat{\boldsymbol{\alpha}}$. Further, if $\widehat{\boldsymbol{\alpha}}$ is a $\sqrt{n}$-consistent estimate, we can estimate the asymptotic variance of $\widehat{\boldsymbol{\theta}}$ by (6.116) if $\pi_{\mathbf{i}2}$ are known or by (6.121) if $\pi_{\mathbf{i}2}$ are unknown and estimated according to (6.118). In other words, the variability of $\widehat{\boldsymbol{\alpha}}$ does not affect the asymptotic variance of $\widehat{\boldsymbol{\theta}}$. This is readily established by a slight modification of the asymptotic expansion in (6.120) (see exercise). These asymptotic properties parallel those of the WGEE estimates for single-response-based regression models discussed in Chapter 4.

**Example 6** (***Linear mixed-effects model***). Consider the FRM for distribution-free inference of the linear mixed-effects model in Example 4 of Section 6.4.1. Under MAR, the estimating equation in (6.109) generally does not provide consistent estimates of $\boldsymbol{\theta}$. As in Example 5 above, a weighted estimating equation must be constructed to ensure valid inference.

As discussed in Example 4, we can use the following modified UGEE to obtain consistent estimates of $\boldsymbol{\theta}$:

$$\mathbf{U}_n(\boldsymbol{\theta}) = \sum_{\mathbf{i}\in C_2^n}\mathbf{U}_{n\mathbf{i}}(\boldsymbol{\theta}) = \sum_{\mathbf{i}\in C_2^n}G_{\mathbf{i}}R_{\mathbf{i}}S_{\mathbf{i}} = \sum_{\mathbf{i}\in C_2^n}G_{\mathbf{i}}R_{\mathbf{i}}(\mathbf{f}_{\mathbf{i}} - \mathbf{h}_{\mathbf{i}}) = \mathbf{0} \qquad (6.122)$$

where $\mathbf{f_i}$, $\mathbf{h_i}$, $S_\mathbf{i}$ and $R_\mathbf{i}$ are defined in Example 4 of Section 6.4.1. In (6.122), $R_\mathbf{i}$ is a $v \times v$ diagonal matrix with the vector $\mathbf{r_i}$ of binary indicators forming the diagonal of the matrix and $v = \frac{1}{2}\left(m^2 + 3m\right)$. The $l$th element of $\mathbf{r_i}$ has the form: $\mathbf{r_{i}}_l = r_{i_1 s} r_{i_1 t} r_{i_2 s} r_{i_2 t}$, where $r_{it}$ is a binary missing data indicator for the $i$th subject defined in Example 4. As in Example 5, $\mathbf{r_{i}}_l$ may be viewed as the missing data indicator for the functional response $\mathbf{f_i}$ of the FRM. Let $\pi_{\mathbf{i}l} = E\left(\mathbf{r_{i}}_l \mid \mathbf{x_i}, \mathbf{y_i}\right)$ and assume again that $\pi_{\mathbf{i}l}$ are known and bounded away from 0 $(1 \leq i \leq n,\ 1 \leq l \leq v)$. Note that unlike Example 5, $\pi_{\mathbf{i}l}$ may also depend on $\mathbf{x}_i$ in addition to $\mathbf{y}_i$. Note also that in this particular example, if $\mathbf{r_{i}}_l = r_{i_1 s} r_{i_1 t} r_{i_2 s} r_{i_2 t}$, then

$$\pi_{\mathbf{i}l} = E\left(r_{i_1 s} r_{i_1 t} \mid \mathbf{x}_{i_1}, \mathbf{y}_{i_2}\right) E\left(r_{i_2 s} r_{i_2 t} \mid \mathbf{x}_{i_2}, \mathbf{y}_{i_2}\right) = \pi_{i_1 st} \pi_{i_2 st} \qquad (6.123)$$
$$\mathbf{i} = (i_1, i_2) \in C_2^n$$

Let

$$\Delta_{\mathbf{i}l} = \pi_{\mathbf{i}l}^{-1} \mathbf{r_{i}}_l, \quad \Delta_\mathbf{i} = \operatorname{diag}_l\left(\Delta_{\mathbf{i}l}\right) \qquad (6.124)$$

where $\operatorname{diag}_l\left(\Delta_{\mathbf{i}l}\right)$ denotes the $v \times v$ diagonal matrix with $\Delta_{\mathbf{i}l}$ on the $l$th diagonal. Define the UWGEE for $\boldsymbol{\theta}$ as follows:

$$\mathbf{U}_n(\boldsymbol{\theta}) = \sum_{\mathbf{i} \in C_2^n} \mathbf{U}_{n\mathbf{i}}(\boldsymbol{\theta}) = \sum_{\mathbf{i} \in C_2^n} G_\mathbf{i} \Delta_\mathbf{i} S_\mathbf{i} = \sum_{\mathbf{i} \in C_2^n} G_\mathbf{i} \Delta_\mathbf{i}\left(\mathbf{f_i} - \mathbf{h_i}\right) = \mathbf{0} \qquad (6.125)$$

where $\mathbf{f_i}$, $\mathbf{h_i}$ and $S_\mathbf{i}$ are defined in (6.122). The estimating equation above is similar to the one defined in (6.122) except for replacing $R_\mathbf{i}$ with $\Delta_\mathbf{i}$ to include the weights in (6.124). Like (6.122), the estimating equation in (6.125) is well defined regardless of whether the components of the functional response $\mathbf{f_i}$ are observed $(\mathbf{i} \in C_2^n)$.

To show that the estimating equation is unbiased, consider the $l$th component $\Delta_{\mathbf{i}l}$, which is given by

$$\Delta_{\mathbf{i}l} = \begin{cases} \left(\pi_{it}\pi_{jt}\right)^{-1} r_{it} r_{jt}\left(f_{1t}\left(\mathbf{y}_i, \mathbf{y}_j\right) - h_{1t}\right) & \text{if } \ 1 \leq l \leq m \\ \left(\pi_{ist}\pi_{jst}\right)^{-1} r_{is} r_{it} r_{js} r_{jt}\left(f_{2st}\left(\mathbf{y}_i, \mathbf{y}_j\right) - h_{2t}\right) & \text{if } \ m+1 \leq l \leq v \end{cases}$$

Note that we switched to the notation $\mathbf{i} = (i, j)$ for clarity in the above. Let $\mathbf{z}_i = \{\mathbf{x}_i, \mathbf{y}_i\}$ denote the collection of both $\mathbf{x}_i$ and $\mathbf{y}_i$. For $1 \leq l \leq m$, it follows from the iterated conditional expectation that

$$\begin{aligned} E\left(\Delta_{\mathbf{i}l} \mid \mathbf{x}_i, \mathbf{x}_j\right) &= E\left[E\left(\left(\pi_{it}\pi_{jt}\right)^{-1} r_{it} r_{jt}\left(f_{1t}\left(\mathbf{y}_i, \mathbf{y}_j\right) - h_{1t}\right) \mid \mathbf{z}_i, \mathbf{z}_j\right)\right] \\ &= E\left[\left(\pi_{it}\pi_{jt}\right)^{-1}\left(f_{1t}\left(\mathbf{y}_i, \mathbf{y}_j\right) - h_{1t}\right) E\left(r_{it} r_{jt} \mid \mathbf{z}_i, \mathbf{z}_j\right)\right] \\ &= E\left[\left(f_{2st}\left(\mathbf{y}_i, \mathbf{y}_j\right) - h_{2t}\right)\right] \\ &= 0 \end{aligned}$$

Similarly, for $m+1 \leq l \leq v$,

$$\begin{aligned}
E\left(\Delta_{\mathbf{i}l} \mid \mathbf{x}_i, \mathbf{x}_j\right) &= E\left[E\left(\frac{r_{is} r_{it} r_{js} r_{jt}}{\pi_{ist} \pi_{jst}}\left(f_{2st}\left(\mathbf{y}_i, \mathbf{y}_j\right)-h_{2t}\right) \mid \mathbf{z}_i, \mathbf{z}_j\right)\right] \\
&= E\left[\left(\pi_{ist} \pi_{jst}\right)^{-1}\left(f_{2st}\left(\mathbf{y}_i, \mathbf{y}_j\right)-h_{2t}\right) E\left(r_{is} r_{it} r_{js} r_{jt} \mid \mathbf{z}_i, \mathbf{z}_j\right)\right] \\
&= E\left[\left(f_{2st}\left(\mathbf{y}_i, \mathbf{y}_j\right)-h_{2t}\right)\right] \\
&= 0
\end{aligned}$$

Thus, the UWGEE in (6.125) is unbiased. It follows from Theorem 2 of Section 6.3 that the estimate $\widehat{\boldsymbol{\theta}}$ from the UWGEE in (6.125) is consistent and asymptotically normal. It is also readily checked that the asymptotic variance of $\widehat{\boldsymbol{\theta}}$ is given by

$$\begin{aligned}
&\Sigma_\theta = 4B^{-1} \Sigma_U B^{-\top} \qquad (6.126) \\
&\Sigma_U = Var\left(E\left(\mathbf{U}_{n\mathbf{i}} \mid \mathbf{y}_{i_1}, \mathbf{x}_{i_1}, \mathbf{r}_{i_1}\right)\right), \quad B = E\left(G_{\mathbf{i}} \Delta_{\mathbf{i}} D_{\mathbf{i}}^{\top}\right)
\end{aligned}$$

A consistent estimate of $\Sigma_\theta$ is readily constructed based on (6.126) (see exercise).

As in the previous example, the weights $\pi_{it} = E\left(r_{it} \mid \mathbf{x}_i, \mathbf{y}_i\right)$ in most applications are unknown and must be estimated before the UWGEE defined in (6.125) can be solved to yield estimates of $\boldsymbol{\theta}$. In light of the identity in (6.123), we can estimate $\pi_{\mathbf{i}l}$ through $\pi_{ist} = \Pr\left(r_{is}=1, r_{it}=1 \mid \mathbf{x}_i, \mathbf{y}_i\right)$. Since the number of assessments $m$ in this example can be $> 2$, $\pi_{ist}$ cannot be directly modeled by logistic regression as in the previous example. In fact, it is generally not possible to model $\pi_{ist}$ without placing some restrictions on the missing data patterns even under MAR. We discussed this issue in Chapters 4 and 5, focusing on the monotone missing data pattern (MMDP) assumption to make this modeling task feasible. Under MMDP, we first model the one-step transition probability $p_{it}$ of the occurrence of missing data and then compute $\pi_{ist}$ for $s \leq t$ as a function of $p_{it}$ using the following relation between the two:

$$\begin{aligned}
\pi_{ist} &= \Pr\left(r_{is}=1, r_{it}=1 \mid \mathbf{y}_i, \mathbf{x}_i\right) = \Pr\left(r_{it}=1 \mid \mathbf{y}_i, \mathbf{x}_i\right) \qquad (6.127) \\
&= \Pr\left(r_{it}=1 \mid H_{i(t-1)}\right) \\
&= \Pr\left(r_{it}=1 \mid r_{i(t-1)}=1, H_{i(t-1)}\right) \Pr\left(r_{i(t-1)}=1 \mid H_{i(t-1)}\right) \\
&= p_{it} \Pr\left(r_{i(t-1)}=1 \mid H_{i(t-1)}\right) = \prod_{s=2}^{t} p_{is} \\
&2 \leq s, t \leq m, \quad 1 \leq i \leq n
\end{aligned}$$

where $H_{it} = \{\mathbf{x}_{is}, \mathbf{y}_{is}; 1 \leq s \leq t-1\}$ denotes the observed data prior to time $t$ $(2 \leq t \leq m,\ 1 \leq i \leq n)$. The one-step transition probability $p_{it}$ is readily modeled by logistic regression as discussed in Chapter 4.

As in Example 5, we can use (6.126) to estimate the asymptotic variance of the UWGEE estimate derived by substituting $\pi_{\mathbf{i}l}(\widehat{\gamma})$ into (6.125), with $\widehat{\gamma}$ denoting the estimated parameter vector of the model for estimating $\pi_{\mathbf{i}l}(\gamma)$ as discussed above. Alternatively, we can derive a version corrected for the variability in the estimated $\widehat{\gamma}$ by applying an argument similar to Example 5 (see exercise).

The considerations above for Examples 5 and 6 are readily extended to the general FRM defined in Section 6.2. The only caveat is that modeling the occurrence of missing data for FRMs can become quite complicated when models are defined by multiple outcomes. In both Examples 5 and 6, we are able to estimate the weight function $\pi_{\mathbf{i}l}$ for the functional response $\mathbf{f}_\mathbf{i}$ through modeling the weight function for individual response $\mathbf{y}_i$, but this is not always possible. For example, if we generalize the FRM for the product–moment correlation in Example 3 of Section 6.4.1 to include cross-variable, cross-time correlations, $\rho_{xyst} = Corr(x_{is}, y_{it})$, and allow $x_{it}$ and $y_{it}$ to have their own missing data patterns, then it is readily checked that the weight function $\pi_{\mathbf{i}l}$ arising from the context cannot be estimated by the weight function for individual responses of the form $\pi_{it}$, but rather by the more complex function $\pi_{ixyst} = \Pr(r_{ixs} = 1, r_{iyt} = 1 \mid \mathbf{x}_i, \mathbf{y}_i)$ (see exercise). As discussed in Chapter 5, we must impose a stronger bivariate monotone missing data pattern (BMMDP) assumption to be able to feasibly model $\pi_{ixyst}$. Models for weight functions can become even more complex when an FRM is defined by more than two outcomes within a longitudinal study setting. Thus, inference for FRM under MAR depends on whether we can model the weight function.

## 6.5 Exercises

**Section 6.1**

1. Consider the random factor analysis of variance (ANOVA) model in (6.3),

(a) Show that the variance of $y_{ki}$ and covariance between $y_{ki}$ and $y_{li}$ are given by (6.4).

(b) Verify (6.6).

2. Show that the dimension of the vector $\mathbf{f}(\mathbf{y}_k, \mathbf{y}_l)$, $3C_4^n$, is also the total number of combinations of $(i,j)$ and $(i',j')$, $\frac{1}{2}\ C_4^n C_2^4$, where $(i,j) \in C_2^{n_k}$ and

$(i', j') \in C_2^{n_l}$ are disjoint with $1 \leq k, l \leq K$.

**Section 6.2**

1. Show that the FRM in (6.31) defines the same model as the GEE defined by (6.29) and (6.30).

2. Symmetrize the functional $\mathbf{f}(y_{ki}, y_{kj})$ defined in (6.15) so that the associated FRM is defined by a symmetric function of $y_{ki}$ and $y_{kj}$.

3. Verify (6.18) under the negative binomial distribution.

4. Verify (6.22) under the parametric zero-inflated model defined in (6.20).

5. Verify (6.25) under the normal based LMM defined in (6.24).

6. Show that $\mathbf{f}$ and $\mathbf{h}$ defined in (6.34) and (6.35) satisfy (6.36) under the model assumptions in (6.32).

7. Verify the identity in (6.45) under the parametric ZIP model in (6.43).

**Section 6.3**

1. Show that (6.52) holds true if $E\left(S_{\mathbf{i}_1,\ldots,\mathbf{i}_K}\right) = \mathbf{0}$.

2. Consider the proof of Theorem 1.

(a) Complete the proof of Theorem 1/1 by showing that the estimate $\widehat{\boldsymbol{\theta}}$ obtained from the UGEE estimating equation (6.51) is consistent.

(b) Verify (6.62).

(c) Verify that $\mathbf{U}_{n\mathbf{i}_1,\ldots,\mathbf{i}_k\ldots,\mathbf{i}_K}\mathbf{U}^{\top}_{n\widetilde{\mathbf{i}}_1(0),\ldots,\widetilde{\mathbf{i}}_k(1),\ldots,\widetilde{\mathbf{i}}_K(0)}$ depends on $\mathbf{i}_k$, $\widetilde{\mathbf{i}}_k(0)$ and $\widetilde{\mathbf{i}}_k(1)$ only through $\mathbf{j}_1, \ldots, \mathbf{j}_K$, where $\widetilde{\mathbf{i}}_k(0)$ and $\widetilde{\mathbf{i}}_k(1)$ are defined in (6.56) and $\mathbf{j}_k$ in (6.57).

(d) Prove Theorem 1/2 under the assumption that $\boldsymbol{\alpha}$ is unknown and replaced by a $\sqrt{n}$-consistent estimate.

3. Consider the UGEE for the generalized three-sample ANOVA for mean and variance defined in (6.63).

(a) Show that the solution to the UGEE is given by (6.64).

(b) Verify (6.65).

4. Find a consistent estimate of the asymptotic variance $\sigma_\theta^2$ of the Wilcoxon signed-rank statistic given in (6.67).

5. Consider the FRM for the generalized $K$-sample Mann–Whitney–Wilcoxon rank sum test with $K = 3$ in Example 7.

(a) Show that under the null hypothesis in (6.69), (6.70) follows from (6.68).

(b) Verify (6.68) and (6.73).

(c) Verify that $H_1$ defined in (6.76) is a symmetric kernel function of $G_1\left(y_{1i_1}, y_{2i_2}, y_{2j_2}, y_{3i_3}, y_{3j_3}\right)$ defined in (6.73).

(d) Find a symmetric $H_k$ for $G_k(y_{1i_1}, y_{2i_2}, y_{2j_2}, y_{3i_3}, y_{3j_3})$ defined in (6.74) and (6.75) for $k = 1, 2$.

6. Show that the estimating equation defined in (6.77) is unbiased, that is, $E(\mathbf{U}_n(\boldsymbol{\theta})) = \mathbf{0}$, if $E(S_{\mathbf{i}}) = E(\mathbf{f}_{\mathbf{i}} - \mathbf{h}_{\mathbf{i}}) = \mathbf{0}$.

7. Consider the proof of Theorem 2.

(a) Show that the U-statistic-like vector $\mathbf{U}_n(\boldsymbol{\theta})$ in (6.77) is asymptotically normal with mean $\mathbf{0}$.

(b) Use (a) to complete the proof of Theorem 2/1 by showing that the estimate $\widehat{\boldsymbol{\theta}}$ obtained from the UGEE in (6.77) is consistent.

(c) Verify (6.80).

(d) Prove Theorem 2/2 under the assumption that $\boldsymbol{\alpha}$ is unknown and replaced by a $\sqrt{n}$-consistent estimate.

8. Consider the FRM for the linear mixed-effects model defined by (6.86) and (6.87).

(a) Develop a UGEE with $G_{\mathbf{i}} = D_{\mathbf{i}} V_{\mathbf{i}}^{-1}$ and $V_{\mathbf{i}}$ determined based on the mean of $\mathbf{f}_{\mathbf{i}}$.

(b) Find the asymptotic variance of the UGEE estimate $\widehat{\boldsymbol{\theta}}$.

(c) Use Theorem 2 to construct a consistent estimate of the asymptotic variance of $\widehat{\boldsymbol{\theta}}$.

9. Consider the FRM for the linear mixed-effects model defined by (6.90) and (6.91).

(a) Develop a UGEE with $G_{\mathbf{i}} = D_{\mathbf{i}}$.

(b) Find the asymptotic variance of the estimate obtained from the UGEE in (a).

(c) Apply Theorem 2 to construct a consistent estimate of the asymptotic variance of $\widehat{\boldsymbol{\theta}}$.

**Section 6.4**

1. Show that $\widehat{\sigma}_2^2$ defined in (6.97) is the usual sample variance estimate calculated based on the observed $y_{i2}$.

2. Show that the estimating equation defined in (6.99) is unbiased, that is, $E(\mathbf{U}_n(\boldsymbol{\theta})) = \mathbf{0}$.

3. Show by solving (6.100) that the estimates of $\mu_t$ and $\sigma_t^2$ at each time are given by the sample mean and sample variance.

4. Consider the estimating equation (6.101).

(a) Show that the solution to the equation is $\widehat{\boldsymbol{\theta}} = \left(\widehat{\mu}_t, \widehat{\sigma}_t^2\right)^\top$ with $\widehat{\mu}_t$ and $\widehat{\sigma}_t^2$ defined by (6.97) and (6.102).

(b) Show that $\widehat{\boldsymbol{\theta}}$ is consistent and asymptotically normal by applying Theorem 1 of Section 6.3.

5. Consider the FRM for the product–moment correlation defined in (6.103) and (6.104).

(a) Verify that the Pearson correlation estimate $\widehat{\rho}_t$ and the sample variance $\widehat{\sigma}_{xt}^2$ of $x_{it}$ and $\widehat{\sigma}_{yt}^2$ of $y_{it}$ satisfy the UGEE for the FRM defined in (6.105).

(b) Verify that the generalized Pearson correlation estimate $\widehat{\rho}_t$ and the sample variance $\widehat{\sigma}_{xt}^2$ of $x_{it}$ and $\widehat{\sigma}_{yt}^2$ of $y_{it}$ given in (6.107) satisfy the UGEE defined in (6.106) in the missing data case.

(c) Generalize the UGEE in (6.106) so that it provides consistent estimates of $\boldsymbol{\theta}$ when $x_{it}$ and $y_{it}$ are allowed to have different missing patterns, that is, $x_{it}$ and $y_{it}$ may not be missing at the same time.

6. Consider the estimating equation (6.112) or more specifically (6.113) for the FRM for the generalized ANOVA used in modeling the variance of a response within the context of pre–post study design. Assume that $\pi_{i2}$ are known.

(a) Verify that this estimating equation is well defined regardless of whether $y_{i2}$ is observed.

(b) Show that the estimates $\widehat{\boldsymbol{\theta}} = \left(\widehat{\sigma}_1^2, \widehat{\sigma}_2^2\right)^\top$ given in (6.115) is the solution to the equation in (6.113).

(c) Show that $\Sigma_\theta$ defined in (6.116) is the asymptotic variance of the UWGEE estimate $\widehat{\boldsymbol{\theta}}$ in (6.115).

(d) Find a consistent estimate of $\Sigma_\theta$.

7. Consider the UWGEE in (6.112) for the generalized ANOVA defined by (6.92) and (6.93). Let $G(\boldsymbol{\alpha})$ be parameterized by a vector $\boldsymbol{\alpha}$ and let $\pi_{i2}(\boldsymbol{\gamma})$ be modeled according to (6.118).

(a) Show that the UWGEE estimate $\widehat{\boldsymbol{\theta}}$ from (6.112) has the following asymptotic expansion:

$$\sqrt{n}\left(\widehat{\boldsymbol{\theta}} - \boldsymbol{\theta}\right) = B\sqrt{n}\left[\mathbf{U}_n + C\left(\widehat{\boldsymbol{\gamma}} - \boldsymbol{\gamma}\right) + D\left(\widehat{\boldsymbol{\alpha}} - \boldsymbol{\alpha}\right)\right] + \mathbf{o}_p(1)$$

where $B$ and $C$ are defined in (6.118) and $D = E\left(\frac{\partial}{\partial \boldsymbol{\alpha}}\widetilde{\mathbf{h}}_1\left(\mathbf{y}_i, \mathbf{r}_i, \boldsymbol{\gamma}\right)\right)^\top$.

(b) Show that $D = E\left(\frac{\partial}{\partial \boldsymbol{\alpha}}\widetilde{\mathbf{h}}_1\left(\mathbf{y}_i, \mathbf{r}_i, \boldsymbol{\gamma}\right)\right)^\top = \mathbf{0}$.

(c) Use the results in (a) and (b) to show that the asymptotic variance of $\widehat{\boldsymbol{\theta}}$ is independent of the variability of $\widehat{\boldsymbol{\alpha}}$ if $\widehat{\boldsymbol{\alpha}}$ is a $\sqrt{n}$-consistent estimate.

8. In Problem 6, assume that $\pi_{i2}$ are unknown and modeled by (6.117).

(a) Verify (6.120) and (6.121).

(b) Find a consistent estimate of $\Sigma_\theta$ based on the expression in (6.121).

9. Consider the UWGEE defined in (6.125) for the FRM in Example 6. Assume that $\pi_{il}$ are known.

(a) Verify (6.126).

(b) Use (6.126) to construct a consistent estimate of the asymptotic variance of the UWGEE estimate $\widehat{\boldsymbol{\theta}}$ from the equation in (6.124).

10. In Problem 9, assume that $\pi_{\mathbf{i}l}$ are unknown and modeled according to (6.127). Let $\boldsymbol{\gamma}$ denote the parameter vector of the model for $\pi_{\mathbf{i}l}$. Assume that the estimate $\widehat{\gamma}$ of $\boldsymbol{\gamma}$ has the asymptotic expansion in (6.119).

(a) Show that the asymptotic variance $\Sigma_\theta$ of the UWGEE estimate $\widehat{\boldsymbol{\theta}}$ has an expression similar to the one in (6.121).

(b) Construct a consistent estimate of $\Sigma_\theta$ based on the expression obtained in (a).

11. Let $\mathbf{z}_{it} = (x_{it}, y_{it})^\top$ denote the continuous bivariate outcome from the $i$th subject at time $t$ from a longitudinal study with $n$ subjects and $m$ assessment times ($1 \le i \le n$, $1 \le t \le m$). Let $\zeta_{xyst} = Cov(x_{is}, y_{it})$ denote the covariance between $x_{is}$ and $y_{it}$. Assume that $x_{it}$ and $y_{it}$ can be missing at different times.

(a) Develop an FRM to model $\zeta_{xyst}$.

(b) Discuss inference under MCAR.

(c) Discuss inference under MAR and show that the weight function for the functional response is of the form $\pi_{ixyst} = \Pr(r_{ixs} = 1, r_{iyt} = 1 \mid \mathbf{x}_i, \mathbf{y}_i)$.

# References


Ash, R.B. (1999). *Probability and Measure Theory.* New York: Academic Press.

Barnhart, H. X., Haber, M. and Song, J. (2002). Overall concordance correlation coefficient for evaluating agreement among multiple observers. *Biometrics* **58**, 1020-1027.

Baron, R. M. and Kenny, D. A. (1986). The moderator-mediator variable distinction in social psychological research: Concept, strategic and statistical considerations. *Journal of Personality and Social Psychology* **51**, 1173-1182.

Bickel, P.J., Klaassen, C.A.J., Ritov, Y. and Wellner, J.A. (1993). *Efficient and Adaptive Estimation for Semiparametric Models.* Baltimore, MD: Johns Hopkins University Press.

Bollen, K.A. (1989). *Structural Equations with Latent Variables.* New York: Wiley.

Casella, G. and Berger, R. (2002). *Statistical Inference.* Pacific Grove, CA: Duxbury Press.

Demidenko, E. (2004). *Mixed Models: Theory and Applications.* New York: Wiley.

Goldstein, H. (1987). *Multilevel Models in Educational and Social Research.* Oxford: Oxford University Press.

Hoeffding, W. (1948). A class of statistics with asymptotically normal distribution. *Annals of Mathematical Statistics* **19,** 293-325.

King, T. S. and Chinchilli, V. M. (2001). A generalized concordance correlation coefficient for continuous and categorical data. *Statistics in Medicine* **20**, 2131-2147.

Kowalski, J. and Powell, J. (2004). Nonparametric inference for stochastic linear hypotheses: Application to high dimensional data. *Biometrika* **91,** 393-408.

Kowalski, J., Blackford, A., Feng, C., Mamelak, A.J. and Sauder, D.N. (2007). Nested, nonparametric, correlative analysis of microarrays for heterogenous phenotype characterization. *Statistics in Medicine* **26,** 1090-1101.

Laird, N. and Ware, J. (1982). Random-effects models for longitudinal data. *Biometrics* **38,** 963-974.

Liang, K.Y. and Zeger, S.L. (1986). Longitudinal data analysis using generalized linear models. *Biometrika* **73,** 13-22.

Lindstrom, M.J. and Bates, D.M. (1988). Newton–Raphson and EM algorithms for linear mixed-effects models for repeated-measures data. *Journal of the American Statistical Association* **83,** 1014-1022.

Mann, H.B. and Whitney, D.R. (1947). On a test of whether one of two random variables is stochastically larger than the other. *Annals of Mathematical Statistics* **18,** 50-60.

McCullagh, P. and Nelder, J. A. (1989). *Generalized Linear Models.* New York: Chapman and Hall.

Nunnally, J.C. and Bernstein, I.H. (1994). *Psychometric Theory.* New York: McGraw Hill.

Prentice, R.L. (1988). Correlated binary regression with covariates specific to each binary observation. *Biometrics* **44,** 1033-1048.

Raudenbush, S.W. and Bryk, A.S. (2002). *Hierarchical Linear Models: Applications and Data Analysis Methods.* Newbury Park, CA: Sage Publications.

SAS Institute Inc. *SAS 9.1.3 Release.* Cary, NC: SAS Institute Inc.

Schroder, K. E. E., Carey, M. P. and Vanable, P. A. (2003). Methodological challenges in research on sexual risk behavior: II. Accuracy of self-reports. *Annals of Behavioral Medicine* **26,** 104-123.

Tsiatis, A.A. (2006). *Semiparametric Theory and Missing Data.* New York: Springer.

Tu, X.M, Feng, C., Kowalski, J., Tang, W., Wang, H., Wan, C. and Ma, Y. (2007). Correlation analysis for longitudinal data: Applications to HIV and psychosocial research. *Statistics in Medicine* **26,** 4116-4138.

van Zyl, J. M., Neudecker, H. and Nel, D. G. (2000). On the distribution of the maximum likelihood estimator of Cronbach's alpha. *Psychometrika* **65,** 271–280.

Verbeke, G. and Molenberghs, G. (2000). *Linear Mixed Models for Longitudinal Data.* New York: Springer-Verlag.

Vuong, Q.H. (1989). Likelihood ratio tests for model selection and non-nested hypotheses. *Econometrica* **57,** 307–333.

Wilcoxon, F. (1945). Individual comparisons by ranking methods. *Biometrics* **1,** 80-83.

Wolfinger, R. (1993). Laplace's approximation for nonlinear mixed models. *Biometrika* **80,** 791-795.

# Subject Index

WILEY SERIES IN PROBABILITY AND STATISTICS

ESTABLISHED BY WALTER A. SHEWHART AND SAMUEL S. WILKS



The ***Wiley Series in Probability and Statistics*** is well established and authoritative. It covers many topics of current research interest in both pure and applied statistics and probability theory. Written by leading statisticians and institutions, the titles span both state-of-the-art developments in the field and classical methods.

Reflecting the wide range of current research in statistics, the series encompasses applied, methodological and theoretical statistics, ranging from applications and new techniques made possible by advances in computerized practice to rigorous treatment of theoretical approaches.

This series provides essential and invaluable reading for all statisticians, whether in academia, industry, government, or research.

† ABRAHAM and LEDOLTER · Statistical Methods for Forecasting
AGRESTI · Analysis of Ordinal Categorical Data
AGRESTI · An Introduction to Categorical Data Analysis, *Second Edition*
AGRESTI · Categorical Data Analysis, *Second Edition*
ALTMAN, GILL, and McDONALD · Numerical Issues in Statistical Computing for the Social Scientist
AMARATUNGA and CABRERA · Exploration and Analysis of DNA Microarray and Protein Array Data
ANDĚL · Mathematics of Chance
ANDERSON · An Introduction to Multivariate Statistical Analysis, *Third Edition*
* ANDERSON · The Statistical Analysis of Time Series
ANDERSON, AUQUIER, HAUCK, OAKES, VANDAELE, and WEISBERG · Statistical Methods for Comparative Studies
ANDERSON and LOYNES · The Teaching of Practical Statistics
ARMITAGE and DAVID (editors) · Advances in Biometry
ARNOLD, BALAKRISHNAN, and NAGARAJA · Records
* ARTHANARI and DODGE · Mathematical Programming in Statistics
* BAILEY · The Elements of Stochastic Processes with Applications to the Natural Sciences
BALAKRISHNAN and KOUTRAS · Runs and Scans with Applications
BALAKRISHNAN and NG · Precedence-Type Tests and Applications
BARNETT · Comparative Statistical Inference, *Third Edition*
BARNETT · Environmental Statistics
BARNETT and LEWIS · Outliers in Statistical Data, *Third Edition*
BARTOSZYNSKI and NIEWIADOMSKA-BUGAJ · Probability and Statistical Inference
BASILEVSKY · Statistical Factor Analysis and Related Methods: Theory and Applications
BASU and RIGDON · Statistical Methods for the Reliability of Repairable Systems
BATES and WATTS · Nonlinear Regression Analysis and Its Applications

*Now available in a lower priced paperback edition in the Wiley Classics Library.
†Now available in a lower priced paperback edition in the Wiley–Interscience Paperback Series.

BECHHOFER, SANTNER, and GOLDSMAN · Design and Analysis of Experiments for Statistical Selection, Screening, and Multiple Comparisons
BELSLEY · Conditioning Diagnostics: Collinearity and Weak Data in Regression
† BELSLEY, KUH, and WELSCH · Regression Diagnostics: Identifying Influential Data and Sources of Collinearity
BENDAT and PIERSOL · Random Data: Analysis and Measurement Procedures, *Third Edition*
BERRY, CHALONER, and GEWEKE · Bayesian Analysis in Statistics and Econometrics: Essays in Honor of Arnold Zellner
BERNARDO and SMITH · Bayesian Theory
BHAT and MILLER · Elements of Applied Stochastic Processes, *Third Edition*
BHATTACHARYA and WAYMIRE · Stochastic Processes with Applications
BILLINGSLEY · Convergence of Probability Measures, *Second Edition*
BILLINGSLEY · Probability and Measure, *Third Edition*
BIRKES and DODGE · Alternative Methods of Regression
BISWAS, DATTA, FINE, and SEGAL · Statistical Advances in the Biomedical Sciences: Clinical Trials, Epidemiology, Survival Analysis, and Bioinformatics
BLISCHKE AND MURTHY (editors) · Case Studies in Reliability and Maintenance
BLISCHKE AND MURTHY · Reliability: Modeling, Prediction, and Optimization
BLOOMFIELD · Fourier Analysis of Time Series: An Introduction, *Second Edition*
BOLLEN · Structural Equations with Latent Variables
BOLLEN and CURRAN · Latent Curve Models: A Structural Equation Perspective
BOROVKOV · Ergodicity and Stability of Stochastic Processes
BOULEAU · Numerical Methods for Stochastic Processes
BOX · Bayesian Inference in Statistical Analysis
BOX · R. A. Fisher, the Life of a Scientist
BOX and DRAPER · Response Surfaces, Mixtures, and Ridge Analyses, *Second Edition*
* BOX and DRAPER · Evolutionary Operation: A Statistical Method for Process Improvement
BOX and FRIENDS · Improving Almost Anything, *Revised Edition*
BOX, HUNTER, and HUNTER · Statistics for Experimenters: Design, Innovation, and Discovery, *Second Editon*
BOX and LUCEÑO · Statistical Control by Monitoring and Feedback Adjustment
BRANDIMARTE · Numerical Methods in Finance: A MATLAB-Based Introduction
† BROWN and HOLLANDER · Statistics: A Biomedical Introduction
BRUNNER, DOMHOF, and LANGER · Nonparametric Analysis of Longitudinal Data in Factorial Experiments
BUCKLEW · Large Deviation Techniques in Decision, Simulation, and Estimation
CAIROLI and DALANG · Sequential Stochastic Optimization
CASTILLO, HADI, BALAKRISHNAN, and SARABIA · Extreme Value and Related Models with Applications in Engineering and Science
CHAN · Time Series: Applications to Finance
CHARALAMBIDES · Combinatorial Methods in Discrete Distributions
CHATTERJEE and HADI · Regression Analysis by Example, *Fourth Edition*
CHATTERJEE and HADI · Sensitivity Analysis in Linear Regression
CHERNICK · Bootstrap Methods: A Guide for Practitioners and Researchers, *Second Edition*
CHERNICK and FRIIS · Introductory Biostatistics for the Health Sciences
CHILÈS and DELFINER · Geostatistics: Modeling Spatial Uncertainty
CHOW and LIU · Design and Analysis of Clinical Trials: Concepts and Methodologies, *Second Edition*
CLARKE and DISNEY · Probability and Random Processes: A First Course with Applications, *Second Edition*
* COCHRAN and COX · Experimental Designs, *Second Edition*

*Now available in a lower priced paperback edition in the Wiley Classics Library.
†Now available in a lower priced paperback edition in the Wiley–Interscience Paperback Series.

CONGDON · Applied Bayesian Modelling
CONGDON · Bayesian Models for Categorical Data
CONGDON · Bayesian Statistical Modelling
CONOVER · Practical Nonparametric Statistics, *Third Edition*
COOK · Regression Graphics
COOK and WEISBERG · Applied Regression Including Computing and Graphics
COOK and WEISBERG · An Introduction to Regression Graphics
CORNELL · Experiments with Mixtures, Designs, Models, and the Analysis of Mixture Data, *Third Edition*
COVER and THOMAS · Elements of Information Theory
COX · A Handbook of Introductory Statistical Methods
* COX · Planning of Experiments
CRESSIE · Statistics for Spatial Data, *Revised Edition*
CSÖRGŐ and HORVÁTH · Limit Theorems in Change Point Analysis
DANIEL · Applications of Statistics to Industrial Experimentation
DANIEL · Biostatistics: A Foundation for Analysis in the Health Sciences, *Eighth Edition*
* DANIEL · Fitting Equations to Data: Computer Analysis of Multifactor Data, *Second Edition*
DASU and JOHNSON · Exploratory Data Mining and Data Cleaning
DAVID and NAGARAJA · Order Statistics, *Third Edition*
* DEGROOT, FIENBERG, and KADANE · Statistics and the Law
DEL CASTILLO · Statistical Process Adjustment for Quality Control
DeMARIS · Regression with Social Data: Modeling Continuous and Limited Response Variables
DEMIDENKO · Mixed Models: Theory and Applications
DENISON, HOLMES, MALLICK and SMITH · Bayesian Methods for Nonlinear Classification and Regression
DETTE and STUDDEN · The Theory of Canonical Moments with Applications in Statistics, Probability, and Analysis
DEY and MUKERJEE · Fractional Factorial Plans
DILLON and GOLDSTEIN · Multivariate Analysis: Methods and Applications
DODGE · Alternative Methods of Regression
* DODGE and ROMIG · Sampling Inspection Tables, *Second Edition*
* DOOB · Stochastic Processes
DOWDY, WEARDEN, and CHILKO · Statistics for Research, *Third Edition*
DRAPER and SMITH · Applied Regression Analysis, *Third Edition*
DRYDEN and MARDIA · Statistical Shape Analysis
DUDEWICZ and MISHRA · Modern Mathematical Statistics
DUNN and CLARK · Basic Statistics: A Primer for the Biomedical Sciences, *Third Edition*
DUPUIS and ELLIS · A Weak Convergence Approach to the Theory of Large Deviations
EDLER and KITSOS · Recent Advances in Quantitative Methods in Cancer and Human Health Risk Assessment
* ELANDT-JOHNSON and JOHNSON · Survival Models and Data Analysis
ENDERS · Applied Econometric Time Series
† ETHIER and KURTZ · Markov Processes: Characterization and Convergence
EVANS, HASTINGS, and PEACOCK · Statistical Distributions, *Third Edition*
FELLER · An Introduction to Probability Theory and Its Applications, Volume I, *Third Edition,* Revised; Volume II, *Second Edition*
FISHER and VAN BELLE · Biostatistics: A Methodology for the Health Sciences
FITZMAURICE, LAIRD, and WARE · Applied Longitudinal Analysis
* FLEISS · The Design and Analysis of Clinical Experiments
FLEISS · Statistical Methods for Rates and Proportions, *Third Edition*

*Now available in a lower priced paperback edition in the Wiley Classics Library.
†Now available in a lower priced paperback edition in the Wiley–Interscience Paperback Series.

† FLEMING and HARRINGTON · Counting Processes and Survival Analysis
FULLER · Introduction to Statistical Time Series, *Second Edition*
† FULLER · Measurement Error Models
GALLANT · Nonlinear Statistical Models
GEISSER · Modes of Parametric Statistical Inference
GELMAN and MENG · Applied Bayesian Modeling and Causal Inference from Incomplete-Data Perspectives
GEWEKE · Contemporary Bayesian Econometrics and Statistics
GHOSH, MUKHOPADHYAY, and SEN · Sequential Estimation
GIESBRECHT and GUMPERTZ · Planning, Construction, and Statistical Analysis of Comparative Experiments
GIFI · Nonlinear Multivariate Analysis
GIVENS and HOETING · Computational Statistics
GLASSERMAN and YAO · Monotone Structure in Discrete-Event Systems
GNANADESIKAN · Methods for Statistical Data Analysis of Multivariate Observations, *Second Edition*
GOLDSTEIN and LEWIS · Assessment: Problems, Development, and Statistical Issues
GREENWOOD and NIKULIN · A Guide to Chi-Squared Testing
GROSS and HARRIS · Fundamentals of Queueing Theory, *Third Edition*
* HAHN and SHAPIRO · Statistical Models in Engineering
HAHN and MEEKER · Statistical Intervals: A Guide for Practitioners
HALD · A History of Probability and Statistics and their Applications Before 1750
HALD · A History of Mathematical Statistics from 1750 to 1930
† HAMPEL · Robust Statistics: The Approach Based on Influence Functions
HANNAN and DEISTLER · The Statistical Theory of Linear Systems
HEIBERGER · Computation for the Analysis of Designed Experiments
HEDAYAT and SINHA · Design and Inference in Finite Population Sampling
HEDEKER and GIBBONS · Longitudinal Data Analysis
HELLER · MACSYMA for Statisticians
HINKELMANN and KEMPTHORNE · Design and Analysis of Experiments, Volume 1: Introduction to Experimental Design, *Second Edition*
HINKELMANN and KEMPTHORNE · Design and Analysis of Experiments, Volume 2: Advanced Experimental Design
HOAGLIN, MOSTELLER, and TUKEY · Exploratory Approach to Analysis of Variance
* HOAGLIN, MOSTELLER, and TUKEY · Exploring Data Tables, Trends and Shapes
* HOAGLIN, MOSTELLER, and TUKEY · Understanding Robust and Exploratory Data Analysis
HOCHBERG and TAMHANE · Multiple Comparison Procedures
HOCKING · Methods and Applications of Linear Models: Regression and the Analysis of Variance, *Second Edition*
HOEL · Introduction to Mathematical Statistics, *Fifth Edition*
HOGG and KLUGMAN · Loss Distributions
HOLLANDER and WOLFE · Nonparametric Statistical Methods, *Second Edition*
HOSMER and LEMESHOW · Applied Logistic Regression, *Second Edition*
HOSMER, LEMESHOW, and MAY · Applied Survival Analysis: Regression Modeling of Time-to-Event Data, *Second Edition*
† HUBER · Robust Statistics
HUBERTY · Applied Discriminant Analysis
HUBERTY and OLEJNIK · Applied MANOVA and Discriminant Analysis, *Second Edition*
HUNT and KENNEDY · Financial Derivatives in Theory and Practice, *Revised Edition*

*Now available in a lower priced paperback edition in the Wiley Classics Library.
†Now available in a lower priced paperback edition in the Wiley–Interscience Paperback Series.

HURD and MIAMEE · Periodically Correlated Random Sequences: Spectral Theory and Practice
HUSKOVA, BERAN, and DUPAC · Collected Works of Jaroslav Hajek—with Commentary
HUZURBAZAR · Flowgraph Models for Multistate Time-to-Event Data
IMAN and CONOVER · A Modern Approach to Statistics
† JACKSON · A User's Guide to Principle Components
JOHN · Statistical Methods in Engineering and Quality Assurance
JOHNSON · Multivariate Statistical Simulation
JOHNSON and BALAKRISHNAN · Advances in the Theory and Practice of Statistics: A Volume in Honor of Samuel Kotz
JOHNSON and BHATTACHARYYA · Statistics: Principles and Methods, *Fifth Edition*
JOHNSON and KOTZ · Distributions in Statistics
JOHNSON and KOTZ (editors) · Leading Personalities in Statistical Sciences: From the Seventeenth Century to the Present
JOHNSON, KOTZ, and BALAKRISHNAN · Continuous Univariate Distributions, Volume 1, *Second Edition*
JOHNSON, KOTZ, and BALAKRISHNAN · Continuous Univariate Distributions, Volume 2, *Second Edition*
JOHNSON, KOTZ, and BALAKRISHNAN · Discrete Multivariate Distributions
JOHNSON, KEMP, and KOTZ · Univariate Discrete Distributions, *Third Edition*
JUDGE, GRIFFITHS, HILL, LÜTKEPOHL, and LEE · The Theory and Practice of Econometrics, *Second Edition*
JUREČKOVÁ and SEN · Robust Statistical Procedures: Aymptotics and Interrelations
JUREK and MASON · Operator-Limit Distributions in Probability Theory
KADANE · Bayesian Methods and Ethics in a Clinical Trial Design
KADANE AND SCHUM · A Probabilistic Analysis of the Sacco and Vanzetti Evidence
KALBFLEISCH and PRENTICE · The Statistical Analysis of Failure Time Data, *Second Edition*
KARIYA and KURATA · Generalized Least Squares
KASS and VOS · Geometrical Foundations of Asymptotic Inference
† KAUFMAN and ROUSSEEUW · Finding Groups in Data: An Introduction to Cluster Analysis
KEDEM and FOKIANOS · Regression Models for Time Series Analysis
KENDALL, BARDEN, CARNE, and LE · Shape and Shape Theory
KHURI · Advanced Calculus with Applications in Statistics, *Second Edition*
KHURI, MATHEW, and SINHA · Statistical Tests for Mixed Linear Models
KLEIBER and KOTZ · Statistical Size Distributions in Economics and Actuarial Sciences
KLUGMAN, PANJER, and WILLMOT · Loss Models: From Data to Decisions, *Second Edition*
KLUGMAN, PANJER, and WILLMOT · Solutions Manual to Accompany Loss Models: From Data to Decisions, *Second Edition*
KOTZ, BALAKRISHNAN, and JOHNSON · Continuous Multivariate Distributions, Volume 1, *Second Edition*
KOVALENKO, KUZNETZOV, and PEGG · Mathematical Theory of Reliability of Time-Dependent Systems with Practical Applications
KOWALSKI and TU · Modern Applied U-Statistics
KVAM and VIDAKOVIC · Nonparametric Statistics with Applications to Science and Engineering
LACHIN · Biostatistical Methods: The Assessment of Relative Risks
LAD · Operational Subjective Statistical Methods: A Mathematical, Philosophical, and Historical Introduction
LAMPERTI · Probability: A Survey of the Mathematical Theory, *Second Edition*

*Now available in a lower priced paperback edition in the Wiley Classics Library.
†Now available in a lower priced paperback edition in the Wiley–Interscience Paperback Series.

LANGE, RYAN, BILLARD, BRILLINGER, CONQUEST, and GREENHOUSE · Case Studies in Biometry
LARSON · Introduction to Probability Theory and Statistical Inference, *Third Edition*
LAWLESS · Statistical Models and Methods for Lifetime Data, *Second Edition*
LAWSON · Statistical Methods in Spatial Epidemiology
LE · Applied Categorical Data Analysis
LE · Applied Survival Analysis
LEE and WANG · Statistical Methods for Survival Data Analysis, *Third Edition*
LEPAGE and BILLARD · Exploring the Limits of Bootstrap
LEYLAND and GOLDSTEIN (editors) · Multilevel Modelling of Health Statistics
LIAO · Statistical Group Comparison
LINDVALL · Lectures on the Coupling Method
LIN · Introductory Stochastic Analysis for Finance and Insurance
LINHART and ZUCCHINI · Model Selection
LITTLE and RUBIN · Statistical Analysis with Missing Data, *Second Edition*
LLOYD · The Statistical Analysis of Categorical Data
LOWEN and TEICH · Fractal-Based Point Processes
MAGNUS and NEUDECKER · Matrix Differential Calculus with Applications in Statistics and Econometrics, *Revised Edition*
MALLER and ZHOU · Survival Analysis with Long Term Survivors
MALLOWS · Design, Data, and Analysis by Some Friends of Cuthbert Daniel
MANN, SCHAFER, and SINGPURWALLA · Methods for Statistical Analysis of Reliability and Life Data
MANTON, WOODBURY, and TOLLEY · Statistical Applications Using Fuzzy Sets
MARCHETTE · Random Graphs for Statistical Pattern Recognition
MARDIA and JUPP · Directional Statistics
MASON, GUNST, and HESS · Statistical Design and Analysis of Experiments with Applications to Engineering and Science, *Second Edition*
McCULLOCH and SEARLE · Generalized, Linear, and Mixed Models
McFADDEN · Management of Data in Clinical Trials, *Second Edition*
* McLACHLAN · Discriminant Analysis and Statistical Pattern Recognition
McLACHLAN, DO, and AMBROISE · Analyzing Microarray Gene Expression Data
McLACHLAN and KRISHNAN · The EM Algorithm and Extensions, *Second Edition*
McLACHLAN and PEEL · Finite Mixture Models
McNEIL · Epidemiological Research Methods
MEEKER and ESCOBAR · Statistical Methods for Reliability Data
MEERSCHAERT and SCHEFFLER · Limit Distributions for Sums of Independent Random Vectors: Heavy Tails in Theory and Practice
MICKEY, DUNN, and CLARK · Applied Statistics: Analysis of Variance and Regression, *Third Edition*
* MILLER · Survival Analysis, *Second Edition*
MONTGOMERY, PECK, and VINING · Introduction to Linear Regression Analysis, *Fourth Edition*
MORGENTHALER and TUKEY · Configural Polysampling: A Route to Practical Robustness
MUIRHEAD · Aspects of Multivariate Statistical Theory
MULLER and STOYAN · Comparison Methods for Stochastic Models and Risks
MURRAY · X-STAT 2.0 Statistical Experimentation, Design Data Analysis, and Nonlinear Optimization
MURTHY, XIE, and JIANG · Weibull Models
MYERS and MONTGOMERY · Response Surface Methodology: Process and Product Optimization Using Designed Experiments, *Second Edition*
MYERS, MONTGOMERY, and VINING · Generalized Linear Models. With Applications in Engineering and the Sciences

*Now available in a lower priced paperback edition in the Wiley Classics Library.
†Now available in a lower priced paperback edition in the Wiley–Interscience Paperback Series.

† NELSON · Accelerated Testing, Statistical Models, Test Plans, and Data Analyses
† NELSON · Applied Life Data Analysis
NEWMAN · Biostatistical Methods in Epidemiology
OCHI · Applied Probability and Stochastic Processes in Engineering and Physical Sciences
OKABE, BOOTS, SUGIHARA, and CHIU · Spatial Tesselations: Concepts and Applications of Voronoi Diagrams, *Second Edition*
OLIVER and SMITH · Influence Diagrams, Belief Nets and Decision Analysis
PALTA · Quantitative Methods in Population Health: Extensions of Ordinary Regressions
PANJER · Operational Risk: Modeling and Analytics
PANKRATZ · Forecasting with Dynamic Regression Models
PANKRATZ · Forecasting with Univariate Box-Jenkins Models: Concepts and Cases
* PARZEN · Modern Probability Theory and Its Applications
PEÑA, TIAO, and TSAY · A Course in Time Series Analysis
PIANTADOSI · Clinical Trials: A Methodologic Perspective
PORT · Theoretical Probability for Applications
POURAHMADI · Foundations of Time Series Analysis and Prediction Theory
POWELL · Approximate Dynamic Programming: Solving the Curses of Dimensionality
PRESS · Bayesian Statistics: Principles, Models, and Applications
PRESS · Subjective and Objective Bayesian Statistics, *Second Edition*
PRESS and TANUR · The Subjectivity of Scientists and the Bayesian Approach
PUKELSHEIM · Optimal Experimental Design
PURI, VILAPLANA, and WERTZ · New Perspectives in Theoretical and Applied Statistics
† PUTERMAN · Markov Decision Processes: Discrete Stochastic Dynamic Programming
QIU · Image Processing and Jump Regression Analysis
* RAO · Linear Statistical Inference and Its Applications, *Second Edition*
RAUSAND and HØYLAND · System Reliability Theory: Models, Statistical Methods, and Applications, *Second Edition*
RENCHER · Linear Models in Statistics
RENCHER · Methods of Multivariate Analysis, *Second Edition*
RENCHER · Multivariate Statistical Inference with Applications
* RIPLEY · Spatial Statistics
* RIPLEY · Stochastic Simulation
ROBINSON · Practical Strategies for Experimenting
ROHATGI and SALEH · An Introduction to Probability and Statistics, *Second Edition*
ROLSKI, SCHMIDLI, SCHMIDT, and TEUGELS · Stochastic Processes for Insurance and Finance
ROSENBERGER and LACHIN · Randomization in Clinical Trials: Theory and Practice
ROSS · Introduction to Probability and Statistics for Engineers and Scientists
ROSSI, ALLENBY, and McCULLOCH · Bayesian Statistics and Marketing
† ROUSSEEUW and LEROY · Robust Regression and Outlier Detection
* RUBIN · Multiple Imputation for Nonresponse in Surveys
RUBINSTEIN and KROESE · Simulation and the Monte Carlo Method, *Second Edition*
RUBINSTEIN and MELAMED · Modern Simulation and Modeling
RYAN · Modern Engineering Statistics
RYAN · Modern Experimental Design
RYAN · Modern Regression Methods
RYAN · Statistical Methods for Quality Improvement, *Second Edition*
SALEH · Theory of Preliminary Test and Stein-Type Estimation with Applications
* SCHEFFE · The Analysis of Variance
SCHIMEK · Smoothing and Regression: Approaches, Computation, and Application
SCHOTT · Matrix Analysis for Statistics, *Second Edition*
SCHOUTENS · Levy Processes in Finance: Pricing Financial Derivatives

*Now available in a lower priced paperback edition in the Wiley Classics Library.
†Now available in a lower priced paperback edition in the Wiley–Interscience Paperback Series.

SCHUSS · Theory and Applications of Stochastic Differential Equations
SCOTT · Multivariate Density Estimation: Theory, Practice, and Visualization
† SEARLE · Linear Models for Unbalanced Data
† SEARLE · Matrix Algebra Useful for Statistics
† SEARLE, CASELLA, and McCULLOCH · Variance Components
SEARLE and WILLETT · Matrix Algebra for Applied Economics
SEBER · A Matrix Handbook For Statisticians
† SEBER · Multivariate Observations
SEBER and LEE · Linear Regression Analysis, *Second Edition*
† SEBER and WILD · Nonlinear Regression
SENNOTT · Stochastic Dynamic Programming and the Control of Queueing Systems
* SERFLING · Approximation Theorems of Mathematical Statistics
SHAFER and VOVK · Probability and Finance: It's Only a Game!
SILVAPULLE and SEN · Constrained Statistical Inference: Inequality, Order, and Shape Restrictions
SMALL and McLEISH · Hilbert Space Methods in Probability and Statistical Inference
SRIVASTAVA · Methods of Multivariate Statistics
STAPLETON · Linear Statistical Models
STAPLETON · Models for Probability and Statistical Inference: Theory and Applications
STAUDTE and SHEATHER · Robust Estimation and Testing
STOYAN, KENDALL, and MECKE · Stochastic Geometry and Its Applications, *Second Edition*
STOYAN and STOYAN · Fractals, Random Shapes and Point Fields: Methods of Geometrical Statistics
STREET and BURGESS · The Construction of Optimal Stated Choice Experiments: Theory and Methods
STYAN · The Collected Papers of T. W. Anderson: 1943–1985
SUTTON, ABRAMS, JONES, SHELDON, and SONG · Methods for Meta-Analysis in Medical Research
TAKEZAWA · Introduction to Nonparametric Regression
TANAKA · Time Series Analysis: Nonstationary and Noninvertible Distribution Theory
THOMPSON · Empirical Model Building
THOMPSON · Sampling, *Second Edition*
THOMPSON · Simulation: A Modeler's Approach
THOMPSON and SEBER · Adaptive Sampling
THOMPSON, WILLIAMS, and FINDLAY · Models for Investors in Real World Markets
TIAO, BISGAARD, HILL, PEÑA, and STIGLER (editors) · Box on Quality and Discovery: with Design, Control, and Robustness
TIERNEY · LISP-STAT: An Object-Oriented Environment for Statistical Computing and Dynamic Graphics
TSAY · Analysis of Financial Time Series, *Second Edition*
UPTON and FINGLETON · Spatial Data Analysis by Example, Volume II: Categorical and Directional Data
VAN BELLE · Statistical Rules of Thumb
VAN BELLE, FISHER, HEAGERTY, and LUMLEY · Biostatistics: A Methodology for the Health Sciences, *Second Edition*
VESTRUP · The Theory of Measures and Integration
VIDAKOVIC · Statistical Modeling by Wavelets
VINOD and REAGLE · Preparing for the Worst: Incorporating Downside Risk in Stock Market Investments
WALLER and GOTWAY · Applied Spatial Statistics for Public Health Data
WEERAHANDI · Generalized Inference in Repeated Measures: Exact Methods in MANOVA and Mixed Models
WEISBERG · Applied Linear Regression, *Third Edition*

*Now available in a lower priced paperback edition in the Wiley Classics Library.
†Now available in a lower priced paperback edition in the Wiley–Interscience Paperback Series.

WELSH · Aspects of Statistical Inference
WESTFALL and YOUNG · Resampling-Based Multiple Testing: Examples and Methods for $p$-Value Adjustment
WHITTAKER · Graphical Models in Applied Multivariate Statistics
WINKER · Optimization Heuristics in Economics: Applications of Threshold Accepting
WONNACOTT and WONNACOTT · Econometrics, *Second Edition*
WOODING · Planning Pharmaceutical Clinical Trials: Basic Statistical Principles
WOODWORTH · Biostatistics: A Bayesian Introduction
WOOLSON and CLARKE · Statistical Methods for the Analysis of Biomedical Data, *Second Edition*
WU and HAMADA · Experiments: Planning, Analysis, and Parameter Design Optimization
WU and ZHANG · Nonparametric Regression Methods for Longitudinal Data Analysis
YANG · The Construction Theory of Denumerable Markov Processes
YOUNG, VALERO-MORA, and FRIENDLY · Visual Statistics: Seeing Data with Dynamic Interactive Graphics
ZELTERMAN · Discrete Distributions—Applications in the Health Sciences
* ZELLNER · An Introduction to Bayesian Inference in Econometrics
ZHOU, OBUCHOWSKI, and McCLISH · Statistical Methods in Diagnostic Medicine

*Now available in a lower priced paperback edition in the Wiley Classics Library.
†Now available in a lower priced paperback edition in the Wiley–Interscience Paperback Series.